Advances in Intelligent Systems and Computing

Volume 934

The series "Advances in Intelligent Systems and Computing" contains publications on theory, applications, and design methods of Intelligent Systems and Intelligent Computing. Virtually all disciplines such as engineering, natural sciences, computer and information science, ICT, economics, business, e-commerce, environment, healthcare, life science are covered. The list of topics spans all the areas of modern intelligent systems and computing such as: computational intelligence, soft computing including neural networks, fuzzy systems, evolutionary computing and the fusion of these paradigms, social intelligence, ambient intelligence, computational neuroscience, artificial life, virtual worlds and society, cognitive science and systems, Perception and Vision, DNA and immune based systems, self-organizing and adaptive systems, e-Learning and teaching, human-centered and human-centric computing, recommender systems, intelligent control, robotics and mechatronics including human-machine teaming, knowledge-based paradigms, learning paradigms, machine ethics, intelligent data analysis, knowledge management, intelligent agents, intelligent decision making and support, intelligent network security, trust management, interactive entertainment, Web intelligence and multimedia.

The publications within "Advances in Intelligent Systems and Computing" are primarily proceedings of important conferences, symposia and congresses. They cover significant recent developments in the field, both of a foundational and applicable character. An important characteristic feature of the series is the short publication time and world-wide distribution. This permits a rapid and broad dissemination of research results.

**** Indexing: The books of this series are submitted to ISI Proceedings, EI-Compendex, DBLP, SCOPUS, Google Scholar and Springerlink ****

More information about this series at http://www.springer.com/series/11156

Jerzy Świder · Sławomir Kciuk ·
Maciej Trojnacki
Editors

Mechatronics 2017 - Ideas for Industrial Applications

Springer

Editors
Jerzy Świder
Faculty of Mechanical Engineering,
Institute of Engineering Processes
Automation and Integrated Manufacturing
Systems
The Silesian University of Technology
Gliwice, Poland

Sławomir Kciuk
Faculty of Mechanical Engineering,
Institute of Theoretical and Applied
Mechanics
The Silesian University of Technology
Gliwice, Poland

Maciej Trojnacki
Industrial Research Institute for Automation
and Measurements PIAP
Warsaw, Poland

ISSN 2194-5357 ISSN 2194-5365 (electronic)
Advances in Intelligent Systems and Computing
ISBN 978-3-030-15856-9 ISBN 978-3-030-15857-6 (eBook)
https://doi.org/10.1007/978-3-030-15857-6

Library of Congress Control Number: 2019934748

This Springer imprint is published by the registered company Springer Nature Switzerland AG
The registered company address is: Gewerbestrasse 11, 6330 Cham, Switzerland

Foreword

There is no doubt that mechatronics is one of the elements conditioning the advent and success of the modern stage of industrial development. Knowledge on this subject is being developed both at worldwide universities and in industrial practice. Collaboration between these two entities is crucial for the results and the level of this development. Recent results of research works conducted by scientists and practitioners on development and application of mechatronics constitute the content of this book, the title of which explicitly underlines the application of mechatronics in industrial practice.

The majority, almost 20 percent of the published works in this book, are devoted to measuring techniques in mechatronic phenomena and related problems. The effectiveness, precision, and reliability of measurements in complex processes of controlling machines and devices often determine the success of these processes. Two equal groups of works, each comprising about 16 percent of the book volume, deal with robotics and design of mechatronic systems. Robotics and robotisation of modern industry are one of the hallmarks and measures of its development level, and the problems of control, sensing, programming, and safe cooperation between a robot and a human are of fundamental importance to mechatronic systems. Fourteen percent of the book's contents are the results of research and applications of mechatronics in medicine and in sport. This is an extremely important area of joint action of doctors and engineers, more and more often providing effects of biomechatronic character, having both scientific and utilitarian, but also deeply humanistic and social dimensions. The last largest group, 12 percent of the works included in the book, deals with modern applications of mechatronics in rapidly changing contemporary mining, which poses strict requirements regarding human safety and the natural environment. The remaining part of the book is devoted to applications of mechatronics in the automotive industry well known to use mechatronics in the design and manufacture of modern cars with exceptional intensity, in the development of the defense technology important in every society protecting its independence, in the extremely demanding aviation industry, in the modern food industry, as well as to teaching of mechatronics at universities and to the paradigm of Industry 4.0.

We hope that this edition of the book will be an important piece of information about the latest findings of researchers, industry representatives, and scientists from universities of technology working in the field of mechatronics, which is an exciting, modern, and fast-growing field of technics and technology, in particular in the context of such important industrial applications.

Acknowledgments. We wish to express our gratitude to Mariola Jureczko, Paweł Bachorz, and Małgorzata Kaliczyńska for their professional and kind help with preparation of the book.

December 2018

J. Świder
S. Kciuk
M. Trojnacki

Contents

About the Editors

Professor Jerzy Świder was awarded the PhD degree of technical sciences in 1981 in the discipline of machine design and maintenance from the Faculty of Energy and Mechanical Engineering of the Silesian University of Technology (SUT), and the DSc (habilitation) degree in the same discipline in 1992. Since 1996 (for two cadences), he was Vice-Dean and since 2005 (for the next two cadences) Dean of the Mechanical Engineering Faculty of the Silesian University of Technology (SUT), Gliwice, Poland. Currently, he is Full Professor (since 2000), Director of the Institute of Engineering Processes Automation and Integrated Manufacturing Systems (since 2005), and Director of doctoral studies in the field of machine design and maintenance (since 2017) at the same university and at the same faculty. Since 2017, he is Chairman of the scientific board of the KOMAG Institute of Mining Technology. He promoted 16 doctors of technical sciences. His research interests focus on graph theory and application, automation of processes, mechatronics, robotics and robotisation of technological processes. His most important projects include such elements as the design of automated production stations, the study of energy consumption by industrial robots, dynamics of systems with couplings, diagnostics of industrial networks, control and diagnostic systems for distributed drives, the use of artificial neural networks in control systems, simulation optimisation of shock absorbers, the mechatronic integrator of vehicle control procedures, the robotic cell for nondestructive quality control of glued connections of car bodies in the General Motors standard, and control of parameters of the scraper conveyor to optimise its working conditions. He is one of the main authors of research, didactics, and laboratories, created at the Mechanical Engineering Faculty of the SUT in the areas of automation, robotics, and mechatronics, in the close cooperation with leading industrial partners in this area.

Sławomir Kciuk, PhD, DSc, Eng. is Prominent Representative of science, who works at the Institute of Theoretical and Applied Mechanics at the Silesian University of Technology, Gliwice. Since 2017, he is Chairman of scientific board of the Research and Development Centre for Mechanical Appliances "OBRUM" Ltd. He is a long-standing, active member of Polish Society of Theoretical and

Applied Mechanics, and currently he is Member of the Board of this renowned society. His research covers a broad scope of issues from the dynamic problems of continuum mechanics and the usage of numerical methods in stereomechanical analysis of machine models to construction of high-speed tracked vehicles.

His science and research issues are mainly connected with the analysis of dynamic phenomena occurring in electromechanical systems. As far as his scientific activities are concerned, he deals with problems concerning the mechanical engineering discipline and, in particular, with analysis of dynamic phenomena in drive systems of mine shaft, theoretical and empirical modal analysis for identification of dynamic properties of machine sets as well as optimization of their dynamic characteristics, analysis of dynamic phenomena occurring in mechatronic sets, vibroisolation systems, and concurrent simulation including real characteristics of dumping elements of suspension systems, including tank suspensions.

Maciej Trojnacki received his MSc Eng. degree in aeronautic control systems in 1999 and PhD degree in machine construction and use in 2003, both from the Faculty of Mechanical Engineering and Aeronautics, Rzeszów University of Technology, Rzeszów, Poland. He received his DSc in mechanics in 2014 from the Faculty of Mechanical Engineering, Silesian University of Technology, Gliwice, Poland. Since 2008, he has been an expert in the National Centre for Research and Development and since 2015 Member of Program Committee of Science and Technology Conference Automation–Innovations and Future Perspectives, organized annually for over 20 years by Industrial Research Institute for Automation and Measurements (PIAP) in Warsaw, Poland.

Currently, he is Associate Professor and Head of Interdisciplinary Research Division in PIAP as well as Member of Scientific Council of PIAP. His research interests include mechanics, control, artificial intelligence, mechatronics, and robotics. In particular, he is a specialist in ground mobile robots (wheeled, legged, tracked, and hybrid ones) and self-driving cars in terms of construction, modeling, and simulation as well as control.

His most important projects realized in PIAP include "Multi Sensor Anti Sniper System (MUSAS)" financed by European Defence Agency (EDA) and Ministry of Science and Higher Education of Poland, "Integrated Mobile System for Support of Anti-terrorist and Anti-crisis Operations (Proteus)" financed from European Regional Development Fund within Innovative Economy Operation Programme, "Dynamics Modeling of a Four-wheeled Mobile Robot and Tracking Control with Limited Slip of Wheels" financed from the means of National Science Centre of Poland and "Study of New Spatial Scanning Head and Its Modification for Mobile Robots for Mapping of the Environment" financed by the National Centre for Research and Development of Poland within the framework of the Program of Applied Research.

Modelling Industrial Robot Fanuc ARC Mate 100iB in LabVIEW

Wacław Banaś and Kamil Lysek(✉)

Institute of Engineering Processes Automation and Integrated Manufacturing Systems, Faculty of Mechanical Engineering, Silesian University of Technology, Konarskiego 18A, 44-100 Gliwice, Poland
kamil.lysek@polsl.pl

Abstract. The purpose of present thesis was to simulate real trajectory of the industrial robot Fanuc ARC Mate 100iB in LabVIEW. For this purpose the simulation of kinematic and dynamic modules were used. The planer generated the trajectory and the robot moved along it. The ending point was made by a human. The program saved the real trajectory, which was compared to the actual trajectory and the comparison was presented in the charts.

Keywords: Modelling · Simulation

1 Introduction

The theme of the work is modelling the industrial robot Fanuc ARC Mate 100iB in LabVIEW. The aim of the work is the most precise representation of the actual work of the robot. For this purpose modules responsible for calculation of the kinematics and dynamics parameters will be used.

The robot should move along the designated trajectory by the planer and the end points entered by a human. The movement trajectory will be written and compared with the real trajectory. The results will be shown in a chart.

2 Building the Model in LabVIEW and Make a Program

At the beginning, the authors made model in LabVIEW from CAD file. All elements was were rebuilt to simple shapes elements and parameters were entered into each part of the model in properties. The model parameters were shown in Tables 1 and 2.

First table showed a centre of gravity in meters for each components (counted from the base), second showed a mass moment of the inertia in $kg \cdot m^2$ for each surface from each components. The mass moment and the centre of the gravity were calculated from the coordinate system embedded in the centre of the geometry.

J. Świder et al. (Eds.): MECHATRONICS 2017, AISC 934, pp. 1–9, 2019.
https://doi.org/10.1007/978-3-030-15857-6_1

Table 1. Means of gravity.

No of component	Centre of gravity $[10^{-3}\ m]$		
	X	Y	Z
1	1.855	0.103	−15.861
2	9.394	1.786	−10.838
3	−1.626	−8.020	0.100
4	31.424	−2.578	13.293
5	−2.155	−2.837	0.005
6	−3.663	0.974	0.308
7	−1.404	0	0

Table 2. Mass parameters of components.

No of component	Mass [kg]	Mass moments of inertia $[10^{-3}\ kg \times m^2]$					
		Ixx	Iyy	Izz	Ixy	Ixz	Iyz
1	68.6	186.32	257.664	525.062	−0.633	31.361	−0.849
2	30.7	352.25	449.067	444.684	4.6403	72.729	−18.078
3	7.7	988.473	960.035	068.797	−0.337	−2.872	109.386
4	20.6	76.949	91.632	109.495	9.682	3.696	−0.783
5	8	29.565	850.569	864.291	−22.844	0.014	−0.002
6	2	4.171	9.259	8.875	−0.103	0.244	−0.017
7	0.4	0.077	0.004	0.004	0	0	0

The model of robot in LabVIEW with simple shapes elements and implemented parameters was shown in Fig. 1.

In order to utilise this robot model, Denavit Hartenberg parameters were determined as was shown in Table 3.

Where:

Θ_i – joint angle;
d_i – link offset;
a_i – link length;
α_i – link twist.

Each calculated homogeneous transformation was shown below (1–7):

$$A_i = \begin{bmatrix} cos\theta_i & -sin\theta_i cos\alpha_i & sin\theta_i sin\alpha_i & a_i cos\theta_i \\ sin\theta_i & cos\theta_i cos\alpha_i & -cos\theta_i sin\alpha_i & a_i sin\theta_i \\ 0 & sin\alpha_i & cos\alpha_i & d_i \\ 0 & 0 & 0 & 1 \end{bmatrix} \quad (1)$$

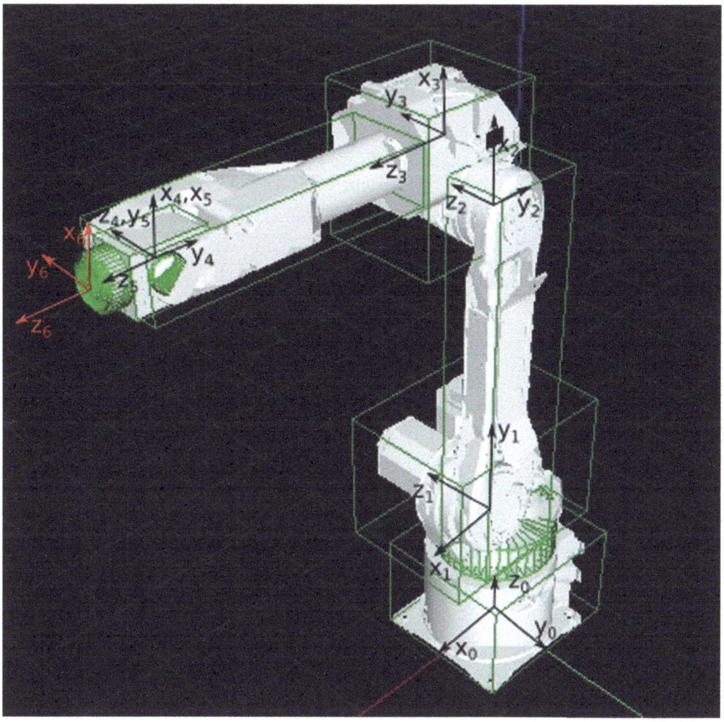

Fig. 1. Model of Fanuc 100iB in LabVIEW with coordinate systems.

Table 3. Denavit–Hartenberg parameters

Number	Θ_i	d_i	a_i	α_i
1	Θ_1	450	150	$\pi/2$
2	$\Theta_2 + \pi/2$	0	600	0
3	Θ_3	0	100	$\pi/2$
4	Θ_4	615	0	$-\pi/2$
5	Θ_5	0	0	$\pi/2$
6	Θ_6	100	0	0

$$A_1 = \begin{bmatrix} cos\theta_1 & 0 & sin\theta_1 & 0.15 * cos\theta_1 \\ sin\theta_1 & 0 & -cos\theta_1 & 0.15 * sin\theta_1 \\ 0 & 1 & 0 & 0.45 \\ 0 & 0 & 0 & 1 \end{bmatrix} \quad (2)$$

$$A_2 = \begin{bmatrix} -sin\theta_2 & -cos\theta_2 & 0 & -0.6 * sin\theta_2 \\ cos\theta_2 & -sin\theta_2 & 0 & 0.6 * cos\theta_2 \\ 0 & 0 & 1 & 0 \\ 0 & 0 & 0 & 1 \end{bmatrix} \quad (3)$$

$$A_3 = \begin{bmatrix} cos\theta_3 & 0 & sin\theta_3 & 0.1 * cos\theta_3 \\ sin\theta_3 & 0 & -cos\theta_3 & 0.1 * sin\theta_3 \\ 0 & 1 & 0 & 0 \\ 0 & 0 & 0 & 1 \end{bmatrix} \quad (4)$$

$$A_4 = \begin{bmatrix} cos\theta_4 & 0 & -sin\theta_4 & 0 \\ sin\theta_4 & 0 & cos\theta_4 & 0 \\ 0 & -1 & 0 & 0.615 \\ 0 & 0 & 0 & 1 \end{bmatrix} \quad (5)$$

$$A_5 = \begin{bmatrix} cos\theta_5 & 0 & sin\theta_5 & 0 \\ sin\theta_5 & 0 & -cos\theta_5 & 0 \\ 0 & 1 & 0 & 0 \\ 0 & 0 & 0 & 1 \end{bmatrix} \quad (6)$$

$$A_6 = \begin{bmatrix} cos\theta_6 & -sin\theta_6 & 0 & 0 \\ sin\theta_6 & cos\theta_6 & 0 & 0 \\ 0 & 0 & 1 & 0.1 \\ 0 & 0 & 0 & 1 \end{bmatrix} \quad (7)$$

The transform of the end of the robot was shown in formulas 8 and 9.

$$T_6^0 = T_5^0 * A_6 = A_1 * A_2 * A_3 * A_4 * A_5 * A_6 \quad (8)$$

$$T_6^0 = \begin{bmatrix} A & B & C & D \\ E & F & G & H \\ I & J & K & L \\ 0 & 0 & 0 & 1 \end{bmatrix} \quad (9)$$

Where:

$$\mathrm{A} = (((-\cos\Theta_1\cos\Theta_2\sin\Theta_3 - \cos\Theta_1\cos\Theta_3\sin\Theta_2)\cos\Theta_4 + \sin\Theta_1\sin\Theta_4)\cos\Theta_5 + (-\cos\Theta_1\cos\Theta_2\cos\Theta_3 + \cos\Theta_1\sin\Theta_2\sin\Theta_3)\sin\Theta_5)\cos\Theta_6 + (-(-\cos\Theta_1\cos\Theta_2\sin\Theta_3 - \cos\Theta_1\cos\Theta_3\sin\Theta_2)\sin\Theta_4 + \sin\Theta_1\cos\Theta_4)\sin\Theta_6,$$

$$B = -(((-\cos\Theta_1\cos\Theta_2\sin\Theta_3 - \cos\Theta_1\cos\Theta_3\sin\Theta_2)\cos\Theta_4 + \sin\Theta_1\sin\Theta_4)\cos\Theta_5 + (-\cos\Theta_1\cos\Theta_2\cos\Theta_3 + \cos\Theta_1\sin\Theta_2\sin\Theta_3)\sin\Theta_5)\sin\Theta_6 + (-(-\cos\Theta_1\cos\Theta_2\sin\Theta_3 - \cos\Theta_1\cos\Theta_3\sin\Theta_2)\sin\Theta_4 + \sin\Theta_1\cos\Theta_4)\cos\Theta_6,$$

$$C = ((-\cos\Theta_1\cos\Theta_2\sin\Theta_3 - \cos\Theta_1\cos\Theta_3\sin\Theta_2)\cos\Theta_4 + \sin\Theta_1\sin\Theta_4)\sin\Theta_5 - (-\cos\Theta_1\cos\Theta_2\cos\Theta_3 + \cos\Theta_1\sin\Theta_2\sin\Theta_3)\cos\Theta_5,$$

$$D = (0.1(-\cos\Theta_1\cos\Theta_2\sin\Theta_3 - \cos\Theta_1\cos\Theta_3\sin\Theta_2)\cos\Theta_4 + 0.1\sin\Theta_1\sin\Theta_4)\sin\Theta_5 - (-0.1\cos\Theta_1\cos\Theta_2\cos\Theta_3 + 0.1\cos\Theta_1\sin\Theta_2\sin\Theta_3)\cos\Theta_5 + 0.615\cos\Theta_1\cos\Theta_2\cos\Theta_3 - 0.615\cos\Theta_1\sin\Theta_2\sin\Theta_3 - 0.1\cos\Theta_1\sin\Theta 2\cos\Theta_3 - 0.1\cos\Theta_1\cos\Theta_2\sin\Theta_3 - 0.6\cos\Theta_1\sin\Theta_2 + 0.15\cos\Theta_1,$$

$$E = (((-\cos\Theta_2\sin\Theta_1\sin\Theta_3 - \cos\Theta_3\sin\Theta_1\sin\Theta_2)\cos\Theta_4 - \cos\Theta_1\sin\Theta_4)\cos\Theta_5 + (-\cos\Theta_2\cos\Theta_3\sin\Theta_1 + \sin\Theta_1\sin\Theta_2\sin\Theta_3)\sin\Theta_5)\cos\Theta_6 + (-(-\cos\Theta_2\sin\Theta_1\sin\Theta_3 - \cos\Theta_3\sin\Theta_1\sin\Theta_2)\sin\Theta_4 - \cos\Theta_1\cos\Theta_4)\sin\Theta_6,$$

$$F = -(((-\cos\Theta_2\sin\Theta_1\sin\Theta_3 - \cos\Theta_3\sin\Theta_1\sin\Theta_2)\cos\Theta_4 - \cos\Theta_1\sin\Theta_4)\cos\Theta_5 + (-\cos\Theta_2\cos\Theta_3\sin\Theta_1 + \sin\Theta_1\sin\Theta_2\sin\Theta_3)\sin\Theta_5)\sin\Theta_6 + (-(-\cos\Theta_2\sin\Theta_1\sin\Theta_3 - \cos\Theta_3\sin\Theta_1\sin\Theta_2)\sin\Theta_4 - \cos\Theta_1\cos\Theta_4)\cos\Theta_6,$$

$$G = ((-\cos\Theta_2\sin\Theta_1\sin\Theta_3 - \cos\Theta_3\sin\Theta_1\sin\Theta_2)\cos\Theta_4 - \cos\Theta_1\sin\Theta_4)\sin\Theta_5 - (-\cos\Theta_2\cos\Theta_3\sin\Theta_1 + \sin\Theta_1\sin\Theta_2\sin\Theta_3)\cos\Theta_5,$$

$$H = (0.1(-\cos\Theta_2\sin\Theta_1\sin\Theta_3 - \cos\Theta_3\sin\Theta_1\sin\Theta_2)\cos\Theta_4 - 0.1\cos\Theta_1\sin\Theta_4)\sin\Theta_5 - (-0.1\cos\Theta_2\cos\Theta_3\sin\Theta_1 + 0.1\sin\Theta_1\sin\Theta_2\sin\Theta_3)\cos\Theta_5 + 0.615\cos\Theta_2\cos\Theta_3\sin\Theta_1 - 0.615\sin\Theta_1\sin\Theta_2\sin\Theta_3 - 0.1\sin\Theta_1\sin\Theta_2\cos\Theta_3 - 0.1\sin\Theta_1\cos\Theta_2\sin\Theta_3 - 0.6\sin\Theta_1\sin\Theta_2 + 0.15\sin\Theta_1,$$

$$I = ((\cos\Theta_2\cos\Theta_3 - \sin\Theta_2\sin\Theta_3)\cos\Theta_4\cos\Theta_5 + (-\cos\Theta_2\sin\Theta_3 - \cos\Theta_3\sin\Theta_2)\sin\Theta_5)\cos\Theta_6 - (\cos\Theta_2\cos\Theta_3 - \sin\Theta_2\sin\Theta_3)\sin\Theta_4\sin\Theta_6,$$

$$J = -((\cos\Theta_2\cos\Theta_3 - \sin\Theta_2\sin\Theta_3)\cos\Theta_4\cos\Theta_5 + (-\cos\Theta_2\sin\Theta_3 - \cos\Theta_3\sin\Theta_2)\sin\Theta_5)\sin\Theta_6 - (\cos\Theta_2\cos\Theta_3 - \sin\Theta_2\sin\Theta_3)\sin\Theta_4\cos\Theta_6,$$

$$K = (\cos\Theta_2\cos\Theta_3 - \sin\Theta_2\sin\Theta_3)\cos\Theta_4\sin\Theta_5 - (-\cos\Theta_2\sin\Theta_3 - \cos\Theta_3\sin\Theta_2)\cos\Theta_5,$$

$$L = 0.45 + (0.1\cos\Theta_2\cos\Theta_3 - 0.1\sin\Theta_2\sin\Theta_3)\cos\Theta_4\sin\Theta_5 - (-0.1\cos\Theta_2\sin\Theta_3 - 0.1\sin\Theta_2\cos\Theta_3)\cos\Theta_5 + 0.615\cos\Theta_2\sin\Theta_3 + 0.615\sin\Theta_2\cos\Theta_3 + 0.1\cos\Theta_2\cos\Theta_3 - 0.1\sin\Theta_2\sin\Theta_3 + 0.6\sin\Theta_2,$$

When the robot parameters were determined, a program was created (it was shown in Fig. 2). The program works according to the algorithms shown in the Fig. 3.

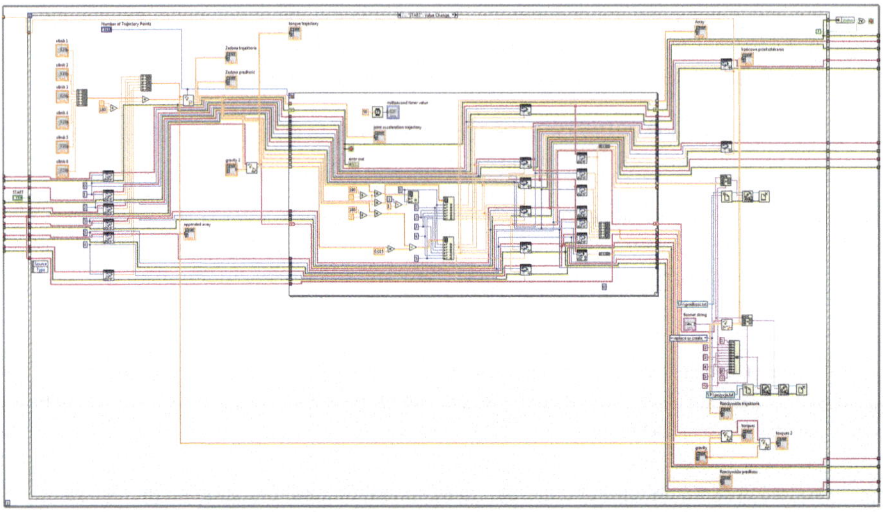

Fig. 2. The created program.

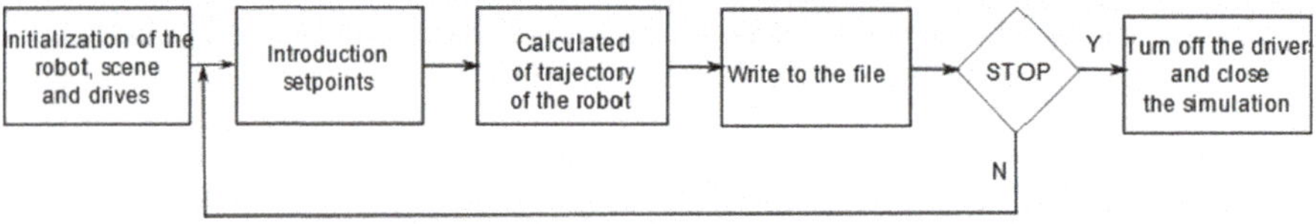

Fig. 3. Algorithm of the program.

The following function blocks were used to create the program:

1. Initialize serial arm VI was used for initialization model of robot arm in LabView environment.
2. Initial motor and Reads motor positions were used for initialization motor model in LabView environment and read motor positions in degrees.
3. Knop was used for introduction setpoint by user.
4. Move motor was used for moving robot arm. Each motor has one move motor function block.
5. Acquire kinematics parameters was used for generating kinematics parameters of robot.
6. Generate joint trajectory – was used for generating joint trajectory for robot from start joint positions, end joint positions and number of intermediate points. This block generated positions and velocities of trajectory.
7. Forward kinematics VI – was used for transforming from joint position trajectory to end effector transform (Cartesian trajectory).
8. Array to spreadsheet string was used to converting an array variable into text
9. Open/Create/Replace file, Write to text file and Close file were used to write measure value into xml file.
10. Close was used to turn off the motor.

3 Tests

Tests were made in program with data from Table 4. As it can be seen, the 15 tests were done with 3 groups of different numbers of interpolated points. Each group was tested with 5 different positions of robot arm.

First test was executed for 36 points of interpolation and end positions as −10° on each of robot arms (which were measured relative to the positioning illustrated in Fig. 1). The result of test was shown in Fig. 4. As it can be seen the calculated value was not the same as real value.

Second and third tests were executed for the same end positions of robot arms, but for different numbers of interpolation points (second with 10 points, third with 50 points). The results were shown in Fig. 5 (2nd test) and Fig. 6 (3rd test). As it can be seen the real value was the closest calculated value during test no 2.

Next three tests were executed for positions as −45° on each of robot arms. The results of tests were shown in Fig. 7, 8 and 9. As it can be seen in these cases also the calculated values were not the same as real values. But in these cases the results of real values were closed to the calculated value.

The last three tests which have been described in that article were executed for position 0° on first arm, −20° on second arm, −30° on third arm, −10° on fourth arm, −20° on fifth arm and −30° on the last arm. The results of tests were shown in Fig. 10, 11 and 12. As it can be seen in these cases, also the calculated values were not the same as real values and the real value was the farthest to calculated value during test no. 8.

Table 4. Data for tests.

No of test	Number of points	Position					
		1	2	3	4	5	6
1	36	−10°	−10°	−10°	−10°	−10°	−10°
2	10						
3	50						
4	36	−45°	−45°	−45°	−45°	−45°	−45°
5	10						
6	50						
7	36	0	−20°	−30°	−10°	−20°	−30°
8	10						
9	50						
10	36	−80°	−15°	−45°	−80°	−15°	−45°
11	10						
12	50						
13	36	−90°	−30°	−50°	−15°	−10°	−45°
14	10						
15	50						

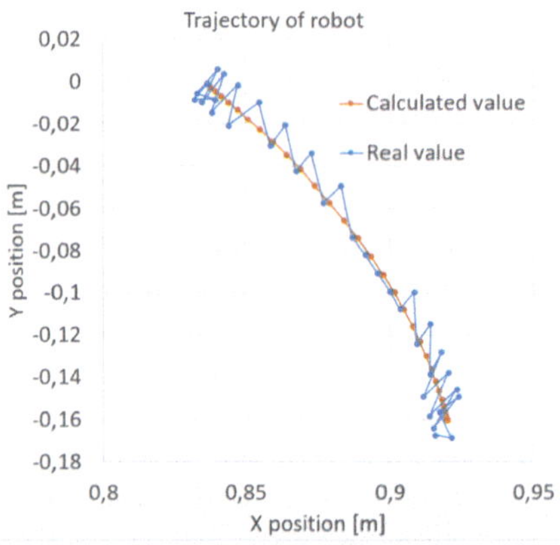

Fig. 4. The results of 1st test.

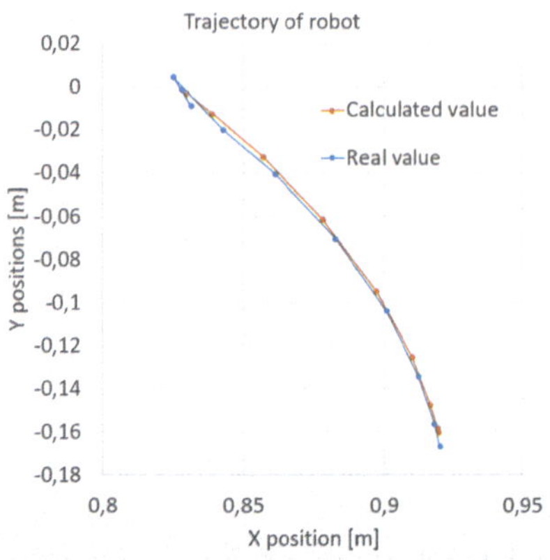

Fig. 5. The results of 2nd test.

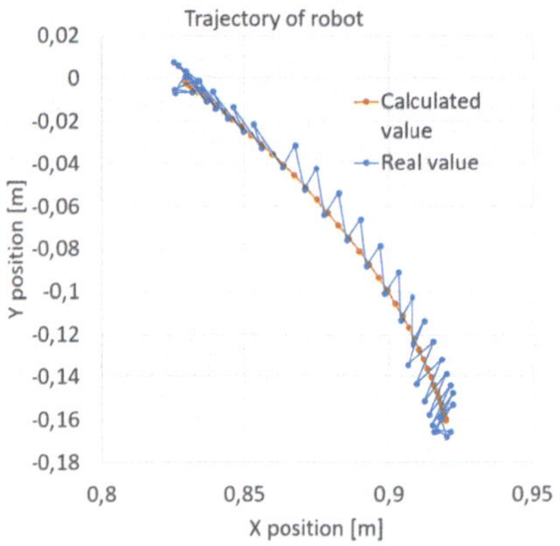

Fig. 6. The results of 3rd test.

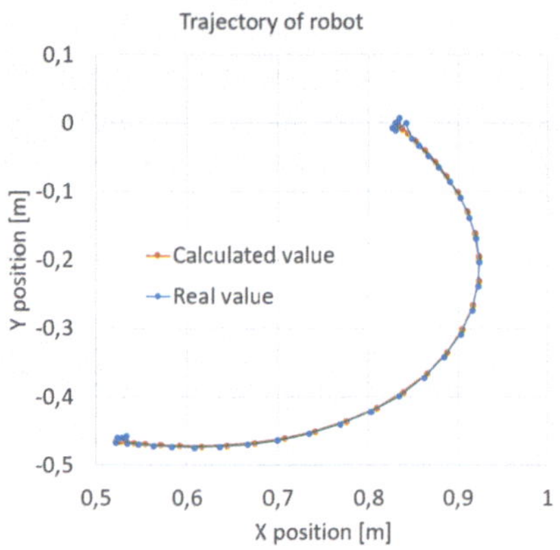

Fig. 7. The results of 4th test.

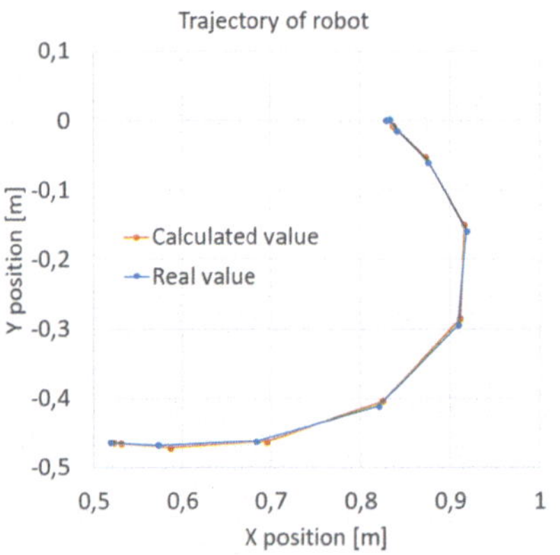

Fig. 8. The results of 5th test.

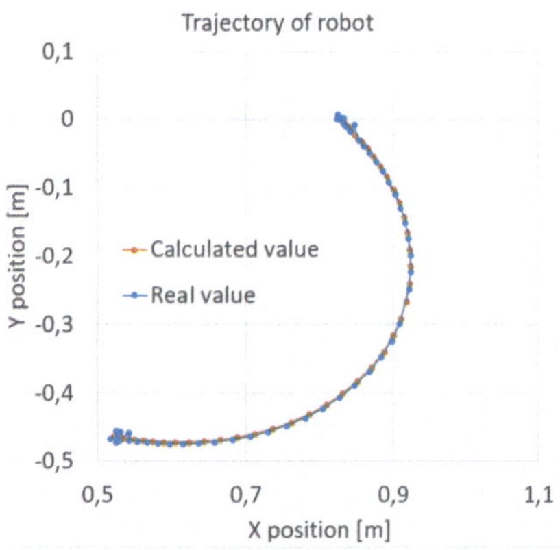

Fig. 9. The results of 6th test.

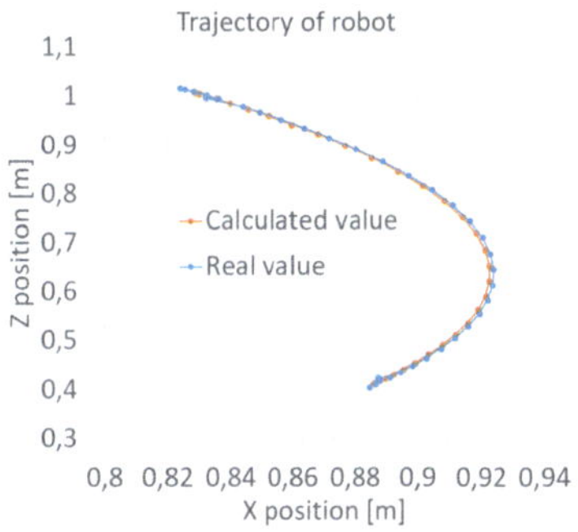

Fig. 10. The results of 7th test.

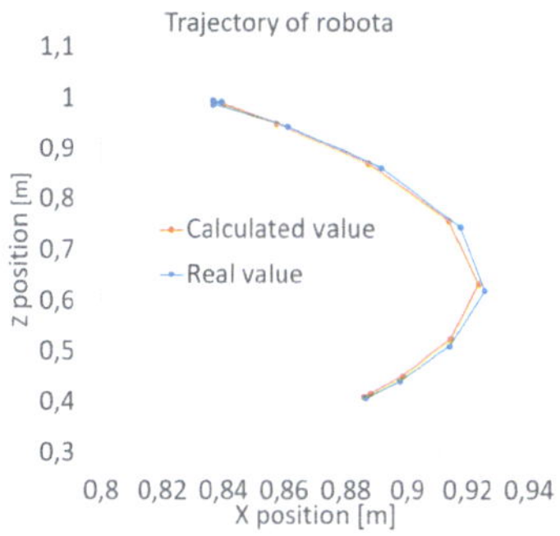

Fig. 11. The results of 8th test.

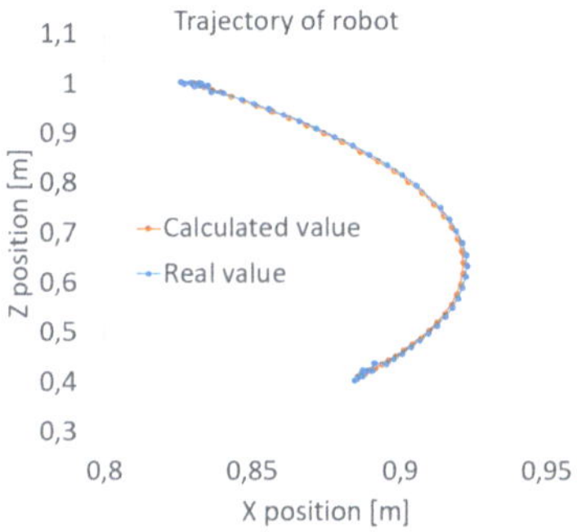

Fig. 12. The results of 9th test.

4 Conclusions

The theme of the work was modelling the industrial robot Fanuc ARC Mate 100iB in LabVIEW. The aim of the work was to obtain the most precise representation of the actual work of the robot. The model was simplified to a simple figure and fixed with drive. The force on the drive was entered by special algorithm based on special modules. These modules for calculation of the kinematics and dynamics parameters were not enough for projection of real robot arm trajectory. The real results had different trajectory than they should have. The reason for these results could be use of too simply algorithm which did not compensate the gravity force sufficiently enough.

References

1. Craig, J.J.: Introduction to Robotics: Mechanics and Control, 4th edn. Pearson, London (2017)
2. Spong, M.W., Vidyasagar, M.: Robots Dynamics and Control, 2nd edn. Wiley, New York (2004)
3. Gwiazda, A.: Modular industrial robots as the tool of process automation in robotized manufacturing cells. In: 2015 IOP Conference (2015)

Recognition of Text Commands and Control of the Mobile Robot Starter Kit 2.0

Wacław Banaś and Bartłomiej Nalepa(✉)

Institute of Engineering Processes Automation and Integrated Manufacturing Systems, Faculty of Mechanical Engineering, Silesian University of Technology, Konarskiego 18A, 44-100 Gliwice, Poland
{waclaw.banas, bartlomiej.nalepa}@polsl.pl

Abstract. This article presents an algorithm for recognizing text commands and how to send information about a text command to a mobile robot Starter Kit 2.0. The first chapters deal with methods of reading image from Kinect sensors, colour palette changes, thresholds, and central moments calculations. The methods for finding closed contours in the analysed image are presented. Then it shows how to divide the text into rows. Each line was also divided into singular letters that were sent to the defined function. Letter recognition was accomplished by creating a new black image and by embedding the characteristic points. Then the image containing the letter was compared to the standard image using the AND operation. The results were compared with the library created with the description of the point's characteristic for each letter. All image processing was done in Python with the Raspbian Jessie. Python has also added libfreenect and OpenCV libraries. The last chapter shows how to communicate between Raspberry Pi3 and the mobile robot Starter Kit 2.0.

Keywords: Robot · LabVIEW · Vision

1 Introduction

Analysed problems associated with robotics [5–9] and digital image processing [1, 10–12] allow to define the need to determine a new algorithm that recognizes the letters forming the command. The main goal of this article was to develop an algorithm for text recognition and describe how to integrate the Microsoft Kinect sensor with the Mobile Starter Kit 2.0.

A man in an unknown place tries to get as much information as possible about the surroundings. Basic data, i.e. the type of environment (forest, room, street), substrate or danger (fire, cliff) are immediately recognized. However, the data collected may not be sufficient for a full description of the environment. Potential hazards may be invisible and information about them can be found on the placards (e.g. "Minefield Note!"). This information enhances the knowledge of the human being about the environment in which it is located or indicates the path in case of movement. Mobile work is

J. Świder et al. (Eds.): MECHATRONICS 2017, AISC 934, pp. 10–19, 2019.
https://doi.org/10.1007/978-3-030-15857-6_2

increasingly being built to replace a human in performing basic tasks. It is essential to develop an algorithm that recognizes the text on different placards in the work environment of the robot.

Text recognition was performed by a prepared set of placards, with commands "turn right" or "turn left". The following devices were used in the project: Mobile Starter Kit 2.0, Raspberry Pi 3 and Microsoft Kinect. The system consisted of two connected control systems. The first (sbRIO-9632) was responsible for controlling the mobile robot. The second (Raspberry Pi 3) implemented a developed text recognition algorithm.

2 Text Recognition

Image processing was performed on a Raspberry Pi 3 and Microsoft Kinect vision sensor. Algorithms were written in Python using the following libraries:

- OpenCV - image processing with documentation [2],
- libfreenect - Microsoft Kinect drivers [3],
- NumPy - operations on matrices with documentation [4].

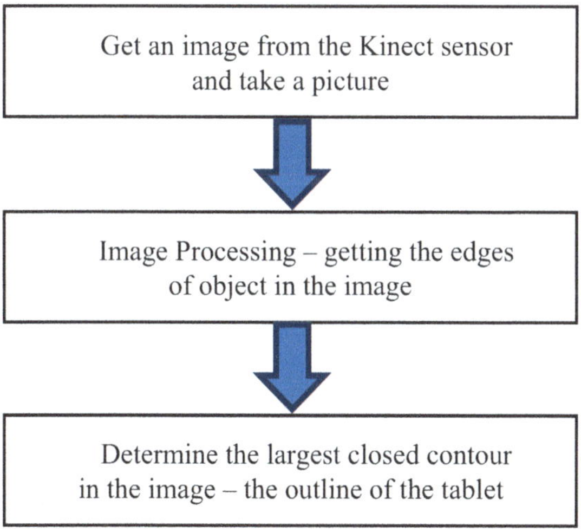

Fig. 1. Image processing procedure part 1 – identification of the largest contour

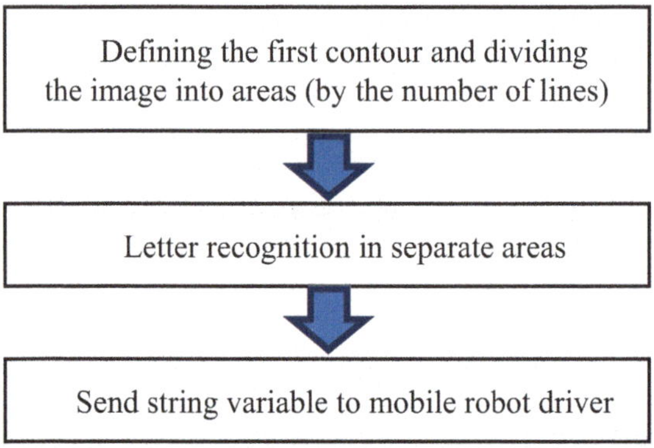

Fig. 2. Image processing procedure part 2 – recognition of letters (compare [2, 13])

A text recognition algorithm has been developed that has been broken down into different stages (see Figs. 1 and 2).

2.1 Picture from the Kinect Sensor

In the first stage of image processing, BGR (Blue Green Red) colour conversion was applied to the HSV (Hue Saturation Value) colour palette and the image threshold was as follows [1, 12]:

$$g(x,y) = \begin{cases} 0, & f(x,y) \leq T_1 \\ 1, & T_1 < f(x,y) \leq T_2 \\ 0, & f(x,y) > T_2 \end{cases} \tag{1}$$

where:

(x, y) – coordinates of object point,
T_1, T_2 – threshold, where $T_1 =$ `color_z_min` and $T_2 =$ `color_z_max`.

As a result of libfreenect [3] and OpenCV [2] libraries, an image was obtained in the BGR color palette. The next step was to convert the BGR image to the HSV color palette using the `cv2.cvtColor` command. Then set the binary range of the image of `color_z_min` and `color_z_max`. The rollback was done using the `cv2.inRange` command.

The main task of this part of the program is binarization and then calculating the central moments of the image obtained. Moment of first order `moments ['m00']` allows you to determine the size of an area that occupies an object whose colour is within a defined area. Exceeding the value defined in the variable area will cause Kinect to take a snapshot and will stop the robot and perform certain imaging operations. In the last step, the program saves the image to a file using the `cv2.imwrite` command.

The first-degree central moments are calculated from the following relationship [1, 10, 11]:

$$\mu_{00} = \sum_x \sum_y f(x, y) \tag{2}$$

where:

$f(x, y)$ – "2-D continuous function":

$$f(x, y) = \begin{cases} 0 \rightarrow \textit{object pixels} \\ 1 \rightarrow \textit{background pixels} \end{cases} \tag{3}$$

Fig. 3. Training image made by Kinect sensor (compare [2, 13])

The Kinect sensor has taken a picture of a sample text command (see Fig. 3). The plate consisted of a printed sheet of paper glued to the wall.

2.2 Preprocessing the Image

In the first part of the program the image is loaded and then processed into a binary form. At first, the image is loaded with `cv2.imread.` The image is subtracted 10 times by the `cv2.resize` function. The following are then performed:

- converting an image from a BGR colour palette to a grayscale image - `cv2.cvtColor`,
- edge detection by determining gradient intensity - `cv2.Canny`.

Based on the processed image, you can determine the outlines in the image (see Fig. 4).

Fig. 4. The effect of the program (compare [2, 13])

2.3 Determine the Greatest Contour

As a result of the image on which the contours are located, the contour with the largest surface area - the plate - should be defined. The program code consists of the following commands:

– defining the parameters of each contour – `cv2.findContours`,
– number of contours - `len (contours)`,
– definition of the matrix containing areas of the individual contours - area,
– first loop for assigning surface contours `(cv2.contourArea)` to individual vector indices,
– second loop for looking for the highest value of the surface area,
– defining the rectangle parameters described on the largest defined contour - `cv2.boundingRect`,
– contour drawing commands `(cv2.rectangle)` and image display (`cv2.imshow`).

The result of the operation of the program is the ability to specify the plaque field and background cut (see Fig. 5).

Fig. 5. The effect of the program (compare [2, 13])

2.4 Defining Text Lines

The next operation is to cut out part of the area. Then the program defines the first line of text and defines its coordinates.

Fig. 6. The effect of the program (compare [2, 13])

Fig. 7. Cut out part of the picture (compare [2, 13])

The presented algorithm allows to cut out one line of text from the image (see Figs. 6 and 7).

2.5 Specifying the Contour Coordinates

As a result of separating the area in which the line of text was located, it was necessary to specify the outer contours of the letters.

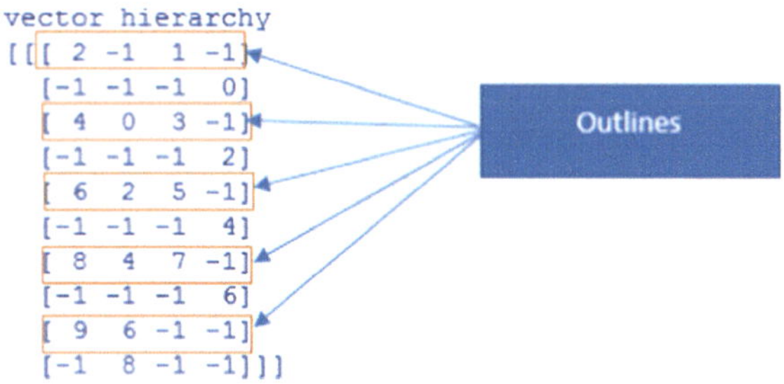

Fig. 8. Print parameters from the console (compare [2, 13])

In the presented printout from the Python console there is a vector hierarchy of contours (see Fig. 8).

2.6 Point Characteristic Method

The contour containing the letter is sent to the function (see Figs. 9 and 10). This function creates a black image. Next, a vector containing characteristic points is created to facilitate the recognition of the letter. The `FOR` loop is added to the black image of the characteristic point. Then the logical AND operation is executed by the `cv2.bitwise_and` command. The logical product components are a binary image containing a letter and a black image with a single point. After performing the algorithm, the area fields are checked for the central moments (`moments ['m00']`). If by logical operation the surface area will be greater than 0 then the vector is written to 1. If the result is a black image then the next vector index is written to 0 (see Figs. 9 and 10).

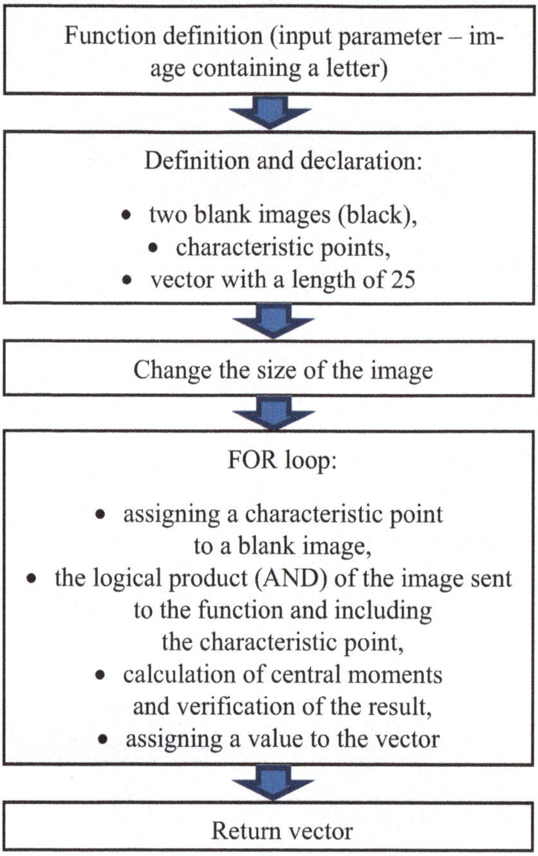

Fig. 9. The algorithm of the function recognizing letters (compare [2, 13])

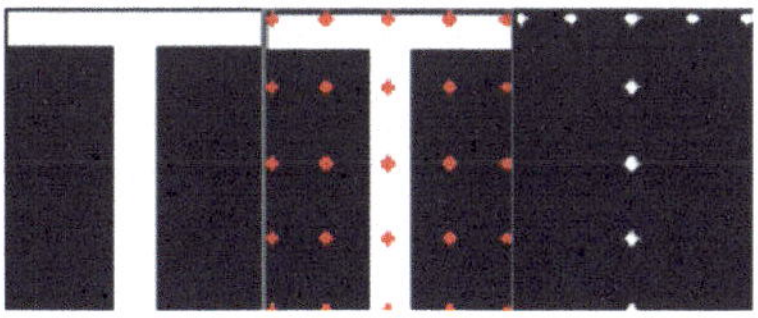

Fig. 10. The effect of the algorithm (compare [2, 13])

Fig. 11. The final result of the algorithm (compare [2, 13])

The end result of the algorithm is to write the letters in the image and to combine them into one string variable (see Fig. 11).

2.7 Microsoft Kinect Integration and Starter Kit 2.0

Image processing was performed in Python on Raspberry Pi 3. The communication between the Raspberry Pi 3 and the mobile robot Starter Kit 2.0 was performed via the RS-232 serial port using a MAX232 based chip. In LabVIEW (robot programming), a program has been implemented with the following components: read the string variable from the serial port (see Fig. 12), processing string variable in SubVI "STRING", comparison of received data from serial port with patterns.

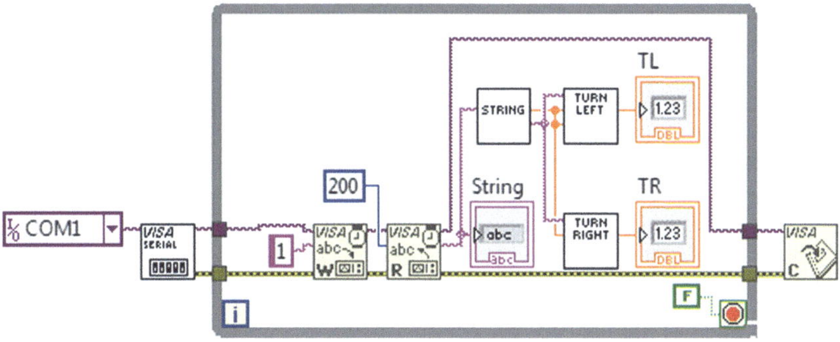

Fig. 12. Block diagram

The TL and TR variables received indicate whether the image is labeled with a right or left turn command. Variables can be used as input to the CASE structure, which will control the rotational speed of the mobile robot engines.

3 Summary and Conclusions

OpenCV, Libfreenect, and basic Python libraries were used for text recognition. The image was processed using binarization and contour definition. An algorithm based on checking the characteristic points of an image is presented. The system of points characteristic of a given letter is written in a specially prepared library. The task of the program was to verify the vector obtained from the analysis of each contour with patterns. The resulting string variable containing the string was sent from Raspberry Pi 3 to the mobile Starter Kit 2.0 via the RS-232 serial port. In Robot LabVIEW, a robot program has been developed to verify the received text with the "turn left" and "turn right" command patterns. In case of preparation of the plate according to the recommendations and if prepared character will occupy a defined surface determined in the work, the algorithm should show high efficiency of operation.

References

1. Gonzalez, R.C., Woods, R.E.: Digital Image Processing, 2 edn. Prentice Hall, Upper Saddle River
2. Documentation of the OpenCV library. http://docs.opencv.org/3.1.0/d6/d00/tutorial_py_root.html. Accessed 27 June 2017
3. Documentation of the libfreenect library. https://github.com/OpenKinect/libfreenect. Accessed 27 June 2017
4. Documentation of the NumPy library. https://docs.scipy.org/doc/numpy-dev/reference/. Accessed 27 June 2017
5. Banaś, W., Ćwikla, G., Foit, K., et al.: Experimental determination of dynamic parameters of an industrial robot. In: IOP Conference Series: Materials Science and Engineering, vol. 227, p. 012012 (2017)
6. Banaś, W., Ćwikla, G., Foit, K., et al.: Modelling of industrial robot in LabView robotics. In: IOP Conference Series: Materials Science and Engineering, vol. 227, p. 012012 (2017)
7. Hryniewicz, P., Banaś, W., Foit, K., et al.: Modelling cooperation of industrial robots as multi-agent systems. In: IOP Conference Series: Materials Science and Engineering, vol. 227, p. 012061 (2017)
8. Banaś, W., Gwiazda, A., Monica, Z., et al.: Analysis of the position of robotic cell components and its impact on energy consumption by robot. In: IOP Conference Series: Materials Science and Engineering, vol. 145, p. 052017 (2016)
9. Banaś, W., Gwiazda, A., Monica, Z., et al.: Modelling and simulation tooling controlled by the PLC in the robot cell in NX. In: IOP Conference Series: Materials Science and Engineering, vol. 145, p. 052016 (2016)
10. Sit, A., Kihara, D.: Comparison of image patches using local moment invariants. IEEE Trans. Image Process. https://doi.org/10.1109/TIP.2014.2315923
11. Huang, Z., Leng, J.: Analysis of Hu's moment invariants on image scaling and rotation. ECU Publications Pre. https://doi.org/10.1109/ICCET.2010.5485542

12. Hamdaoui, F., Sakly, A., Mtibaa, A.: An efficient multi level thresholding method for image segmentation based on the hybridzation of modified PSO and Otsu's method. Springer International Publishing Switzerland. https://doi.org/10.1007/978-3-319-11017-2_29
13. Recognizing digits with OpenCV and Python. https://www.pyimagesearch.com/2017/02/13/recognizing-digits-with-opencv-and-python/. Accessed 06 Apr 2018

System for Positioning of the Roadheader in Roadways of Hard Coal Mines

Sławomir Bartoszek[1(✉)] and Gabriel Kost[2]

[1] KOMAG Institute of Mining Technology, Gliwice, Poland
sbartoszek@komag.eu

[2] Institute of Engineering Processes Automation and Integrated Manufacturing Systems, Silesian University of Technology, Gliwice, Poland

Abstract. The system for positioning of the roadheader operating in hard coal mines roadways developed in KOMAG is presented. The system, which uses propagation of radio and ultrasonic waves is not typical as regards known systems which mainly use laser technology. Design of the system, measuring technology as well as the method for determination of roadheader's position and orientation in roadways is described. Two versions of the system configuration are presented. In the first one, geometry of arrangement of ultrasound receivers on a roadheader enables determination of its orientation. In the latter one, one of the parameters describing roadheader's orientation is determined from the position of ultrasound receivers, and two other parameters are determined from readings of inclinometer. Components of physical model of the positioning system, which was tested in mine experimental roadway are presented. Results of tests as well as accuracy of the system readings are given.

Keywords: Mine workings · Roadheader · Directional navigation · Localization · Orientation

1 Introduction

Development of roadways in hard coal mines is the main process in coal extraction. The process opens coal seams for mining, enabling ventilation and run-of-mine transportation. The roadways are driven by mining/cutting method with roadheader transverse or longitudinal cutter head (Fig. 1). Conditions of roadheader operation in front of the roadway are extremely difficult. There are high concentrations of dust as well as natural hazards e.g. rock and gas (methane) bursts. That is why time of work of the personnel servicing the roadway complex in the face front, including roadheader's operators should be maximally reduced.

Automation of cutting operation is one of the method for solving the problem [7, 8, 10]. It enables the roadheader operator and other personnel from the area of face front withdrawal. Automation concerns mainly cutting process, which is realized in turn with roof supporting and stabilizing process, which secures the roof against roof fall. However, the roof support process is still realized manually.

J. Świder et al. (Eds.): MECHATRONICS 2017, AISC 934, pp. 20–29, 2019.
https://doi.org/10.1007/978-3-030-15857-6_3

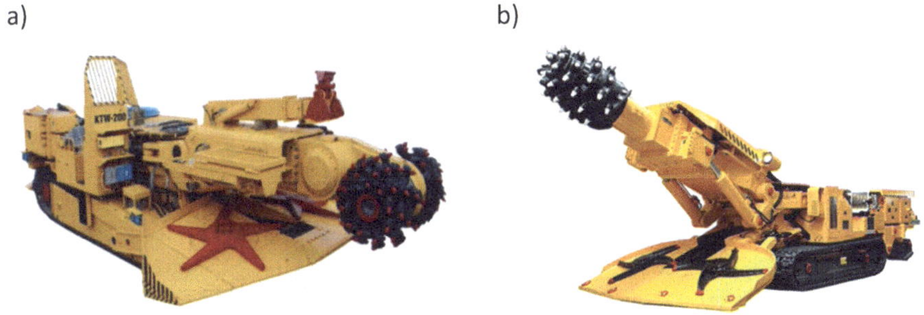

Fig. 1. Roadheaders: (a) KTW-200 made by KOPEX with transverse cutter head [11], (b) EBZ200 made by XCMG with longitudinal cutter head [12]

Roadheader operation in autonomous or remote control mode requires its accurate positioning, i.e. determination of its current position and orientation in roadway.

Positioning of the roadheader in a hard coal mine roadway is a difficult task and it has not been not fully solved so far. The present systems, which use laser technology appeared to be not useful due to many limitations.

The positioning system, which uses propagation of radio and ultrasonic waves, developed in KOMAG is described in the following chapters.

2 Design and Principle of Operation

Before starting the project on development of new system for positioning the roadheaders, currently used roadheaders' positioning systems, GPS as well as the methods used in indoor positioning system (IPS) [9] and in mobile robot positioning systems [6] were analysed. Taking into account underground mine conditions and special character of roadheaders operation, the method for determination of the roadheader's position in relation to the fixed reference system in the roadheader surrounding without necessity of referring to its movement was selected. Measuring technology and available measuring instruments were analysed [4]. Tools such as ultrasound transducers, rangefinders, scanners, laser total stations, radio communication modules and radio units for distance measurements were divided into nine groups. Devices belonging to each group were analysed as regards their possible use in hard coal mines in the roadheader positioning system [4]. As the result, technology using the ultrasound waves was selected. Measurements with use of ultrasound waves require additional radio communication modules. Detailed information about the analysis and selected transceivers can be found in [4].

The principle of operation as well as main components of the developed roadheader's positioning system are presented in Fig. 2. Position of the roadheader is determined in relation to transmitters (beacons) installed on roadway walls or roof. In the case of roadway of hard coal mine the transmitters will be fixed to arches of the ŁP roof support. The transmitters are independent, active devices having the element generating ultrasound waves, radio module, power supply source and magnetic holder which enables its attachment to the roadway support arches.

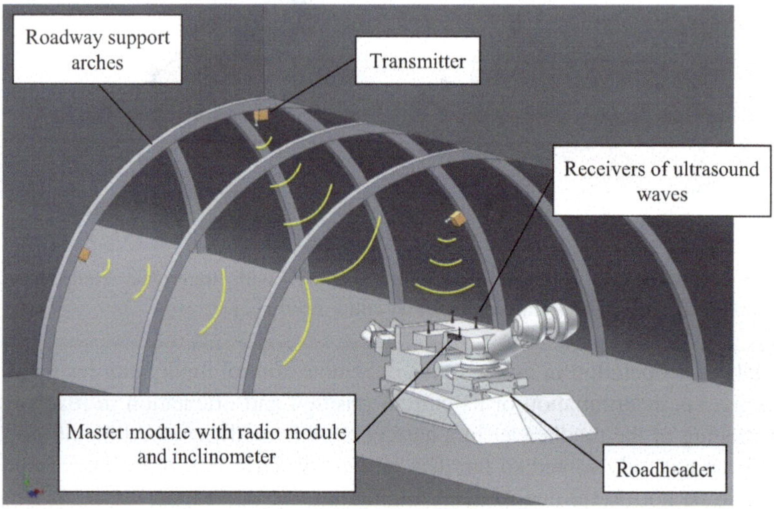

Fig. 2. Concept of the method for positioning the roadheader [1–4]

Receivers of ultrasound waves, master module with radio module and inclinometer are installed on the roadheader.

Coordinates of the transmitting points on the roadway support are the input data for calculation of position and orientation of roadheader. With the advance of roadway development, new sets of transmitters will be installed. Calibration procedures enabling automatic determination of newly installed transmitters, using the position of previous transmitters and elements installed on the roadheader, are planned in the positioning system.

3 Measuring Technology

Transmission of radio and ultrasound waves was used in measuring the distance between receivers on the roadheader and transmitters on the roadway support. Measured distance will be proportional to time of ultrasound waves propagation between those components. High speed of radio waves propagation and relatively small distances in a roadway were the reason of using the ultrasound waves to increase measurements accuracy. Radio waves play a role of synchronizing and communication signals. The idea of measurements is presented in Fig. 3.

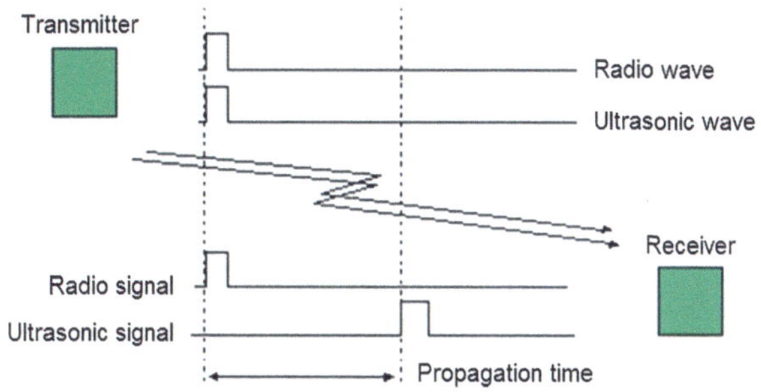

Fig. 3. Measurements of the distance between the transmitter and receiver with use of ultrasound and radio waves [1–4]

The measurements are realized according to TOA (Time of Arrival) rule. At the same time radio waves and ultrasound waves are emitted from the transmitter. Time from the moment of receiving the radio wave in the receiver until detection of ultrasound wave is recorded. The distance between the transmitter and receiver is calculated from the following relationship (1) [4]:

$$d = \Delta t \frac{v_{RF} \cdot v_{US}}{v_{RF} - v_{US}} [m] \quad (1)$$

where: Δt – time difference in detecting both signals in the receiver (in seconds), v_{US} – speed of ultrasound wave propagation (in m/s), v_{RF} – speed of radio wave propagation (in m/s).

As speed of radio waves is about 10^6 times higher than ultrasound wave, it has been assumed that time of propagation of radio wave can be neglected, which made the relationship (1) simpler and can be expressed as [4]:

$$d = v_{US} \cdot \Delta t \ [m] \quad (2)$$

The system emits short beams of ultrasound waves of duration of few ms. To minimize disadvantageous phenomena accompanying the propagation of ultrasound waves i.e. reflection from barriers in a roadway, the moment of detecting the wavefront which passed the shortest way is recorded – so called LOS (Line of Sight) conditions.

4 Position and Orientation of the Roadheader

Position of the roadheader in a roadway is described by two parameters: position and orientation. Reference coordinate system in a roadway (A) as well as local coordinate system (B) on the roadheader begun at the place of installation of one of receivers of ultrasound wave are determined (Fig. 4).

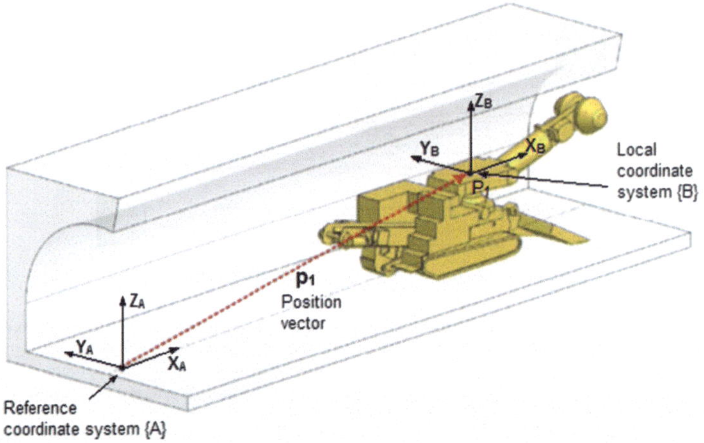

Fig. 4. Coordinate system associated with a roadway and a roadheader [4]

Roadheader's position is defined by the vector of displacement from the beginning of the coordinate system (B) in relation to the reference system (A). Trilateration rule is used in determination of the position coordinates (Fig. 5).

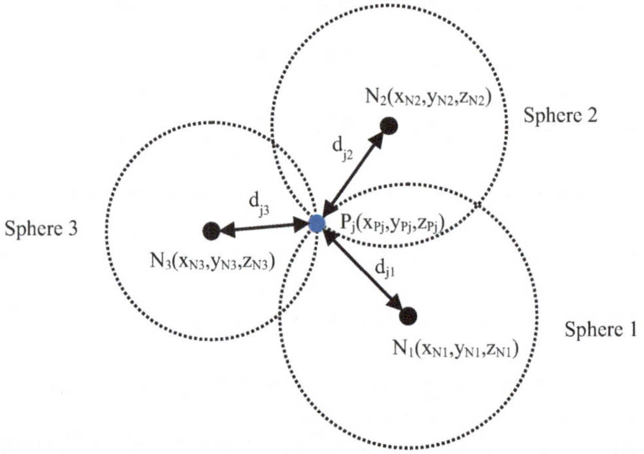

Fig. 5. Trilateration – parameters required for determination of object position (simplified view – projection on 2D plane) [4]

Spheres around transmitters N_i, which radiuses are equal to distances between transmitters installed on the roof support and receiver P_j on the roadheader are defined according to the above mentioned rule. The distance is determined by the following formula (3).

$$d_{ji} = \sqrt{\left(x_{Pj} - x_{Ni}\right)^2 + \left(y_{Pj} - y_{Ni}\right)^2 + \left(z_{Pj} - z_{Ni}\right)^2}[\mathrm{m}] \tag{3}$$

Installation of three transmitters is required for determination of object's position. The following system of three equations should be solved to determine coordinates of object's position (4):

$$\begin{cases} \left(x_{Pj} - x_{N1}\right)^2 + \left(y_{Pj} - y_{N1}\right)^2 + \left(z_{Pj} - z_{N1}\right)^2 = d_{11}^2 \\ \left(x_{Pj} - x_{N2}\right)^2 + \left(y_{Pj} - y_{N2}\right)^2 + \left(z_{Pj} - z_{N2}\right)^2 = d_{12}^2 \\ \left(x_{Pj} - x_{N3}\right)^2 + \left(y_{Pj} - y_{N3}\right)^2 + \left(z_{Pj} - z_{N3}\right)^2 = d_{13}^2 \end{cases} \tag{4}$$

Increasing the number of transmitters and using suitable algorithms improve calculations accuracy [4] which can be advantageous due to the possibility of overshadowing the measuring lines to the selected transmitters by obstacles present in the roadway.

Orientation of the roadheader is described by the following three parameters: roll γ, pitch β and yaw α angles. This describes position of axis of local roadheader's coordinates system (B) in relation to the reference coordinates system (A). Determination of roadheader's orientation was realized in two ways. In the first version, three receivers of ultrasonic waves were installed on the roadheader. Their known spatial arrangement (on one plane) enables determination of the roadheader orientation by use of simple geometrical relationships (Figs. 6 and 8), having determined receivers position. In the second version two receivers were installed on the roadheader, which enabled determination of one parameter describing the orientation (yaw angle α) (Fig. 9). Two other parameters (roll angle γ and pitch angle β) were determined by means of two-axial inclinometers (Fig. 7).

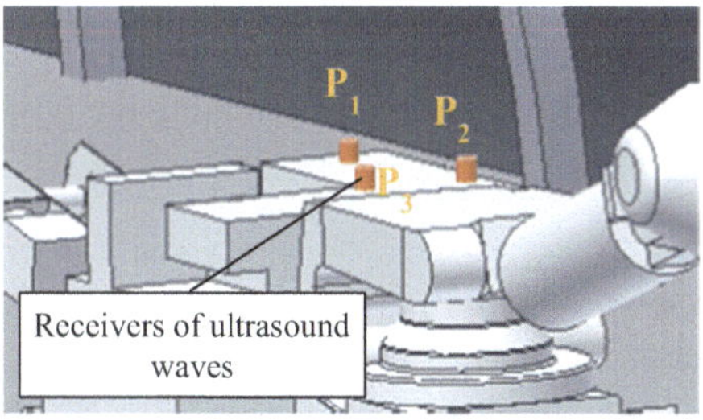

Fig. 6. Components used in determination of roadheader's orientation – variant 1 [4]

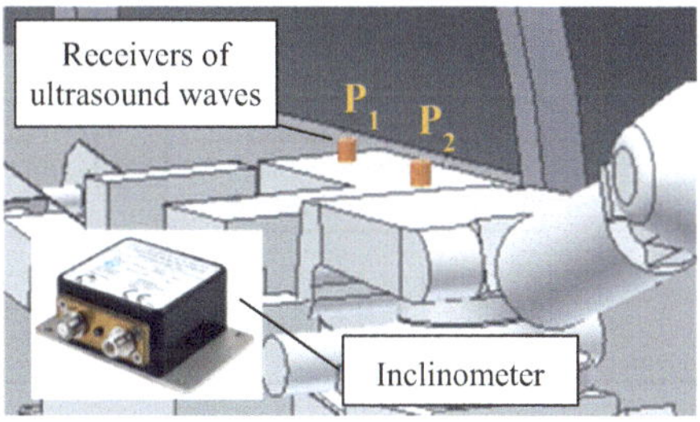

Fig. 7. Components used in determination of roadheader's orientation – variant 2 [4]

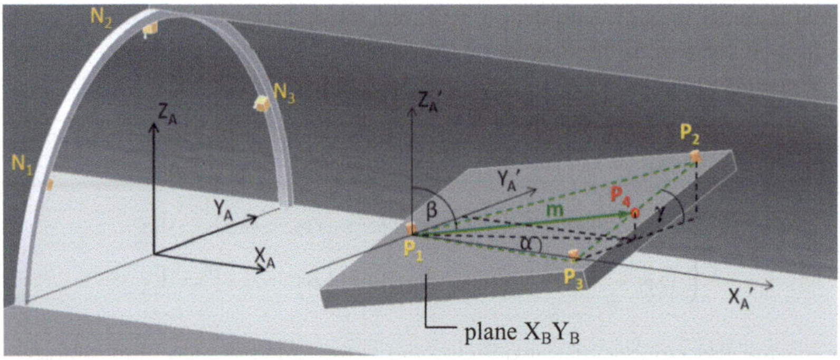

Fig. 8. Parameters describing the roadheader's orientation – variant 1 [4]

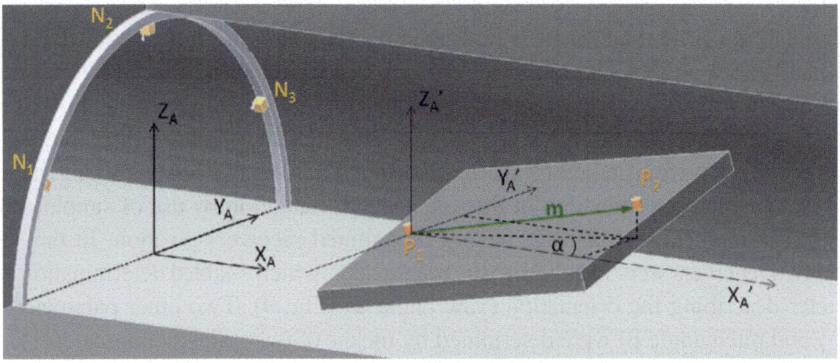

Fig. 9. Parameters describing the roadheader's orientation – variant 2 [4]

Algorithm for a cycle of measurements determining the position and orientation of the roadheader is presented in Fig. 10.

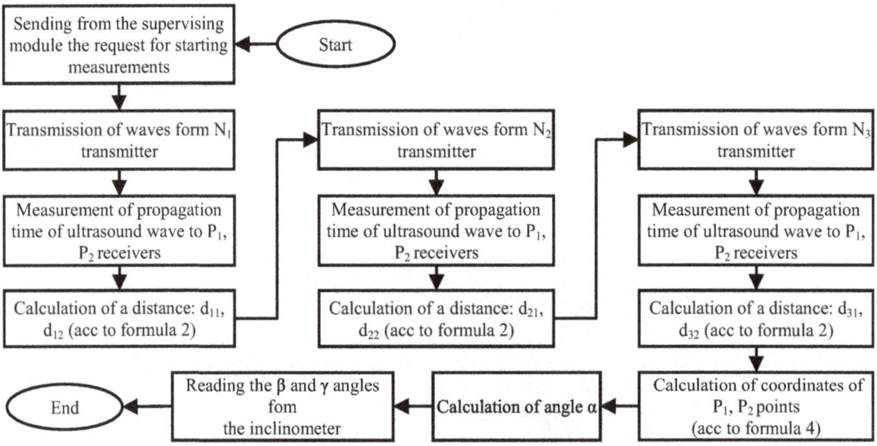

Fig. 10. Algorithm for determination of position and orientation of the roadheader according to the second variant [4]

5 Testing the Physical Model of the Positioning System

Tests of operation of the positioning system's (Fig. 11) were carried out in an experimental model simulating the real mine roadway built of arches of ŁP12 roadway support with the use of real roadheader.

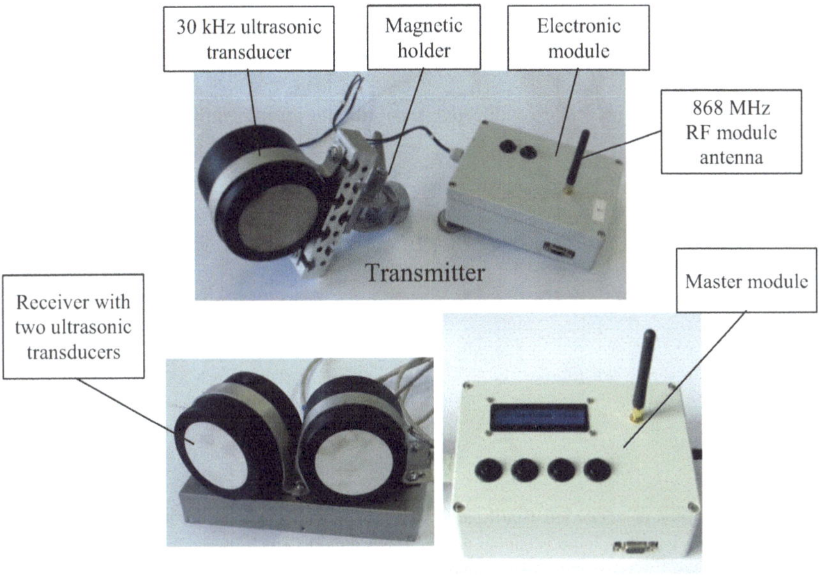

Fig. 11. Components of the positioning system [5]

Measurement results of roadheader's position during its run through the tunnel are presented in Figs. 12, 13 and 14. Each recorded coordinate of the roadheader's position is presented on the separate diagram.

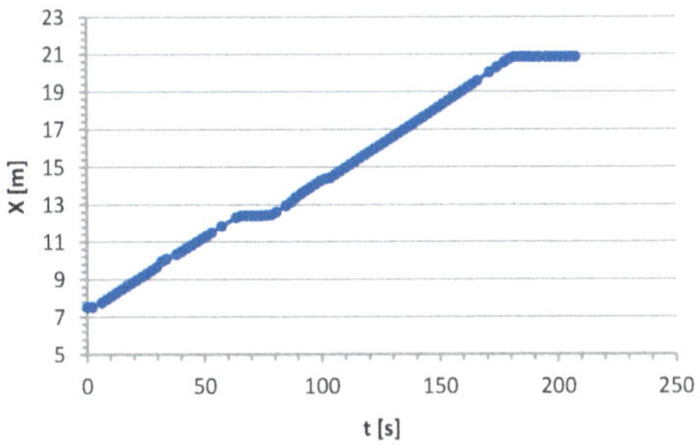

Fig. 12. Curve of recorded roadheader's coordinate during run – coordinate X [5]

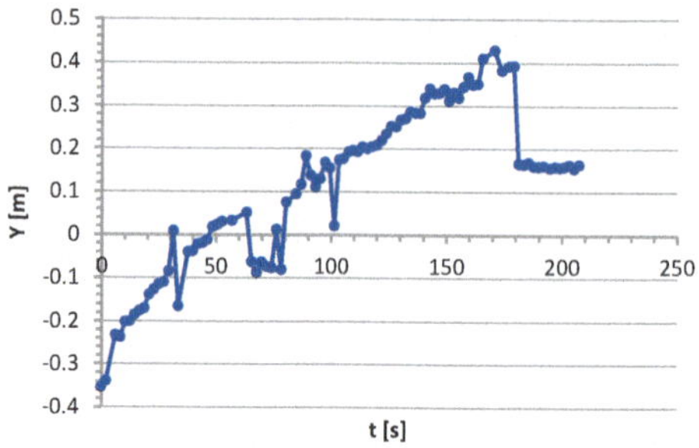

Fig. 13. Curve of recorded roadheader's coordinate during run – coordinate Y [5]

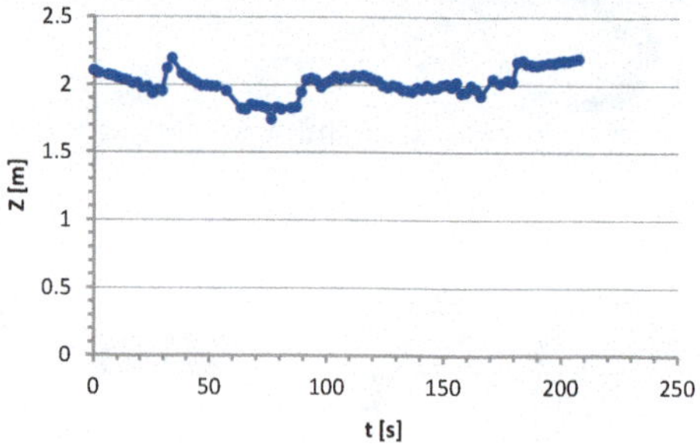

Fig. 14. Curve of recorded roadheader's coordinate during run – coordinate Z [5]

Position of the roadheader was determined in a distance 7–21 m from the arch of roadway support, on which transmitters were installed. Maximum distance resulted from the tunnel length. At distances shorter than 7 m, the system did not receive signals from all the transmitters at the same time, which resulted from their arrangement as well as from limited angle of ultrasound waves transmission. Accuracy of determination of roadheader's position coordinates was verified at the selected places of the tunnel, see Table 1. The obtained values were compared with the position coordinates measured manually. Due to technical reasons only two coordinates X and Y were verified.

Table 1. Comparison of the system readings with manual measurements [5]

Item	Expected value (measured manually)			Value determined by the system			Relative error	
	X [m]	Y [m]	Z [m]	X [m]	Y [m]	Z [m]	ΔX[m]	ΔY[m]
1	7.37	−0.37	–	7.39	−0.55	2.15	0.02	0.18
2	10.76	−0.12	–	10.80	−0.25	2.20	0.04	0.13
3	12.40	0.05	–	12.43	−0.10	1.85	0.03	0.15
4	20.81	0.36	–	20.87	0.16	2.18	0.06	0.20

6 Conclusions

Roadheader's positioning system developed by KOMAG in collaboration with Silesian University of Technology is presented. Ultrasound and radio waves were used in determining the current position and orientation of roadheader during its operation for development of roadway in a hard coal mine.

The system was verified in the experimental tunnel, where position of moving roadheader was measured. It was found that accuracy of determination of each coordinate is different. Accuracy of determination of coordinate X was equal to few centimetres, but in the case of coordinate Y it was more than 10 cm. Accuracy of determination of coordinate Z was not verified due to the lack of proper measuring instrument. More detailed information regarding the system accuracy can be found in [4]. Measuring range of the system with one set of transmitters is 7–21 m.

Proper positioning of the roadheader during cutting operation enables increasing automation of the process. That will enable the personnel withdrawal, including the operator, or it will reduce exposure time of miners to harmful conditions during cutting. That fact will significantly improve work safety by elimination or reduction impact of natural hazards and harmful environment.

References

1. Bartoszek S.: Pozycjonowanie kombajnu chodnikowego w wyrobisku korytarzowym. Maszyny Górnicze **1**, 22–35 (2016). ISSN 0209-3693
2. Bartoszek, S.: System pozycjonowania maszyn mobilnych w wyrobiskach korytarzowych. Praca wewnętrzna ITG KOMAG, Gliwice, 2012–2015 (materiały niepublikowane)
3. Bartoszek, S.: Metoda pozycjonowania górniczych maszyn mobilnych w wyrobiskach korytarzowych. W: Innowacyjne techniki i technologie dla górnictwa. Bezpieczeństwo – efektywność – niezawodność. KOMTECH 2012, Instytut Techniki Górniczej KOMAG, Gliwice, pp. 387–399 (2012)
4. Bartoszek, S.: Metoda pozycjonowania kombajnu chodnikowego w wyrobisku korytarzowym. Rozprawa doktorska, Politechnika Śląska, Gliwice (2017)
5. Bartoszek, S.: i inni: System pozycjonowania kombajnu chodnikowego. Praca wewnętrzna ITG KOMAG, Gliwice (2016, materiały niepublikowane)
6. Borenstein, J., Everett, H.R., Feng, L.: Where am i - systems and methods for mobile robot positioning. The University of Michigan (1996)
7. Jasiulek, D., Świder, J.: Mechatronic systems in mining roadheaders – examples of solutions. Pomiary Automatyka Robotyka **1** (2013). ISSN 1427-9126, Indeks 339512
8. Jasiulek, D., Stankiewicz, K., Świder, J.: An adaptive control system of roadheader with intelligent modelling of mechanical features of mined rock. J. KONES Powertrain Transp. **18** (2), 197–203 (2011)
9. Lukianto, C., Sternberg, H.: Overview of current indoor navigation techniques and implementation studies. In: FIG Working Week (2011)
10. Rogala-Rojek, J.: System doradczy wspomagający operatora kombajnu chodnikowego. Rozprawa doktorska, AGH Kraków (2010)
11. KOPEX GROUP Homepage. www.kopex.com.pl. Accessed 24 May 2017
12. XCMG Homepage. www.xcmg.com. Accessed 24 May 2017

Biomechatronic Simulator for Fencing Training Using Virtual Reality Technology

Andrzej Bieniek[1(✉)], Anna Szczygioł[2], Miłosz Chrzan[1], Piotr Wodarski[1], Michał Morys[2], Bogdan Bacik[2], Grzegorz Juras[2], Robert Michnik[1], Katarzyna Paszek[1], and Marek Gzik[1]

[1] Department of Biomechatronics, Faculty of Biomedical Engineering, Silesian University of Technology, Zabrze, Poland
{andrzej.bieniek,milosz.chrzan,piotr.wodarski, robert.michnik,marek.gzik}@polsl.pl, paszekkatarzyna@wp.pl

[2] Department of Human Motor Behavior Jerzy Kukuczka, Academy of Physical Education, Katowice, Poland
{a.szczygiol,m.morys,b.bacik,g.juras}@awf.katowice.pl

Abstract. Fencing is a combat sport in which fencers compete in three weapons: foil, saber and epee. They differ from each other in blade type, valid target and scoring technique. Although the fundamental role of speed, coordination and sensory-motor skills in fencing is indisputable, merely rapid and correct accomplishment of a movement does not ensure success. Outright bout against an opponent requires high level of perceiving, information processing and accurate decision-making. Thus, in fencing, as a 'game of anticipation and deception', technical abilities are just a tool used to execute chosen strategy and tactics. The repeatability of joints' movements relevant to the lunge performance was measured. Shoulder, elbow and wrist joint of the upper arm, hip joint and knee leg movements were checked in order to match and determine the number of necessary recordings. The study covered 10 athletes, 6 saber and 4 epee fencers, members of Polish National Team. There is a repeatability in measured movements. The weak and negative correlation coefficient is observed just between several subjects. The strong correlation coefficient was noted across trials within a single subject. This result implies, that frequently repeated movements create stable movement patterns which is individual for each of the subjects. According to individual character of performed actions, animation database should contain possible highest number of recorded athletes. Just widening of database can provide expected level of diversity to fencing training and prevent the effect of 'learning' just one opponent.

Keywords: Virtual reality · Fencing · Simulator · Motion capture · Biomechatronics

1 Introduction

Fencing is a combat sport in which fencers compete in three weapons: foil, saber and epee, which differ from each other in blade type, valid target and scoring technique (Chen et al. 2017). Although the fundamental role of speed, coordination and

J. Świder et al. (Eds.): MECHATRONICS 2017, AISC 934, pp. 30–37, 2019.
https://doi.org/10.1007/978-3-030-15857-6_4

sensory-motor skills in fencing is indisputable, merely rapid and correct accomplishment of a movement does not ensure success. Outright bout against an opponent requires high level of perceiving, information processing and accurate decision-making. Thus, in fencing, as a 'game of anticipation and deception', technical abilities are just a tool used to execute chosen strategy and tactics (Czajkowski 2009; Borysiuk and Waskiewicz 2008).

Biomechanical studies of fencing evaluates its technique, mostly concerns basic movements such as advance, retreat, fleche and lunging. Each of these actions start from en garde position, where both of athlete's feet are situated on the ground at an angle of around 90°. Perpendicular feet placement enables fencer to use entire sole surface of rear leg, and in result produces the greatest power and velocity in the propulsive phase of most common attack -lunge (Gresham-Fiegel et al. 2013; Guilhem et al. 2014). While the role of non-leading leg is to forcefully push the body forward (Sinclair et al. 2013), front leg extensors contributes in breaking phase while its foot strikes against the floor (Guilhem et al. 2014).

Chosen parameters of lower limbs kinematics significantly affect variables of the armed arm and in result touché performance (Bottoms et al. 2013; Gholipour et al. 2008). Great values of range of motion at rear leg's knee and front leg's hip flexion increase sword velocity during the lunge. Moreover, the weapon reaches the highest values of velocity in the moment of touché (Hassan and Klauck 1998). When armed arm motion is initiated together with foot off phase, its average speed significantly decreases due to associated anticipatory postural adjustment (Do and Yiou 1999).

While the kinematics of fencing is well known in the literature, the influence of different performing conditions on its parameters is still limited. According to Klinger, Adrian and Dee (1983) attack with lunge is performed swifter and longer under competitive conditions than during practice. While the kinematics of front leg differ slightly, the rear leg's drag extended up to 5 times in competitive condition. Additionally, uncertainty, caused with target change during attack with lunge, do not affect its precision, horizontal velocity of the center of mass at the end of the acceleration phase and participation of body segments in elite fencers. However, increased number of solutions significantly extends information processing (Williams and Walmsley 2000; Gutiérrez-Dávila et al. 2014; Gutierrez-Davila et al. 2013). Further research of the same authors shows, that when target change was delayed and occur at the end of advance phase, technical execution of attack has changed, leading to more mistakes (Gutiérrez-Cruz et al. 2016).

Just highly controlled environment, which allows to repeat chosen movements in realistic and reproducible conditions, enables to collect reliable data of athletes' performance. With technological progress, virtual reality (VR) became one of the most efficient ways to analyse complex interactions between players in different sporting scenarios. Additionally, VR simulators allow for numerical simulations in immersive environment and help understanding perception based on information in natural user interface.

Creating VR application for sport training involves capturing sport-specific motion. Obtained data is essential for creating virtual animation, and what is more enables comparison of athlete's performance in real and immersive world. Third step includes presentation of virtual environment which contains perception or perception-action tasks (Bideau et al. 2009).

VR simulators are already in common use for sports training or rehabilitation (Gzik et al. 2017). Applications mostly evaluate sensorimotor skills in ball games (Miles et al. 2012), for example goalkeeper's behavior during curved free kicks in soccer (Craig 2013) and while throwing in handball, reaction times in table tennis or passing skills in rugby. In another applications, athletes were asked to anticipate deceptive movement of the opponent (Bideau et al. 2010) or foresee the action played by opponent team (STRIVR). There is also VR batting simulator (using EON engine) designed to improve the batter's pitch recognition and performance. Each of VR training system has its own limitations. Firstly, optical motion capture systems provide the possibility of obscuring the markers and in result the loss of data and incorrect avatar's movement (Miles et al. 2012). Secondly, commonly used data projectors do not enable full immersion in comparison to head-mounted displays (HMD) or cave systems. It is also worth noticing, that most of the above mentioned simulators do not provide real-time 3D tracking system and are able to obtain just limited amount of data. There was an attempt to create virtual simulator for fencing. Nevertheless, the motions are visualised in 2D space so the elaborated visualisation is not fully interactive or spatial. These kinds of systems do not reflect motion in a real world in natural way.

2 Aim

The aim of this study was to create VR simulator which enables personalized fencing training with virtual opponent in fully immersive environment. Additionally, using inertial motion tracking system we aim to obtain real-time 3D kinematics and reaction times for visual stimulus produced by application.

3 Materials and Methods

In order to develop VR simulator, it was necessary to define its structure, obtain kinematics data and acquire the fencing-specific movements in various tactical situations. The created simulator has multilayer structure and contains hardware, visual track, application and visual layer. Hardware layer is responsible for real-time acquisition of position and orientation of body segments in space. Due to necessity of real-time tracking it was required to use system which enables online analyses. Such systems are based on electromagnetic field interferences, IMU and MIMU sensors or depth mapping. The hardware system used in this study was inertial motion tracking system (MVN Biomech) using MIMU sensors. The proposed technology is much more cheaper and has wider range than electromagnetic systems and shows higher accuracy than depth mapping. MVN Biomech (XSENS, Netherland) enabled to measure full-body 3D real-time kinematics at sampling frequency 120 Hz and with accuracy of 1 mm and 0.5°. Visual track layer presents generated graphics to athlete exercising. It is important to generate the stereoscopic image of combat area with appropriate frequency and quality, allowing athlete full-immersion in a computer generated environment. In this study we used HMD Oculus DK2 which allows for stereoscopic image generation that makes the athlete move around on the piste. The visual layer was limited to cut off additional visual stimuli (pic.). Application is responsible for receiving data from the hardware layer, and in addition selecting and activating previously recorded movements in the form of a virtual opponents.

The repeatability of joints' movements relevant to the lunge performance was measured. Shoulder, elbow and wrist joint of the upper arm, hip joint and knee leg movements were checked in order to match and determine the number of necessary recordings. The study covered 10 athletes, 6 saber and 4 epee fencers, members of Polish National Team. The fencers were instructed to perform fencing lunge as rapidly as possible in three different tactical situations. Firstly as an isolated movement, secondly preceded with step forward and backward. Each of movement sequence was recorded 5 times in two starting condition: self-paced and after fencing-specific visual stimuli (opponent's step forward with flexed arm). To check repeatability the correlation coefficient between each pair of measurement was calculated (pic.). These results are presented as correlation matrices (pic.). The values in the matrices were set at level 1 for correlations above 0.5, which means strong correlation and at level 0.5 for correlation between 0.5–0.3 which means weak correlation. Analogical scale was used for negative correlations.

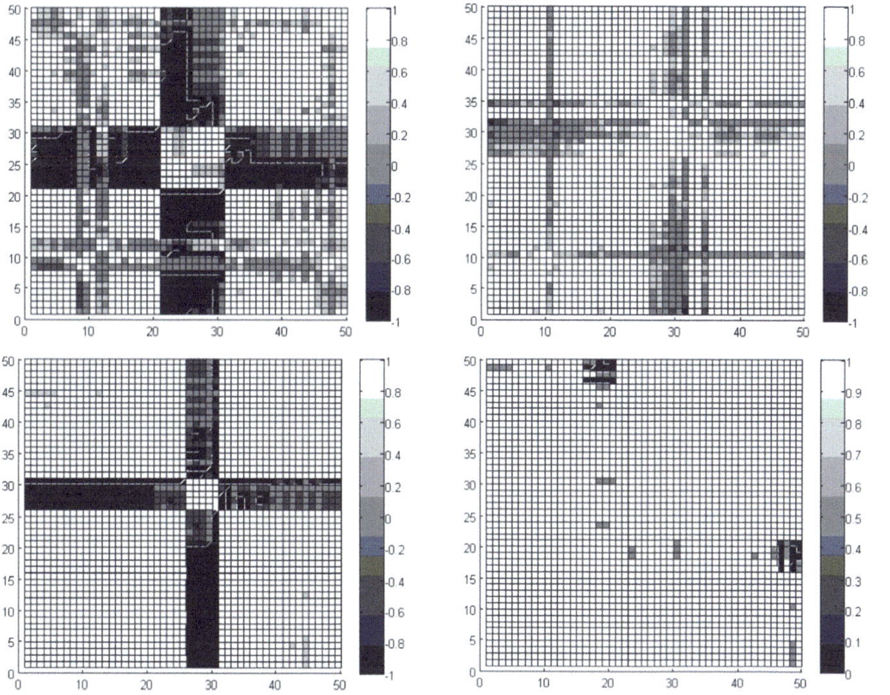

Fig. 1. Exemplary correlation table, each row and column corresponds to one repeating of movement and every 5 rows and columns correspond to one subject.

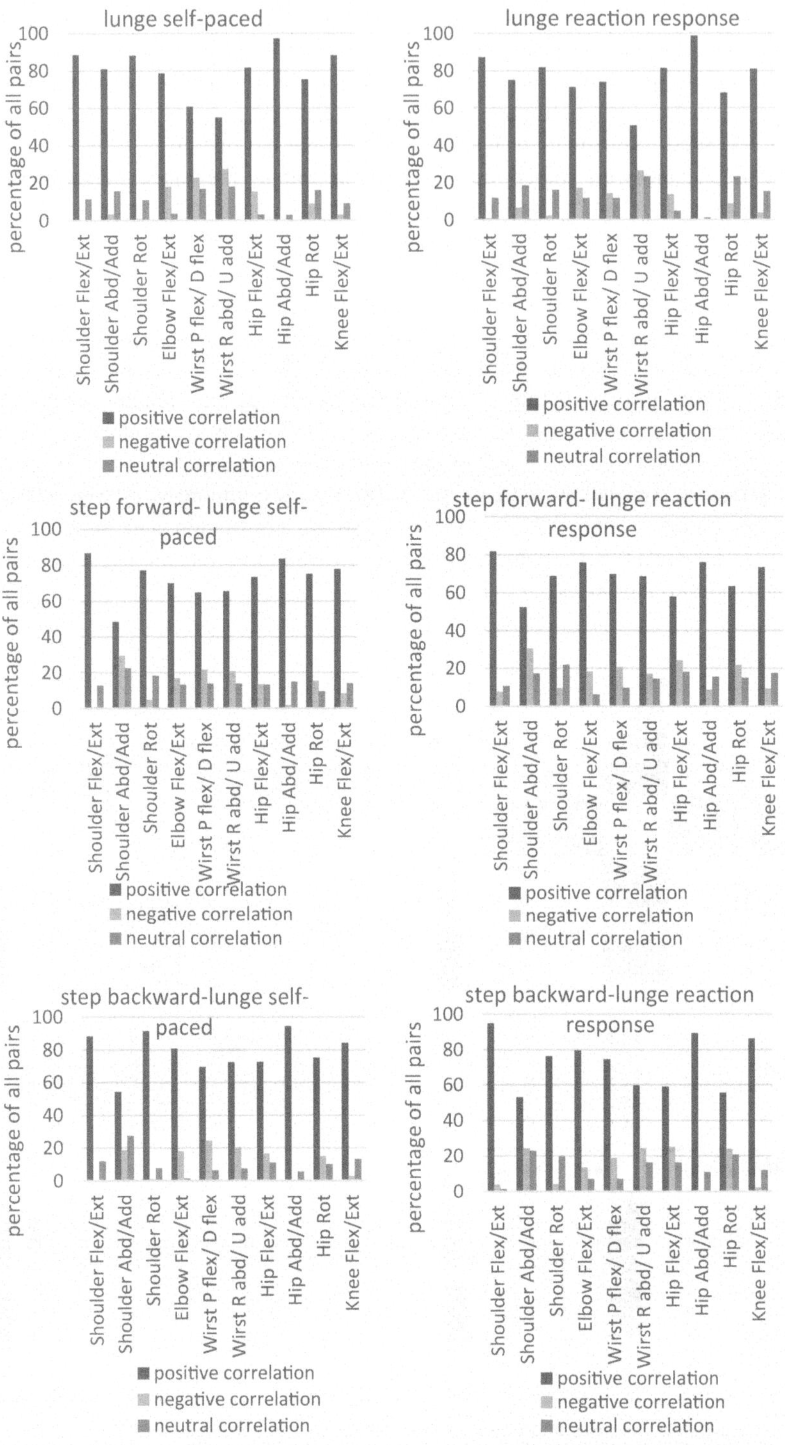

Fig. 2. Comparison of strong, weak and neutral correlations for all of the subjects.

For each tactical situation and starting condition, the number of positive, negative and neutral correlations were calculated and presented on bar charts as a percentage of all pairs (Fig. 2).

4 Discussion

Designed simulator allows to measure the kinematic data related to temporal joints' angular values in fencing-specific movements, as in previous publications (Gresham-Fiegel et al. 2013; Guilhem et al. 2014; Bottoms et al. 2013; Gholipour et al. 2008). The simulator's realism allows the virtual environment to contain perceptions or perception-like tasks (Bideau et al. 2009). According to Miles (2009) real-time avatar tracking enables immediate fencer movements' correction and in result more efficient training. A large amount of obtained data provides thorough analysis.

The repeatability of angular values for all 8 subjects was found (strong positive correlation) (Fig. 1.). The weak and negative correlation coefficient is observed just between 2 subjects. The strong correlation coefficient was noted across trials within a single subject. This result implies, that frequently repeated movements creates stable movement patterns which is individual for each of the subjects. According to individual character of performed actions, animation database should contain possible highest number of recorded athletes. Just widening of database can provide expected level of diversity to fencing training and prevent the effect of 'learning' just one opponent.

In Fig. 1 we can observe that one of the fencers executes the movement in chosen join with strong negative correlation in comparison to others. This type of information is extremely valuable in the process of developing motion patterns for a simulator for fencing. Single deviations at angular values and, at the same time, positive correlation of other movements, indicate a strong repeatability of the selected fencing action within individual. Due to he above mentioned fact, it is crucial to enlarge the simulator database and record as many possible fencers performing as many actions as possible.

It is also worth emphasizing, that in this study the comparison of fencing lunge was made in the laboratory environment. It may be assumed, that in real fencing bout, where the target area is changeable and individual for each fencer, the correlation coefficients between the kinematic measurements of the armed arm will be weak while among lower extremities we predict strong correlation coefficients.

5 Conclusions

Creating a simulator for contact sports, such as fencing, is extremely difficult because just a small part of the training can be done in the simulator. Therefore, it can play only supporting role in training of selected fencing actions. The simulator implementation must be done in close cooperation with fencing experts.

In order to create simulator the biomechanical studies of human movement are inevitable as in papers (Michnik et al. 2016). The simulator designed in this study should be equipped with large fencing movements database and enable an athlete to select virtual opponent. It is important to understand, that the selection of virtual opponent means the choice of properly recorded animation for which the correlation between the player's behavior and the animation is positive. Accordingly, a fencer will be able to learn correct movements pattern on the basis of recorded movement sequence.

References

Bideau, B., et al.: Using virtual reality to analyze sports performance. IEEE Comput. Graph. Appl. **30**(2), 14–21 (2010)

Bideau, B., et al.: Virtual reality, a serious game for understanding performance and training players in sport. IEEE Comput. Graph. Appl. **30**, 14–21 (2009)

Borysiuk, Z., Waskiewicz, Z.: Information processes, stimulation and perceptual training in fencing. J. Hum. Kinet. **19**(1), 63–82 (2008)

Bottoms, L., Greenhalgh, A., Sinclair, J.: Kinematic determinants of weapon velocity during the fencing lunge in experienced epee fencers. Acta Bioeng. Biomech. **15**(4), 109–113 (2013)

Chen, T.L.-W., et al.: Biomechanics of fencing sport: a scoping review. PLoS One **12**(2), e0171578 (2017). Barbosa, T.M. (ed.)

Craig, C.: Understanding perception and action in sport: how can virtual reality technology help? Sport. Technol. **6**(4), 161–169 (2013)

Czajkowski, Z.: The essence and importance of sense of timing in fencing. Stud. Phys. Cult. Tour. **16**(3), 241–247 (2009)

Do, M.C., Yiou, E.: Do centrally programmed anticipatory postural adjustments in fast stepping affect performance of an associated "touche" movement? Exp. Brain Res. **129**(3), 0462–0466 (1999)

Gholipour, M., Tabrizi, A., Farahmand, F.: Kinematics analysis of lunge fencing using stereophotogrametry. World J. Sport. Sci. **1**(1), 32–37 (2008)

Gresham-Fiegel, C.N., House, P.D., Zupan, M.F.: The effect of nonleading foot placement on power and velocity in the fencing lunge. J. Strength Cond. Res./Natl. Strength Cond. Assoc. **27**(1), 57–63 (2013)

Guilhem, G., et al.: Mechanical and muscular coordination patterns during a high-level fencing assault. Med. Sci. Sport. Exerc. **46**(1), 341–350 (2014)

Gutiérrez-Cruz, C., Rojas, F.J., Gutiérrez-Davila, M.: Effect of defence response time during lunge in foil fencing. J. Sport. Sci. **34**(7), 651–657 (2016)

Gutierrez-Davila, M., et al.: Response timing in the lunge and target change in elite versus medium-level fencers. Eur. J. Sport. Sci. **13**(4), 364–371 (2013)

Gutiérrez-Dávila, M., et al.: Effect of uncertainty during the lunge in fencing. J. Sport. Sci. Med. **13**(1), 66–72 (2014)

Gzik, M., Wodarski, P., Jurkojć, J., Michnik, R., Bieniek, A.: Interactive system of enginering support of upper limb diagnosis. In: Gzik, M., Tkacz, E., Paszenda, Z., Piętka, E. (eds.) Innovations in Biomedical Engineering. AISC, vol. 526, pp. 115–123. Springer, Cham (2017). https://doi.org/10.1007/978-3-319-47154-9_14

Hassan, S., Klauck, J.: Kinematics of lower and upper extremities motions during the fencing lunge: results and training implications. In: ISBS-Conference Proceedings, pp. 170–173 (1998)

Klinger, A.K., Adrian, M.J.: Foil target impact forces during the fencing lunge. Biomech. VIII **4B**, 882–888 (1983)

Michnik, R., Jurkojć, J., Wodarski, P., Gzik, M., Jochymczyk-Woźniak, K., Bieniek, A.: The influence of frequency of visual disorders on stabilographic parameters. Acta Bioeng. Biomech. **18**(1), 25–33 (2016). https://doi.org/10.5277/ABB-00201-2014-04

Miles, H.C., et al.: A review of virtual environments for training in ball sports. Comput. Graph. **36**(6), 714–726 (2012)

Sinclair, J., Taylor, P.J., Bottoms, L.: The appropriateness of the helical axis technique and six available cardan sequences for the representation of 3-D lead leg kinematics during the fencing lunge. J. Hum. Kinet. **37**(June), 7–15 (2013)

Williams, L.R., Walmsley, A.: Response timing and muscular coordination in fencing: a comparison of elite and novice fencers. J. Sci. Med. Sport./Sport. Med. Aust. **3**(4), 460–475 (2000)

Application of Industrial Automatic Systems for Operating Parameters Identification of Mining Machines

Jarosław Brodny(✉) and Magdalena Tutak

Silesian University of Technology, Akademicka 2A, 44-100 Gliwice, Poland
jaroslaw.brodny@polsl.pl

Abstract. Mechatronics, as an interdisciplinary technique finds increasingly more and more practical application, also in mining industry. In particular, this applies to mining machines that are part of a mechanized longwall complex.

In mechatronic category, as whole it can be treated as advanced mechatronic system (model). It consists of a complex of integrated machines and devices, which together constitute a specific mega-machine. It is equipped with advanced industrial automation systems, dozens of sensors that monitor their work and proper data warehouse management software, being interpreter for behaviors and events for each systems.

In the paper, proposals for use of the data to analyze the work of selected machines of the complex (heading machines and longitudinal conveyor) were presented. An adaptation was proposed for mining conditions, for acknowledged method of effective estimation of the whole machine (OEE). It captures not only observed today area of critical events (which results in longer shutdowns), but also: unobserved micro-disruptions, fluctuations in the field of utilization (and hence productivity) of particular machines and events resulting in losses in the field of quality. However, the model resulting from cause-effect analyzes accompanying the adaptation of the method must often require correlated effects of several parameters. They identify both the technological phases, as well as give n images of behavior of machines in these phases. In this way, additional range of information dedicated primarily to production and maintenance services is acquired. Developed software should be a significant element of "mechatronic model" as longwall complex is recognized for. It should facilitate also its operation. Fundamental aim of the paper is to identify significant, from the point of view of exploitation, operational parameters of the machine, based on recorded data acquired by industrial automation system.

Keywords: Mechatronics system · Mining machines · OEE · Automatics system

1 Introduction

Industrial automation is currently one of most dynamically developing branches of control systems engineering and electrical engineering. Its fundamental task is automation of manufacturing and technological processes, and consequently optimization of

J. Świder et al. (Eds.): MECHATRONICS 2017, AISC 934, pp. 38–51, 2019.
https://doi.org/10.1007/978-3-030-15857-6_5

production processes. In order for an automation process to be effective full integration of production, control, control-measurement devices, communication and data transfer systems, as well as specialized software are necessary. All these elements form less or more complicated system of industrial automation. Industrial information technology forms a very significant part of this system connecting issues from information technology and manufacturing technology. It also includes monitoring and modelling of production processes, data acquisition and processing, information and knowledge acquisition from that data.

It can be assumed that currently applied industrial automation systems are kind of mechatronics systems. They connect fields of mechanics, theory of machines, electronics, information technology, automation, robotics and control engineering. In both systems all components have to cooperate closely to each other.

Industrial automation and mechatronics are currently practically applied in each domain of economy. It includes also mining industry. Especially it refers to machines applied in hard coal mining. These machines, due to specific conditions in which they work, are endangered for action of series of factors, not occurring in plants of other branches (so called closed facilities). Mines and environment in which mining machines work can be treated as open facility. External factors, connected with random character of natural environment interaction (in this case rock mass, gases, temperature etc.), cause that not always it is possible to predict conditions in which these machines will work. That is why they are characterized by much higher security factors and technical parameters than their equivalents used in surface production facilities.

Due to this difficult unequivocal identification of working conditions it becomes necessary to determine operation parameters and effectiveness of application of these machines. Carrying out such actions results also in the necessity of accommodation to very competitive energy market, where mining enterprises in order to be competitive, are forced to take actions to reduce production costs. A method which can effectively reduce these costs is to use production measures owned by these enterprises. Practice shows that incomplete use of machines is currently one of most important problems of mining enterprises.

In order to improve use of mining machines it is necessary to carry out reliable assessment of their operation. Fundamental meaning in this process has the assumption of proper research methodology, which has to be based on reliable data about their operation. It seems reasonable to apply this assumption in this range system of industrial automation. Particularly it concerns this part of the system which includes measurements, registration, transmission and archiving of operational parameters of machines.

Mechanized longwall complex assigned for mining of rock mass and transport of output from the face zone, was subjected to studies, which results are presented in the paper. This complex consists of longwall-shearer, armoured face conveyor, beam stage loader, crusher and powered support. This complex is used directly for exploitation of hard coal seams. In mechatronic category, as whole it can be treated as advanced mechatronic system (model). It consists of a complex of integrated machines and devices, which together constitute a specific mega-machine. It is equipped in advanced industrial automation systems, dozens (and increasingly hundreds) of sensors that monitor their work and proper data warehouse management software, being interpreter for behaviors and events for each systems [3, 5]. Great problem in operation of this system is proper integration of registered data. In further stages it is important also elaboration of conclusions and recommendations for particular groups of stakeholders.

In the systemic approach to this point, undoubtedly the obstacle is a lack of repeatability in the exploitation process of such system. It results from different and not foreseeable conditions of its work (mining and geological conditions), on each exploited longwall and technologically conditioned phase of the exploitation cycle.

The range of information located in the massive data cloud is a great source of information and, after the use of proper processing and knowledge, can be used directly in an exploitation process of a longwall which lasts for months.

In the paper, proposals for use of the data to analyze the operation of selected machines of the complex (shearer and face conveyors) were presented. An adaptation of acknowledged method of overall effective estimation of machines (OEE) was proposed for mining conditions [4, 8, 9]. It captures not only observed today area of critical events (which results in longer shutdowns), but also: unobserved micro-disruptions, fluctuations in the field of utilization (and hence productivity) of particular machines and events resulting in losses in the field of quality. In many cases, these values are available to diagnose only on the basis of proper processing and analysis of recorded data by the system.

However, the model resulting from cause-effect analyses accompanying the adaptation of the method often must require correlated effects of several parameters. They identify both the technological phases, as well as give images of behavior of machines in these phases. Additional effect of multiplicity of recorded parameters is the ability to observe also other indicators appropriate to maintain production. In this way, additional range of information dedicated primarily to production and maintenance services is acquired.

It was assumed that primary goal of this paper is to identify the operational parameters, significant from the exploitation point of view, basing on the data registered by industrial automation system. For this purpose, in a significant range there was used operational process supervisory system of SCADA type (Supervisory Control And Data Acquisition). Obtained great amount of data, depending on needs and expectations, can be used in very wide range of applications. Presented paper concentrates on the determination of effectiveness of use of two machines included in the complex. The examples of more wide use of registered parameters to monitor operation state of shearer was also presented.

Obtained and presented results unequivocally indicate that application of industrial automation systems enables identification of series of operational parameters of machines. Received data can be used for many analyses in order to optimize the production process [11]. The examples presented in the paper, in the Authors' opinion are only a fragment of possibilities of use of the data.

Authors have developed also computational tool intended to carry out such analyses. The software, as a result of the works carried out, and giving mentioned range of information, can be concluded as another component that extends existing "mechatronic model" of the longwall complex, which may decide about significant improvement of its utilitarian values. In the Authors' opinion the developed methodology of acquisition of data on machines operation and their analysis is so universal that it should find practical application also in many other fields.

2 Characteristics of the Examined System

As mentioned before, to the studies were performed on two machines forming the mechanical longwall complex. They were longwall shearer and armored face conveyor. A task of the shearer is to directly loosen the (coal) rock from the rock mass and its loading on armored face conveyor (scraper). However, armored face conveyor serves to transport the output from longwall heading to bottom gate. In further step, output is transported by conveyors (scraper and belt type) to the shaft area and further to the surface. Both machines which are part of the complex were working in longwall heading during hard coal mining. The diagram of longwall heading with marked machines and direction of transport of the output is presented in Fig. 1.

The fundamental technical parameters of longwall shearer and armored face conveyor are presented in Table 1. The parameters characterizing the longwall, where two machines were working are presented in Table 2. These parameters have significant meaning, because they determine operating parameters of machines working at this longwall.

During operation of both machines series of parameters of their operation were registered. In a case of shearer 145 values were registered but three in a case of conveyor. Selected operation parameters of both machines through industrial automation system are presented in Table 3.

Registration of all parameters was possible thanks to operation of industrial automation system. View of control panel used in this process is presented in Fig. 2, whereas Fig. 3 presents view of electrical installation box with block of JNA controller (main control unit of machine) [12]. All devices used in underground mining institutions must have proper safety certificate. These devices must be intrinsically safe. They belong to group I, M1 category and are intended for application in underground mines and installations on the surface which are in danger of firedamp and/or flammable dusts and which must be used also when explosive atmosphere is present. Devices which do not fulfill safety requirements cannot be used in these conditions.

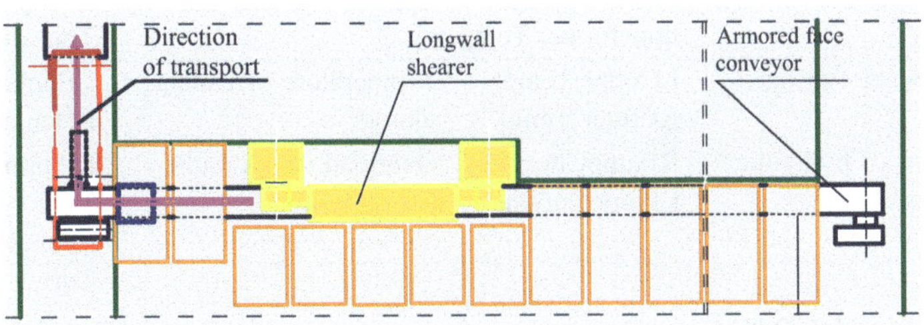

Fig. 1. Diagram of longwall heading with marked machines and direction of transport of the output.

Table 1. Technical parameters of longwall shearer and armored face conveyor.

Longwall shearer			
Parameter	Value	Parameter	Value
Cutting range	3400 mm	Haulage pull	2 × 320 kN
Maximum power installed			
• on cutting drums • on haulage drive • on hydraulic system	2 × 285 kW 2 × 60 kW 2 × 13 kW	Control of hydraulic circuits Cutting drum diameter	Through solenoid valve 24 V DC 1600 mm
Supply voltage	1000 V, 50 Hz	Haulage speed (operating)	0–12 m/min
Velocity of cutting drum	47 rev/min	Haulage speed (maneuvering)	0–20 m/min
Cutting drum web	800 mm	Weight	about 47.5 t
Armored face conveyor			
Maximum power	3 × 400 kW	Conveyor's chain	2 × ϕ38 × 126 mm
Supply voltage	3.3 kV	Peak output	Ok. 1200 t/h
Conveyor chain speed	1.11 m/s	Conveyor's length	150 m

Table 2. Parameters characterizing the longwall, where two machines was working.

Seam thickness	To 2.4 m do 3.4 m	Height of the longwall	To 3.0 m
Seam inclined	About 6°	Length of the longwall	Medium 139 m
Live of face	746.0 m	Roof control	Roof caving
Longitudinal seam inclined	Medium 1°	Transverse seam inclined	Medium 6°

Table 3. Selected operation parameters of both machines through industrial automation system.

Parameter	Designation	Parameter	Designation
Longwall shearer			
Velocity of haulage	VelHaulage	Position of shearer	Mach_shield pos
Currents of the cutting engines	RCutter current LCutter current	Currents of the haulages	LHaulage current RHaulage current
Power transformer temperature	Power transformer Temp	Supply voltage	Voltage Transducer
Temperature of cutting motors	LCutter Temp RCutter Temp	Temperature of haulage motors	LHaulage Temp RHaulage Temp
Temperature of hydraulic pump motors	RPumpCurrent LPumpCurrent	Overload of hydraulic pump motors	RPumpThermal Trip LPumpThermal Trip
Armoured face conveyor			
Currents of the main engines	RCarier cutter LCarier cutter	Currents of the auxilliary engines	JNP

Presented set of machines equipped for industrial automation system was submitted to test during the normal operation in hard coal mining.

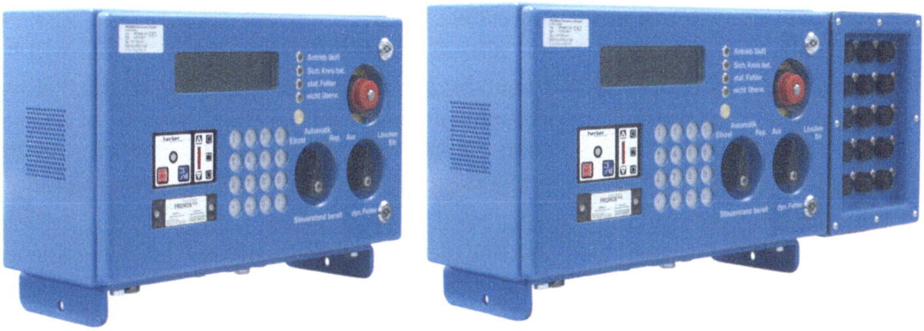

Fig. 2. View of control panel type Promos Plus [12].

Fig. 3. View of electrical installation box with block of JNA controller (main control unit of machine).

3 Research Methodology

From the entire technological chain including underground exploitation of hard coal, a set of machines forming the mechanized longwall complex was selected for studies. This is one of the many sets functioning in the field of coal production that can be distinguished in this process. Authors assumed that this set is the most representative and most important from the final beneficiary point of view. Such assumed research model based on the workplace and functionality implicated a lot of practical problems which, in order to reach assumed aim, had to be solved. The aim was to use the industrial automation systems to identify the parameters of studied machines and further to use the acquired information and knowledge to improve the efficiency of their utilization and the entire production process.

Besides the research object, it was also assumed that firstly obtained data will be used to determine the efficiency of the investigated machines. It was assumed that the overall efficiency model OEE will be applied for this purpose. In order to adapt this model to the specificity of underground mining, it was subjected to significant modification [2, 10]. In the further stages, the use of recorded data for diagnostic purposes and full control of the operation are expected.

Generally, it can be assumed that the developed methodology of utilization of industrial automation systems to identify machine parameters is based on the method, innovative for hard coal mining practice in the country, using advanced IT tools. These tools are relational databases and data warehouses. At the next stage of construction of IT system for monitoring the state of machines, the creation IT platform is expected.

In the presented paper, one focused mainly on obtaining objective data about the parameters of machines operation and their utilization to evaluate the efficiency of using these machines. Finally, a methodology was developed to use the OEE model to analyze the work of these machines. A general diagram of this process is shown in Fig. 4. Ideological scheme of operation and application of OEE model for operation analysis of mining machines.

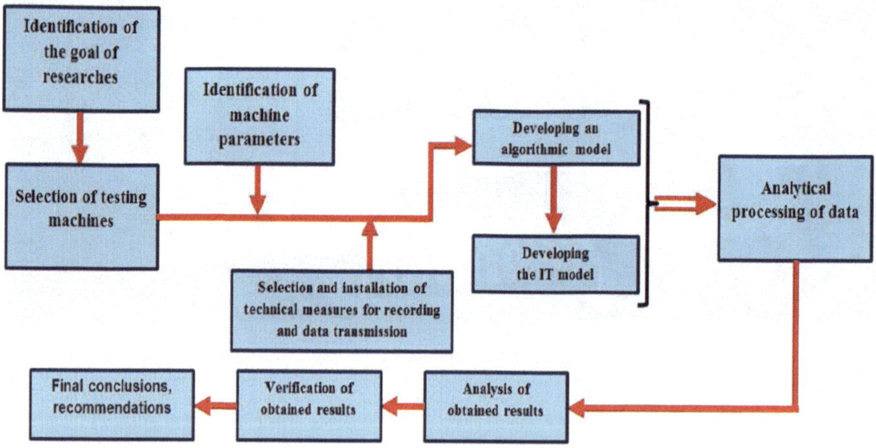

Fig. 4. Ideological diagram of functioning and application of the model OEE for operational analysis of mining machines.

Generally, the operations connected with assumed target can be divided into three basic phases: concept and planning, implementation and verification. On the other hand, particular actions enclosed in the developed methodology include:

- selection of testing machines,
- identification of the goal of research (e.g. assessment of effectiveness),
- developing an algorithmic and IT model,
- identification of machine parameters (important for the project goal)
- selection and installation of technical measures for recording and data transmission,
- registration and archiving of obtained data on machine performance parameters (database creation),
- pre-processing of acquired data (stratification, data cleaning, etc.)
- basic analytical processing of data (according to assumed algorithm)
- analysis of obtained data (acquirement of machine operation information)
- verification of obtained results.

The OpenEye system was used to manage the entire IT system, particularly for the acquired data [13]. This tool uses the components of the SCADA process control system. This system works with relational databases of Oracle, MS SQL and PostgreSQL types, and enables very fast data analysis and visualization of the obtained results. It should also be emphasized that the developed and used IT tool is integrated with the commonly used mining management system SZYK2. This system is the original EAM/CMMS solution and is part of the ERP system [1, 6, 7]. In general, its task is to manage the current maintenance of the means of production (preventive measures to prevent failures - reviews, repairs, adjustments, replacement of items, maintenance).

It can be assumed that the accepted research methodology and developed IT tool integrated with other applied systems and using industrial automation systems to obtain data guarantee the reliability and objectivity of the obtained results.

The final idea of the presented methodology is to develop a tool for systematic control of the efficiency of the production process, aimed to limit the number of failures and faults and reduce downtime. This tool has to indicate the places and possibilities for improvement or elimination of "bottlenecks" in the hard coal mining process.

Running specific actions, as a derivative of the OEE analysis, must already be a part of the ongoing management process of production process. A systematic study of OEE indicators tree, built for individual machines and the whole system, should also provide a way to visualize the degree of improvements in the manufacturing process which are implemented in the enterprise.

It should be also emphasized that the effectiveness of the implemented solution requires full co-operation of mine services responsible for proper management of production resources and compatibility of used IT systems.

Therefore, a significant meaning for the presented system has creation of uniform platform for registration of operational data of machines. Thanks to this platform it will be possible to optimally use machines and devices, and thereby to obtain positive effects in a field of production management and maintenance. This in turn, will translate into measurable economic benefits of the mines.

4 Test Results

The main parameters of the machines, which were used in the analysis, were the time courses of current intensity, absorbed by their motors. In a case of a longwall shearer, these courses were related to velocity of its feed and position in the longwall. It enabled to perform the analysis of operation of shearer and conveyor in relation to a single cycle. This cycle, depending on applied mining technology, includes a few phases. In tested mine, the one-way mining technology was applied. In this case, this cycle consisted of mining phase, return movement and sumping. Real course of operation of shearer in a time function is presented in Fig. 5. Besides the position of shearer in the longwall, also the time courses of change of current intensity consumed by the motors of the mining equipment of shearer and velocity of its feed are presented in this chart.

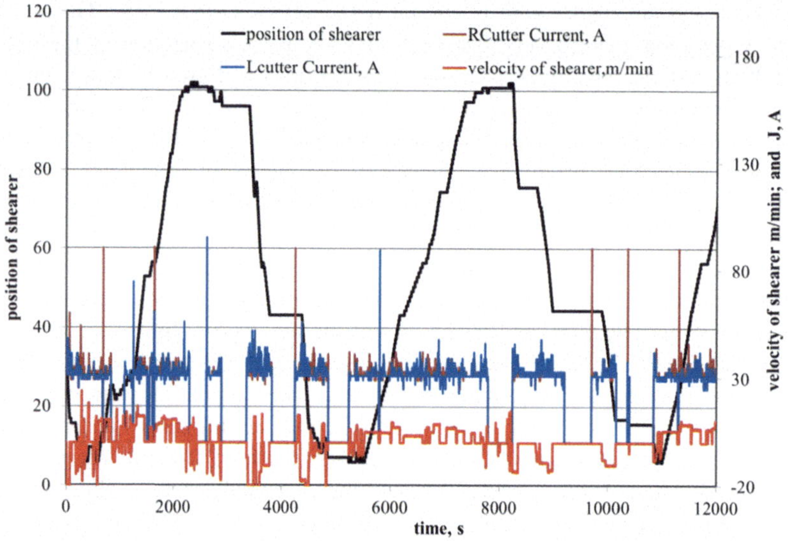

Fig. 5. Temporal waveforms of the velocity and position of the longwall shearer for one cycle of its operation.

Based on the presented courses all planned and unplanned interruptions in the operation of shearer can be determined. These interruptions determine times in which position of shearer has not changed and velocity of its feed equals to zero. Also in these periods the most often values of intensity current charged by the mining motors of the shearer equals to zero. Similar courses determining the changes in value of current intensity consumed by the motors were recorded for the armoured face conveyor.

On the base of such registered courses the availability indexes for shearers and conveyor for particular working shifts were determined. The analysis included one month of operation of the set, that is 63 working shifts (360 min each). Availability index was determining the ratio of real operation time of machine to the normative time (360 min).

Based on the determined availability indexes of both machines availability indexes of the whole set and their average values were also determined. Obtained results are presented in Table 4. Changes in the availability values of both machines for particular working shifts are presented in Fig. 6.

Table 4. Results of analysis of availability of tested machines.

	Availability, %		
	Average	Maximum	Minimum
Longwall shearer	58.26 ± 8.19	77.23	44.12
Armoured face conveyor	65.23 ± 6.89	81.37	54.21
Set of machines	61.75 ± 8.31	81.37	44.12

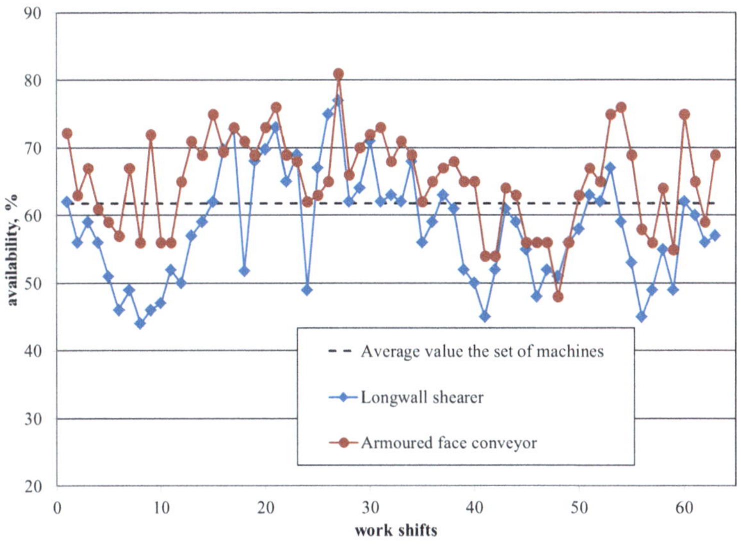

Fig. 6. The changes in the availability values of both machines for particular working shifts.

Determination of availability index of tested machines is a base for determination of indicator of effectiveness of their utilization. Therefore, it is necessary to determine the productivity of machines and quality of output, treated in this case as a final product. In the presented example determination of this indicator was omitted.

Next stage of analysis of obtained data is to determine the structure of resisted interruptions and determine their causes. Only after this step, the appropriate steps in order to optimize the production process can be taken. In this case, also data recorded by the industrial automation system was used. Their identification was able thanks to specially developed module registering the unplanned interruptions in machines' operation. This module has been developed to be integral part of commonly applied management production system SZYK. Basing on the data registered in this system the analysis of time of lasting of unplanned operation interruptions of machines for investigated period was carried out. Time structure of interruptions is presented in Table 5, whereas their reasons are listed in Fig. 7. During determination of causes of interruptions, obtained results for their part, were proportionally calculated for the whole.

Table 5. The structure of the registered breaks.

	Duration of the breaks, min					
	0.5–1	1–5	5–10	10–30	30–60	Above 60
Longwall shearer	757.5	1136.5	1420.5	2935.0	2083.0	1137.0
Armoured face conveyor	394.5	867.5	1340	2366	2050	868

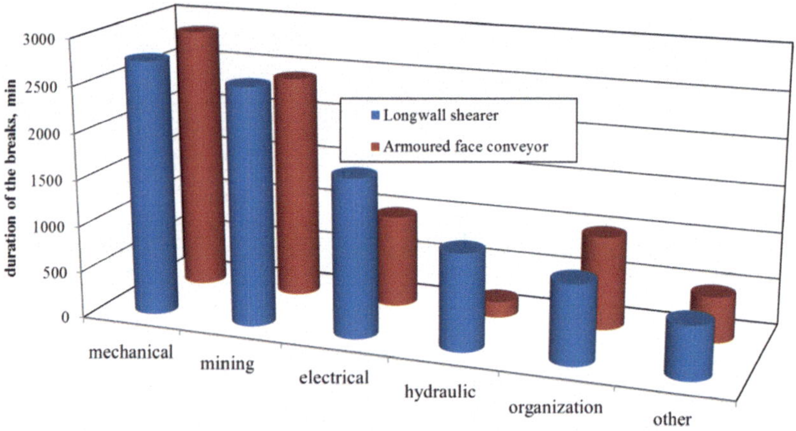

Fig. 7. Reasons for unplanned breaks in the studied machines' work.

Registration of many operational parameters, especially for longwall shearer enabled also for example the analysis of temperature of its mining motors and power transformer. In Fig. 8 example time courses of the temperature change is presented.

Analyzing registered data and time course presented on their base, one can state that currently applied sensor systems enable very precise monitoring of the operational state of particular machines. But for this data to be practically used, it is necessary to apply adequate computational tools for their archiving, analysis and presentation of results. Industrial automation systems in connection with tools developed and presented in this paper create such possibilities.

Results presented in the paper, being an effect of application of developed data registration and analysis system are only a small part of information which can be obtained from this system.

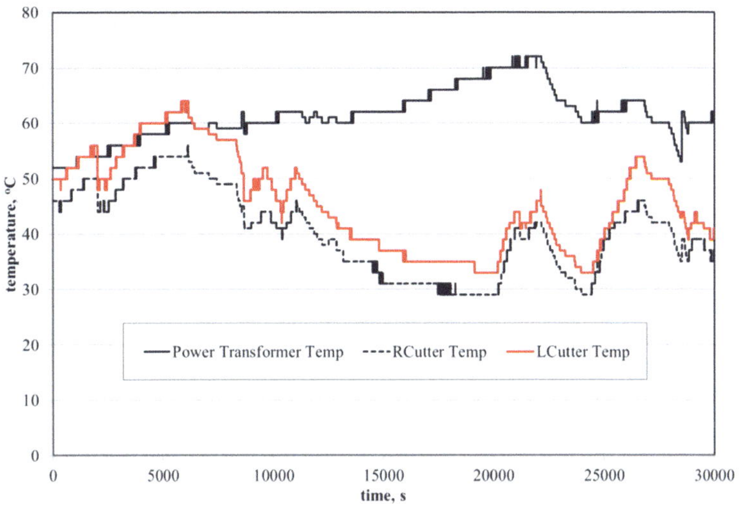

Fig. 8. Time courses of the temperature change of mining motors and power transformer.

5 Conclusions

Very fast development of science causes that its effects are in greater degree introduced into the economy. Increasing speed of this process is the result of great competition in global economy, which in greater degree is based on knowledge.

So called fourth industrial revolution, lasting from some time, forces these changes increasingly. Development of technology, especially information technology ideally fits in this trend. One more important region in which changes take place very fast, is machine industry. In range of state of technology, its development has the key meaning. Whereas, in range of interdisciplinary scientific researches mechatronics can be considered as a discipline which had the most influence and also currently influences the development of machine industry. These changes are sensible in all branches of economy, including the mining engineering. Ineffective and low mechanized mining engineering can turn out to be one of the greater beneficiaries of changes taking place in technology.

Methodology of utilization of industrial automation systems in identification of operational parameters of mining machines and their further analysis, presented in the paper, creates great possibilities of optimization of production process in mines.

This in turn should add to improvement of economic effectiveness of mining enterprises. The topic covered in the paper refers to two areas related to mechatronics. In the first one, more general, it should be assumed that entire longwall complex, applied to mining of coal is a complex mechatronic system.

In the second one, more detailed approach, the industrial automation system together with developed IT tool is a specific component of this system. The concept of utilization of these two areas, presented in the paper, according to the authors, brings very positive effects.

Realization of accepted assumptions has enabled obtaining of large amount of data on operation of machines. Processed data has become source of information and knowledge about the state of tested machines.

In the presented example, we focused on the analysis of operation of two machines, whereas the developed methodology and IT tools enable analysis of much greater number of machines.

It can include the whole mechanized complex, as well as all machines located in the technological series of coal production. Very significant advantage of developed methodology is reliability of the input data.

Obtained data (marked as raw) practically is not affected by any human error. It causes the fact that registered parameters of machine operation, which in subsequent stages are the base for the assessment of their operation condition and utilization, can be treated as objective measures in this range. Based on that, reliable and objective information and knowledge about operations of these machines can be obtained.

Therefore, it is reasonable to conclude, that without application of proper advanced systems of recording, transmission, archiving and analysis of data, it would be impossible to obtain such knowledge.

In the presented example of determination of efficiency of machines' utilization very important meaning has defining of structure and reasons of unplanned interruptions in their work.

Thanks to the applied tools it was possible to identify these causes. This information should be used by the maintenance services and management of companies to limit these interruptions. Subsequently, it should improve effectiveness of the production process.

Other registered diagnostic parameters of machines enable full monitoring of their condition. Basing on this, with assumption of appropriate criterial values the degree of utilization, consumption and proximity to the critical state can be determined. This information enables to make proper decisions regarding the inspection, control or maintenance of machine components.

Therefore, it is reasonable to conclude that advanced mechatronic systems together with the appropriate systems and algorithms used for analysis and inference enable effective control of production systems. Subsequently, it should have an impact on the increased security and economic efficiency of these systems. In the case of mining industry it can have a decisive meaning for its further existence in a highly competitive energy market.

Acknowledgments. This article is the result of the research project No. PBS3/B6/25/2015 "Application of the Overall Equipment Effectiveness method to improve the effectiveness of the mechanized longwall systems work in the coal exploitation process" realized in 2015–2018, financed by NCBiR.

References

1. Bingi, P., Sharma, M.K., Godla, J.K.: Critical issues affecting an ERP implementation. Inf. Syst. Manag. **16**(3), 7–14 (1999)
2. Brodny, J., Tutak, M., Michalak, M.: The use of the TGŚP module as a database to identify breaks in the work of mining machinery. In: 13-th International Conference Beyond Databases, Architectures and Structures BDAS 2017, Beyond Databases, Architectures and Structures. Towards Efficient Solutions for Data Analysis and Knowledge Representation, pp. 441–452 (2017). https://doi.org/10.1007/978-3-319-58274-0_35
3. Einicke, G.A., et al.: Longwall mininig automation. An application of minimum-variance smoothing. IEEE Control Syst. Mag. **6**, 28–37 (2008)
4. Elevli, S., Elevli, B.: Performance measurement of mining equipments by utilizing OEE. Acta Mont. Slovaca **15**(2), 95–101 (2010)
5. Gosiewski Z., Osiecki J.: Panasiuk J.: Elementy Mechatroniki, Wydawnictwo Wojskowej Akademii Technicznej. Warszawa (2007)
6. Lech, P.: Zintegrowane systemy zarządzania ERP/ERPII. Difin, Warszawa (2003)
7. Loska, A.: Scenario modeling exploitation decision-making process in technical network systems. Maint. Reliab. **19**(2), 268–278 (2017)
8. Mazurek, W.: Wskaźnik OEE – Teoria i praktyka.. Neuron (2014). www.neuron.com.pl. Accessed 30 Sept 2017
9. Nakajima, S.: Introduction to TPM. Total Productive Maintenance. Productivity Press, Portland (1988)
10. Stecuła, K., Brodny, J., Tutak, M.: Application of the TPM strategy to analyze the effectiveness of using a set of mining machines. In: 16th International Multidisciplinary Scientific GeoConference SGEM, Conference Proceedings, Book1, vol. 2, pp. 65–72 (2016). https://doi.org/10.5593/sgem2016/b12/s03.009
11. Szewczyk, K.: Efektywność wyposażenia jako czynnik wzrostu wartości przedsiębiorstwa. Zeszyty Naukowe Uniwersytetu Szczecińskiego **685** (2011)
12. Becker Homepage. http://www.becker-mining.com.pl/phdi.html. Accessed 30 Sept 2017
13. Wasko Homepage. https://www.wasko.pl/monitorowanie-i-kontrola-instalacji…/openeye-scada. Accessed 30 Sept 2017

Collection of Essential Methods Among the Beams Analysis as an Introduction into the Dynamic Reverse Task Solution of Bending Vibration within Mechatronic System

Andrzej Buchacz(✉)

Institute of Engineering Processes Automation and Integrated Manufacturing Systems, Silesian University of Technology, Konarskiego Street 18 A, 44-100 Gliwice, Poland
andrzej.buchacz@polsl.pl

Abstract. Purpose of this paper is analysis of vibrating beam by the exact and approximate methods and creating the hypergraphs of the beam in case of two methods of analysis. Approach was to nominate the relevance or irrelevance between the characteristics obtained by considered methods – especially concerning the relevance of the natural frequencies-poles of characteristics of beam. The main subject of the research is the continuous free beam as a subsystem of vibrating beam-system with constant cross sections. Findings this approach is fact, that approximate solutions fulfill all conditions for vibrating beams and can be introduction to synthesis of these systems modeled by hypergraphs. Research limitation is that linear continuous transverse vibrating free beam is considered. Practical implications of this study is the main point is the introduction to synthesis of transverse vibrating continuous beam-systems. Originality of this approach relies on application approximate methods of analysis of beam and modeling the one of transformed hypergraphs.

Keywords: Dynamical flexibility · Mechanical or mechatronic complex systems · Exact and approximate methods of analysis

1 Introduction

In the Gliwice Research Centre the multiple problems of different models of vibrating beam systems analysed by the structural numbers methods modelled by means of the graphs and hypergraphs[1] have been solved (e.g. [3–8]). The discrete - continuous torsionally and flexibly vibrating mechanical and mechatronic systems were considered

[1] The problems of synthesis of electrical systems [1] and of selected class of continuous, discrete - continuous discrete mechanical systems [6] and active mechanical systems concerning the frequency spectrum has been made (e.g. [8]). Other diverse problems have been modeled by different kind of graphs [2] next they were examined and analyzed in (e.g. [16–18]).

J. Świder et al. (Eds.): MECHATRONICS 2017, AISC 934, pp. 52–60, 2019.
https://doi.org/10.1007/978-3-030-15857-6_6

[9–12, 15]. In comparison to dynamical flexibilities only for mechanical torsionally vibrating bar and flexibly vibrating beam, as a parts of complex mechatronic systems, exact method and approximate methods were used [8–14, 19]. In this paper the hypergraphs methods have been used for modeling of mechanical subsystems – vibrating beams – of simply mechatronic subsystems of complex mechatronic systems.

On the base of the obtained formulas, which were determined by the exact and approximate method, it is possible to make the analysis of the considered vibrating system by only approximate method. Taking into consideration other boundary conditions of mechanical or mechatronic systems and other kinds of their vibrations it is necessary to achieve other researches review in this paper. The problems will be presented in future works, because necessary conditions to synthesis of transverse vibrating mechanical or/and complex mechatronic systems must be obtained.

2 The Algorithm of Exact Method Obtaining Within the Solution of Transverse Vibrating Beam

In the global case the equation of motion of the beam is considered as:

$$EJ_z y(x,t)_{,xxxx} + \rho A y(x,t)_{,tt} = 0 \tag{1}$$

where: $y(x,t)$ is deflection at the time moment t of the lining beam section within the distance x from the beginning of the system, E - Young modulus, ρ - mass density of material of the beam, J_z - polar inertia moment of the beam cross section, A – area of the beam cross section.

The boundary conditions subsystem of beam system are known and suitably equal within the beneath equation:

- free (F)

$$y_{,xx}(x,t) = 0, \; y_{,xxx}(x,t) = 0, \tag{2}$$

- clamped (C)

$$y(x,t) = 0, \; y_{,x}(x,t) = 0, \tag{3}$$

- pinned (P)

$$y(x,t) = 0, \; y_{,xx}(x,t) = 0, \tag{4}$$

- sliding (S)

$$y_{,x}(x,t) = 0, \; y_{,xxx}(x,t) = 0. \tag{5}$$

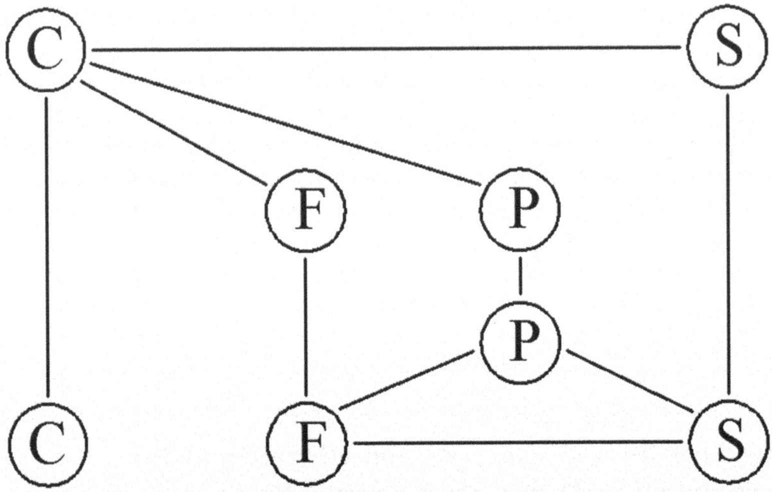

Fig. 1. Graphical representation of combinations of boundary conditions

In the Fig. 1 possible combinations of boundary conditions were presented. The boundary condition are combinations of formulas (2–5).

The solution of Eq. (1) is the function in form of (E.g. [13–15]).

$$y(x, t) = X(x) \sin \omega t \tag{6}$$

Substituting of derivatives of (6) to (1) is obtained

$$X^{(\mathrm{IV})}(x) - k^4 X(x) = 0 \tag{7}$$

where: k depends on boundary conditions (E.g. [13–15]).

The solution $X(x)$ of (7) takes the form

$$X(x) = A_1 \sin kx + A_2 \cos kx + A_3 \sinh kx + A_4 \cosh kx \tag{8}$$

After substitution of (8) into combinations of boundary conditions (2–5) for considered beam and after transformations of own functions the relationships between constants $A_1, \ldots, A_4$ receive the form

$$X_n = f\left(\frac{z_n}{l} x\right), n = 1, 2, 3, \ldots. \tag{9}$$

where: z_n depends on boundary conditions (E.g. [13–15]).

The problem of determining the own function concerning the discrete – continuous and transverse vibrating mechatronic systems is presented in (e.g. [12]).

3 The Algorithm of Exact Method Obtaining the Dynamical Flexibility of Transverse Vibrating Beam

Taking into consideration, that mechanical subsystem of transverse vibrating complex system is extorted with harmonic force in form $P(t) = P_0 \sin \omega t$, therefore one boundary condition in (2) or (5) takes form

$$EJ_z y(x,t)_{\prime xxxx} + \rho A y(x,t)_{\prime tt} = P(t) \tag{10}$$

The set of equations, after transformations, were obtained in matrix form as

$$\mathbf{WA} = \mathbf{F}, \tag{11}$$

where: **W**, **A**, **F** - are the matrices received from one of the boundary conditions of elementary beam.

To qualify constants $A_1,\ldots,A_4$, we should count combinations of following determinants (e.g. [12]).

$$|\mathbf{W}|, |\mathbf{W}_1|, \ldots, |\mathbf{W}_4|. \tag{12}$$

Substituting (11) and (12) to (7) and taking into account (6) deflection beam is considered

$$y(x,t) = f(E,\ J_z, k, l)P_0 \sin \omega t \tag{13}$$

According to the definition of dynamic flexibility, on the basis of (11), it takes form

$$Y = f(E,\ J_z, k, l). \tag{14}$$

Apart from that detailed transformations and final sentences the combinations of boundary conditions have been presented in (E.g. [13–15]).

4 The Algorithms of the Approximate Methods Obtained Within the Dynamical Flexibilities of Transverse Vibrating Beam

It has to be considered that if the beam is under the action of moment with continuous factorization threw the beam length with the value $F(x) \sin \omega t$ on the length unit – then the equation of motion of the element with length dx lining in the point x takes form:

$$EJ_z y_{,xxxx} \mathrm{d}x + \rho A y_{,tt} \mathrm{d}x = F(x) \sin \omega t \, \mathrm{d}x \tag{15}$$

To determine the dynamic flexibility of the factors, which are compatible to concentrate loading $F \sin \omega t$, which works in point z have to be found. The loading can be considered as a limit of concentrate loading threw the length - as follows:

$$P(x) = \begin{cases} \frac{P}{h} \text{ when } z - h \leq x \leq z, \\ 0 \text{ in other section} \end{cases} \tag{16}$$

and the Eq. (15) takes form:

$$EJ_z y'_{xxxx} + \rho A y'_{tt} = P_0 \sin \omega t \tag{17}$$

where: $P_0 = F/h$, h - range of considerations.

Using approximate Galerkin's method the solution of Eq. (17) is given in form

$$y(x,t) = \sum_{n=1}^{\infty} y_n(x,t) = \sum_{n=1}^{\infty} A^{(n)} f(k,x) \sin \omega t \tag{18}$$

Substituting the following derivatives of (18) to (17), the amplitude value $A^{(n)}$- after transformations - takes form:

$$A^{(n)} = P_0 f(a,k,\omega) \tag{19}$$

where: $a = \sqrt{EJ_z/\rho A}$.

In general the dynamical flexibility at the arbitrary point of the beam gets shape of [12]

$$Y_{xl} = \sum_{n=1}^{\infty} Y_{xl}^{(n)} \tag{20}$$

For sum k = 0,1,2,3 the plot of absolute value of dynamical flexibility is defined by expression (25) for combinations of boundary conditions, that means suitably (C)-(S) and (F)-(P) in [12, 15].

By solving the equation of (17) and using orthogonalization method, the deflection $y(x,t)$ is a harmonic function that is considered to be:

$$y(x,t) = \sum_{n=1}^{\infty} S_n(t) X_n(x) \tag{21}$$

where: $S_n(t) = \frac{1}{\gamma_n^2} \int_0^l y(x,t) X_n(x) \mathrm{d}x$, $\gamma_n^2 = \int_0^l X_n^2(x) \mathrm{d}x$, z_n- are roots of equation for its own value (e.g. [12]).

As a result of orthogonalization the equation movement beam in form (20) has been received

$$EJ_z \left[y_{,xxx} X_n - y_{,xx} X_n' + y_{,x} X_n'' - y X_n''' \right]_0^l + EJ_z \int_0^l y(x,t) X_n^{(IV)} \mathrm{d}x + \rho A \int_0^l y_{,tt} X_n \mathrm{d}x = 0 \tag{22}$$

Taking into consideration the boundary conditions (F)-(P) and/or (C)-(S) (e.g. [14]) the following extortion and the conditions of own functions, has been received

$$\ddot{S}_n + \omega_n^2 S_n = -\frac{P(t)}{\rho A \gamma_n^2} = -\frac{P_0}{\rho A \gamma_n^2} \sin \omega t \tag{23}$$

The solution of Eq. (23) is following

$$S_n(t) = -\frac{P_0}{\rho A \gamma_n^2} \frac{1}{\omega_n^2 - \omega^2} \sin \omega t \tag{24}$$

where: ω_n - the n own frequency.
Deflection of beam is equal

$$S_n(t) = -\frac{P_0}{\rho A} \sum_{n=1}^{\infty} \frac{X_n(x)}{\gamma_n^2 (\omega_n^2 - \omega^2)} \sin \omega t = \sum_{n=1}^{\infty} Y_n \sin \omega t \tag{25}$$

On the basis of (25) the dynamic flexibility is given as

$$Y_n = \frac{X_n(x)}{\rho A \gamma_n^2 (\omega^2 - \omega_n^2)}. \tag{26}$$

Comparing the plots obtained by the exact and approximate methods it can be noticed that the poles of flexibilities have the same values (e.g. [14]).

5 Graphs Different Category's Concerning Models of Vibrating Beams

The couple

$${}^kX = \left({}_1X, {}_2^kX\right) \tag{27}$$

is called a hypergraph, where: is the set as in (27), and ${}_2^kX = \left({}_2^kX^{(i)}/i \in \mathrm{N}\right)$, (k = 2,3, … ∈N) is a collection of subsets of set ${}_1X$; the collection ${}_2^kX$ is called a hypergraph over ${}_1X$ as well, and ${}_2^kX = \left\{{}_2^kX^{(1)}, {}_2^kX^{(2)}, \ldots, {}_2^kX^{(m)}\right\}$ is a set of edges [2], called hyperedges or blocks, if

$$\begin{cases} {}_2^kX \neq \varnothing (i \in \mathrm{I}), \\ \bigcup\limits_{i \in \mathrm{I}} {}_2^kX^{(i)} = {}_2X. \end{cases} \tag{28}$$

If a subset from the collection of subsets of vertices with $n_z \leq n$, is distinguished from hypergraph ${}^kX$ with n vertices, then the complete graph of hypergraph ${}^kX$ changes into graph X_Z. In this graph each pair of vertices is incident, and graph X_Z has $m = \binom{n}{2}$ edges.

Graphs X and hypergraphs ${}^{k}X$ have been shown in their geometrical representation. Sets of edges ${}_{2}X$ have been marked by lines, subsets of family ${}_{2}^{k}X$ (hyperedges or blocks) - two-dimensional continuum with enhanced vertices, in the shape of circles.

In this paper hypergraphs - graphs of category k - ${}^{k}X$ (k = 2, 3) were used. It will be clearly mentioned each time, also according to graphs X, called also graphs of the first category - ${}^{1}X$ (see for example [4, 5]).

The basic notions have been written in italics and are presented in (e.g. [12, 15]).

In the paper the flexibly vibrating of the free-free beam (i) with constant cross-section and constant flexibly rigidity $(EJ_z)^{(i)}$ (where $E^{(i)}$- Young's modulus of the beam, $J_z^{(i)}$ - polar moment of inertia of cross-section of the beam [look Eq. (1)] as well as length has been considered. The model in the form of a determined and continuous system was introduced.

Presented model takes the root also and - this beam as well as every studied in more far draught of beam in figure of arrangement limited. In this model, generalized displacements – deflections ${}_{1}s_{1}^{(i)}$ and ${}_{1}s_{2}^{(i)}$ correspond to its extreme points. Moreover the extreme points of beam were subordinated and generalized within - the slopes of the beam - ${}_{1}s_{3}^{(i)}$ and ${}_{1}s_{4}^{(i)}$. These general displacements are measured in the inertial system of reference. Moreover, the origin of the inertial system of reference has generalized coordinate ${}_{1}s_{0}^{(i)} = 0$ assigned to it. So a set of the generalized displacements of a flexibly vibrating beam can be formulated as follows: ${}_{1}S^{(\mathrm{i})} = \left\{ {}_{1}s_{0}^{(i)}, {}_{1}s_{1}^{(i)}, {}_{1}s_{2}^{(i)}, {}_{1}s_{3}^{(i)}, {}_{1}s_{4}^{(i)} \right\}$.

Making mutually one-to-one transformation imitation

$$f : {}_{1}S^{(i)} \to {}_{1}X^{(i)} \tag{29}$$

in this way, that

$$f\left({}_{1}s_{j}^{(i)}\right) = {}_{1}x_{j}^{(i)}, \tag{30}$$

where: ${}_{1}s_{j}^{(i)} \in {}_{1}S^{(i)}, {}_{1}x_{j}^{(i)} \in {}_{1}X^{(i)}$, j = 0,…,4,

the five-vertex hypergraph as a model of flexibly vibrating beam with constant cross-section is obtained

$${}^{2}\underset{f}{X}^{(i)} = \left[{}^{2}X^{(i)}, f \right] \tag{31}$$

where: ${}_{2}^{k}X^{(i)}$ - one-element family - five-element subset of vertices ${}_{1}X^{(i)}$.

In case of bending vibrating beam, clamped on the left end and free on the right one, the set of the generalized displacements of the beam can be described in form: ${}_{1}S^{(\mathrm{i})} = \left\{ {}_{1}s_{0}^{(i)},\ {}_{1}s_{1}^{(i)} = 0,\ {}_{1}s_{2}^{(i)},\ {}_{1}s_{3}^{(i)} = 0,\ {}_{1}s_{4}^{(i)} \right\}$. After transformations of (30) and (31) in case of the flexibly vibrating clamped – free beam with constant cross-section graphical representation is shown in (e.g. [12, 15]).

6 Conclusion

The problems presented in this paper concerning the analysis of transverse vibrating beam were the introduction to the solution of dynamic inverse task - that means the synthesis of transverse vibrating complex mechanical and mechatronic systems with assumed frequency spectrum. In future research works these problems shall be discussed.

Acknowledgement. This work has been conducted as a part of research project PBS3/B6/37/2015 (PST-41/RMT2/2015) from 2015 until 2018.

References

1. Bellert, S., Woźniacki, H.: The Analysis and Synthesis of Electrical Systems by Means of the Method of Structural Numbers. WNT, Warszawa (1968). (in Polish)
2. Berge, C.: Graphs and Hypergraphs. American Elsevier Publishing Co., Inc./North Holland Publishing Co., New York/Amsterdam (1973)
3. Białas, K., Buchacz, A., Dzitkowski, T.: Synthesis of active mechanical systems with dumping in view of polar graphs and structural numbers. Silesian University of Technology Press, Gliwice (2009). (in Polish) Monograph No. 230
4. Buchacz, A.: Modelling, synthesis and analysis of bar systems characterized by a cascade structure represented by graphs. Mech. Mach. Theory **30**(7), 969–986 (1995)
5. Buchacz, A.: The synthesis of vibrating bar-systems represented by graph and structural numbers. Sci. Lett. Silesian Univ. Technol., Mech **104** (1991). (in Polish) Silesian University of Technology Press
6. Buchacz, A., Dymarek, A., Dzitkowski, T.: Design and Examining of Sensitivity of Continuous and Discrete-Continuous Mechanical Systems with Required Frequency Spectrum Represented By Graphs and Structural Numbers. Silesian University of Technology Press, Gliwice (2005). (in Polish) Monograph No. 88
7. Buchacz, A.: The expansion of the synthesized structures of mechanical discrete systems represented by polar graphs. J. Mater. Process. Technol. **164–165**, 1277–1280 (2005). Complete Elsevier
8. Buchacz, A., Żurek, K.: Reverse Task of Dynamic of Active Mechanical Systems by Means the Graphs and Structural Numbers Methods. Silesian University of Technology Press, Gliwice (2005). (in Polish). Monograph No 81
9. Buchacz, A.: Exact and approximate analysis of mechanical and mechatronic systems. J. Achieve. Mater. Manuf. Eng. **33**(1), 47–52 (2009)
10. Buchacz, A.: Modelling, synthesis, modification, sensitivity and analysis of mechanic and mechatronic systems. J. Achieve. Mater. Manuf. Eng. **24**(1), 198–207 (2007). International OCOSCO World Press
11. Buchacz, A.: Dynamical flexibility of discrete-continuous vibrating mechatronic system. J. Achieve. Mater. Manuf. Eng. **28**(2), 159–166 (2008). International OCOSCO World Press
12. Buchacz, A.: Transformations of hypergraphs of beams as models in synthesis of flexibly vibrating continuous mechanical systems, pp. 35–38. Vestnik Krymskoy Akademii Nauk – Sevastopolskie otdelenie. Release 1, Sevastopol (2010)

13. Buchacz, A.: Orthogonalization method of analysis of mechanical and mechatronic systems. In: Les Problèmes Du Technoshère Et De La Formation Des Cadres D'ingénieurs Contemporains, Recueil des exposés des participants du III Séminaire international scientifique et méthodique, Novembre 2009 à Sousse (Tunisie), Donetsk, pp. 85–88 (2009)
14. Buchacz, A.: Comparison of characteristics of subsystems and systems as introduction to synthesis of torsionally vibrating mechatronic systems. In: Les Problèmes Du Techno- Shère Et De La Formation Des Cadres D'ingénieurs Contemporains. Recueil des exposés des participants du IV Séminaire International scientifique et méthodique, Novembre 2010 à Hammamet (Tunisie), Donetsk, pp. 85–88 (2010)
15. Buchacz, A.: The supply of formal notions to synthesis of the vibrating discrete-continuous mechatronic systems. J. Achieve. Mater. Manuf. Eng. **44**(2), 168–178 (2011). International OCOSCO World Press
16. Callahan, J., Baruh, H.: Vibration monitoring of cylindrical shells using piezoelectric sensors. Finite Elem. Anal. Des. **23**, 303–318 (1996)
17. Kurnik, W.: Damping of mechanical vibrations utilizing shunted piezoelements. Mach. Dyn. Probl. **28**(4), 15–26 (2004)
18. Lu, P., Lee, K.H., Lim, S.P.: Dynamical analysis of a cylindrical piezoelectric transducer. J. Sound Vib. **259**(2), 427–443 (2003)
19. Wojnarowski, J., Buchacz, A.: Use of Hypergraphs and complete structural numbers in the analysis of vibrating beam systems with non-linearly changing cross-sections. Vibr. Eng. **3** (4), 593–598 (1989). Hemisphere Publishing Corp.

Tunable Model of a Servo Hydraulic Tester for Shock Absorbers Vibrational Evaluation

Piotr Czop[1], Mariusz Hetmańczyk[2(✉)], Grzegorz Wszołek[2], and Jakub Słoniewski[1]

[1] AGH University of Science and Technology, Mickiewicza St. 30, 30-059 Cracow, Poland
[2] The Silesian University of Technology, Konarskiego 18A St., 44-100 Gliwice, Poland
mariusz.hetmanczyk@polsl.pl

Abstract. The aim of the paper is to propose a method of tuning first-principle model of a general-purpose servo-hydraulic test-rig. The servo-hydraulic test-rig dynamics is a significant contributor considering shock absorber vibrational analyses and tests. Therefore, the derived model reproduces the important operational parameters of the servo-hydraulic test-rig applicable in laboratory tests. The paper provides a system identification case study in the time domain based on the experimental data.

Keywords: Shock absorber · System identification · Nonlinear model · First-principle model

1 Introduction

The paper presents the architecture of the first-principle model combining mechanical-hydraulic sub-systems and briefly describes each component of the model. The model provides a combination of first-principle equations and data-driven approach [1]. The advantage of a first-principle approach to modeling is the ability to gain a physical insight into the underlying system. On the other hand, the advantage of the data-driven approach is the speed in which an accurate model can be developed, as well as the confidence arising from the fact that the model comes from data measured on an actual system.

Nevertheless, disadvantages of this approach are the need to handle multiple data sets in order to cover the entire range of operating conditions and the inability to extrapolate the best-fit curve to operating conditions outside the range covered by the measurement data. Investigation of servo-hydraulic dynamics requires detailed models comprising sub-models which represent particular components. These models are based on first-principle equations (e.g. mass, momentum and force balance) that involve phenomenological correlations, such as discharge coefficients. Such models are commonly utilized to gain an understanding of physical processes as well as in process efficiency optimization [2]. These models are knowledge models, thanks to which process dynamics can be understood. The proposed model applies three categories of parameters, geometrical, physical and phenomenological.

J. Świder et al. (Eds.): MECHATRONICS 2017, AISC 934, pp. 61–69, 2019.
https://doi.org/10.1007/978-3-030-15857-6_7

2 Updating Model Parameters

The process of servo-hydraulic tester model calibration consists of: selection of representative data sets, including a sufficiently broad operating range that the model proved could work correctly using a data fit measure and adjustment of model parameters to fit a model response to data. The model is represented by a set of nonlinear state-space equations formulated in the continuous-time domain. The objective of the estimation is to minimize the error function between the measured signals and model responses by means of an iterative numerical technique. The function describing the error has to be a positive and decreasing function. The procedure for updating model parameters consists of two in-a-loop phases: simulation of a model by solving differential equations numerically in Simulink [3], and numerical minimization in the parameter space with respect to an error-related criterion function using the MATLAB Optimization Toolbox [4].

After each simulation of the model for fixed-length input signals, the simulated output data are sampled and the criterion function is re-evaluated to calculate a new set of model parameters. A general form of the model in the continuous domain is as follows:

$$\begin{aligned} \tfrac{d}{dt}x(t) &= f(t, x(t), u(t), w(t); \theta) \\ y(t) &= h(t, x(t), u(t), v(t); \theta) \\ x(0) &= x_0 \end{aligned} \tag{1}$$

where vector f(.) is a nonlinear, time-varying function of the state vector x(t) and the control vector u(t), while vector h(.) is a nonlinear measurement function; w(t) and v(t) are sequences of independent random variables and θ denotes a vector of unknown parameters. In nonlinear systems, the state vectors and the measurements vectors are not Gaussian distributed. The predictor resulting from the model (1) takes the form:

$$\hat{y}(t|\theta) = g\left(t, Z^{t-1}; \theta\right) \tag{2}$$

while the prediction error equation has the form:

$$\varepsilon(t, \theta) = y(t) - g\left(t, Z^{t-1}; \theta\right) \tag{3}$$

The sum of squared errors is used as an error criterion. This problem is known in numerical analysis as "the nonlinear least-square problem" [5]. The objective of the estimation is to minimize the $V_N(\theta)$ error function by means of an iterative numerical technique. The error function $V_N(\theta)$ has the form:

$$V_N\left(\theta, Z^N\right) = \frac{1}{N} \cdot \sum_{t=1}^{N} \frac{1}{2} \cdot \varepsilon^2(t, \theta) \tag{4}$$

The standard Gauss-Newton method was considered and used for adjusting model parameters in this paper.

3 First-Principle Data-Driven

The case study demonstrates estimation of parameters of a continuous-time model based on the servo-hydraulic test rig [6]. The model is developed for the Hydropuls MSP25 servo-hydraulic test-rig, equipped with an IST8000 electronic controller. The functional closed-loop layout of the system is presented in Fig. 1.

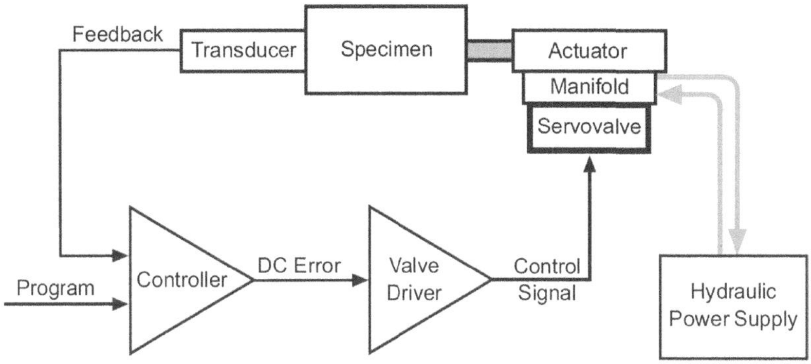

Fig. 1. Closed-loop servo-hydraulic test system signal flow layout.

The test rig was used to load a shock absorber and capture its dynamic characteristics, i.e. displacement vs. force. Data acquisition was performed with an 8-channel ICP amplifier manufactured by LMS. The test rig is equipped with an oil supply system (the so-called servo-pack) that provides a pressure of 28 MPa at a flow-rate of 90 l/min. The actuator provides a 25 kN force at the rod, while the maximum stroke is 250 mm at the maximum achievable velocity of 3 m/s. The actuator rod is coupled to the adapter, which transfers the force to a shock absorber mounted on a test rig. The main components of the servo-hydraulic system are the hydraulic actuator with integrated displacement transducer in a piston-rod assembly (IST-Schenk) and the three-stage servo-valve system. The test rig is equipped with a PID controller. The excitation signal is converted into a voltage applied to the servo-valve, which controls the amount of oil supplied to the chambers of the actuator. The simulation first-principle model of the Hydropuls MSP25 test-rig consists of a servo-valve and a hydraulic actuator model. The Moog G761 series industrial servo-valve consists of a polarized electrical torque motor and two stages of hydraulic power amplification. This valve dynamics is approximated by a transfer function with an input as a voltage and output as the spool displacement. The electromagnetic torque motor, driving the flapper, is controlled by an electrical current input from a current amplifier. A valve control input u (voltage) is converted into a current, with a current amplifier gain Ke.

The hydrodynamic forces acting on the spool are neglected. Therefore, the spool position depends on the input voltage, angular natural frequency ω_n, servo and damping ratio of the servovalve ζ:

$$G_{servo} = \frac{x_{spool}(s)}{x_{PID}(s)} = \frac{K_e \cdot K_{spool}}{1 + \left(\frac{2\zeta}{\omega_{n,servo}}\right)s + \left(\frac{s}{\omega_{n,servo}}\right)^2} \tag{5}$$

The pressure drop among the test rig chambers (A and B), input (P) and output (T) are as follows:

$$\begin{aligned} \Delta p_{PA} &= p_P - p_A \\ \Delta p_{BT} &= p_B - p_T \\ \Delta p_{AT} &= p_A - p_T \\ \Delta p_{PB} &= p_P - p_B \end{aligned} \tag{6}$$

The flows at all servo-valve restrictions are determined by orifice equations based on pressure drops and spool displacement:

$$\begin{aligned} q_{PA} &= Cd_{PA} \cdot b \cdot \left|\Delta x_{spool}\right| \cdot \sqrt{2 \cdot |\Delta p_{PA}|/\rho} \cdot sgn(\Delta p_{PA}) \\ &\quad if\ \Delta x_{spool} = x_{spool} - x_{spool,0} \geq 0 \\ q_{BT} &= Cd_{BT} \cdot b \cdot \left|\Delta x_{spool}\right| \cdot \sqrt{2 \cdot |\Delta p_{BT}/\rho|} \cdot sgn(\Delta p_{BT}) \\ &\quad if\ \Delta x_{spool} = x_{spool} - x_{spool,0} \geq 0 \\ q_{AT} &= Cd_{AT} \cdot b \cdot \left|\Delta x_{spool}\right| \cdot \sqrt{2 \cdot |\Delta p_{AT}|/\rho} \cdot sgn(\Delta p_{AT}) \\ &\quad if\ \Delta x_{spool} = x_{spool} + x_{spool,0} < 0 \\ q_{PB} &= Cd_{PB} \cdot b \cdot \left|\Delta x_{spool}\right| \cdot \sqrt{2 \cdot |\Delta p_{PB}|/\rho} \cdot sgn(\Delta p_{PB}) \\ &\quad if\ \Delta x_{spool} = x_{spool} + x_{spool,0} < 0 \end{aligned} \tag{7}$$

In case of a hydraulic tester, the oil density can be assumed as a constant since for the maximal load conditions (15 MPa) the hydraulic oil density change of 1% gives 0.5% change in flow. For typical load conditions during shock absorber test (1 MPa), the change in flow is smaller than 0.034%. A model of a hydraulic actuator is derived assuming the pressures in the actuator chambers and displacement of the rod as the state variables. The bypass flow, controlled by the throttle valve, is defined by the following formula:

$$q_{AB} = Cd_{AB} \cdot A_{leak} \cdot \sqrt{2 \cdot \frac{|\Delta p_{AB}|}{\rho}} \cdot sgn(\Delta p_{AB}) \tag{8}$$

The total balance of the inlet and outlet flows is determined as follows:

$$\begin{aligned} q_A &= q_{PA} - q_{AT} - q_{AB} \\ q_B &= q_{PB} - q_{BT} - q_{AB} \end{aligned} \tag{9}$$

The pressures p_A and p_B are determined based on the mass equilibrium equations, where the volumes are determined based on the initial oil volume in the actuator and the position of the piston-rod:

$$\begin{aligned} \frac{dp_A}{dt} &= \frac{\left(q_A - \dot{V}_A\right) \cdot B}{V_{A0} + A_A \cdot x} \\ \frac{dp_B}{dt} &= \frac{\left(q_B + \dot{V}_B\right) \cdot B}{V_{B0} + A_B \cdot x} \end{aligned} \tag{10}$$

where changes of chamber volumes:

$$\dot{V}_A = A_A \cdot \dot{x} \cdot \dot{V}_B = A_B \cdot \dot{x} \tag{11}$$

Finally, the displacement of the actuator x is calculated based on the force equilibrium equation:

$$\ddot{x}_m + F_{fric}(\dot{x}) = p_A \cdot A_A - p_B \cdot A_B - F_{ext}(x, \dot{x}) \tag{12}$$

where: AA, AB are the pressure acting areas of two sides of the piston; p_A, p_B are the pressures in the chambers. The dry friction force F_{fric} between the piston and cylinder and rod and seal is modelled as follows:

$$F_{fric = F_{fric\_max}} \cdot tanh\left(\frac{\dot{x}_{tube} - \dot{x}_{rod}}{V_{ref}}\right) \tag{13}$$

The friction force F_{fric} depends on the direction of the rod travel x, not on velocity nor piston position. The maximal friction force F_{fric_max} is obtained from experimental tests. The applied friction model uses the hyperbolic tangent function tanh(.) which approximates measurement data and the reference velocity v_{ref} which provides a smooth (realistic) switch of friction force similar to measurement data. A model of the electronic controller uses a program command and sensor feedback to control the servovalve. A test rig controller uses a group of gain controls, i.e. proportional (P), integral (I) and derivative (D) gains. A transfer function representation of a PID controller is given as follows:

$$G_{PID} = \frac{u_{PID}(s)}{\Delta x(s)} = K_P \cdot \left(1 + \frac{1}{s \cdot T_I} + s \cdot T_D\right) = K_P + \frac{K_I}{s} + K_D \cdot s \tag{14}$$

where

$$\Delta x = x_{exp} - x \tag{15}$$

The output of the PID controller is used as input of servovalve transfer function. The equations have been implemented in the MATLAB/Simulink package.

4 Tuning of the Model

A physical model can conveniently be represented as a set of non-linear state-space equations formulated in the continuous-time domain. The objective of the estimation is to minimize the error function between the measurement signals and model responses in the time-domain by means of an iterative numerical technique constituting "the non-linear least-squares problem" [5]. The function describing the error has to be a positive and decreasing function. The procedure of model tuning consists of two in-a-loop phases: (i) simulation of a model by solving differential equations numerically in Simulink, and (ii) numerical minimization in the parameter space with respect to an error-related criterion function using MATLAB Optimization Toolbox [3]. The interested reader may find more information concerning the available methods and algorithms that support identification of first-principle models in [2]. The simulation and optimization settings used in the parameter adjustment process are presented in Table 1.

Table 1. Simulation and optimization settings.

Simulation		Optimization (minimization)	
Option	Value	Option	Value
Solver	ode23tb (stiff/TR-BDF2)	Gradient type	Basic
Max step size	Auto	Algorithm	lsqnonlin
Min step size	Auto	Cost type	SSE
Zero crossing control	Disable all	DiffMaxChange	0.1
Relative tolerance	Auto	DiffMinChange	1E-08
Absolute tolerance	Auto	Large scale	true
		MaxIter	28
		RobustCost	False
		TolCon	1E-6
		TolFun	1E-6

The pink signal was used as a standard excitation signal in order to perform shock absorber tests (Fig. 2).

The Newton-Gauss method, lsqnonlin(.) routine implemented in the Optimization Toolbox of MATLAB [4], was used to minimize the function describing the error in the measurement signals and model responses in the frequency domain. The model performance is evaluated either by visual inspection of the plot of the measured data y vs. model response $\hat{y}$ or by analyzing the value of:

$$R^2 = 1 - \frac{\frac{1}{N} \cdot \sum_{i=1}^{N} |y(i) - \hat{y}(i)|^2}{\frac{1}{N} \cdot \sum_{i=1}^{N} |y(i)|^2} \tag{16}$$

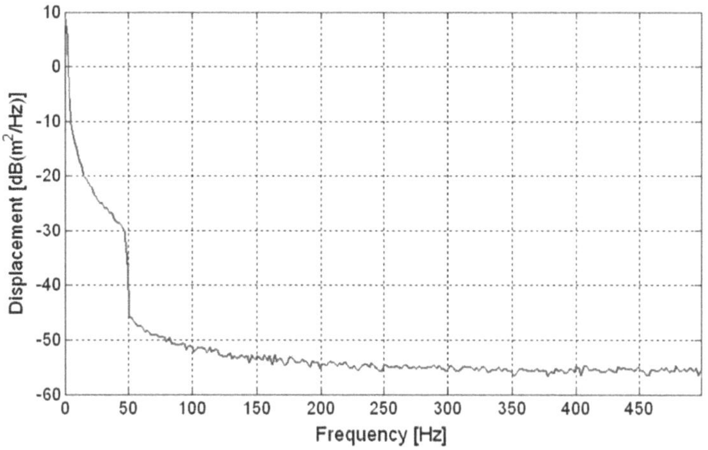

Fig. 2. Frequency response of the servo-hydraulic tester measured at the top of the rod actuator.

where N is the number of samples, and i is the discrete time. The signals were filtered using pass-band filter in the range 0–200 Hz.

The known geometrical and physical parameters of the servo-hydraulic tester model were shown in Table 2, while the unknown free parameters (phenomenological) were shown in Table 3. The tuning scenario considers estimation of four free parameters simultaneously. The first parameter is the leakage between the upper and bottom chamber of the hydraulic actuator, the second is the piston friction force, the third is the discharge coefficient of the servo-valve, and the fourth is the damping ration of the servo-valve.

Table 2. Values of fixed model parameters.

Configuration parameters	Values
PID controls	
PID proportional gain KP	400 [V/m]
PID integral gain KI	0 [V·s/m]
PID derivative gain KD	0 [V/(m·s)]
Hydraulic actuator	
Oil density	960 [kg/m^3]
Chamber volume VA and VB	480 [ccm]
Pressure acting area AA and AB	1920 [mm^2]
Piston-rod assembly mass m	45 [kg]
Fluid bulk modulus	1.5e3 [MPa]
Reference friction velocity vref	1 [mm/s]
Servovalve	
Amplifier gain Ke	1 [A/V]
Natural frequency ωn, servo	43 [Hz]
Spool gain Kspool	0.0024 [m/A]
Servovalve channel width b	0.5 [mm]

Table 3. Estimated values of the free model parameters.

Configuration parameters	Estimated values
Piston friction force F_{fric_max}	53 [N]
Product $Cd_{AB\ Aleak}$	3.2863e-3 [mm^2]
Discharge coefficient Cd	0.2191 [–]
Damping ratio ζ	0.3701 [–]

The input to the model is the demanded displacement of the rod of the actuator, while the output is the measured (simulated) displacement. The boundary conditions were imposed on the estimating parameters values according to physical limitations. The iterative convergence process was run for a couple of minutes (15 min) to minimize the error objective function.

The error was computed comparing reference system output (rod acceleration) with predicted model output after each iteration. The evaluated accuracy in the time domain for a random pink noise signal (Fig. 2) was 87% using the measure expressed by Eq. 12. Figure 3 shows the system identification results in the frequency domain since the qualitative accuracy is important to excite the shock absorber in the specific range of frequencies.

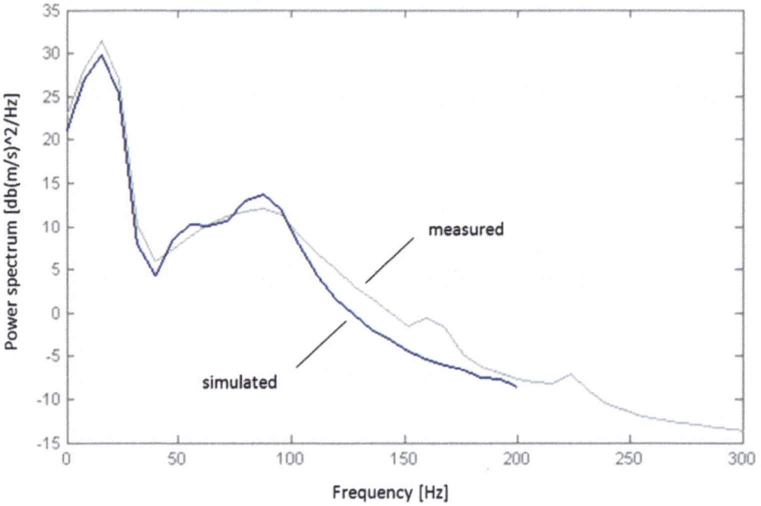

Fig. 3. Frequency response of the servo-hydraulic tester measured at the top of the rod actuator.

Table 2 specifies values of the physical and geometrical parameters used in the simulation model, while Table 3 specifies phenomenological ones to be estimated.

5 Summary

This work shows a model adjustment process based on the measured output acceleration signal and the input displacement signal in order to achieve the high correlation data fit between the test-rig model response and the measured experimental data. The achieved modeling accuracy was better than 87% regarding the high-frequency vibration content in the frequency range 0–200 Hz.

References

1. Czop, P.: Development of First-Principle Data-Driven Models for Design and Assessment of Technical Systems. Institute for Sustainable Technologies - PIB, Radom (2011)
2. Bohlin, T.P.: Practical Grey-box Process Identification. Theory and Applications. Springer, London (2006)
3. Ljung, L.: System Identification Toolbox for Use with MATLAB. The MathWorks Inc., Natick (2007)
4. Natick, L.: Global Optimization Toolbox. User's Guide. The MathWorks Inc., Natick (2015)
5. Ljung, L.: System Identification: Theory for the User, 2nd edn. Prentice Hall, Upper Saddle River (1999)
6. Kost, G., Wszołek, G., Czop, P., Jakubowski, D., Wlodarczyk, T.: Parameter estimation of first-principle models formulated using nonlinear ordinary differential equations. Silesian University Publishing House, Gliwice (2013)

Modelling and System Identification of a Monotube Shock Absorber

Piotr Czop[1], Mariusz Hetmańczyk[2(✉)], Grzegorz Wszołek[2], and Jakub Słoniewski[1]

[1] AGH University of Science and Technology, Mickiewicza 30 St., 30-059 Cracow, Poland
[2] The Silesian University of Technology, Konarskiego 18A St., 44-100 Gliwice, Poland
mariusz.hetmanczyk@polsl.pl

Abstract. This paper demonstrates linear and nonlinear system identification methods based on a first-principles monotube shock absorber model. Two models were considered, namely low- and high-content a priori knowledge first-principles models. Operational data collected during testing of a shock absorber on a servo-hydraulic test rig were used in order to validate the proposed system identification methods and evaluate their accuracy. The proposed approach in system identification has potentials for a wide range of engineering applications such as control of active and semi-active suspension system.

Keywords: Shock absorber · System identification · Nonlinear model · First-principles model

1 Introduction

This paper aims to develop of system identification methodology intended for adjusting the parameters of low- and high-content a priori knowledge shock absorber models. The low-content models are based on autoregressive linear and nonlinear black-box models, while high-content model are based on first-principle equations (e.g. conservation of mass, momentum, force and energy) that involve phenomenological correlations, like friction or flow coefficients. The taxonomy of models regarding available a priori knowledge is discussed by Bohlin [1] or Ljung [2].

The a priori knowledge about a given phenomenon comes from the analysis, comprising of finding all possible connections to other phenomena and physical laws, which precedes the modeling. The a priori knowledge is of key importance in modeling although its availability is always limited by the complexity of the physical system. Even if the governing physical principia are known, it is usually difficult to formulate the specific relationships and obtain particular values of the parameters. Availability of the a priori knowledge and the modeling purpose determine the following: (i) the final type of the model, (ii) the accuracy requirements, (iii) the type of specific modeling procedure, (iv) the complexity of the model and lastly, (v) the method and the cost of its realization. According to the degree to which the a priori knowledge is available, then either a first-principle or a data-driven model, or a combination of both, can be

J. Świder et al. (Eds.): MECHATRONICS 2017, AISC 934, pp. 70–80, 2019.
https://doi.org/10.1007/978-3-030-15857-6_8

applied. High-content a priori knowledge models use understanding of the system underlying physics to derive its mathematical representation. These models are expensive in development since expertise in the area of knowledge at the advanced level is required to derive equations from physical laws, while low-content a priori knowledge models use system test data to derive its mathematical representation. The advantage of the former approach is the depth of the insight into the behavior of the system and thus ability to predict the performance, while the advantage of the latter is the speed in which an accurate model can be created and confidence gained thanks to the use of the data obtained from the actual system. The difficulty of the former approach lies in the determination of the phenomenological parameters like the friction or the heat transfer coefficient. High-content models are frequently adjusted by trial-and-error, which can lead to non-optimal results. On the other hand, the disadvantage of low-content models is the need to handle multiple data sets in order to cover the range of system operation.

First-principle lumped-parameter models of shock absorbers were originally discussed by Lang [3] and Talbot and Starkey [4]. The developed models were intended to describe the damping force performance however they also included the detailed analysis of valve system models. There have been several other studies about shock absorber. Reybrouck [5] proposed a mathematical model of the mono-tube gas-charged shock absorber. Simms et al. [6] investigated the influence of shock absorber properties on vehicle dynamic behaviour through the simulation and test. Liu et al. [7] reported the characteristics of non-linear dynamic response for the twin-tube hydraulic shock absorber. Baracat [8] discussed the principia of shock absorber design. Duym et al. [9–11] developed several simulation models for mono and twin-tube shock absorbers. Basso [12] described a method to predetermine the characteristic diagram of a shock absorber by means of numerical simulation of the piston shim valve. Boggs [13] proposed an intuitive empirical nonlinear dynamic shock absorber model for simulation studies. Other authors presented their works focusing on specific design aspects, i.e. durability, noise and vibration reduction. Cherng et al. [14] reported the effect of noise of shock absorber using the acoustic index method. Yamauchi [15] developed the model of a shock absorber to understand and reduce the self-excited vibration in the piston-rod assembly. Kruse [16] developed the model focusing on structural noise in order to improve the vehicle interior acoustic properties. Lee and Moon [17] developed the lumped-parameter model to reproduce the operation of the displacement-sensitivity (i.e. stroke-dependent) shock absorber. Fukushima at el. [18] used theoretical analysis and simulation to obtain the shock absorber damping coefficients in order to meet the suspension requirements under driving conditions. Sikora [19] studied fluid induced pressure fluctuations inside the shock absorber flow passages. Subramanian et al. [20] discusses the introduction of a nonlinear shock absorber model to be used for low-frequency NVH applications in body-on-frame vehicles. Liu [21] developed a double tube shock absorber model using neural network in order to analyze the influence of relative speed to shock absorber's force. The model was then embedded into the whole vehicle model, and the optimization was performed to improve the vehicle riding comfort. Wang et al. [22] developed the parametric shock absorber model with actual in-service parameters, such as variable viscous damping or comprehensive stiffness and a small mounting clearance was developed for a rail vehicle's axle-box shock absorber.

The obtained and experimentally validated model coupled all of the main factors that had significant impact on the damping characteristics, so that the model could be useful in more extensive parameter effects analysis, optimal specification and product design optimization of shock absorbers for track-friendliness, ride comfort and other high-speed transit problems [22].

2 Nonlinear Models Applicable to Reproduce Shock Absorber Dynamics

2.1 Low-Content of a Priori Knowledge Models

Low-content of a priori-knowledge models can be represented using a Linear Time-Invariant (LTI) models. These models are formulated using discrete-time difference equations [23]:

$$y(i) = G\left(z^{-1}\right) \cdot u(i) + H\left(z^{-1}\right) \cdot e(i) = \frac{B(z^{-1})}{A(z^{-1}) \cdot F(z^{-1})} \cdot u(i) + \frac{C(z^{-1})}{A(z^{-1}) \cdot D(z^{-1})} \cdot e(i) \tag{1}$$

where $G(z^{-1})$ and $H(z^{-1})$ are discrete-time transfer functions of z^{-1}operator and represent the input-to-output dynamics and the disturbance-to-output dynamics, respectively. Moreover, y(i), u(i), and e(i) denotes the output, input, and disturbance input respectively. The polynomials A, B, C, D, and F contain adjustable coefficients and have orders nA, nB, nC, nD, and nF respectively. They general form is as follows: $A(z^{-1}) = 1 + a_1 z^{-1} + a_2 z^{-1} + \ldots + a_{nA} z^{-1}$. The other representation is a state-space formulation, where the relationship among the input, the disturbance and the output signals is expressed in the form of a system of first order difference equations and parameterized as the State-Space Innovations Form (SSIF):

$$\begin{aligned} x(i+1) &= Ax(i) + Bu(i) + Ke(i) \\ y(i) &= Cx(i) + Du(i) + e(i) \end{aligned} \tag{2}$$

where A, B, C, D are system matrices, K is the disturbance matrix, y(i) is the output, u (i) is the input, and e(i) is the disturbance input. Nonlinear systems represent the majority of physical structures, systems and installations; therefore there is an interest to develop nonlinear models and adequate methods to estimate their parameters. Nonlinear models have strong justifications since real systems frequently do not follow the superposition rule producing signals with harmonics corresponding to an input signal and its inherent properties (natural frequencies). A shock absorber is one of the most nonlinear systems where mass, damping, and stiffness can vary according to applied excitation by means of excitation amplitude and frequency. There are available semi-nonlinear models convertible to linear ones and inherently nonlinear models elaborated in this paper. The first group requires a priori knowledge regarding the type of the nonlinearity model. Hamerstein and Wiener models allow to take into account the input- and output-related static nonlinearities.

The Hamerstein-Wiener model has the following general structure:

$$y(z^{-1}) = \frac{B(z^{-1})}{A(z^{-1})} \cdot f_N \cdot \left[u(z^{-1})\right] \tag{3}$$

while, the Wiener model has the following structure:

$$y(z^{-1}) = f_N \cdot \left[\frac{B(z^{-1})}{A(z^{-1})} \cdot u(z^{-1})\right] \tag{4}$$

where polynomial A, B represent the input-output dynamics as in Eq. (1), while $f_N[.]$ is an arbitrary static nonlinear function. The second group, namely inherently nonlinear models, applies a general nonlinear mapping between the model inputs and outputs including delayed data samples as follows [24]:

$$y(i+1) = g \cdot [y(i+d-1), K, y(i+1-nA), u(i), K, u(i-nB)] \tag{5}$$

where g[.] represents the neural network structure, nA, nB represent the number of delayed samples of y, u signals, respectively. The following model functional symbols referred to in this paper are available based on the introduced model structures:

$$\begin{gathered}
Eq.\,1\,ARX:\ ARX(nA,nB):\ linear\,ARX\,model\\
Eq.\,2:\ SS(nN):\ linear\,State-Space\,Innovation\,Form\,model\\
Eq.\,3:\ NLWH(f_N,nA.nB):\ Hammerstein-Wiener\,model\\
Eq.\,4:\ NLW(f_N,nA,nB):\ Wiener\,mode\\
Eq.\,5:\ NARX(nA,nB,d):\ nonlinear\,ARMAX\,model
\end{gathered} \tag{6}$$

The paper provides system identification results for all structures referred to in Eq. (6).

2.2 High-Content of a Priori Knowledge Models

High-content of a priori knowledge models are commonly nonlinear models that capture all relevant nonlinearities, which frequently result from technical limits that follow the product or system specification or intend operating range. The parameters' ranges are therefore set as the hard boundaries to state variables and parameters, which are required to use only the allowed variability range due to model stability, nonlinearity (root square functions), sensor boundary ranges, other type of saturation (e.g. clearances), or unsigned variables values (e.g. absolute pressure). Initial parameters' and state variables' values are assigned using the known state variables measurements, e.g. internal pressure, rod displacement. There is also possibility to use a heuristic approach or equivalent measurements. The following a priori information is acquired and applied to increase the first-principle content in the model: (i) physical laws represented by conservation equations (Newton's laws of motion, first law of thermodynamics, etc.), (ii) phenomenological laws represented by constitute equations (e.g. piston-cylinder friction model [24], the Bouc-Wen hysteresis model [25], density,

Hook's law, ideal gas law), (iii) parameters ranges (minimum and maximum values), (iv) state variables ranges (minimum and maximum values), (v) initial values of model parameters (rough calculation, the best guess, etc.), (vi) initial values of model states (measurements, previous simulations, etc.), (vii) deterministic or stochastic disturbance models, e.g. linear trend, white or filtered noise. High-content of a priori knowledge models are mostly formulated as a set of nonlinear state-space equations in the continuous-time domain:

$$\begin{aligned} \tfrac{d}{dt}x(t) &= f(t, x(t), u(t), w(t); \theta) \\ y(t) &= h(t, x(t), u(t), v(t); \theta) \\ x(0) &= x_0 \end{aligned} \tag{7}$$

where $f(.)$ and $h(.)$ are nonlinear time-varying functions of the state vector $x(t)$, the control vector $u(t)$, the process disturbance vector $w(t)$, the measurement disturbances vector $v(t)$, and the unknown parameters vector θ.

3 Shock Absorber Model and Experimental Results

In order to demonstrate a system identification approach for low and high content a priori knowledge models, the mono-tube shock absorber model developed in [26] was used. The model is capable to reproduce the experimental measurements conducted at servo-hydraulic tester described in [26]. The Sect. 3.2 shows a low-content knowledge model system identification approach where the parameters do not represent a physical meaning variables, while the Sect. 3.3 shows a high-content knowledge model system identification approach where two parameters are part of the friction equations of the monotube shock absorber model.

3.1 Shock Absorber Model

The monotube shock absorber model developed in [26] includes a friction equation representing the resultant friction force F_{fric} between the piston-rod assembly and the pressure tube, and the rod guide:

$$F_{fric} = F_{fric\_max} \cdot tanh\left(\frac{\dot{x}_{tube} - \dot{x}_{rod}}{V_{ref}}\right) \tag{8}$$

where F_{fric_max} is the maximal friction force, v_{ref} is the reference velocity, $\dot{x}_{rod}$ is the piston-rod assembly velocity if it is fixed in the top mount, $\dot{x}_{tube}$ is the shock absorber body velocity. There are two unknown parameters in Eq. (8), namely F_{fric_max} and v_{ref} which required to be obtained. The parameters value are obtained from experimental tests while the excitation signal has relatively low velocity v = 0.01 m/s (stroke S = 90 mm) and hydraulic restrictions are as low as feasible (i.e. the valves are fully opened). The mono-tube shock absorber schematic is presented in Fig. 1.

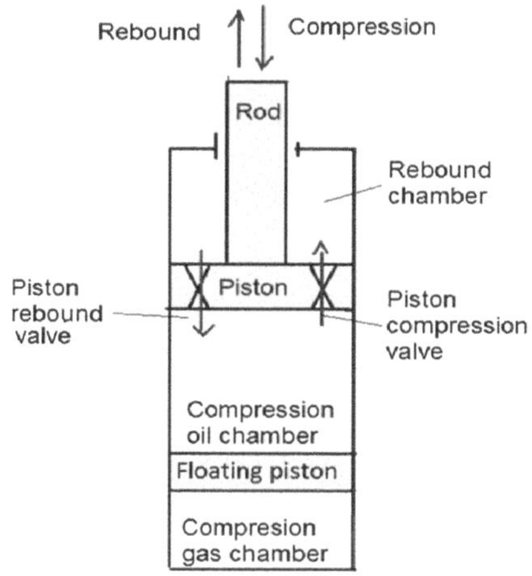

Fig. 1. Mono-tube shock absorber scheme.

The proposed friction model uses the hyperbolic tangent function tanh(.) which provides a smooth switching of friction force, similar to the measurement data. The friction between floating piston and pressure tube is neglected since it is significantly smaller than F_{fric}. The next Sect. 3.2 applies a low-content knowledge model (i.e. black-box model) to capture overall shock absorber dynamic including friction effects, while Sect. 3.3 applies the high-content knowledge shock absorber model from [26] where F_{fric_max} and v_{ref} parameters are adjusted (estimated) to achieve the best agreement between operational data and the model response. Low-content of *a priori* knowledge model validation results

3.2 Low-Content of a Priori Knowledge Model Validation Results

Linear and nonlinear model structures were used to capture dynamics of the hydraulic shock absorber. The model applied structures were, i.e. ARX, Wiener-Hamerstain, and NARX as presented in Table 1. Model structures were developed using inputs and outputs specified in Table 2, while the simulation and optimization parameters are presented in Table 3. The Levenberg-Marquardt (trust-region) algorithm was used to estimate the parameter.

Table 1. Validation results

Model structure	Orders of model parameters (Eq. 6)	Time domain fit [%]
ARX	nA = 6; nB = 6	86.50
SS	nN = 6	86.82
NLHW	nA = 6; nB = 6	96.26
NARX	nA = 6; nB = 6, d = 1	93.11

Table 2. First-principle model inputs, states, outputs, and parameters.

Inputs	Shock absorber housing displacement	[m]
	Shock absorber housing velocity	[m/s]
Outputs	Force at the piston-rod assembly	[N]
	Pressure in the rebound chamber/compression	[N/m2]

Table 3. Simulation and optimization settings [27].

Simulation		Optimization	
Solver type in MATLAB [27]	ode23tb	Search method	Nonlinear least square
Relative tolerance	1E-3 [s]	Criterion	Trace
Absolute tolerance	1E-6 [s]	Max iteration	20
		Tolerance	0.01 [–]

Linear models, as expected, do not correctly reproduce a nonlinear behavior of the hydraulic shock absorber. The predicted model response is affected by presence of a nonlinear content in the model. The operational measurement shows distorted measured output signal (i.e. force) due to static friction effect as presented in Figs. 2 and 3.

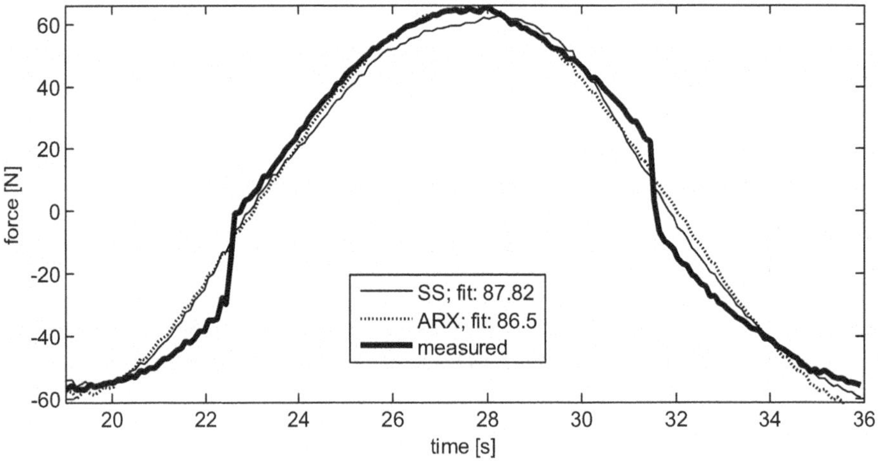

Fig. 2. Force-time characteristic of the hydraulic shock absorber (Part 1).

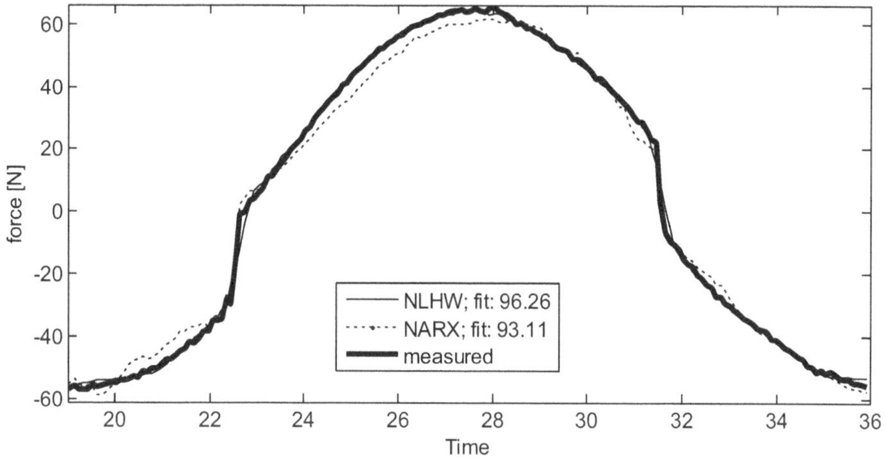

Fig. 3. Force-time characteristic of the hydraulic shock absorber (Part 2).

3.3 High-Content of a Priori Knowledge Model Validation Results

The grey box model structure (Eq. 7) available in System Identification Toolbox (MATLAB [27]) was used to represent the simulation model developed in [26]. The model has two inputs which are the signals of displacement and velocity and one output which is the piston-rod assembly force. It is therefore classified as MISO structure. The adjustment algorithm simulates the model several times trying various parameter values to reduce the prediction error. The following algorithm properties can affect the quality of the results (i) solver settings, (ii) optimization method, (iii) gradient options, and (iv) other specified algorithm properties. The first-principle (Eq. 7) model takes into account the inputs, states, outputs, and parameters as presented in Table 4. The simulation and optimization parameters are presented in previous Table 3.

Table 4. First-principle model inputs, states, outputs, and parameters.

Inputs	Shock absorber housing displacement	[m]
	Shock absorber housing velocity	[m/s]
States	Oil mass	[kg]
Outputs	Force at the piston-rod assembly	[N]
Parameters	Maximal velocity (Eq. 8) F_{fric_max}	[N]
	Reference velocity (Eq. 8) v_{ref}	[m/s]

The model response Fig. 4 was obtained only for the known input signal. A single shock absorber cycle consisting of the compression and rebound stroke was used to adjust the model with the simulation and optimization algorithm settings from Table 3.

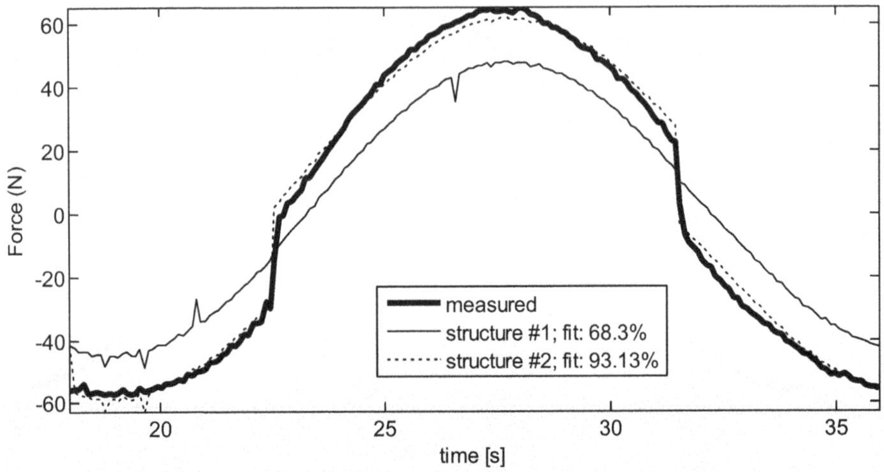

Fig. 4. Force-time characteristic of the hydraulic shock absorber.

Table 5. Adjusted Mmodel parameters.

	Parameter range		Preadjusted parameters (result #1) time domain fit 68.03 [%]	Adjusted parameters (result #1) time domain fit 93.13 [%]
	Min	Max		
F_{fric_max}	0.1	100	0.1	14.84
v_{ref}	1E-6	1	1E-5	1.53E-5

The unknown parameters were initially preadjusted based on trial-and-error approach and reported in Table 5 (result #1). Then the model parameters were finally adjusted using optimization technique, i.e. Levenberg-Marquardt while the results reported in Table 5 (result #2).

Table 6. Ranking of models.

	Model structure	Time domain fit [%]
Low-content model	ARX	86.50
	NLHW	96.26
	NARX	93.11
	SS	86.82
High-content model (Table 5)	Result #1	68.03
	Result #2	93.13

4 Summary and Conclusions

Low-content of a priori knowledge (black-box) and high-content of a priori knowledge (first-principle) models were compared using operational data of a hydraulic shock absorber setup. Linear autoregressive model (ARX) provided a good fit accuracy results, however it does not reproduce qualitatively a nonlinear behavior of the hydraulic shock absorber. The linear model response is a sine-form in which the nonlinear affect is only roughly approximated. The nonlinear "black-box" models approximate well the 'true' model response; however, they required higher number of parameters to cover the full operating range.

The adjusted first-principle models showed good fit to operational data required only two physical parameters compared to a few of neural network models (NLHW, NARX), see Table 6.

References

1. Bohlin, T.P.: Practical Grey-box Process Identification. Theory and Applications. Springer, London (2006)
2. Ljung, L.: Approaches to identification of nonlinear systems. In: Proceedings of the 29th Chinese Control Conference, pp. 1–5. IEEE Xplore Digital Library, China (2010)
3. Lang, H.H.: A study of the characteristics of automotive hydraulic dampers at high stroking frequencies. Ph.D. thesis, The University of Michigan, Michigan (1977)
4. Talbott, M., Starkey, J.: An experimentally validated physical model of a high-performance mono-tube damper. In: Proceedings of the 2002 SAE Motorsports Engineering Conference and Exhibition, Indianapolis, pp. 382–402 (2002)
5. Reybrouck, K.: A non linear parametric model of an automotive shock absorber. SAE Technical Paper 940869, Indianapolis (1994)
6. Simms, A., Crolla, D.: The influence of damper properties on vehicle dynamic behaviour. SAE Technical Paper 2002-01-0319, Indianapolis (2002)
7. Liu, Y., Zhang, J., Yu, F., Li, H.: Test and simulation of nonlinear dynamic response for the twin-tube hydraulic shock absorber. SAE Technical Paper 2002-01-0320, Indianapolis (2002)
8. Baracat, D.E.: A proposal for mathematical design of shock absorbers. SAE Technical Paper 931691, Sao Paulo (1993)
9. Duym, S.W.R.: Simulation tools, modelling and identification, for an automotive shock absorber in the context of vehicle dynamics. Int. J. Veh. Mech. Mob. **33**(4), 261–285 (2010)
10. Duym, S.W., Reybrouck K.G., Baron, G.V., Stiens, R.: Physical modeling of the hysteretic behaviour of automotive shock absorbers. In: Steering and Suspensions Technology, pp. 125–137. SAE, Warrendale (1997)
11. Duym, S., Stiens, R., Reybrouck, K.: Evaluation of shock absorber models. Veh. Syst. Dyn. **27**, 109–127 (1997)
12. Basso, R.: Design of a single-tube shock absorber with a pre-established characteristic diagram. Int. J. Heavy Veh. Syst. **17**, 179–195 (2010)
13. Boggs, C., Ahmadian, M., Southward, S.: Efficient empirical modelling of a high-performance shock absorber for vehicle dynamics studies. Veh. Syst. Dyn. **48**, 481–505 (2010)

14. Cherng, J.G., Ge, T., Pipis, J., Gazala, R.: Characterization of air-borne noise of shock absorber by using acoustics index method. In: SAE Conference Proceedings: Noise and vibration, p. 342. SAE (1999)
15. Yamauchi, H., Sugahara, T., Mishima M., Noguchi, E.: Theoretical analysis and proposition to reduce self-excited vibration of automotive shock absorber. SAE Paper 2003-01-1471, Michigan (2003)
16. Kruse, A.: Characterizing and reducing structural noises of vehicle shock absorber systems. SAE Technical Paper 2002-01-1234, Michigan (2002)
17. Lee, C.T., Moon, B.Y.: Simulation and experimental validation of vehicle dynamic characteristics for displacement-sensitive shock absorber using fluid-flow modelling. Mech. Syst. Signal Process. **20**, 373–388 (2006)
18. Fukushima, N., Hidaka, K., Iwata, K.: Optimum characteristics of automotive shock absorbers under various driving conditions and road surfaces. Int. J. Veh. Des. **4**, 463–472 (1983)
19. Sibielak, M.: Study of flow-induced vibration phenomena in automotive shock absorbers. Solid State Phenom. **248**, 204–210 (2016)
20. Subramanian, S., Surampudi, R., Thomson, K.: Development of a nonlinear shock absorber model for low-frequency NVH applications. SAE Paper 79-84, Michigan (2003)
21. Liu, S.W., Li, G., Hao, L.: Optimization research of vehicle damper matching parameters. In: Proceedings 2013 International Conference on Mechatronic Sciences, Electric Engineering and Computer (MEC), China, pp. 3284–3289 (2013)
22. Wang, W., Yu, D., Zhou, Z.: In-service parametric modelling a rail vehicle's axle-box hydraulic damper for high-speed transit problems. Mech. Syst. Signal Process. **62**, 517–533 (2015)
23. Ljung, L.: System Identification: Theory for the User, 2nd edn. Prentice Hall, Upper Saddle River (1999)
24. Czop, P., Slawik, D.: A high-frequency first-principle model of a shock absorber and servo-hydraulic tester. Mech. Syst. Signal Process. **25**, 1937–1955 (2011)
25. Spencer Jr., B., Dyke, S., Sain, M., Carlson, J.D.: Phenomenological model for magnetorheological dampers. J. Eng. Mech. **123**, 230–238 (1997)
26. Kost, G., Wszołek, G., Czop, P., Jakubowski, D., Wlodarczyk, T.: Parameter estimation of first-principle models formulated using nonlinear ordinary differential equations. Silesian University Publishing House, Gliwice (2013)
27. Ljung, L.: System Identification Toolbox for Use with MATLAB. The MathWorks Inc., Natick (2007)

Particle Image Velocimetry Technique Applied to Flow Evaluation Through a Shock Absorber Intake Valve

Piotr Czop[1], Mariusz Hetmańczyk[2(✉)], Grzegorz Wszołek[2], Jakub Słoniewski[1], Damian Gąsiorek[2], and Zbigniew Buliński[2]

[1] AGH University of Science and Technology, Mickiewicza 30, 30-059 Cracow, Poland
[2] The Silesian University of Technology, Konarskiego 18A, 44-100 Gliwice, Poland
mariusz.hetmanczyk@polsl.pl

Abstract. This paper provides a feasibility study regarding experimental PIV (Particle Image Velocimetry) measurements in order to determine a velocity map of the flow developed across the hydraulic shock absorber intake valve. The velocity map is required to validate and calibrate a CFD/FSI model, which is intended to assess the risk of occurrence of aeration/cavitation phenomena. The model allows to reduce the risk at the shock absorber design stage by means of virtual prototyping. The paper presents the developed hydraulic experimental setup and image processing algorithms in order to obtain a velocity map. The initial test results are briefly reported.

Keywords: PIV · CFD · FSI · Shock absorber · Fluid flow model · Image processing · Intake valve

1 Introduction

Vehicle handling and comfort are negatively affected by common problems observed in hydraulic shock absorbers, namely the presence of gas pockets (aeration) and vapour cavities ("bubbles" or "voids") formation in the hydraulic liquid due to severe operation conditions (i.e. multiple high strokes). Gas and vapour pockets have negative contributors to the low- and high-frequency ride performance One of the most important negative contributors to the low- and high-frequency ride performance of hydraulic shock absorbers used in vehicles is the combined aeration and cavitation effect [1, 2]. The low-frequency effect is the delay in the build-up of damping force, or equivalently, slower than expected increase of pressure in the chambers, and the hysteresis loop in the force-velocity response attributable to abnormal fluid compressibility, which is caused by the existence of either gas (aeration) or liquid vapour phase (cavitation) at certain stages of the stroking cycle. The high-frequency effect manifesting itself as presence of excessive vibrations and emission of noise [3, 4], is caused by abrupt, catastrophic collapse of cavities entrapped in the hydraulic liquid, and is attributable to the aeration and cavitation phenomena as well [5, 6].

J. Świder et al. (Eds.): MECHATRONICS 2017, AISC 934, pp. 81–90, 2019.
https://doi.org/10.1007/978-3-030-15857-6_9

Aeration/cavitation phenomena occurrence risk can be evaluated applying CFD (Computational Fluid Dynamics) or FSI (Fluid-Structure Interactions) simulation approach which allow to determine the velocity profiles through shock absorber restrictions in valve systems [7]. An accurate CFD model to reproduce physical phenomena qualitatively and quantitatively in shock absorber under operational conditions is required [8]. Experimental validation and calibration of a CFD model is feasible with the use of PIV (Particle Image Velocimetry) measurements and image processing technique. The experimental measurements allow to confirm CFD/FSI model assumption regarding laminar, transient, or fully developed flow model. They also provide valuable knowledge about asymmetry of the intake disc spring valve opening, while the velocity map allows direct comparison to CFD/FSI simulation results and, if required to verify assumptions regarding simplification of the valve flow passages by means of mesh accuracy and the valve interior geometry, e.g. wall chamfer, rounding radius or surface roughness.

2 Particle Image Velocimetry

The PIV technique is the most recent development in non-invasive flow measurement methods, within the class of optical ones. Fundamental idea of this measurement technique is straightforward. The flow of interest is seeded with small tracer particles, then their images are recorded with a digital camera. The fluid velocity field is determined by finding displacements of tracer particles between two subsequent images if time lag between them is known. Obviously, it is assumed that tracer particles follow the fluid flow closely. This assumption can be fulfilled be proper choice of tracer particles. Most commonly Stokes number criterion is used to verify correctness of this assumption. Particles are acceptable if their characteristic Stokes time τ_p is small compared to relaxation time of the flow. The particle Stokes time is defined as:

$$\tau = \frac{\rho_p d_p^2}{18\mu} \tag{1}$$

where p_p is particle density, d_p refers to mean particles diameter and μ is dynamic viscosity of fluid. For most commercial seeding commonly used in water applications, the characteristic Stokes number is between 1–10 μs. The most important advantages of the PIV method are its non-invasiveness and capability of capturing the whole velocity field in the region of interest (ROI). This makes this technique very well suited to gather experimental data for the purpose of validation of the CFD models. During more than 30 years of intensive work in the subject different varieties of the PIV technique have been developed. PIV installations can be distinguished depending on the number of dimensions of the ROI and number of reconstructed velocity components. Most typical commercial installations are often referred to as 2D-2c PIV systems. PIV installations designed to obtain all three component of the velocity vectors in the selected plane are available on the market, such installations are often referred to as stereographic PIV or 2D-3c PIV. Also, very advanced setups, called holographic PIVs, are available to obtain full 3-dimensional (3D) velocity fields. Such installations are however expensive and thus impractical in industry applications. Other classification of the PIV systems relays on the tracer particles density in the flow. Basically, three density levels are distinguished, low seeding density when images of individual particles can be detected and particles can be tracked across subsequent images. Often this situation is referred to as the Particle Tracking Velocimetry (PTV). Medium level density of the seeding particles describes situation when individual particles can be detected, however it is no longer possible to track particles between subsequent images. In this situation images are evaluated statistically with use of the cross-correlation. Usually this situation is addressed as a PIV method. Finally, in case of high level density of tracer particles, they cannot be individually detected and subsequent images are statistically correlated only. This situation is usually referred to as a Laser Speckle Velocimetry (LSV). Other criterion distinguishing different PIV approaches is the recording mode. It can vary from multi-exposed single frame approach up to multiple frames single exposed. Comprehensive review of the PIV technique itself is discussed in [9]. Introductory treatment of applications of the PIV technique to investigate structures of the fluid flow containing gas cavities is presented in [10]. Variants of the PIV are also used relying on different tracing algorithms as well as different methods called Particle Tracking Velocimetry (3D-PTV). An interested reader will find a good overview of experimental methods in [11]. PIV application in shock absorber investigations is reported in [12, 13]. As mentioned above, in this measurement method the flow under investigation is seeded by tracer particles with physical properties i.e. density and mean diameter chosen appropriately to the investigated flow type. Utilization of many different seeding particles is reported in literature like polyester particles, glass spheres, oil droplets, titanium oxide powder, liquid crystals, oxygen or hydrogen bubbles or smoke. Properly chosen tracer particles are moving with the same velocity as the fluid and thus can be used to reconstruct velocity field of interest thanks to position recording of reflective images of tracer particles. In most PIV systems double-pulsed high power Nd:YAG lasers are employed to highlight particles. Wide range of these devices is available on the market with double-impulse repetition rate varying from 5 Hz to 20 kHz in case of time-resolved PIV systems. To record light scattered by moving particles digital image CCD or CMOS cameras are used. In traditional PIV, depending on the exposing mode (whether multi-exposed single frame or single-exposed multi frame is applied) images are processed using auto-correlation or cross-correlation algorithms to obtain vector maps. If the PIV system works in the multi-exposure single frame mode, the autocorrelation method is used. Alternatively, if the PIV system works in the single-exposure multi frame mode, the cross correlation method is most commonly used. The data processing starts with division of the image into the so called interrogation windows. Their size may vary from 16×16 (or even 8×8) up to 256×256. The pair of corresponding interrogation windows taken from two subsequent images is correlated with use of the cross correlation function:

$$R(\delta_\xi, \delta_\eta) = \sum_\xi \sum_\eta I_1(\xi, \eta) I_2(\xi + \delta_\xi, \eta + \delta_\xi) \tag{2}$$

where ξ and η are pixel centers coordinates in the image coordinate system and I_1, I_2 stand for the grayscale intensities for the first and second image respectively. The summation in the above equation is done over all pixels in the interrogation window. Maximization of this function results in the relative displacements (ξ_1, ξ_η) of the corresponding interrogation windows from the two successive images. Having these displacements and interframe time between the two images the velocities can easily be calculated. Obviously, they will be expressed in the image coordinates; therefore they need to be transformed to the object coordinates. Most often Eq. (2) is solved in the frequency domain using the Fast Fourier Transform (FFT). It is much faster solution method than searching (ξ_1, ξ_η) pixel by pixel. However, this results in the fact that the interrogation window size can by multiplied by 2 only. Unfortunately the formulation given by Eq. (2) is very prone to the non-uniform illumination of the image background. In such conditions very often false displacements are produced. Hence, usually the normalized correlation function is used in practice:

$$R(\delta_\xi, \delta_\eta) = \frac{\sum_\xi \sum_\eta I_1(\xi, \eta) I_2(\xi + \delta_\xi, \eta + \delta_\eta)}{\sum_\xi \sum_\eta I_1^2(\xi, \eta) \cdot \sum_\xi \sum_\eta I_2^2(\xi + \delta_\xi, \eta + \delta_\eta)} \quad (3)$$

where the denominator is the product of the intensity standard deviations of the first and the second image interrogation windows, respectively. Equation (3) produces better results than the standard cross correlation approach (2). Other common method of vector estimation is the least square function, it is especially well suited in the non-uniform illumination conditions. The interrogation windows displacements (ξ_1, ξ_η) are calculated by minimization of the least square functional:

$$R(\delta_\xi, \delta_\eta) = \sum_\xi \sum_\eta \left[I_1(\xi, \eta) - I_2(\xi + \delta_\xi, \eta + \delta_\eta)\right]^2 \quad (4)$$

The PIV technique is a very powerful non-invasive flow measurement technique. However, it has many crucial parameters that may influence final measurement result and need to be chosen appropriately to specific application. Starting with the parameters of the illumination source, digital camera, flow seeding and ending with the images processing methods. However, the method allows error correction algorithms to be applied.

The major disadvantage of the classical PIV approach is the fixed segmentation of analysed images. Replacement of the similarity analysis (correlation) by the optical stream (optical flow) method allows the displacement of each image element (pixel) to be found between image sequences which significantly improves the accuracy. In general, the accuracy of the classical PIV method depends on multiple factors among which the most important are the camera resolution, selection of the correlation sequence and frequency as well as the choice of seeding particles. Basic parameters influencing PIV accuracy are also pixel intensity (light intensity), resolution and sensitivity of the optical system, size of the imaging window, maximal size of registered displacement, flow characteristics and selection of the time interval between subsequent images.

3 Experimental Setup

An experimental setup [14] was designed and built for the purpose of PIV measurements of the flow through intake valve assembly reproducing shock absorber operational conditions. The base valve assembly (Figs. 1 and 2) is mounted into a transparent tooling which allows the flow through the valve to be observed. The flow is directed in such a manner that opens the sliding intake disc D1 supported by the coil spring (Fig. 1).

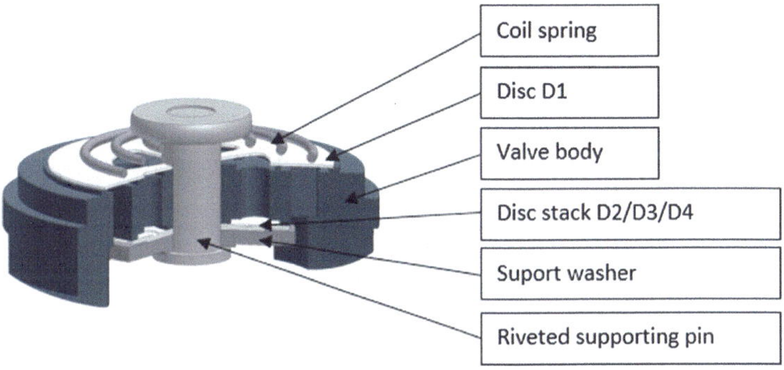

Fig. 1. Transparent valve casing and valve magnification showing top-view of the valve.

Fig. 2. Valve assembly components.

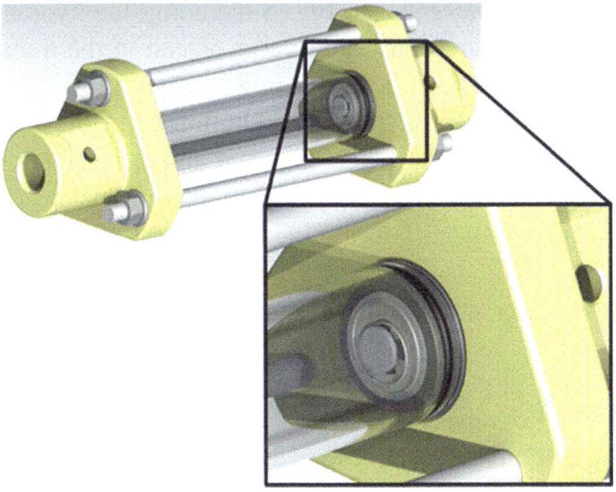

Fig. 3. Transparent valve casing and valve magnification showing top-view of the valve.

A polycarbonate pipe is installed in inlet and outlet elements allowing flow images to be captured (Fig. 3). The valve assembly is secured in the inlet element of the tooling, which is further sealed and attached to a frame of the PIV installation.

A hydraulic supply device utilizes the servo-hydraulic test rig MTS 858 (Fig. 4) to ensure desired mass-flow rate of the flow through valve.

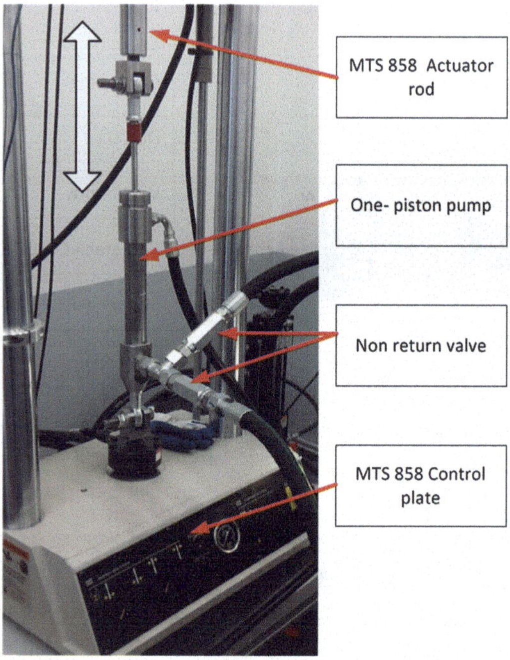

Fig. 4. Hydraulic supply device based on MTS 858.

The hydraulic circuit is presented in Fig. 5 including the hydraulic supply device, hydraulic lines, pressure sensor, and tested valve placed in the transparent tube.

The entire equipment has been designed in a manner allowing image the velocity field of the liquid phase of the flow to be taken by means of the digital Particle Image Velocimetry (PIV) (cf. Fig. 6). The schematic view and the picture of the experimental setup are shown in Fig. 5. The PIV system used in measurements consisted of the high speed digital camera Phantom v9.1 from Vision Research. It has high quality CMOS sensor with resolution 1632 × 1200 pixels and 14-bit image depth. It is capable of recording 1000 frames per second at full resolution. The recording speed can be significantly increased up to over 150000 fps at reduced resolution. This camera is equipped with electronic global shutter allowing exposition time down to 2 μs and on board 16 GB DRAM. The camera was equipped with the high quality Nikkor lens with 50 mm focal length. The investigated flow was seeded with hollow glass spheres with average size equal to 10 μm and density close to the liquid density. High power green (532 nm) CW DPSS (Continuous Wave Diode-Pumped Solid State) laser was used to illuminate the flow. Green laser are favourable in the PIV application as most of the digital image sensors have maximum quantum efficiency around the wave length equal to 532 nm. The equipment applied to PIV measurements is listed in Table 1.

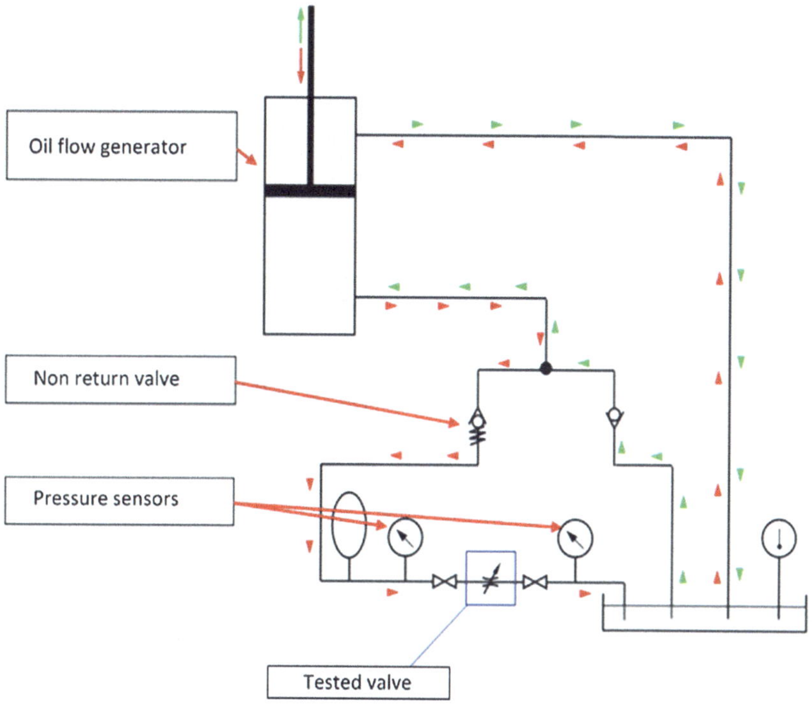

Fig. 5. Schematic representation of the experiment.

Table 1. Equipment applied to experimental PIV measurements.

Device type	Technical specification
Servo-hydraulic tester MTS 858	Load cell 7 kN; Max. stroke 200 mm
Camera - Phantom v9.1	Resolution: 1632 × 1200 - 14bit Max. frame rate – 1k fps to 153.846k Memory 6 GB–24 GB
Lens - Nikon AF	AF-S Nikkor 50 mm f/1.4G
Green laser - constant wave	Output power P = 18 mW Wavelength l = 532 nm
Hydraulic fluid	Density 15 °C 855 kg/m^3 Viscosity 20 °C 22.45 mm^2/s Viscosity 40 °C 12.55 mm^2/s
Pressure sensors	HBM-PM3B 100 bar Uout = 0–10 V HBM-P2VA1 80 bar Uout = 0–10 V

The gathered images were processed using the open source toolbox written in the MATLAB software package. The velocity vectors were calculated using hierarchical cross correlation method. In this approach vectors are calculated iteratively starting with big interrogation window and gradually the window size is decreased with subsequent iteration of the algorithm.

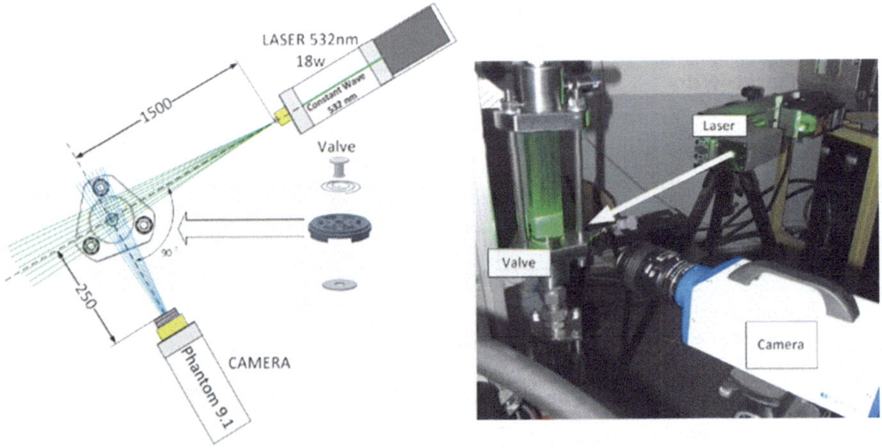

Fig. 6. Particle Imaging Velocimetry (PIV) 2D-2c setup schematic (left) and picture of the actual installation (right).

In present work initial window size equal to 64 × 64 was used and five iterations of the algorithm were carried out with final window size equal to 16 × 16 pixels. At each iteration the displacement map was estimated with cross correlation function in the form described by Eq. (4).

4 Measurement Results

Figure 7 shows an exemplary image generated by the PIV technique. Left pane is a raw image from the CCD camera with red squares indicating unwanted reflexes from an external light source. A magnification on the right-bottom part is used to indicate the base-line of the solid geometry (yellow line). The same boundary is shown on the PIV-processing result – a velocity (scalar) map computed by the PIV processing algorithms described in the previous section. Tracer particles and the plane of the light sheet are clearly visible. Analyzing image shown in Fig. 7 indicates that the light sheet used in the PIV method does not contain the axis of symmetry. Closer analysis of the light ray path through air-glass-oil interfaces that each deviation from angle 0 (measured with respect to the normal to the glass surface) causes the refraction to occur making the light sheet not pass through the axis of symmetry of the valve casing. Figure 8 shows final results of image processing, namely the velocity map (the z-axis is calibrated in m/s).

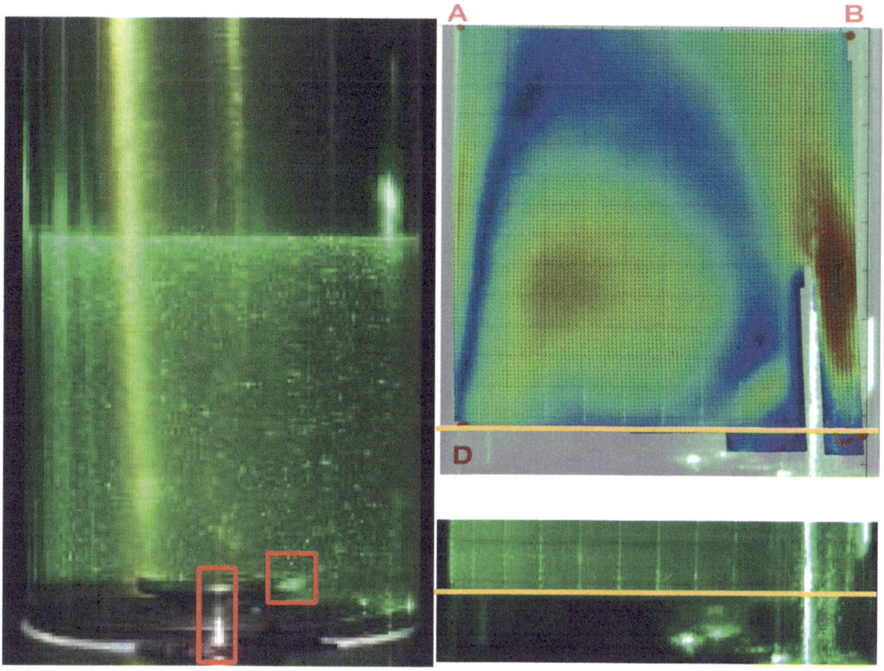

Fig. 7. Direct image from the CCD camera used in the PIV setup (left pane), the base-line of the flow in the image (yellow line) and a screen-shot of the scalar velocity field computed from the PIV measurements (upper right pane); red squares denote unwanted light reflexes (bottom right pane).

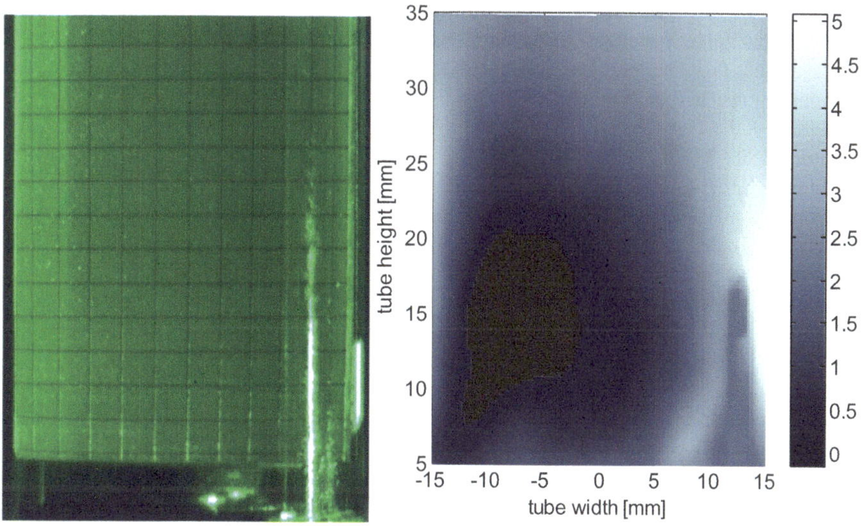

Fig. 8. PIV calibration grid (left pane) and PIV-obtained velocity image (right pane)

5 Summary

The work described in this paper proves the method to be feasible and capable of providing information useful in validating CFD and FSI simulation models. Post-processing has been created to extract appropriate mesh elements, their volumes as well as velocity vector coordinates and to compute projections onto a plane defined in the PIV physical experiments by the sheet of light. Visualization algorithms were created along.

References

1. Duym, S.W., Baron, G.V., Stiens, R., Reybrouck, K.G.: Physical modeling of the hysteretic behaviour of automotive shock absorbers. SAE Technical Paper 970101, pp. 125–137 (1997)
2. Dixon, J.C.: The Shock Absorber Handbook. Society of Automotive Engineers (SAE), Warrendale (1999)
3. Ceccio, S.L., Brennen, C.E.: Observations of the dynamics and acoustics of travelling bubble cavitation. J. Fluid Mech. **233**, 633–660 (1991)
4. Czop, P., Sławik, D.: A high-frequency first-principle model of a shock absorber and servo-hydraulic tester. Mech. Syst. Signal Process. **25**, 1937–1955 (2011)
5. Brennen, C.E.: Cavitation and Bubble Dynamics. Cambridge University Press, Cambridge (2013)
6. Czop, P., Sławik, D., Włodarczyk, T., Wojtyczka, M., Wszołek, G.: Six Sigma methodology applied to minimizing damping lag in hydraulic shock absorbers. J. Achieve. Mater. Manuf. Eng. **49**, 243–250 (2011)
7. Czop, P., Śliwa, P., Gniłka, J., Gąsiorek, D., Wszołek, G.: A computational fluid flow analysis of a disc valve system. J. KONES **18**(1), 112–117 (2011)
8. Świder, J., Wszołek, G., Czop, P., Jakubowski, D., Gąsiorek, D., Sławik, D., Skrobol, A., Włodarczyk, T., Buliński, Z., Gniłka, J.: Model-based Approach Applied in Optimization of Hydraulic Valve Systems. Jacek Skalmierski Publishing House, Gliwice (2013)
9. Cavazzini, G.: The Particle Image Velocimetry – Characteristics, Limits and Possible Applications. InTech, France (2012)
10. Adrian, R.J.: Particle-imaging techniques for experimental fluid mechanics. Annu. Rev. Fluid Mech. **23**, 261–304 (1991)
11. Schwer, L.E.: An overview of the ASME V&V-10. Guide for verification and validation in computational solid mechanics. In: 20th International Conference on Structural Mechanics in Reactor Technology (SMiRT 20), pp. 1–10. The American Society of Mechanical Engineers, Finland (2010)
12. Guzzomi, F.: Investigation of damper valve fluid-structure interaction through the application of experimental visualisation techniques. Ph.D. thesis, The University of Western, Australia (2007)
13. Guzzomi, F., O'Neill, P., Tavner, A: Investigation of damper valve dynamics using parametric numerical methods. In: 16th Australasian Fluid Mechanics Conference (AFMC), pp. 1123–1130. School of Engineering, The University of Queensland, Queensland (2007)
14. Czop, P., Wszołek, G., Hetmańczyk, M., Dawid J.: Method for the determination of working parameters of the passive valves, in particular the disk valves of hydraulic attenuators by PIV method (granted patent). Silesian Technical University, PL222810 (B1) (2014)

Concept, Physical Design and Simulator of IRYS Social Robot Head

Przemysław Dąbek(✉), Maciej Trojnacki, Piotr Jaroszek, and Karolina Zawieska

Industrial Research Institute for Automation and Measurements (PIAP), Warsaw, Poland
{pdabek,mtrojnacki,pjaroszek,kzawieska}@piap.pl

Abstract. The paper describes concept behind IRYS social robot and a system of accompanying tools for application in the education and entertainment areas. Some important developments in these areas are pointed out. The concept of a multi-faceted solution of the system is described, which includes the robot and its high fidelity computer simulator. Main assumptions for the robot high-level control system are given, including the concept of lesson scenarios. Because the system is already under development, its current state is described and in particular the current state of the robot head, neck and upper trunk (a bust), where important mechatronic aspects are highlighted. Some illustrative examples of emotions expressed by means of the developed robot bust and its simulator are presented. Possible directions for further development are indicated.

Keywords: Social robot · Robot head · Emotion expression · Physical design · Simulator

1 Introduction

Among many definitions of social robots, this paper follows this one: "Social Robotics is the study of robots that interact and communicate among themselves, with humans, and with the environment, within the social and cultural structure attached to their roles" [1]. The International Federation of Robotics categorizes social robots as a subcategory of personal and domestic robots, which are a subcategory of service robots – one of two main branches of robotics, another one being the industrial robots. In Europe, service robots are seen as one of the means that are going to help relieve the problem of ageing society. It is envisaged that elderly people will keep a robot at home as a support of or substitute for elderly care personnel. Also, robots at home can perform many tedious everyday tasks to relieve people from doing them, and thus giving them more free time. As discussed for example in [2], the social robots may also play the role of a workmate. Yet another area of application is education and entertainment, where robots to some extent take the role of a teacher (including the use of several robots under supervision of one human educator). It has been proved that robots can make learning more attractive to children, thus boosting up their motivation to acquire new knowledge. In education, there is a particular focus on children with different degrees of disability such as autism, where robots serve as effective means of therapy.

J. Świder et al. (Eds.): MECHATRONICS 2017, AISC 934, pp. 91–100, 2019.
https://doi.org/10.1007/978-3-030-15857-6_10

In general, robotic tools have been increasingly applied in education, from inserting them to the existing curricula, for example, as SMART (Science, Mathematics, Art, Robot and Technology), to creating distinctive educational programs known for example as "r-Learning" or "Robotics in Education" (RiE). There has been a growing tendency to employ robots not only to pursue "Learning about Robots" but also "Learning with Robots". This is where non-technical subjects increasingly come into play as well as social robots do. A choice of a particular robotic platform, such as that of a social robot, implies also a choice of specific educational roles for robots – from tools, peers to tutors – as well as of pedagogical paradigms (for a detailed review see, for example, [3–5]). As educational robotics advances, different studies investigate not only the impact of and the interaction of robots with learners, but also the perception of educational robots on the part of teachers [6, 7] and parents [8].

Robots are used as teaching support at different educational levels, in particular in pre-school and primary education. Social robots are particularly useful in a situation where young children prefer robots that take the role of peers or tutors rather than mere tools [3], and they generally anthropomorphise robots more than adults do [9]. This in line with the assumption that education may come together with entertainment, a relationship that has been well-captured by the term "edutainment". It is worth noting that edutainment robots are often social or humanoid robots, and they can be used to teach both technical and non-technical subjects, for example, story-telling [10] or music [11]. Social robots have also been used to support teaching of these and other subjects in education and therapy of children with autism [12, 13]. For the purposes of this research it is interesting to note that robots have often been used in language education. This includes using robots to improve vocabulary skills, in pre-school education as well as to teach foreign languages, in particular English language. Such a type of education has been sometimes addressed as Robot-Assisted Language Learning (RALL) [14]. In addition to teaching particular school subjects, educational robots allow forming interactive relationships with learners [15] and development of different cross-disciplinary skills (problem-solving and team-work), as well as social skills [6]. The latter is particularly relevant to education of children with autism [12, 13]. Regardless of the school subject and robot design, there is a wide consensus on the fact that robotics is an excellent motivational tool for students with different abilities.

It is possible to distinguish two major styles of appearance of social robots: anthropomorphic – similar to human, and zoomorphic – similar to animals. Yet another style is associated with imaginary creatures like Furby or Photon (qualities of animals and machines, e.g. wheels).

Taking into account the fidelity of head features, those styles can be further divided into: realistic that map features in the most realistic way and iconic that map features in intentionally overemphasized way (comic/cartoon, caricatural). Iconic solutions are most often anthropomorphic. Realistic anthropomorphic solutions include Albert HUBO [16], whilst iconic cover Nao [17] and EMYS [18]. Zoomorphic solutions include PABI [19] with penguin-like appearance.

Finally, existing solutions can be considered in terms of various ways of technical realization of individual components of the head, which are used to express the emotions of the robot. They may include: mechanical solutions in which movement of physical components e.g. eyes or mouth occurs, and electronic solutions in which particular elements of robot's face are displayed on LCD displays, projectors (displaying part or complete face on the appropriate mask like in Furhat robot [20]) or LEDs. The simplest solutions use just LCD displays as a robot face. The most complex are mechanical solutions, especially those that involve the mapping of human facial muscles, as in the case of the ROMAN robot [21].

The long-term aim of the present research is development of technology for personal robots intended for educational and home use. The objective of the article is to describe concept behind the social robot and system of accompanying tools for education and entertainment purposes called IRYS (Intelligent Robot for Young Society) and its actual state of realization. Some examples of expression of emotions are shown and compared in case of physical system and its computer simulator.

2 System Concept

This article discusses the head of social robot called IRYS. The head has been designed with educational use for pre-school and early-school children in mind, but also some needs of autistic children were considered. The application of the robot in question is intended to increase children interest in the learning process and improve knowledge acquisition.

The IRYS social robot will have a modular design with the following main components:

- bust (a head with upper part of a trunk),
- mobile platform,
- other components (e.g. manipulators).

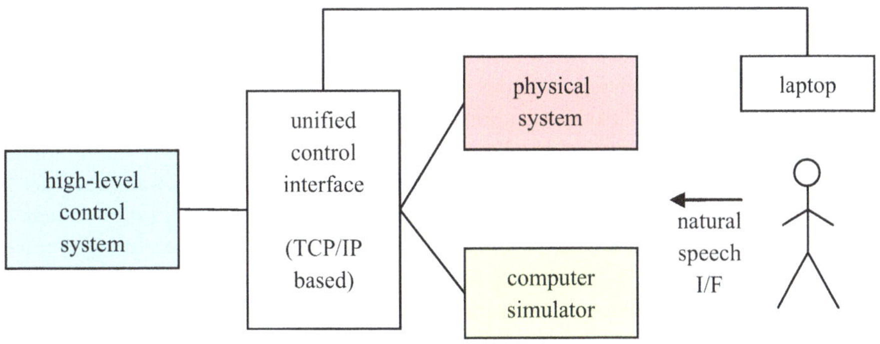

Fig. 1. Overview of the system

The concept of the system assumes discerning physical system, computer simulator, the high-level control system and connection of all modules by means of a unified interface (Fig. 1). Operational control of the system is possible via human-machine interfaces based on natural speech or using a traditional terminal (e.g. laptop, tablet).

3 Physical Head with Upper Trunk

As a result of works conducted to date, the demonstrator of the bust of IRYS social robot (Fig. 2) was created. The demonstrator of the robot bust was developed in Industrial Research Institute for Automation and Measurements (PIAP), Poland in collaboration with students from Warsaw University of Technology, Poland. Much of the work associated with design of the robot head and assumptions for software was carried out within the BEng theses of Pytko [22] and Jargot [23]. The automatic speech recognition system for the Polish language for the robot based on the CMU Sphinx library was developed as a result of work of Zygadło [24, 25]. Currently, robot software can also use an alternative commercial solution available in this area, that is, Google Cloud Speech API, because of large vocabulary support.

Development of the bust was carried out according to methodology associated with the V-Model, with clearly distinguished phases of architecture and requirements formulation, concept studies, detailed design of modules, software implementation and manufacturing, integration and system validation.

As part of the initial works, various concepts of the robot head were considered. They assumed different design styles and various ways of technical realization of individual components of the head (used to express the emotions of the robot). The final concept accepted for implementation was chosen as a result of a survey research conducted among potential robot users, including teachers of early-school education and teachers working in schools with integration classes. The adopted solution includes features of anthropomorphic and zoomorphic styles, fidelity of features is iconic, whereas the technical solution is hybrid (mechanical-electronic).

The robot has to allow verbal and nonverbal means of communication with human. Verbal communication includes speech analysis (supported by machine vision) and speech synthesis. Speech analysis is related to the processing of human speech into text. The synthesis of speech consists in converting the text into human-understandable audio wave. Machine vision includes analysis of lips expressions, but also face detection or recognition, text recognition, and so on.

Nonverbal communication involves simulating the expression of main human emotions, and recognizing emotions of people. Examples of such emotions considered in IRYS include happiness, sadness, fear and anger. Each emotion is programmed in the form of the appropriate head pose, that is, the configuration of the ears, eyebrows, eyelids, irises, eyes, cheeks, mouth and neck. All configuration items are coordinated in software to provide a consistent message to the user.

The demonstrator of the bust consists of three main components: head, neck and upper part of a trunk. The trunk is the base of the robot in which the following components are assembled: control computer, electrical and electronic systems, loudspeakers and microphone. The trunk allows the robot to perform such functions as speaking and listening. The neck allows the robot head to rotate around four different axes (lower pitch, roll, upper pitch and yaw).

The IRYS robot head contains the following controlled elements that allow to express different emotions:

- eyes that include eyelids, eyebrows, and eyeballs (3 drives for pitch and independent yaw), inside installed cameras and irises mounted in the form of LED rings,
- lips displayed on the LCD,
- ears that may stand or fold,
- cheeks that are made in the form of LED rings.

To increase adaptability to user's preferences, face covers are easily exchangeable.

Table 1 shows information about the components of the robot demonstrator, including the functions they perform. It can be noticed that the most complex parts of the robot head are its eyes. In turn, the nose has only aesthetic function and is not controlled in any way. The lips, apart from the function of expressing emotions, will move in the rhythm of pronounced words by the robot – appropriate works are in progress.

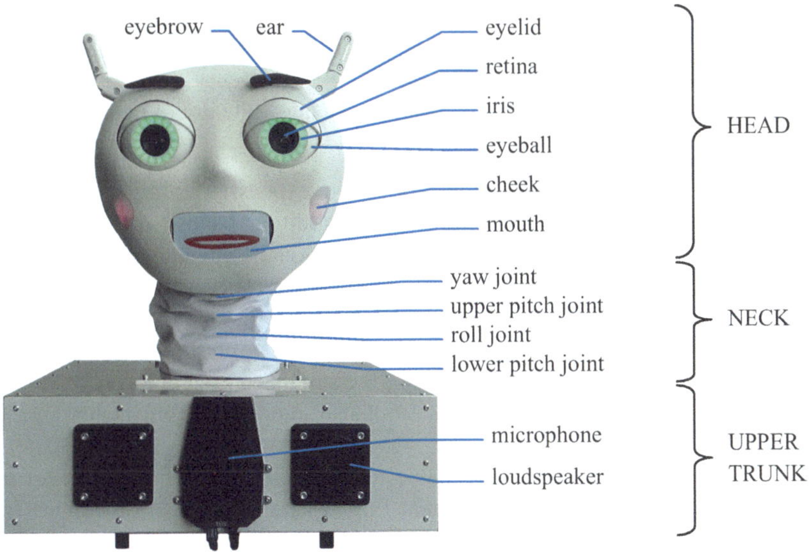

Fig. 2. Demonstrator of the bust of IRYS social robot.

Table 1. Breakdown of components of demonstrator of the robot bust and their functions

Component	Subcomponents and their functions	Modules	Implementation method
Head	2 eyes for seeing and expressing emotions	3 joints	3 servos
		2 pupils	2 cameras
		2 irises	2 LED rings
		2 eyelids	2 servos
		2 eyebrows	2 servos
	Nose for aesthetic function only		Aesthetic shape
	Mouth for expressing emotions		LCD display
	2 ears for expressing emotions		2 servos
	Color of 2 cheeks for expressing emotions		2 LED rings
Neck	4 joints for head movements		4 servos
Trunk	Voice		2 speakers
	Hearing		1 microphone

4 Simulator

For the head of IRYS a simulator was developed. The simulator is intended to accomplish the following objectives:

- eliminating the need for real robot at the stage of development of its software and creating scenarios for it,
- replacing the real robot with a virtual solution to allow the use of multiple robots by a wider group of pupils.

From the user's point of view, instead of a more expensive real robot, they can use a cheaper virtual version with similar functions. In addition, the use of larger number of virtual robots could enable personalization of educational content and its adaptation to the needs, interests and abilities of particular children.

The main function of the simulator is as realistic as possible visualization of the appearance of the robot head and its behavior, resulting from the interaction with the user. The specific, anthropomorphic nature of the simulated system forces not only the need for correct visualization of static poses but also smooth and convincing transitions between them.

There are several technologies available for simulators: Gazebo, V-Rep for more technical approach, and e.g. Unity, Unreal Engine and Cry Engine for more visual aspects. The simulator was made using the Unreal Engine. This choice was based on the criteria such as having one of the best visualization possibilities, a fully open source code, and fast and convenient prototyping of the entire system, based on a graphical state editor.

5 High-Level Control System and Unified Control Interface

The high-level control system is responsible for behavior which is appropriate to situation. It interprets the sensory input, determines emotional state of the robot, and finally the appropriate response.

The entry point of human-robot interaction is the supervisor, where the user selects one mode of operation from command control and predefined scenario. When the current mode of operation is left, the control is passed back to the supervisor, where selection of the mode of operation can be repeated.

In the command control mode, the user issues commands by means of voice or keyboard input. This mode is mainly for the purpose of testing robot functions and therefore engagement of the high-level interpretation is limited.

An important element underlying the robot concept is a *scenario* that defines the way the robot works and the rules of interacting with people. When using a robot to teach children, the scenario describes the scope and process of realization of a lesson. It is intended to allow the education of children in a topic chosen by the operator (teacher).

In the predefined scenario mode the user chooses a scenario for following interaction. An example is the answer-game scenario, where the robot asks educative questions, a user answers them and then the robot shortly comments on the answer. If the answer is wrong, the robot becomes angrier, otherwise it becomes happier. The tone of the short comment depends on the actual state of emotion of the robot.

The scenarios will eventually contain elements of machine learning, which will increase the flexibility of realization of scenarios by the robot. The use of machine learning will allow the content of lesson to be adapted to the progress of students (learners). The material less assimilated by the children will be repeated more often.

The Unified Control Interface is the interface between the high-level control system and a terminal. The terminal can be either the physical system or the simulator. The purpose of this architecture is separation of the high-level control system from the controlled object (physical or virtual). An explicit benefit is a possibility of programming new scenarios on a computer where only Windows system is installed, because the high-level control system runs within the Robot Operating System environment, which is Linux based.

6 Expressing Emotions of a Robot Head

An important issue is the perception of human emotions expressed by the robot in the case of physical and simulated solutions. Figure 3 compares four selected emotions expressed within both contexts.

It was possible to achieve a good agreement of poses for the simulated and physical solutions, but some work is necessary to implement drawing of lips based on a series of connected points on the physical system, as it is done on the simulator.

When analyzing the way of expressing emotions one can draw the conclusion that the anger Fig. 3e–f would be better expressed if the eyes had also the bottom eyelid. This observation is even truer in case of expression of disgust (not shown).

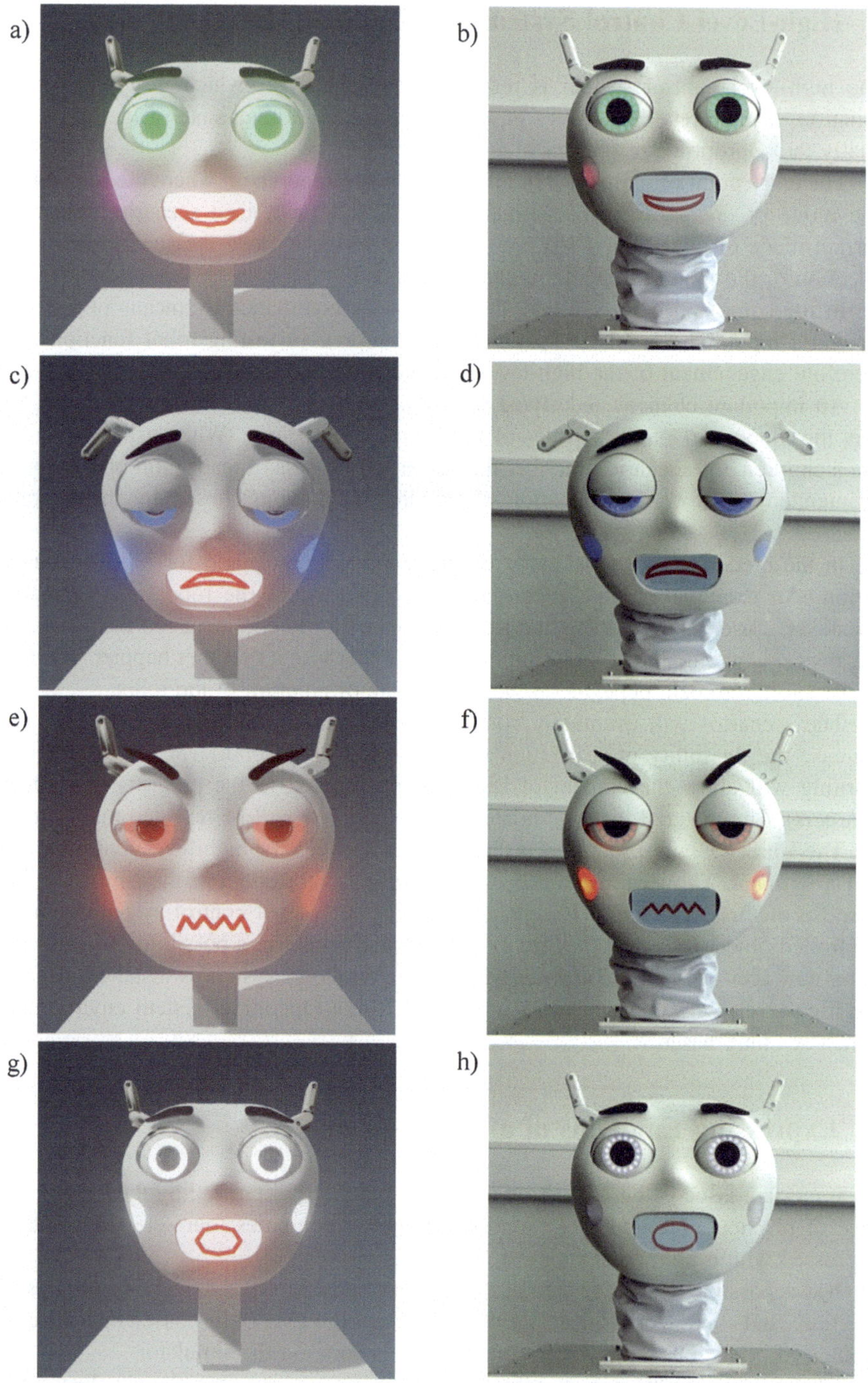

Fig. 3. Emotions expressed by simulated (left) and physical (right) robot head: a, b – happiness, c, d – sadness, e, f – anger, g, h – fear.

7 Conclusions and Future Works

On the one hand, this research is part of a larger, relatively well-established research area that focuses on the development of social robots capable of interacting and communicating with human users. On the other hand, the design of the IRYS robot seems to bring novel contribution to the field of social robotics on a national and international level. The present paper gives an update on progress of works on PIAP first social robot. The work leads to the following preliminary conclusions:

- It is known from other research that young children are more likely to anthropomorphise robots, and anthropomorphisation often goes hand in hand with play. In view of that, anthropomorphic design along with multimodal communication through facial expressions, head gestures and speech, with its emphasis on emotions, makes the IRYS robot particularly suitable for primary education. This is because emotions have been often explicitly employed in teaching strategies and learning processes.
- Given a variety of its forms as well as the adaptability of some of its functions (e.g. face covers) as well as its friendly appearance, the robot has the potential to become accepted by different groups of end users, not only children, but also, for example, the elderly or persons with autism.

The use of IRYS in the form of physical system and simulator opens the door to diversified forms of training provided to both robot operators and users.

Future works will include:

- Development of different scenarios for human-robot interaction, with the focus on lesson scenarios.
- Further testing with potential users.
- Research on possibility of safe and responsible artificial intelligence introduction.

Acknowledgements. The work was financed from Industrial Research Institute for Automation and Measurements (PIAP), Poland statutory and own research funds.

References

1. Ge, S.S., Matarić, M.J.: Preface. Int. J. Soc. Robot. **1**(1), 1–2 (2009)
2. Piçarra, N., Giger, J.-C., Pochwatko, G., Możaryn, J.: Designing social robots for interaction at work: socio-cognitive factors underlying intention to work with social robots. J. Autom. Mob. Robot. Intell. Syst. **10**, 17–26 (2016)
3. Mubin, O., Stevens, C.J., Shahid, S., Al Mahmud, A., Dong, J.-J.: A review of the applicability of robots in education. J. Technol. Educ. Learn. **1**(209-0015), 13 (2013)
4. Hong, N.W.W., Chew, E., Sze-Meng, J.W.: The review of educational robotics research and the need for real-world interaction analysis. In: 2016 14th International Conference on Control, Automation, Robotics and Vision (ICARCV), pp. 1–6 (2016)
5. Benitti, F.B.V.: Exploring the educational potential of robotics in schools. Comput. Educ. **58** (3), 978–988 (2012)

6. Alimisis, D.: Educational robotics: open questions and new challenges. Themes Sci. Technol. Educ. **6**(1), 63–71 (2013)
7. Fridin, M., Belokopytov, M.: Acceptance of socially assistive humanoid robot by preschool and elementary school teachers. Comput. Hum. Behav. **33**, 23–31 (2014)
8. Lin, C.H., Liu, E.Z.F., Huang, Y.Y.: Exploring parents' perceptions towards educational robots: gender and socio-economic differences. Br. J. Educ. Technol. **43**(1), E31–E34 (2012)
9. Belpaeme, T., Baxter, P., De Greeff, J., Kennedy, J., Read, R., Looije, R., Neerincx, M., Baroni, I., Zelati, M. C.: Child-robot interaction: perspectives and challenges. In: International Conference on Social Robotics, pp. 452–459 (2013)
10. Kory, J., Breazeal, C.: Storytelling with robots: learning companions for preschool children's language development. In: 2014 RO-MAN: The 23rd IEEE International Symposium on the Robot and Human Interactive Communication, pp. 643–648 (2014)
11. Han, J.-H., Kim, D.-H., Kim, J.-W.: Physical learning activities with a teaching assistant robot in elementary school music class. In: 2009 Fifth International Joint Conference on the INC, IMS and IDC, NCM 2009, pp. 1406–1410 (2009)
12. Fridin, M.: Storytelling by a kindergarten social assistive robot: a tool for constructive learning in preschool education. Comput. Educ. **70**, 53–64 (2014)
13. Robins, B., Dautenhahn, K., Te Boekhorst, R., Billard, A.: Robotic assistants in therapy and education of children with autism: can a small humanoid robot help encourage social interaction skills? Univ. Access Inf. Soc. **4**(2), 105–120 (2005)
14. Han, J.: Robot-aided learning and r-learning services. In: Human-Robot Interaction (2010)
15. Han, J.: Emerging technologies. Robot assisted language learning. Lang. Learn. Technol. **16** (3), 1–9 (2012)
16. Oh, J.-H., Hanson, D., Kim, W.-S., Han, Y., Kim, J.-Y., Park, I.-W.: Design of android type humanoid robot Albert HUBO. In: 2006 IEEE/RSJ International Conference on Intelligent Robots and Systems, pp. 1428–1433 (2006)
17. Shamsuddin, S., Ismail, L.I., Yussof, H., Zahari, N.I., Bahari, S., Hashim, H., Jaffar, A.: Humanoid robot NAO: review of control and motion exploration. In: 2011 IEEE International Conference on Control System, Computing and Engineering (ICCSCE), pp. 511–516 (2011)
18. Kędzierski, J., Muszyński, R., Zoll, C., Oleksy, A., Frontkiewicz, M.: EMYS-emotive head of a social robot. Int. J. Soc. Robot. **5**(2), 237–249 (2013)
19. Dickstein-Fischer, L., Alexander, E., Yan, X., Su, H., Harrington, K., Fischer, G.S.: An affordable compact humanoid robot for autism spectrum disorder interventions in children. In: Proceedings of the Annual International Conference of the IEEE Engineering in Medicine and Biology Society, EMBS, pp. 5319–5322 (2011)
20. Al Moubayed, S., Beskow, J., Skantze, G., Granström, B.: Furhat: a back-projected human-like robot head for multiparty human-machine interaction. In: Cognitive behavioural systems, pp. 114–130 (2012)
21. Hirth, J., Schmitz, N., Berns, K.: Emotional architecture for the humanoid robot head ROMAN. In: 2007 IEEE International Conference on Robotics and Automation, pp. 2150–2155 (2007)
22. Pytko, K.: Opracowanie konstrukcji mechanicznej głowy robota społecznego dla dzieci. BEng thesis, Warsaw University of Technology, Warsaw (2016) (in Polish)
23. Jargot, D.: Opracowanie systemu komunikacji niewerbalnej z człowiekiem dla głowy robota społecznego. BEng thesis, Warsaw University of Technology, Warsaw (2016) (in Polish)
24. Zygadło, A., Janicki, A., Dąbek, P.: System rozpoznawania mowy polskiej dla robota społecznego. Pomiary Autom. Robot. **20**(4), 27–36 (2016)
25. Zygadło, A.: System rozpoznawania mowy polskiej dla robota. BEng thesis, Warsaw University of Technology, Warsaw (2016) (in Polish)

Cognitive Maintenance and Polymorphic Production as the Leading Industry 4.0 Paradigms

Marek Drewniak(✉) and Marek Gabryś

Aiut Ltd., Wyczółkowskiego Str. 113, 44-100 Gliwice, Poland
{marek.drewniak,marek.gabrys}@aiut.com

Abstract. The increasing interest in the integration of informatics with automation control systems results in the creation of a brand new philosophy of production. The philosophy defined as Industry 4.0 revolutionizes manufacturing in the form as it is understood nowadays. The classical methods of optimization of a production process are insufficient for current requirements and therefore need totally new assumptions and should be reconstructed. Two significant paradigms of Industry 4.0 are presented in the paper: Cognitive Maintenance and Polymorphic Production. Polymorphic Production is an innovative organizational concept that enables dynamic switching between numerous production variants (frameworks). The idea is based on the gathering and analysis of various elements: production orders, current production process status, operational Key Production Indicators, resources and energy availability or priorities set by management. As a result, dynamic, real-time switching between production paths is possible in order to manufacture in the most optimal and efficient way. Due to increased system complexity and flexibility, effective maintenance becomes a key factor for successful implementation of Industry 4.0. Therefore, the second paradigm is the reinforcement of the Cognitive Maintenance concept by the use of knowledge and resources management, which are related to the industrial installation. The purpose of the approach is to gather the resources at every step of an installation life span and to use them effectively so they can support the maintenance. The technical supplement for the approach are methods like technological modeling, remote support and the use of augmented and virtual reality. By the use of such approaches like Polymorphic Production or Cognitive Maintenance, the significant improvement of efficiency of industrial plants is possible. The process can be performed not only by the optimization of production paths but also thanks to modern maintenance of industrial installations. The reason for stressing the significance of those two paradigms results from their complexity: on one hand they affect the whole concept of Industry 4.0, and on the other they combine many aspects that are usually considered separately or which are only partially treated together.

Keywords: Industry 4.0 · Internet of Things · Key Performance Indicator · Shop floor management · Polymorphic Production · Cognitive Maintenance · Technological modelling · Augmented Reality

J. Świder et al. (Eds.): MECHATRONICS 2017, AISC 934, pp. 101–110, 2019.
https://doi.org/10.1007/978-3-030-15857-6_11

1 Introduction to Industry 4.0

The trend which more and more often enters industrial manufacturing is the concept of Industry 4.0. Defined as the Fourth Era or Fourth Industrial Revolution, after steam power, electrification and automation, it is the next significant step towards a brand new philosophy of manufacturing. Its superior target is to integrate fields from automation control with information systems in order to make production faster, cheaper and more efficient [1].

The model of selling products has changed in recent years. More and more often the customer selects and configures the product. Such an approach gradually leads to the total customization of a product and manufacturing in small series. This phenomenon generates the necessity to adapt the production environment to simultaneously manufactured products and forces the process of its maintenance to be defined in parallel to the design of a production process. The implementation of Industry 4.0 will enable the production of short, dedicated and customized series and therefore the required time needed for the start-up of a production will be shortened. In addition, the approach will have to allow the easy reconfiguration of processes through the continuous management of changes at the production level [1–3].

The philosophy of Industry 4.0 can be implemented in various ways. The first one is continuous monitoring and gathering of data from process and administrative-business layers, which are related to the activity of an enterprise. The process layer can be analyzed thanks to Cyber-Physical Systems (CPS) and Internet of Things (IoT), which grant access to process data acquired from the current operation of end-point and control devices [4].

The second way integrates ERP (Enterprise Resource Planning), EAM (Enterprise Asset Management) and MES (Manufacturing Execution Systems). The approach delivers knowledge about the use of materials and operational staff, information about logistic systems and data about currently processed production orders.

Thanks to the use of data warehouses and mechanisms of Big Data along with artificial intelligence and data mining methods it is possible to assign complicated dependencies which may occur between production parameters. The indication of those dependencies was previously impossible owing to limited computational power and capabilities of storing large amounts of data. Currently, those calculations are possible and can be used as one of the ways of implementing Industry 4.0 [1, 2].

The robotics and automation control devices are equipped by the manufacturers with more and more advanced technological solutions and functionalities. Classical robots are being progressively replaced by adaptive and self-learning collaborative robots, which are able to cooperate with human beings.

Devices of intelligent logistics are used not only to supply the production line with parts and components needed for manufacturing but often replace conventional transportation mediums, allowing the product to move freely on the production path [5].

The processes of designing of production stands are supported by virtual reality which can be combined with simulations of kinematics and 3D scenes. This approach reduces the time and costs of their production and start-up.

Finally, additive 3D printing used for prototyping becomes indispensable element on product life span and significantly reduces its release time to the destination market.

Although the examples mentioned already present the general trends towards which the current industrial manufacturing market is heading, they do not exhaust the Industry 4.0 concept. Due to the fact that Industry 4.0 cannot be achieved by the use of classical ways of optimization, it requires a total reforming of most of the currently used approaches to the implementation of industrial solutions. Cognitive Maintenance and the Polymorphic Production – two paradigms which may be the basis for the successful start-up of Industry 4.0 and which are the combination of factors mentioned above – are introduced and extended.

2 Maintenance in Today's Industry

One of the most important factors influencing the efficiency of the manufacturing process being carried out is the upkeep of the industrial plant in a state of the highest efficiency [6]. In order to make the reaction to occurring abnormalities in production effective, it is necessary to use the knowledge about specific cases and diagnostic methods which match those cases. Moreover, this term does not only mean that the anomaly needs to be diagnosed quickly. The maintenance is gradually and more intensely supported by mechanisms of continuous monitoring of aspects such as state of end-point devices, degradation of their components or consumed energy. The analysis and efficiency assessment of values which result from those observations permit the assessment of the condition of the production stand and, as a result, estimation and forecast of events which may affect the continuity of the stand's work. This idea, realized as a "prevent instead of react" paradigm, is defined as Predictive and Preventive Maintenance [6–9].

The second type of support for the maintenance of an industrial plant is the systematization and management of procedures used for various types of maintenance and intervention actions. Computerized-Maintenance Management Systems (CMMS) permit the administering of the equipment, plan inspections and service actions and manage the documentation, procedures and work schedule of technical staff [10]. Thanks to the processing of selected fragments of knowledge related to the work of production stands, those systems are capable of providing schedules for repairs and machine services. Additionally, they support the planning of the production capabilities of an industrial plant.

3 Cognitive Maintenance

Mechanisms for managing resources and procedures become the standard used in supporting the maintenance of an industrial installation and the whole plant. It is a natural consequence that those mechanisms can be combined together in order to create common correlations and to form up the knowledge. Additionally, the experience (documents, knowledge, procedures) gathered during the work with one case (e.g. an intervention) can be transferred and used during an occurrence of other events of this type. This concept, described as the support of maintenance by the use of knowledge along with parallel understanding of its sources, was defined as Cognitive Maintenance and was first proposed by IBM in 2016. Its main goal is, firstly and specifically, to improve an action area of the local maintenance services which are responsible for maintaining the infrastructure. Secondly and more generally, it directly affects the efficiency of the performed manufacturing process [11].

When analyzing the needs and demands of the current industrial manufacturing market it is possible to move one step forward and expand Cognitive Maintenance. Knowledge about the technology and the process can be gathered not only as technological resources, e.g. as ready-to-use solutions, technical documentations or manuals, but can also be supported by expert knowledge from specialists from specific disciplines and which would be gathered with all steps of a production stand's life span.

Current approaches for the implementation of the production stand are based on the flow of technical data through all stages: from the request, technical offer and specifications, through the design and start-up, to the maintenance and service of the production stand. Resources which are generated during one of those steps feed the following ones, e.g. the technical description of the process is used to specify the technological solutions which will allow its carrying out.

The first possibility to extend Cognitive Maintenance is to gather the knowledge related to technology and resources of an installation during each step of its life span. The knowledge can be systematized and its components can be linked internally in order to support designing and start-ups of current and consecutive, new industrial installations. Besides providing information about the ways in which the process can be performed, it also allows a transfer of the technology which can be used in comparable or similar operating conditions. Technological resources which are generated during each step of an installation life span, their flow through those stages and possibilities of using them thanks to the transfer of knowledge are presented in Fig. 1. The approach minimizes the possibility of making expensive mistakes related to the start-up of detailed solutions, e.g. the use of improper material for the production of components of an end-point device, resulting in a waste of time spent on its design and manufacturing in a mechanical workshop.

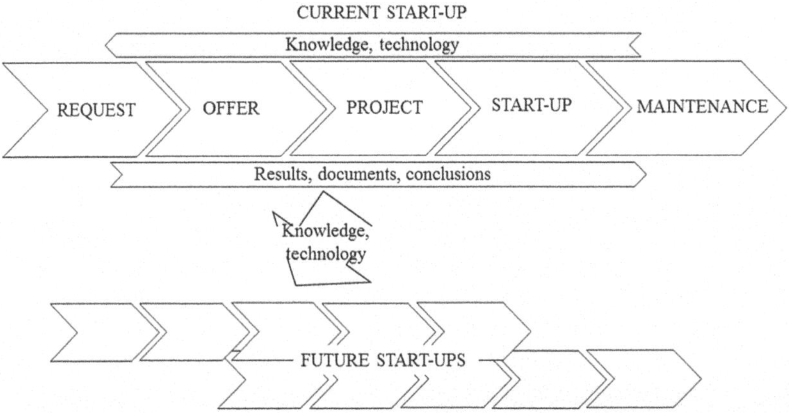

Fig. 1. The knowledge on various stages of an installation life span

The second possibility of extending Cognitive Maintenance is to use the knowledge gathered in the stage of designing an installation for its effective maintenance. The mechanical engineering is more and more often supported by mechanisms of Virtual Commissioning and Virtual Reality. Transferring some of their resources to the area of maintenance can greatly support current actions related to the service of end-point devices. Detailed 3D models processed and used e.g. by Augmented Reality can hasten the diagnosis and therefore more quickly detect sources of anomalies. Additionally, the simulation of kinematics can help with the verification of correctness of the realized technological process. An example of the use of Augmented Reality for processing of 3D model at the industrial plant is presented in Fig. 2.

Fig. 2. 3D model of collaborative robot YUMII by ABB, processed by Augmented Reality

The third approach which extends Cognitive Maintenance is enhancing the knowledge about the industrial installation with a technological model, combining not only the installation physical structure but also the information about roles and interfaces of its components along with all the dependencies and correlations which occur between them. Technological modelling unifies the knowledge from various disciplines and eliminates one of the major problems related to the maintenance of an industrial installation – the necessity to use specialist expert knowledge in case of an occurrence of a discipline-oriented problem. The properly created technological model allows for an interdisciplinary combination of a component of an industrial installation with all disciplines it interacts with [12, 13]. An example is a pneumatic actuator, which, besides its own resources related to mechanical (manufacturer documentation, 3D model) and pneumatical engineering (pneumatic schema, pipelining and control valves) is also related to electrical engineering (sensors for detection of the position of piston) and control layer (control signals, control modules).

Standards which appear on the industrial manufacturing market and which are used for the description and technological modelling of an installation are more and more often supported by leading manufacturers of ready-to-use solutions and software. A CAEX (Computer Aided Engineering Exchange) standard is used for the hierarchical description and object-oriented representation of the structure, e.g. factory. Universal COLLADA (Collaborative Design Activity) format supports interactive solutions in

3D modelling. Open programming standards such as PLCOpen XML grant access to the content of an installation control software and give a possibility of linking its fragments with the components of an installation [12, 13].

The technological model can either be partially generated automatically – using tools provided by manufacturers of ready-to-use solutions and software – or manually by integrators of automation control solutions. The model can form up the repository of knowledge which can then be used by technical staff or with the support of remote experts during intervention actions.

4 Knowledge-Integrated Manufacturing

Although the recent observation of a process and an analysis of data which is generated at its level was successfully used e.g. in mechanisms of the Predictive and Preventive Maintenance, they slowly become insufficient for complex analyses and assessments of a factory efficiency. Today's industrial manufacturing market expects tools which would be capable not only of a simple dedicated assessment of the quality and efficiency of a production. In order to provide complete information about the manufacturing process it is necessary to integrate the analysis with the knowledge of a production process and the technology in which it is performed. Such an approach allows one not only to explain phenomena which may influence the production parameters but also, and most importantly, optimizes the manufacturing.

There are three general sources which feed the knowledge base: process data, technological data and business data. Sources where the information is generated come out from various layers of a production, not necessarily from the horizontal division (e.g. end-point, measurement or control devices). It is possible to continuously develop this knowledge by reaching vertical layers, e.g. systems for product ordering, managing resources, managing worktime and personnel availability or managing the logistics responsible for supplying the production. The last, additional supplement for knowledge base are procedures, which can be used as a tool for affecting those sources.

The interaction and cross-referencing of information from various layers and their simultaneous combination allow the creation of indicators defined as KPIs (Key Performance Indicators). KPIs are usually formulated as a set of rules, in which the change of one parameter may affect not only the KPI itself but also other parameters which form this KPI. Those values are used for fuzzy assessment of efficiency and quality of a production and can be used for the detection and diagnosis of a source which generates temporary worsening of the KPI. In addition, their use is bound to a hierarchy of managing the structure of an enterprise: KPIs for lower layers formulate KPIs for upper ones. This fact allows widening of the field of interest of specific groups of users, who equipped with dedicated operational cockpits can work only with data which is important to them (the personnel managing the production line is interested in the efficiency of single production stands while the manager of the factory is interested in the general production efficiency of the whole industrial plant) [14, 15].

5 Polymorphic Production

Two of the most important targets which the current industrial manufacturing market turns to is, first, to perform production processes in parallel and second, to achieve flexibility in their realization. Such solutions grant the possibility of producing simultaneously and as an effect, of the faster manufacturing and launching a product to the target market [16, 17]. They also allow the full configurability of a process by the definition of production paths which are used for the manufacturing.

The first factor which illustrates the heading towards those targets is equipping of end-point devices with unified interfaces. The Internet of Things permits the accessing of most currently produced end-point devices, to recalibrate them, diagnose and cancel anomalies. Production stands gradually become more and more configurable by equipping them with the functionality to dynamically add or remove devices used within the production process.

The second aspect is the nearly-free possibility of a customer to configure the product. Manufacturers of such products need to cope with problems of large possibilities to configure a process which needs to be realized at the production level. It is more and more often necessary to equip the factories with tools capable of flexible adaptation of end-point devices to such production but also with informatic tools for managing such dynamically changing orders.

The third symptom is the continuous attempt to separate the manufacturing technology from the physical structure in which the production is realized. The classical production line usually does not allow the same process to be performed on two stands simultaneously. It is also uncommon that the production can be planned freely between production stands. Instead, it is realized in line, one stand after another. Industry 4.0 redefines this approach thanks to factors which were mentioned in the previous points and chapters. Mechanisms for calculation of KPIs and for an analysis of the current state of a production permits the continuous recalibration of stands according to the occupancy of a line. The product can be moved to a specific production path which is calculated and set on the basis of the current production state. As an effect, the maximal use of production stands and maximal efficiency of a production can be achieved. Such dynamic switching of production path, due to its variety, can be defined as Polymorphic Production. Not only does it grant the possibility of producing many independent variants and, as a result, optimize the process but also allows for a quick diagnosis in case of the occurrence of an anomaly and the possibility of continuing production despite the exclusion of a stand or a group of stands. The example of a mechanism which regulates the flow of a product on various production paths is presented in simplified form in Fig. 3.

Although the example presented in Fig. 3 is a simplification of industrial reality, it shows one more important aspect which is counted as a phenomenon that increases flexibility in the manufacturing. Transportation mediums used by now, e.g. belt or chain conveyors, rail trolleys and carts, skids or blisters do not grant a free flow through the production path. The first step towards increasing the latitude in the transportation were AGVs (Automated Guided Vehicles), navigated by magnetic or optical tapes, and which are now mainly used for supplying the production line with parts needed for

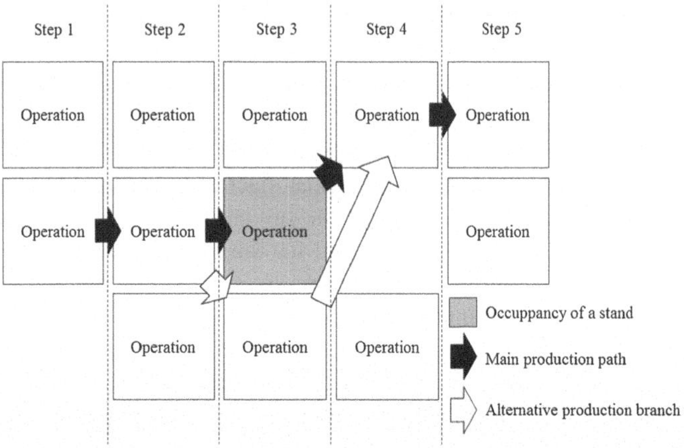

Fig. 3. The simplified example of the dynamic recalculation of a production path according to the occupancy of a production stand

production. Recently, the AGV concept has been changing dynamically and the main direction of those changes is visible in the use of natural navigation, which is based on the observation of the surrounding and making decisions in order to reach the destination point. Such an approach opens up a brand new chapter in industrial logistics: it will permit the free movement of a product through the production process and will crown the implementation of totally flexible production.

6 Summary

Industry 4.0 will not only change the implementation of new technologies diametrically but the whole concept of manufacturing. This approach redefines what the current industry looks like and, as a consequence, reforms the methods which are currently used. In a world that delivers more and more innovative solutions it is important to understand how those solutions may impact all aspects of the production from the relation between the customer and the manufacturer, through the technology and the process of design and manufacturing of a product to the life span of a product. Two of Industry 4.0 paradigms are especially capable of increasing this awareness i.e. Cognitive Maintenance and Polymorphic Production.

Nowadays, manufacturers need to meet more and more increasing quality requirements. Therefore, the classic idea of observing the process and making a simple analysis of a production becomes insufficient for the effective management of a process. For this reason, the process layer needs to be supplied with the knowledge of the technology, manufacturing process and procedures which are related to it. This demand allows the Cognitive Maintenance concept to be used as a complex approach to the problems of design, start-up and the maintenance of industrial installations. The concept – extended and supported by remote maintenance based on mechanisms such as augmented reality – can expedite the process of gathered knowledge in order to make interventions and the work of local maintenance services easier and more efficient.

Systems which are used for the support, maintenance and management of production processes are often narrowed down to a single layer or interact only with the neighboring ones. Industry 4.0 will gradually force an interaction between those solutions and will eventually lead to a complete integration of industrial plant management, from the production to the business layer. The knowledge gathered at various production stages and enterprise layers will serve to determine the indicators that are important for the assessment of the quality and efficiency of industrial plants. Indicators, technology and process data will result in a new quality of manufacturing which will be realized by Polymorphic Production. This, along with the elastic configuration of a process, will allow the dynamic recalibration of production stands in order to manufacture faster, cheaper and in the most optimal way.

Acknowledgement. This work was supported by Polish National Centre of Research and Development from the project ("Knowledge integrating shopfloor management system supporting preventive and predictive maintenance services for automotive polymorphic production framework" (grant agreement no: POIR.01.02.00-00-0307/16-00). The project is realized as Operation 1.2: "B+R sector programs" of Intelligent Development operational program in years 2014–2020 and co-financed by European Regional Development Fund.

References

1. Bundesministerium fur Bildung und Forschung, Industrie 4.0 – Innovationen fur die Produktion von morgen, Die neue high-tech Strategie - Innovationen fur Deutschland, Berlin (2017)
2. Industrie 4.0 (2017). http://www.plattform-i40.de/
3. Cupek, R., Ziebinski, A., Huczala, L., Erdogan, H.: Agent-based manufacturing execution systems for short-series production scheduling. Comput. Ind. **82**, 245–258 (2016)
4. Fell, M., Melin, H.: Roadmap for the emerging "Internet of Things. Clayton MacKenzie (2014)
5. Ullrich, G.: Automated Guided Vehicle Systems. Springer, Heidelberg (2015)
6. Center for Intelligent Maintenance Systems (2017). http://www.imscenter.net/
7. Samdani, S.: The Predictive Production System: The Future of Making. McKinsey & Company, New York (2017)
8. Borgi, T., Hidri, A., Neef, B., Naceur, M.S.: Data analytics for predictive maintenance of industrial robots. In: IEEE International Conference on Advanced Systems and Electric Technologies (2017)
9. Fossier, S., Robic, P.-O.: Maintenance of complex systems – from preventive to predictive. In: IEEE, 12th International Conference on Live Maintenance (2017)
10. Labib, A.: Computerised maintenance management systems. In: Complex System Maintenance Handbook, pp. 417–435, Springer, Heidelberg (2008)
11. IBM. https://www.ibm.com/blogs/internet-of-things/equipment-maintenance-cognitive/. Accessed 10 June 2016
12. Drath, R.: Datenaustausch in der Anlagenplanung mit AutomationML – Integration von CAEX, PLCopen XML und COLLADA. Springer, Heidelberg (2010)
13. AutomationML Consortium: <AutomationML/> - The Glue for Seamless Automation Engineering, Whitepaper AutomationML, Part 1 – Architecture and General Requirements (2014)

14. Weber, A., Thomas, R.: Key Performance Indicators – Measuring and Managing the Maintenance Function. Ivara Corporation, Burlington (2005)
15. Parmenter, D.: Key Performance Indicators: Developing, Implementing and Using Winning KPIs. Wiley, Hoboken (2015)
16. Dima, I.C.: Industrial Production Management in Flexible Manufacturing Systems. Engineering Science Reference, Hershey (2013)
17. Chryssolouris, G., Manufacturing Systems. Theory and Practice, Springer Texts in Mechanical Engineering, Springer, Heidelberg (2013)

Concept of Coupling the Rehabilitation Treadmill with Foot Pressure Sensors

Sławomir Duda(✉) and Grzegorz Gembalczyk

Silesian University of Technology, Gliwice, Poland
{Slawomir.Duda, Grzegorz.Gembalczyk}@polsl.pl

Abstract. The main aims presented in the work constitute conceptual solutions of treadmill control systems with foot pressure sensors feedback. The article also contains results of further experimental studies regarding the device.

Keywords: Mechatronic device · Gait rehabilitation · Rehabilitation treadmill · Real-time operations

1 Introduction

Strokes are one of the reasons for remaining in the disabled condition for a longer period of time. The restoration of the ability to move or retain balance in the upright position is significant for the everyday life of a patient. The deficit connected with the lack of the ability to walk pertains to almost 70% of the patients, who have experienced a stroke. It leads to a sedentary lifestyle and further fall of mobility and the ability to keep balance as a consequence. Such people lose the possibility of full participation in social life [1, 2]. Physiotherapy constituting a conventional therapy of rehabilitation is a basic kind of therapy meant for patients with a walking deficit. Having in mind relieving the work of physiotherapists, various devices are created causing the acceleration of motorics capabilities regeneration during the period of hospital care (Fig. 1) [3, 4].

Fig. 1. Example of a rehabilitation device [5].

J. Świder et al. (Eds.): MECHATRONICS 2017, AISC 934, pp. 111–118, 2019.
https://doi.org/10.1007/978-3-030-15857-6_12

A mechatronic device for locomotor training (see Fig. 2) used for walking exercises in unloading conditions [6] has been created at the Institute of Theoretical and Applied Mechanics within an implemented project financed by the National Centre for Research and Development.

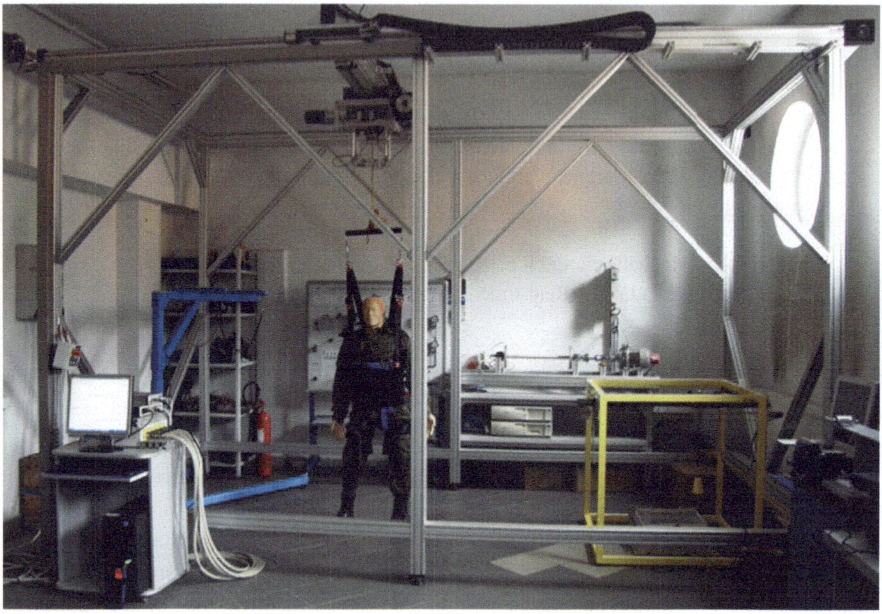

Fig. 2. Mechatronic device for locomotor training developed at Silesian University of Technology.

The models presented in this work are the result of further research on the device constituting various solutions conceptually. The idea of the system involves using foot pressure sensors against the ground placed in shoe insoles in order to steer the working mode of a training treadmill, i.e. start – stop. As a result, this solution can be designated as an indirect steering system of a mechatronic device for walking re-education, for example, by detecting emergency conditions or supporting driver steering.

2 The Set Up of the Test Bench

The presented device basically consists of two systems. The mechanical one comprises a commercial treadmill (DOMYOS SOFT WALK 8), and the electronic one contains a PC computer equipped with a real time card connected with a USB connector and MATLAB software. The systems are coupled with each other creating a mechatronic system by signals coming from the insoles placed in shoes (sensors) and the system of a speed regulation of treadmill tape transport (actuator) (Fig. 3).

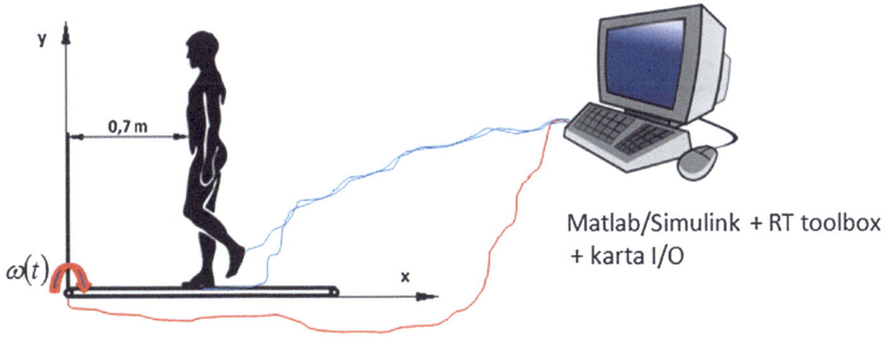

Fig. 3. The concept of coupling a treadmill with insoles registering the pressure of a foot against the ground.

Steering the speed of treadmill tape transport can be useful in the course of rehabilitation of people who learn to walk again. The tape movement will be adjusted to the current walking abilities of a patient being in a full interaction with the behaviour of the person working out, e.g. if they stop, the tape will also be stopped. A separate task is the possibility of using a relevant algorithm, making it possible "to adjust" the speed of treadmill tape transport to the real speed of walking. The step change of a tape speed with a time delay between particular settings can be an obstacle. It results from the fact that the device has been pinned up to the current system of a speed change of a treadmill tape transport.

3 Integration of Components

The described devices work in the mode of real time, and steering is implemented with the help of MATLAB/Simulink software installed at a stationary computer. Coupling between a training treadmill and a PC computer is implemented with the help of RT-DAC4/PCI cards [7]. Measurement signals are conditioned with a specially designed interface. The signals that steer the work of the treadmill are generated by the computer [8]. A training treadmill equipped with an asynchronous engine with the power of 0.8 kW that is available on the market has been used in the developed version of the device. The communication between the treadmill and the computer takes place through a RT-DAC/USB2 card of real time and an additionally developed electronic system (Fig. 4).

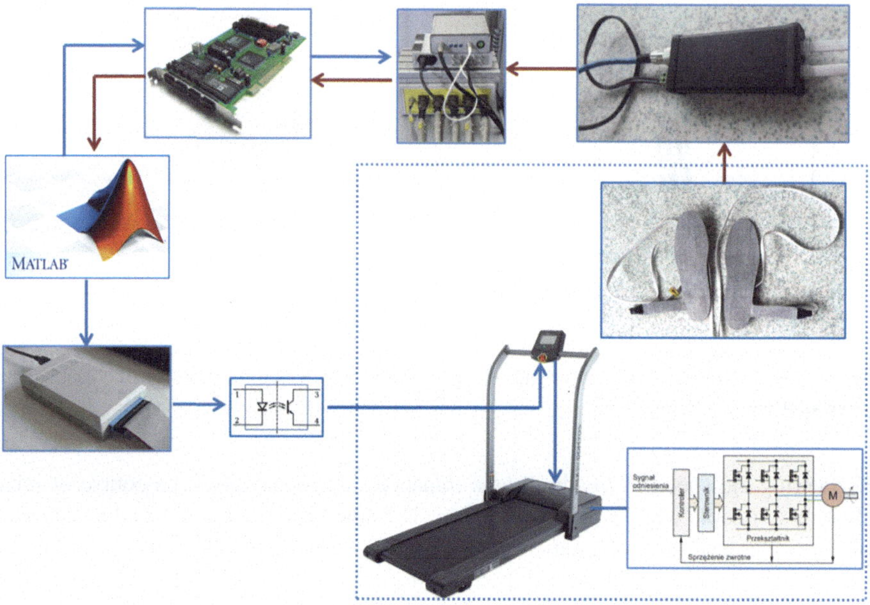

Fig. 4. Components cooperation in mechatronic treadmill for gait reeducation.

In order not to interfere with the drive system of the treadmill, its steering panel has been modified by a parallel connection of an additional drive whose task is to simulate the work of buttons placed in the steering panel of a treadmill. The developed system has been built with the use of optocouplers steered with digital signals generated by a steering system in the PC computer. The change of a digital signal from a high condition to a low one (and the return to the high condition is performed after 0.1 s) is equal to pressing a button on the treadmill steering panel.

4 Algorithm of the System Operation

As a result of analyzing the distribution of foot strain, a measurement system based on baroresistant sensors has been developed. As it is presented in the figure, there are four pressure sensors on each insole. The sensors of this type are characterized by a significant resistance change that decreases with pressure rise.

In order to achieve a measurement signal which would inform about foot pressure against the ground, a system converting a measurement signal has been made suiting the baroresistant pressure sensors used in shoe insoles (Fig. 5). The system consists of a signal amplifier, a power supply, and cabling. Each insole is equipped with RJ-11 connections pinned up to the amplifier system. The amplifying system generates voltage signals in the scope of 0–10 V, registered by the real time card. It should be added that the converting system summarizes the values of the signals from the pairs of sensors placed on the front and back of the insoles. In connection with the above, four signals corresponding to particular foot pressure zones reach the data acquisition card:

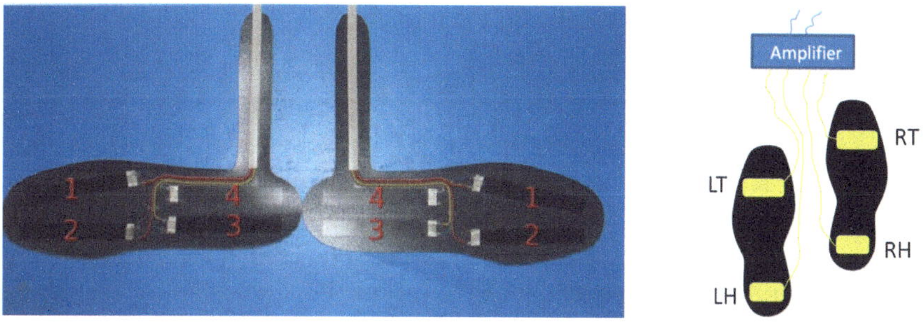

Fig. 5. The measurement system of foot pressure.

the front part of the right foot (RT), the back part of the right foot (RH), the front part of the left foot (LT), the back part of the left foot (LH). The determination of a reference value of a measurement signal at which it is possible to state that a given foot part exerts pressure on the ground has required carrying out a series of experimental research. The tests have been carried out while performing various activities e.g. walking, simulating inclining or standing upright. The results have been registered for two people of various weight.

As the result of the carried out research, the course of a pressure change has been obtained during walking for the front and back part (toes-heel) of the right and left foot. An excerpt of the obtained chart is presented in Fig. 6.

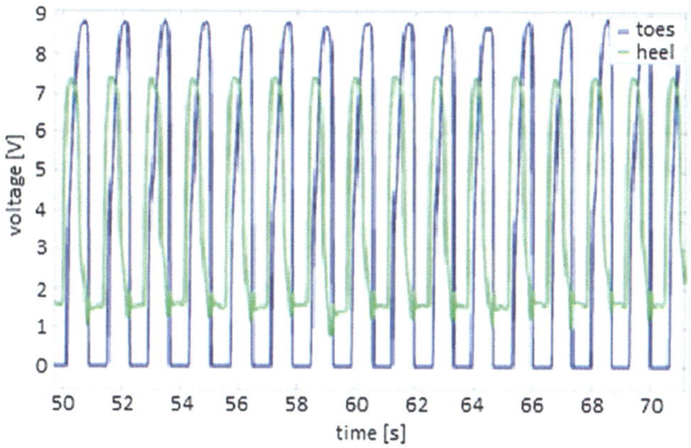

Fig. 6. A fragment of an original chart of a strain insole.

It has been assumed that there is a contact of particular foot zones with the ground for voltage higher than 4.5 V. These conditions have been used for developing the detection of a contact: foot-ground in the form of a zero-one function (Fig. 7).

Let the f function be expressed with the formula:

$$f(t) = \begin{cases} 1, & U \geq 4,5 \\ 0, & U < 4,5 \end{cases} \quad (1)$$

where: t – time [s], U – voltage [V].

The *f* function for a single measurement point assigns the value of 0 – no support or 1 – the appearance of support. The indicator of pressure, used in the value assessment and referring to the voltage signal, does not include the average value of registered analogue signal [9]. Thus, a column matrix has been obtained in which there is information about the appearance of support or its lack in single cells.

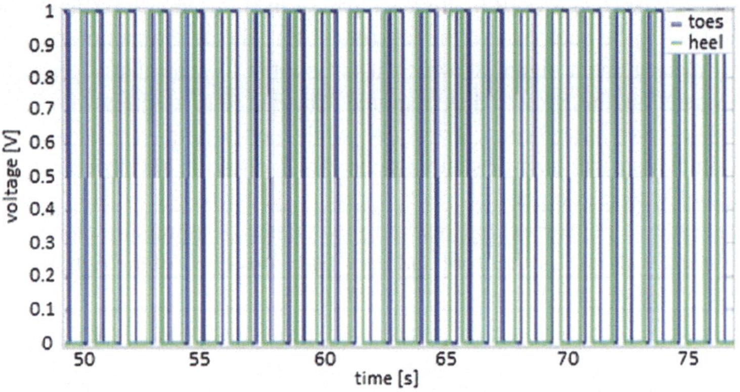

Fig. 7. A fragment of a chart of a strain insole after unification.

If both feet are on the ground (signal 1 from all sensors), then the treadmill is switched off in the position; if one foot is raised (signal 0 from the pressure sensors of the heel H and toes T), the treadmill is switched on. The treadmill tape moves with the same speed of v = 0.5 km/h. There is an obvious situation while walking that both feet stick to the ground in the support stage; if such a period lasts no longer than 0.6 s, it is not treated as a person's lack of movement- thus the treadmill is switched on.

5 Test of an Operation System Accuracy

The use of computer possibilities of modelling and testing designed systems is a significant aspect of mechatronic testing. Testing software with the participation of existing system elements carried out in real time belongs to such possibilities. This testing involves connecting the real elements of a system built with a real time card and software. The advantage of such approach is the possibility of simulating the system operation without the necessity of implementing this system into a device [10].

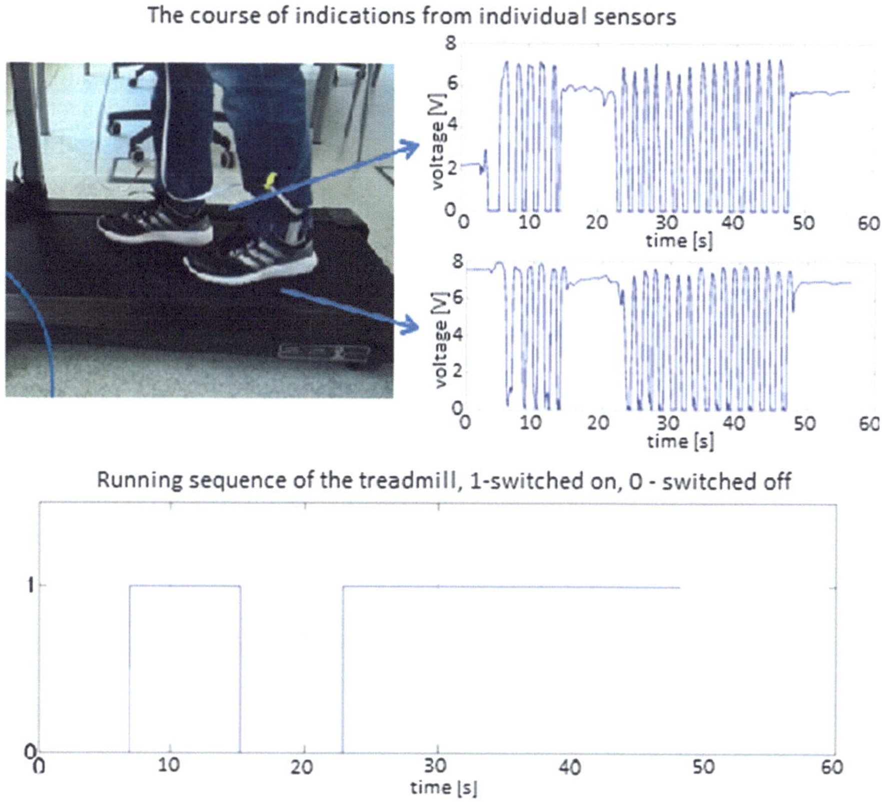

Fig. 8. A figure caption is always placed below the illustration. Short captions are centered, while long ones are justified. The macro button chooses the correct format automatically.

Testing the algorithm described in Sect. 4 has been carried out in the MATLAB/Simulink environment. Input signals constitute readings from baroresistant sensors from the insoles made within real time. Testing the system has involved walking and stopping on the treadmill (Fig. 8).

6 Conclusions

The mechatronic system presented in this thesis has assumed switching a treadmill on and off depending on the fact, if a patient is walking on it or standing. The simple algorithm presented in thesis has been based on the signals coming from baroresistant pressure sensors placed in the shoe insoles of a rehabilitated person meets the presented assumption. The accuracy of the system operation has been confirmed with the tests carried out.

Further work will be focused on the expansion of the algorithm by introducing the function of increasing and decreasing the tape transport speed, depending on the speed of walking; the development of an adaptation algorithm. Such a functionality of a treadmill will aid the rehabilitation of patients learning to walk again by the automatic adjustment of a tape transport speed to the individual walking capabilities of each patient.

References

1. Feigin, V.L., Roth, G.A., Naghavi, M., Parmar, P., Krishnamurthi, R., Chugh, S., Mensah, G.A., Norrving, B., Shiue, I., Ng, M., Estep, K., Cercy, K., Murray, C.J.L., Forouzanfar, M. H.: Global burden of stroke and risk factors in 188 countries, during 1990–2013: a systematic analysis for the Global Burden of Disease Study 2013. Lancet Neurol. **15**(9), 913–924 (2016)
2. Lisabeth, L.D., Baek, J., Morgenstern, L.B., Zahuranec, D.B., Case, E., Skolarus, L.E.: Prognosis of midlife stroke. J. Stroke Cerebrovasc. Dis. **27**(5), 1153–1159 (2018)
3. Koceska, N., Koceski, S.: Review: robot devices for gait rehabilitation. Int. J. Comput. Appl. **62**(13), 1–8 (2013)
4. Sommers, J., Wieferink, D.C., Dongelmans, D.A., Nollet, F., Engelbert, R.H., van der Schaaf, M.: Body weight-supported bedside treadmill training facilitates ambulation in ICU patients: an interventional proof of concept study. J. Crit. Care **41**, 150–155 (2017)
5. BIODEX. http://www.biodex.com. Accessed 28 Dec 2017
6. Duda, S., Gąsiorek, D., Gembalczyk, G., Kciuk, S., Mężyk, A.: Mechatronic device for locomotor training. Acta Mechanica et Automatica **10**(4), 310–315 (2016)
7. Kopetz, H.: Real-Time Systems Design Principles for Distributed Embedded Applications. Springer, Wien (2011)
8. Duda, S., Gembalczyk, G., Jurkojć, J., Kciuk, S., Michnik, R.: Mechatronic solution of components cooperation in the device for gait reeducation. In: Proceedings of the ECCOMAS Thematic Conference on Multibody Dynamics 2015, pp. 799–805. International Center for Numerical Methods in Engineering (CIMNE), Barcelona (2015)
9. Paszkowski, W., Loska, A.: The use of neural network model in the assessment of annoyance of the industrial noise sources. In: Burduk, A., Mazurkiewicz, D. (eds.) Intelligent Systems in Production Engineering and Maintenance – ISPEM 2017. Advances in Intelligent Systems and Computing, vol. 637, pp. 428–439. Springer, Cham (2017)
10. Kciuk, S., Duda, S., Mężyk, A., Świtoński, E., Klarecki, K.: Tuning the dynamic characteristics of tracked vehicles suspension using controllable fluid dampers. In: Nawrat, A. (ed.) Innovative Control Systems for Tracked Vehicle Platforms, pp. 243–258. Springer, Cham (2014)

Forecasting of Methane Hazard State in the Exploitation Wall Using Neural-Fuzzy System

Dariusz Felka[1(✉)] and Jarosław Brodny[2]

[1] Institute of Innovative Technologies EMAG, Katowice, Poland
dariusz.felka@ibemag.pl
[2] Silesian University of Technology, Gliwice, Poland
jaroslaw.brodny@polsl.pl

Abstract. In hard coal mines the series of natural hazards take place, one of the most dangerous are gaseous hazards. However, the most commonly occurring gaseous hazard is methane hazard. Methane is odorless and colorless, lighter than air organic gas, of which large quantities are emitted during an exploitation of hard coal. It is inflammable gas which with appropriate air concentration becomes a very dangerous explosive gas. In order to ensure safety of the staff, the critical concentration of this gas cannot be allowed to occur in mining atmosphere. Therefore, it is reasonable to apply mechatronic systems which task is to register current methane concentrations in mining atmosphere, also to develop analytical systems in order to forecast these concentrations. Application of advanced artificial intelligence systems provides opportunities for effective forecasting of methane concentrations in underground headings. This forecast is based on the registered measurement data and determined ventilated parameters. In the paper, an example of application of artificial intelligence methods to build an analyzing methane hazard model in the region of exploited wall was presented. The way of formation of neural-fuzzy system based on numerical data was described. Hybrid system was built to determine current level of methane hazard and forecast its future concentration in the region of longwall during normal mining operations. For determination of system parameters automatically, methods of knowledge recovery (based on fuzzy clustering of measured data) and adaptive learning algorithm of neural-fuzzy system were used. Developed tool gives a lot of opportunities for practical application.

Keywords: Mechatronic system · Methane hazard · Neural-fuzzy system · Forecasting

1 Introduction

Methane hazard is defined as an occurrence of excessive amounts of methane causing presence of dangerous state for staff and property of mine [1, 2, 17–19, 21]. Presence of methane in mining atmosphere is caused by its release from coal deposits as a results of its mining (exploitation).

J. Świder et al. (Eds.): MECHATRONICS 2017, AISC 934, pp. 119–133, 2019.
https://doi.org/10.1007/978-3-030-15857-6_13

Among the basic criteria of assessment of methane hazard, one can distinguish ventilation-, absolute- and criteria-methane-bearing capacity. Currently, due to the way of collection of data necessary for risk assessment (averaging of results) and detailed elaboration of measurements, these forecasts are characterized by a great delay.

Therefore, it is necessary to look for the new methods, which will quickly and reliably determine the methane hazard state. The monitoring and forecasting system for methane concentration in mining headings developed by Authors meets the requirements above mentioned. Its base are credible measuring data registered by modern sensors and intelligent analytical system processing these data. The idea of proposed solution is based on application of fuzzy logic and neural networks for evaluation and forecasting of methane hazard state.

Theory of fuzzy sets was created as alternative for classical concepts of sets and logic theory. It describes phenomena which are difficult to define with ordinary mathematics, because they are defined in non-precise and out-of-focus way. In fuzzy logic there is no sharp border between elements belonging and do not belonging to the set. Systems based on the fuzzy logic are applied to solve problems and advice with use of implemented data base.

Models based on fuzzy logic are intelligent models, enabling modeling very complex phenomena, which often are difficult to describe in a conventional way.

Applications of techniques of automatic knowledge extraction enable to build fuzzy model, which can analyze on a regular base e.g. the level of methane hazard in the region of longwall. The input values of such model can be measuring data of physical values measured during the exploitation process [12].

However, neural networks are structures which can model nonlinear relationships basing on input/output data. They are built of neurons creating next layers of the network. Knowledge of neural networks is written in the form of weights of interconnections between the particular neurons, modified during the learning process based on the examples. Taught structure associates acquired knowledge and responses based on the input data. The way of connection of particular neurons decides about the architecture of network [14].

Neural-fuzzy systems are formed through artificial connection of neural network and fuzzy systems. Created fuzzy model is transformed to neural-fuzzy network, which weights corresponds to parameters of memberships function in premises and to polynomial coefficients in conclusions of fuzzy model rules [6].

While creating a model representing selected issue, the fundamental problem is to acquire the knowledge necessary to organize the structure of model. In fuzzy systems two methods of acquiring knowledge are being used.

The first of them is to get the knowledge from an expert. Structure and parameters are determined arbitrarily on the basis of expert's knowledge. Problem of such approach can be difficulty to precisely determine parameters in a case of complex representation. Therefore, a frequent solution is to create a preliminary model which then is precisely tuned on the base of measuring data [10].

Second method is an automatic extraction of rules based on numerical data. Effective action of such algorithm enables to determine system parameters together with organization of its structure. It is also possible to connect both methods. In such a case some knowledge will come from the expert and some from the automatic selection of rules.

Currently for forecasting of methane hazard also computer modeling (CFD) based on the finite volume method is used [2, 17, 18, 20, 22].

In the presented paper concept of application of neural-fuzzy model for improvement of safety of underground mining exploitation is shown.

Application of advanced artificial intelligence systems provides opportunities for effective forecasting of methane concentrations in underground headings. This forecast is based on the registered measurement data and determined ventilated parameters. In the paper, an example of application of artificial intelligence methods to build an analyzing methane hazard model in the region of exploited wall was presented. The way of formation of neural-fuzzy system based on numerical data was described. Hybrid system was built to determine current level of methane hazard and forecast its future concentration in the region of longwall during normal mining operations. For determination of system parameters automatically, methods of knowledge recovery (based on fuzzy clustering of measured data) and adaptive learning algorithm of neural-fuzzy system were used.

The reliable measuring system is significant for quick and effective actions. It is complex mechatronic system providing continuous data flow to intelligent analytical system.

Forecasting methodology of methane hazard state presented in the paper and developed tool create a lot of opportunities for practical application. It subsequently should have an effect on improvement of safety of work in mining companies.

2 Characteristics of Monitoring, Assessment and Methane Hazard Prediction System

Measuring systems applied in mines register many parameters used to determine the methane and fire hazards and characteristics of the ventilation network [11, 16]. For this purpose, temperature, barometric pressure, airflow velocity and methane content sensors are being used. Particularly important is monitoring of regions of longwalls, where the most dangerous phenomena of random nature often take place.

For this reason in these regions the most measuring devices are concentrated. Results of these measurements are the base for analysis of safety state in these regions. In order to obtain reliable and effective analysis it is necessary to use appropriate measuring devices and systems for analysis of acquired data.

In the chapter, measuring devices used for monitoring the methane hazard state were presented and the concept of creating a system for assessment and prediction of this hazard in hard coal mines was discussed. The basic task of the developed system is rapid and effective forecasting of methane hazard state in investigated region. Results of these forecast will be a base for taking actions in order to eliminate or/and reduce the hazard state.

2.1 Measuring Devices of Methane Hazard Monitoring System

The measurement of methane concentration was carried out using MM-4 methane meter, which is being used in Polish hard coal mines. This device is characterized by good metrological properties and short response times. Thermocatalytic transducers are used to measure the concentration of methane.

To measure the velocity of air flow in mining headings the AS-3 anemometer was used (Fig. 1). Measurement with this device is done using a temporary ultrasonic method. An integrated methane drainage sensor ZCO (Fig. 2) was used to measure the amount of methane content in methane drainage process. It measures physical parameters (pressure, differential pressure on the orifice, temperature) and concentration of methane in the pipeline. On the base of the obtained results, the amount of pure methane in the pipeline is determined by simplified or iterative method.

a) b)

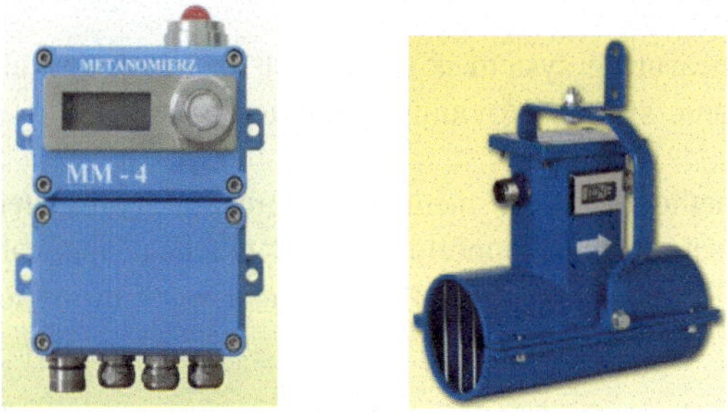

Fig. 1. Methane meter type MM-4 (a) and anemometer type AS-3 (b).

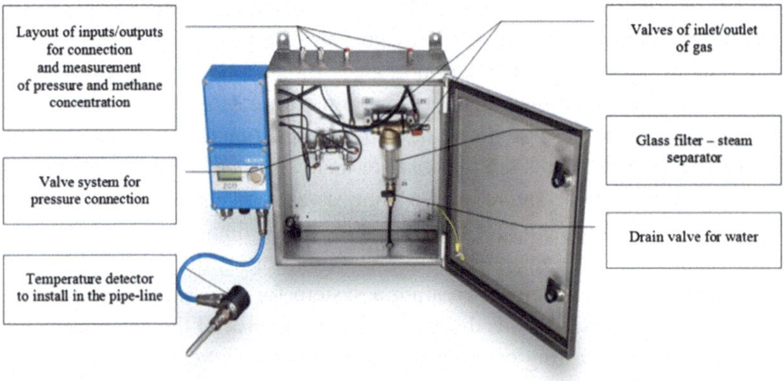

Fig. 2. Integrated methane drainage sensor ZCO.

2.2 Concept of Methane Hazard Assessment and Prediction of Methane Hazard

Developed methane hazard assessment and prediction system in mining headings is based on the data registered through the measured device and analysis of that data in the neural-fuzzy model. The system operates on two levels. The first includes the definition of heading and creation of inference model and methane hazard degree and the second focuses on the current operation of the developed system.

The first level has a design and configuration character. It includes mainly identification of the headings parameters (way of ventilation, sensors location of the monitoring system together with cross section of heading in the measured place).

Based on the defined structure of the heading and uploading learning data, a fuzzy model defining the level of methane hazard in the area of the exploited wall is created, basing on the current measurements. The output files of this process are files with heading structure and infer model. In the modeling process, the following parameters are determined:

- fuzzy division on input/output spaces (clustering) into subspaces with simultaneous determination of centers of defined groups (clusters),
- base of rules of infer mechanism,
- creation and optimization of parameters of model membership function,
- definition of infer method system (aggregation, implication, method of determining output value of model).

The second level has a character of current system operation. A structure consisted of two models determines the level of methane hazard according to the rules described in instructions [9]. Fuzzy model determines current level of methane hazard on the base of measured data vector obtained from the sensors of monitoring system. Determined levels of hazard are achieved in a form of moving timeframe. Based on recorded values in archive and values of current adaptation neural-fuzzy net the forecast hazard is determined (ANFIS) [8]. At fixed intervals of time the network is tuned based on recorded data in an archive.

Values of current level assessment of methane hazard and forecast value are the result of the layer operation. The concept of assessment system and prediction of methane hazard in the region of exploited longwall is presented in Fig. 3.

Structure of heading and formation of fuzzy system is designed on the base of numerical data set in *Wyrobisko & Model Learning* software. Calculating algorithms are available by means of the graphical user interface (GUI). Designed user interface significantly facilitates process of formation of the fuzzy model from uploading numerical data to generation of the final model. The window of program with clustered results is presented in Fig. 5 in Sect. 3.2.

Functionality of the program can be divided into two groups. The first of them is design of heading for which system must operate. Window of program contains three panels (*Anemometry*, *Metanomierze* i *Odmetanowanie*), that allow defining devices operating in the region of exploited wall. Additionally, structure of heading is defined by a choice of ventilation system (U, Z or Y) and conducting the methane drainage

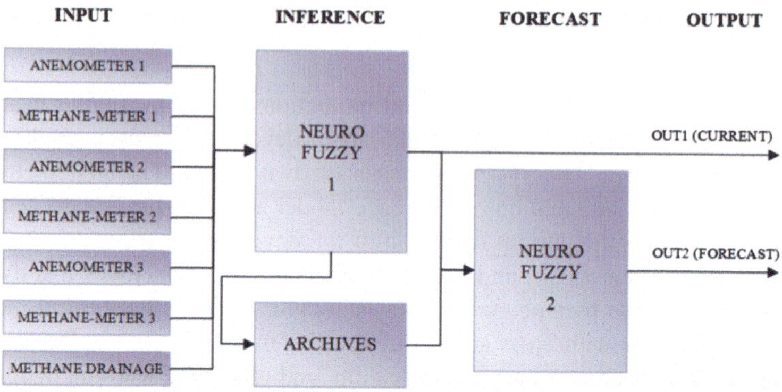

Fig. 3. Conception of assessment system and prediction of methane hazard in the region of exploited longwall.

process. After defining all parameters, in *Wyrobisko* panel, the structure of heading with location of measuring devices is displayed.

The second feature includes creation of fuzzy system on the base of numerical data. In *Tworzenie modelu* panel fundamental parameters of this process and uploads set of learning data are defined.

Then, in *Wykres danych numerycznych* panel there are data, which can be edited using pop-up menu. Additionally, after completed clustering process in the Chart, there are presented centers of determined clusters, recorded in the *Środki klasterów* panel.

The most important, from the point of view of system creation, is *Klasteryzacja danych* panel. This panel enables to select clustering parameters on the base of uploaded set of numerical data. In the software four clustering methods are implemented. For each of them edition of particular parameters is possible. After finish grouping process in *Wyniki klasteryzacji* panel, the implemented indicators of quality clustering and function of measurement time are displayed, which enables to estimate effectiveness of the process of division inlet/output spaces.

Current state of methane safety and also forecasting of future state in investigated region is determined using the *Methane Prediction* software. Numerical algorithms similarly as in case of *Wyrobisko & Model Teaching* window are available through the graphical user interface (GUI). Designed user interface allows for analysis of safety state together with data visualization. The window of the software with real data is shown in Fig. 7 in Sect. 3.2.

The most important panels of the software are *Model rozmyty, ANFIS, Uczenie ANFIS* and *Predykcja* panels. In the *Model rozmyty* panel enables preview of model structure, and also edition of the rule base and parameters of membership function and visualization of the inferring way. After uploading the model, its membership functions for each of inputs is displayed. During operation of the system, in the edition field current value of calculated index is displayed.

ANFIS panel enables to preview the structure of model, edit the rule base, membership function parameters and preview the inferring way, together with displaying the inputs membership function. During operation of the system in the edition field current forecast value is displayed. In the *Uczenie ANFIS* panel a visualization of teaching process of neural-fuzzy net is presented, whereas in the charts teaching data and levels of effectiveness of teaching and testing are displayed. *Predykcja* panel includes a visualization of current indicator of methane hazard calculated by the system, and the prediction value, together with prediction error. In the edition fields current value of network teaching and testing errors are displayed.

3 Characteristics of Neural-Fuzzy System for Prediction of Methane Hazard in Region of Longwall

3.1 Identification of the Research Area

The tests were carried out in longwall together with adjacent headings in one of hard coal mines. This region is characterized by high methane content. In longwall ventilated in "Y" system active methane drainage is carried out [15, 16]. For this wall value of criteria-methane-bearing capacity was determined, taking into account efficiency of methane drainage at the level of 45%.

Real distribution of headings in investigated region (an extract from the original map) together with indicated directions of ventilation and placed sensors are presented in Fig. 4 [4, 13].

3.2 Characteristics of the Developed System

The base for the developed system is data registered through the measurement system at the time of normal operation of the longwall. Based on it, a current level of methane hazard was determined. The measurement included:

- velocity of air flow (anemometer AS038 - heading N-2), m/s,
- velocity of air flow (anemometer AS099 - heading N-3), m/s,
- velocity of air flow (anemometer AS072 – rise gallery N-3), m/s,
- methane concentration (methane-meter MM137 – heading N-2), %,
- methane concentration (methane-meter MM116 – heading N-3), %,
- methane concentration (methane-meter MM104 – rise gallery N-3), %,
- volumetric flowrate of methane in methane drainage pipeline (sensor WM46), m^3/min.

Output value for such developed model is methane hazard indicator, determined as the ratio of absolute-methane-bearing capacity to the criteria-methane-bearing capacity in range of values from "0" to "1". Value "0" means the highest level of safety, and value "1" the real methane hazard, which requires appropriate action in order to reduce the methane amount in ventilated air [6].

The process of creating the whole system can be divided into several steps. At the beginning the properties of headings in region of the longwall, location of measurement sensors, then properties of teaching data *i.e.* number of measurements and their dimensions (number of inputs and outputs) were determined. After loading, measured data set was subjected to fuzzy clustering. Thanks to that, system parameters describing

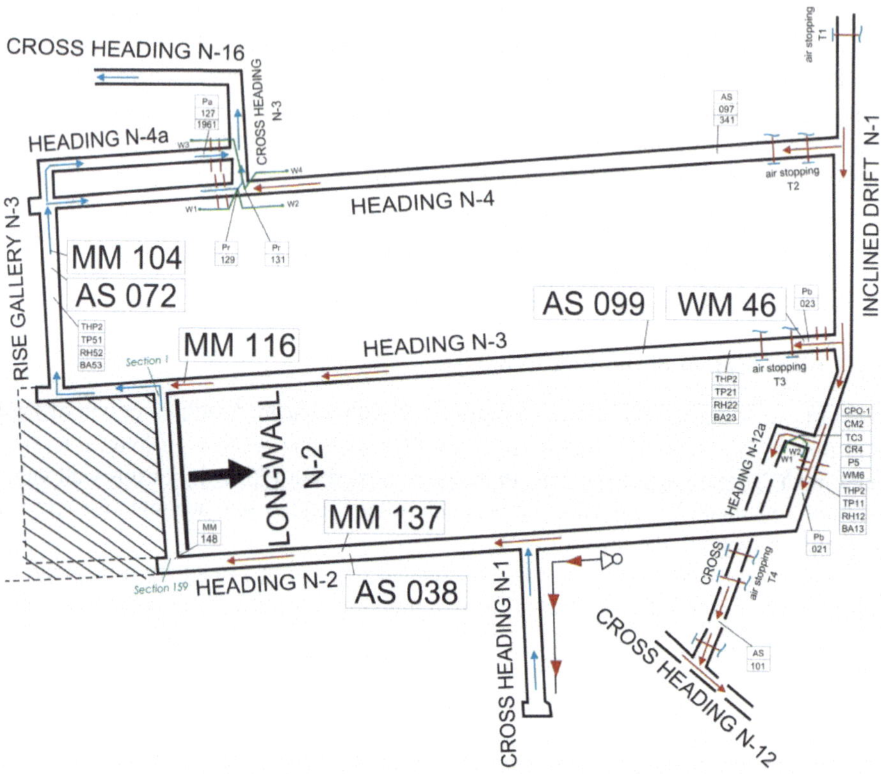

Fig. 4. Real distribution of headings in investigated region (an extract from the original map) together with indicated directions of ventilation and placed sensors [4].

its structure and way of acting were obtained. Algorithm returned the position of centers of clusters and optionally the number of clusters. These parameters are very significant because the number of clusters equals to number of rules of data base, and coordinates of centers of clusters correspond to the location of centers of affiliation function of the fuzzy model.

As the result of performed clustering of measurement data set, 9 clusters were obtained. Two methods determined this parameter in clustering process, whereas in remaining ones the number of clusters was declared on the base of trial tests carried out for different number of clusters. For applied methods, localization of clusters is similar. Coordinates of centers were projected on the axes of particular input data, in order to determine the parameters of membership function. These coordinates correspond to maximum values of the function. For further description the Gauss function was chosen, due to its shape and description using two parameters.

Figure 5 presents window of *Wyrobisko & Model Teaching* software after defining parameters of headings and caring out fuzzy clustering process of measurement data.

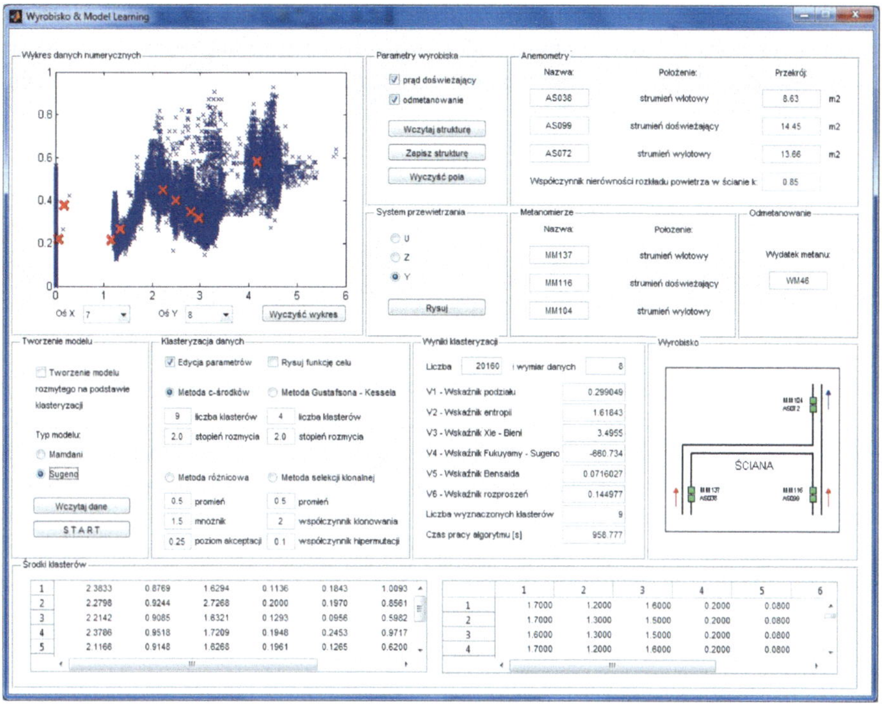

Fig. 5. View of the window of *Wyrobisko & Model Teaching* software after defining parameters of headings and carried out fuzzy clustering process of measurement data.

Based on the obtained results in clustering process the structure of fuzzy model was created. Tha type of fuzzy model (*Sugeno*), the number of inputs (7) and outputs (1) were determined. As a mechanism of fuzzy inference the implication (AND) and aggregation rules (OR) were assumed. Then, a list of fuzzy rules was created. Membership functions located in the input space were assigned to appropriate rules, and in output space the function parameters were assigned respectively as well. Structure and parameters of fuzzy model built on the base of measurement data were presented in Fig. 6.

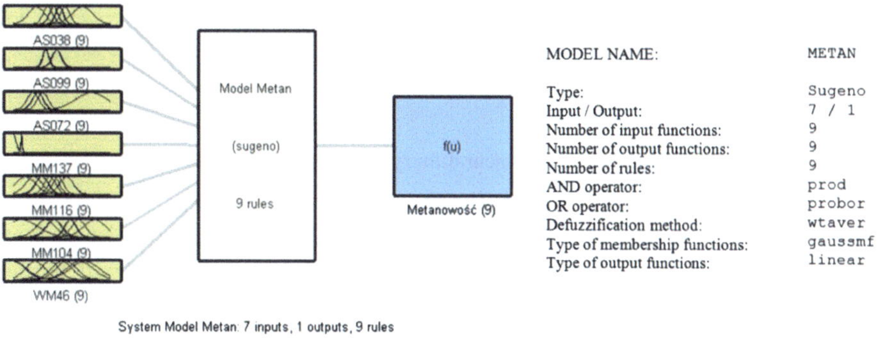

Fig. 6. Structure and parameters of fuzzy model built on the base of measurement data.

The creation process of ANFIS networking in application for forecasting the level of methane hazard is based on the creation of matrix of teaching vectors and vectors testing the network [3, 5, 7]. For the purposes of test the teaching vectors were generated in a form:

$$w(t) = [x(t-(n-1)\Delta) \quad \cdots \quad x(t-2\Delta) \quad x(t-\Delta) \quad x(t) \quad x(t+\Delta)] \tag{1}$$

where:

- t – current measurement time, min,
- $(t \pm \Delta)$ – measurement times shifted by Δ, min,
- n – dimension of the input vector.

In the next step, preliminary structure of network was created, basing on division of grid numerical data collection. Default number of bell-shaped membership function of input model equals to 2, while on the outputs linear functions are generated. ANFIS network created in this way was submitted to tuning in order to optimize inference parameters. In the process of teaching the network a hybrid method, which is combination of the least squares method and backward error propagation, was used. Adaptive structure of neural-fuzzy network after teaching process is presented in Fig. 7.

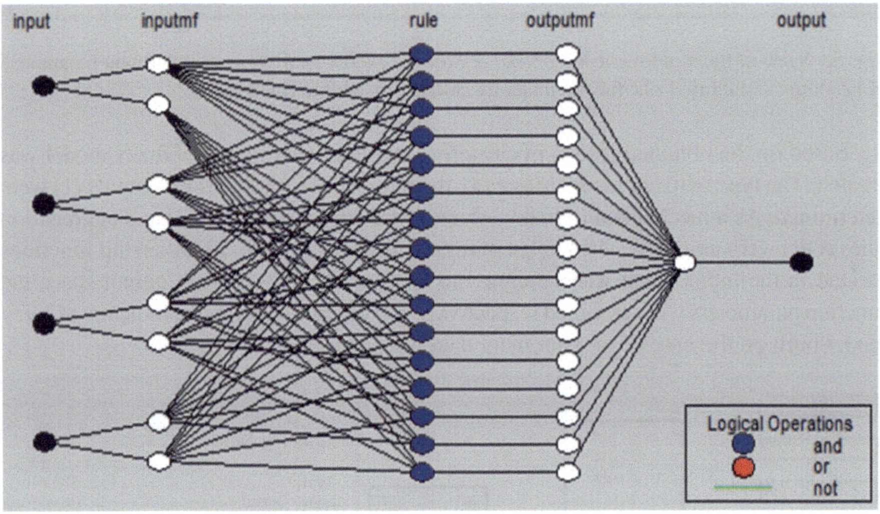

Fig. 7. Adaptive structure of neural-fuzzy network after teaching process.

Figure 8 presents the view of *Methane Prediction* window during operation. It shows the current measurements form sensors of monitoring system. Based on this program current value of methane-bearing-capacity is calculated and shown. In the *Model rozmyty* panel, current level of methane hazard indicator, and in the *ANFIS* panel the predicted index value are displayed. Visualization of values in time is located on the

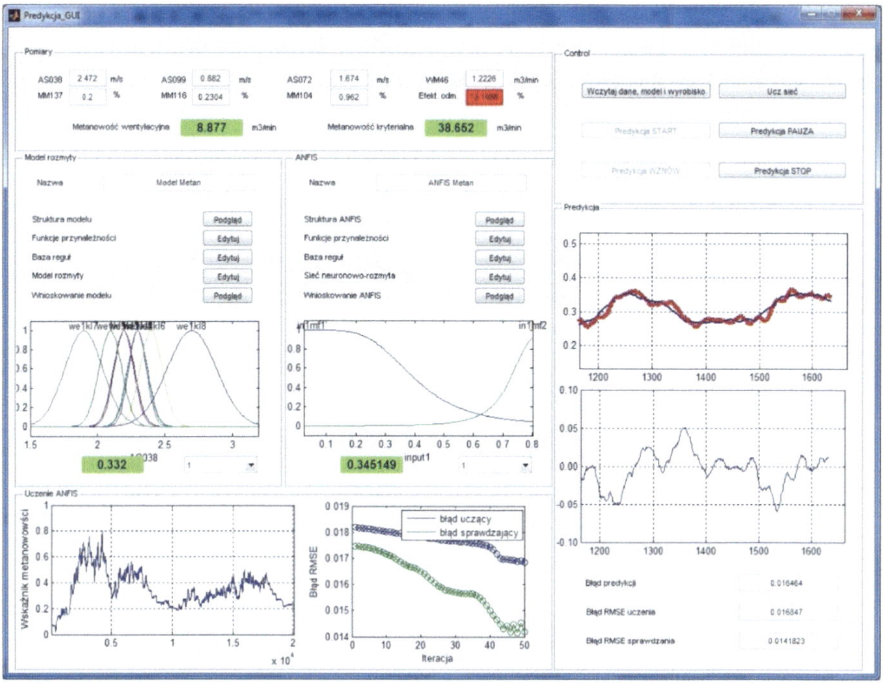

Fig. 8. View of *Methane Prediction* window during operation.

graph in the *Predykcja* panel. Additionally, prediction error is also presented. The tests of system carried out in wide range, showed the good representation of phenomena and small prediction error.

4 Test Results

Developed system was used to determine the methane hazard level in a region of longwall during exploitation. Two different sets of real registered measurements during exploitation in tested period of time have been prepared for the need of system testing. Teaching set used in tuning process of parameters and testing set simulating the registration of indications of measurement devices during the operation of the longwall were created. Based on them, system determines the current values of methane hazard indicator. Additionally, neural-fuzzy structure determines the short period prediction on the base of data recorded and updated in the archive. Based on the tests carried out, an effectiveness of performance of system was verified i.e. prediction values were compared with the real ones and written in tables in a form of the following parameters:

- Maximum Prediction Error,
- Mean Absolute Error (MAE),
- Mean Absolute Percentage Error (MAPE),
- Root-Mean-Squared Error (RMSE).

Table 1 presents the results of simulations for different values of forecast horizons. Iterations number of the process was set as 50, while dimensions of input vector as 4. The obtained results indicated the increase of error together with increasing horizon.

Table 1. Results of simulations for different values of forecast horizons.

No.	Learning process				System operation			
	Horizon, measurement	Error train	Error check	Learning time, s	Max error of prediction	MAE	MAPE [%]	RMSE
1	1	0.0010	0.0009	20.52	0.0040	0.0007	0.2153	0.0009
2	2	0.0019	0.0017	20.46	0.0067	0.0011	0.3589	0.0015
3	5	0.0045	0.0036	20.46	0.0149	0.0025	0.7694	0.0033
4	10	0.0090	0.0069	20.22	0.0250	0.0051	1.5654	0.0066
5	15	0.0130	0.0108	20.26	0.0422	0.0077	2.3803	0.0102
6	20	0.0168	0.0142	20.23	0.0542	0.0101	3.0931	0.0133
7	25	0.0201	0.0215	20.27	0.0837	0.0145	4.4296	0.0195
8	30	0.0258	0.0351	20.16	0.1274	0.0206	6.2356	0.0286
9	40	0.0314	0.0546	20.25	0.2502	0.0314	9.4494	0.0457
10	50	0.0349	0.0817	20.42	0.2214	0.0366	10.8798	0.0481

Table 2 presents the results of simulation for different length of input data vector. Amount of process iteration was established as 50, however forecast horizon as 10. The obtained results indicate that accuracy increases gradually together with increase of input vector. However, it should be noted that together with an increase of input vector the time of learning increases significantly.

Table 2. Results of simulation for different length of input data vector.

No.	Learning process				System operation			
	Vector	Error train	Error check	Learning time, s	Max error of prediction	MAE	MAPE [%]	RMSE
1	2	0.0097	0.0072	2.8760	0.0267	0.0051	1.5560	0.0066
2	3	0.0094	0.0071	6.2940	0.0241	0.0050	1.5436	0.0065
3	4	0.0090	0.0069	20.22	0.0250	0.0051	1.5654	0.0066
4	5	0.0084	0.0071	94.50	0.0286	0.0050	1.5425	0.0066
5	6	0.0073	0.0071	507.87	0.0238	0.0049	1.5127	0.0064
6	7	0.0058	0.0067	4508.29	0.0262	0.0046	1.4171	0.0060

Table 3 shows the results of simulation for different values of teaching net epoch. Horizon of forecast was set as 10, however dimension of input vector as 4. Together with increasing value of iteration number, learning time increases as well. Also, it can be observed that above 40–50 iterations the efficiency of teaching increases, despite executing following cycles of teaching of network.

Table 3. Results of simulation for different values of teaching net epoch.

No.	Learning process				System operation			
	Number of epochs	Error train	Error check	Learning time, s	Max error of prediction	MAE	MAPE [%]	RMSE
1	5	0.0092	0.0071	3.59	0.0258	0.0053	1.6416	0.0069
2	10	0.0092	0.0071	5.42	0.0258	0.0053	1.6416	0.0069
3	15	0.0091	0.0071	7.28	0.0258	0.0053	1.6416	0.0069
4	20	0.0091	0.0071	9.23	0.0307	0.0052	1.5975	0.0068
5	30	0.0091	0.0070	12.85	0.0280	0.0052	1.5956	0.0067
6	40	0.0091	0.0070	16.62	0.0250	0.0051	1.5727	0.0066
7	50	0.0090	0.0069	20.29	0.0250	0.0051	1.5654	0.0066
8	60	0.0090	0.0070	24.08	0.0250	0.0051	1.5639	0.0066
9	80	0.0090	0.0070	31.42	0.0250	0.0051	1.5639	0.0066
10	100	0.0090	0.0070	38.93	0.0250	0.0051	1.5639	0.0066

5 Conclusions

The subject matter presented in the article regards application of artificial intelligence method for analyzing of the level of methane hazard in underground mining in region of longwall. This issue has a great practical meaning as methane hazard is one of the most dangerous in coal mining. In order to monitor this hazard a tool for predicting its state was developed.

Application of advanced artificial intelligence systems provides opportunities for effective forecasting of methane concentrations in underground headings. This forecast is based on the registered measurement data and determined ventilated parameters. Necessary credibility of the data is achieved thanks to application of advanced mechatronic systems for measurement, registration, archiving and transmission of registered parameters. Application of automatic measurement system enables very fast and reliable data acquisition. This is a big progress over the current system, which is based on the data collection by mining staff. The data is averaged, which causes that results of analyzes carried out are characterized by different credibility and long delay.

Thanks to use of automatic acquisition data techniques, it was possible to create a neural-fuzzy system representing the methane hazard problem. In analyzed case the hazard is understood as a relation of absolute- and criteria-methane-bearing capacity in the region of longwall. However, operation of the developed system is based on the real data obtained from the measurement system.

In the paper, methodology for construction of fuzzy model based on clustering of measured data was presented. Also concept of the system, which is responsible for the current determination of methane hazard based on registered measurements by sensors, as well as short-term prediction on methane hazard level was discussed.

The obtained data clearly indicate, that assessment of methane hazard, thanks to applied advanced methodology, provides greater control of parameters, particularly in the situations of dynamic changes and exceeds of extreme values.

Therefore, it can be assumed that artificial intelligence methods could be an effective complement to conventional methods of acquiring knowledge and processing measurement data. The developed methodology and tools create great opportunities for practical application and improvement of the occupational safety in mining industry.

References

1. Brodny, J., Tutak, M.: Analysis of methane emission into the atmosphere as a result of mining activity. In: 16th International Multidisciplinary Scientific GeoConference SGEM 2016, SGEM Vienna GREEN Extended Scientific Sessions, SGEM2016 Conference Proceedings, Book 4, vol. 3, pp. 83–90 (2016). https://doi.org/10.5593/sgem2016/hb43/s06
2. Brodny, J., Tutak, M.: Determination of the zone endangered by methane explosion in goaf with caving of operating longwalls. In: 16th International Multidisciplinary Scientific GeoConference SGEM 2016, SGEM 2016 Conference Proceedings, Book 1, vol. 2, pp. 299–306 (2016). https://doi.org/10.5593/sgem2016/b12/s03.039
3. Dokumentacja projektu europejskiego AVENTO (Zaawansowane narzędzia do kontroli wentylacji i emisji metanu), Instytut Technik Innowacyjnych EMAG, Katowice 2014–2015 (niepublikowana)
4. Dokumentacja ściany N-2 w pokładzie 404/2 w KWK Pniówek, Pawłowice (2013)
5. Felka, D.: Metody budowy inteligentnych modeli na bazie danych numerycznych. Monografia: Młodzi Innowacyjni. Innowacyjne rozwiązania z dziedzinie automatyki, robotyki i pomiarów, Wydawnictwo PIAP, Warszawa (2012)
6. Felka, D., Brodny, J.: Application of neural-fuzzy system in prediction of methane hazard. In: Intelligent Systems in Production Engineering and Maintenance - ISPEM 2017. Advances in Intelligent Systems and Computing, vol. 637, pp. 151–160. Springer, Cham (2018)
7. Fuzzy Logic Toolbox User's Guide. Revised for Version 2.2.24 (Release 2016b), The MathWorks, Inc. (2016)
8. Jang, J.-S.R., Sun, C.-T., Mizutani, E.: Neuro-Fuzzy and Soft Computing. A Computational Approach to Learning and Machine Intelligence. Prentice-Hall, Upper Saddle River (1997)
9. Krause, E., Łukowicz, K.: Zasady prowadzenia ścian w warunkach zagrożenia metanowego. Instrukcja nr 17 GIG, Katowice - Mikołów (2004)
10. Łęski, J.: Systemy neuronowo – rozmyte, Wydawnictwo Naukowo – Techniczne, Warszawa (2008)
11. Miśkiewicz, K., Wojaczek, A., Wojtas, P.: Systemy Dyspozytorskie kopalń podziemnych i ich integracja - Wybrane problemy. Monografia, Wydawnictwo Politechniki Śląskiej, Gliwice (2011)
12. Mrozek, B., Felka, D.: Inteligentny model wskaźnika zagrożenia pożarowego w kopalni węgla. PAR Pomiary Automatyka Robotyka **02**, 540–545 (2012)
13. Mróz, J., Felka, D., Broja, A., Małachowski, M.: Devices for measuring ventilation parameters and methane concentration as well as concept of complex monitoring of methane hazard in longwall area. Min. - Inform. Autom. Electr. Eng. **01**(529), 7–18 (2017)
14. Osowski, S.: Sieci neuronowe w ujęciu algorytmicznym. Wydawnictwo Naukowo-Techniczne, Warszawa (1996)
15. Pawiński, J., Roszkowski, J., Strzemiński, J.: Przewietrzanie kopalń. Śląskie Wydawnictwo Techniczne, Katowice (1995)
16. Roszczynialski, W., Trutwin, W., Wacławik, J.: Kopalniane pomiary wentylacyjne. Wydawnictwo Śląsk, Katowice (1999)

17. Tutak, M., Brodny, J.: Assessment of hydrodynamics of gas flow through homogeneous and heterogeneous porous rock media. In: 17th International Multidisciplinary Scientific GeoConference SGEM 2017, vol. 17, no. 14, pp. 539–546 (2017). https://doi.org/10.5593/sgem2017/14
18. Tutak, M., Brodny, J.: Influence of auxiliary ventilation devices on a distribution of methane concentration at the crossing of longwall and ventilation roadway. In: 17th International Multidisciplinary Scientific Geoconference SGEM (2017), vol. 17, no. 13, pp. 437–444 (2017). https://doi.org/10.5593/sgem2017/13
19. Tutak, M., Brodny, J.: Analysis of influence of goaf sealing from tailgate on the methane concentration at the outlet from the longwall. In: IOP Conference Series: Earth and Environmental Science, vol. 95, p. 042025 (2017). https://doi.org/10.1088/1755-1315/95/4/042025
20. Tutak, M., Brodny, J.: Determination of particular endogenous fires hazard zones in goaf with caving of longwall. In: IOP Conference Series: Earth and Environmental Science, vol. 95, p. 042026 (2017). https://doi.org/10.1088/1755-1315/95/4/042026
21. Tutak, M.: Analysis of varying levels of methane emissions from coal mines in Poland. In: SGEM 2017 Vienna GREEN Conference Proceedings, vol. 17, no. 43, pp. 301–308 (2017). https://doi.org/10.5593/sgem2017h/43/s19.038
22. Tutak, M.: Assessment of hydrodynamics of gas flow through the porous rock structures. In: 17th International Multidisciplinary Scientific GeoConference SGEM 2017, SGEM2017 Vienna GREEN Conference Proceedings, vol. 17, no. 15, pp. 53–60 (2017). https://doi.org/10.5593/sgem2017h/15/s06.007

Modular Approach to the Planning of the Robot's Tasks in the Context of Holons and Graph-Based Methods

Krzysztof Foit(✉)

Institute of Technological Processes Automation and Integrated Manufacturing Systems, Silesian University of Technology, Gliwice, Poland
krzysztof.foit@polsl.pl

Abstract. The philosophy of "Industry 4.0" assumes that the contribution of the computer science in the modern industry will intensively grow. The manufacturing tasks are no longer just programs for digitally controlled machines, but are a part of the overall process that includes transport, machine-to-machine communication, and human-machine interfaces that allow the ongoing control of the whole process. Under such circumstances, describing the whole system – or just a part of it – needs to adjust the level of information detail. The key is to find the right level of detail that will guarantee a comprehensible description of the activity performed by the machine, but concomitantly will not contain any information that is redundant at the given level of generality. Robots are special kinds of machines, because apart from the typical handling and manipulation tasks, they also often perform the work that is the part of the machining or assembly process. This involves the higher level of complexity of the code written in the robot's programming language, but such code is hard to understand for people without the ability of robots' programming. The paper presents the other way of the description of tasks performed by robots during the manufacturing process that is the modular approach, based on classic, structural programming methods in connection with the holon-based and graph-based methods. Taking into account the dynamic development of modern ways of robots' programming, the presented method may provide the starting point for further research in the context of task-level programming or programming by demonstration.

Keywords: Manufacturing system · Task planning · Holon · Petri nets

1 Introduction

The "Fourth Industrial Revolution" (in short "Industry 4.0") is a kind of philosophy that determines the modern trends in the industry. Lasi et al. [1] list some major concepts that describe the "Industry 4.0" compatible factory:

- Smart Factory – the factory equipped with the "smart technology", where the machines have certain level of independence and decision making ability, the digitalized model of products and factories are used; the term "Smart Factory" is also tightly bonded with ubiquitous computing;

J. Świder et al. (Eds.): MECHATRONICS 2017, AISC 934, pp. 134–143, 2019.
https://doi.org/10.1007/978-3-030-15857-6_14

- cyber-physical systems – the merge of physical and digital level of systems; in the other words, such system is an integration of software, networking and physical layer;
- decentralization and self-organization – the machines and devices have some level of machine intelligence and the centralized management of the manufacturing process has been dropped; the machines are able to automatically form structures, according to the requirements of the processed element;
- adaptation to the human needs – the process of designing the new manufacturing systems should take into consideration human needs as the major criterion.

There is another important property of the modern industry, which is modularity. It is indirectly connected with decentralization and self-organization, because modular systems have the ability to adapt to the dynamically changing environment. Such system could be built from the available modules, according to the current needs, based on seasonal fluctuations or changes in the product requirements. The top-level modules, e.g. the one that forms the production line, consists of lower-level modules, which in turn consists of another lower-level modules, and so on. Because we are considering the machines and devices as the cyber-physical systems, the modularity can occur in two different domains: hardware domain (type of the machine, the equipment, tools etc.) and software domain (programs, subroutines, keywords, variables etc.).

Referring specifically to the robots domain, it should be emphasized, that there is no standard programming language, which could be used on different machines. Norman et al. [2] and Smith et al. [3] point out that the major problem of beginner programmer is to formulate the problem that he thinks about in a way, which is understandable for computer. They claim that there are two possible solutions of this problem. The first one is "to move the user closer to the system" [3], which is connected with learning a programming language. The second is to "move the system closer to the user" [3], where the programmer uses human-understandable programming language. Both of the solutions lead to the conclusion that the domain of robotics needs the better and simpler way of programming.

The aim of this paper is to present the modular approach to the robot's task planning. Such method can encompass the different, known techniques of robots' programming, like "programming by example/demonstration" [3–5], task-level programming [6–9] and any of the classic methods. In the described case, the holons philosophy is used as the base for the modular approach. In order to "bind" the modules together, the proper method of actions modelling and presenting their sequence should be chosen. One of the possibilities are methods based on graphs or networks, because by using them it is possible to show the flow of the program in graphical way that is intuitive and understandable even by people, which have nothing in common with programming. The method presented in the paper is theoretical but functional and could be seen as the origin for the further works in the field of advanced methods of robots' programming.

2 The Comparison of Classic and Advanced Methods of Planning of the Robotic Tasks

This section provides a brief description of selected classic and advanced methods of robots' programming. It will allow highlighting the most important features and properties of each technique and will also characterize the level of generality of described methods.

2.1 The Classic Form of Robots' Programming

Considering the classic form of robots' programming methods, it is worth to mention about the general classification into off-line and on-line forms of programming. The off-line programming is done without the presence of the real robots, using personal computer and the proper software. The on-line programming requires direct programmer-robot relation, the physical presence of the programmer and the robot in the same place.

The off-line programming is done using dedicated computer application. The program is often created in the code editor, which is assisted by graphics simulation of the manipulator and the emulation of robot's controller. The code has a text form that is very similar to the high-level programming languages, like BASIC or Pascal [10]. The programming languages, used in the field of robotics, have limited capabilities in comparison with the languages developed for computers. Although there are some manifestations of structural programming possibilities, a great number of the languages based on BASIC syntax still offer the command that realizes the direct jump to the particular line of the code. Use of this statement does not allow the programmer to create the good quality code, based on the structural programming philosophy and thus modularizing the program. The essence of modularization manifests itself in the special design of code, where the repetitive fragments can be defined only once, and used repeatedly.

The on-line programming of the robot involves direct interaction with the robot. The programmer uses robot's interface to create the sequence of movements that are saved in the robot's controller memory. Such movements could be defined in two, different ways. The first one is to move the robot's manipulator to the desired position and save the coordinates to the controller's memory. The program is created by repeating this action until all required movements will be done. Such method of programming is called "point-to point" (PTP), because the operator only set the desired points and their orientation, while the control system plans the path between two subsequent points. The second manner of the on-line programming is called "continuous path" (CP). In this method, the operator moves the wrist of the manipulator along the desired path. The position of the arm is sampled by the control system with certain frequency and coordinates are stored in the memory. Both of the on-line programming methods could be mixed with the off-line programming – such approach is called hybrid programming and is often used in order to minimize the period of excluding the robot from the manufacturing process.

As it was mentioned earlier, the classic techniques of robots' programming could be used for creation of program that has modular form, but at least structural programming model is required. Although such technique could be supported by the robot's controller, it may be unavailable in on-line programming mode.

2.2 Programming by Example/Demonstration

"Programming by example" or "programming by demonstration" is a term, which comes from the field of personal computer programming. Today, the term "programming by demonstration" is used in reference to the programming of robots, while the former refers mainly to the computer programming field.

It is worth to mention that programming by demonstration operates on the higher level of abstraction than the classic methods of programming. The one of the manners of realizing the PbD is to perform the task by the user without the presence of the robot. It could be processed using virtual reality, vision systems, special markers or a sensoric outfit. Another way could be seen as quite similar to the Continuous Point method of programming, because it involves operating of the robot using special interfaces or simply move the effector in the desired way. It allows the end-user to teach robots new actions by simply demonstrating the movement of the manipulator. The main difference between CP programming and programming by demonstration is that PbD requires multiple demonstrations of movements, what lets the robot to generalize the problem and act properly in the unseen situations. The ability of generalization is one of the key issues in this technique of programming. Muench et al. [11] pay attention to the role of the operator during the generalization process, because the software cannot decide about the importance of recorded action and recognize if it represents a part of demonstration, a correction of the previous error or the inaccuracy. The whole process of PbD is schematically illustrated in Fig. 1. The authors also divide the learning task into three parts:

- learning of structural knowledge (program structure, subprograms etc.),
- learning of parameters knowledge (criteria, conditions and action parameters),
- learning of skill knowledge.

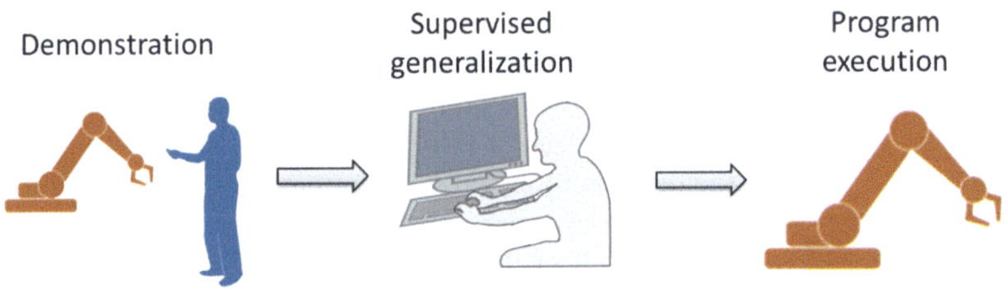

Fig. 1. The phases of programming by demonstration process.

Further, the authors describe the process of learning of the structural knowledge as a classification task that divides the program into modules [11]. One of the aims is to find repeating activities and assign them to subroutines. In turn, Billard et al. [12]

divide the demonstrated task into segments that are a sequence of state-action-state transitions. Such approach could be usable in the further processing of the task using graph-based methods.

2.3 Task-Level Programming

The task-level programming is another advanced method of programming the tasks performed by robots. This technique uses the higher level of abstraction than programming by demonstration and could be seen as a form that gets closer to the natural language. This manner of programming is strictly goal-oriented, and it is the software problem to convert a command issued in a quasi-natural language into the form interpretable by the robot's controller. For example, the command "paint the box red" is understandable by human if somebody knows what box should be painted, where it is and – what is more important – how to paint a box. Of course the problem "how to paint a box" is also defined on high level of generality, because it is connected with the knowledge about the painting process of any box. This is the key to understand what the task-level programming is. This approach has different consequences. On the one hand it is very comfortable for the end-user, who may quickly start programming the robot, without the need to learn complicated programming language. On the other hand, most of the conversion work from task-level language to the robot's programming language is done by the software and the operator should be accustomed to assist the conversion software. Simmons and Apfelbaum [13] raise the problem of debugging of such computer-generated code, because some operations may be done concurrently: for example the robot may do the planning and executing at the same time. Moreover, Somani et al. [14] points out that the hardware processing of the task-level program must take into account different constraints, which are the result of dynamically changing environment. These may be, for example, different ways of grasping the same object, where the control system should choose one of the possibilities, or the differences in properties of manipulated object (e.g. empty container and the same container filled-up with the liquid).

The task-level programming is still in the early phase of development. The development of this method is very difficult due to the complexity of the problem, which requires considerable resources and computing power. The work is also hampered by the lack of standardization of robotic programming languages – because of the possibility of synthesizing the program only for particular type of robots, the application of the results of performed scientific research is much more difficult.

Referring to the issue of modularization in relation to task-level programming, it should be mentioned that the division of the program into modules plays a similar role to the classification task in the case of programming by demonstration. However, the problem here is different, because all of the modules (subprograms) must be at first synthesized and then bonded together into program.

3 Holons

3.1 Introduction

The term "holon" has been introduced by A. Koestler in the field of philosophy, popularized in his book "The Ghost in the Machine". The holon has a relation with the hierarchical model of nature. Koestler, in [15] has written that "*Holon is a system of relations which is represented on the next, higher level as a unit, that is, a relatum*". He also claimed that holon is intelligent, self-organizing part of structure that may be triggered by the signal from higher level of hierarchy. The activated holon acts according to its inner "pattern of activity" – there is no need to send any special kind of signal, which will specify what the holon should do.

The hierarchy of holons is called holarchy. The holon can be a member of more than one holarchy at the same time and its membership can change during the time, so the holarchy is regarded as the dynamic structure. The examples of such structures could be the living organisms, where the organs acts as holons, or a flock of birds, where the particular birds are the holons, but the flock acts as unit – holon that may be triggered by the signal from the environment.

The holons theory has also gained popularity in software engineering and manufacturing management. Holonic approach is often used in the mentioned fields along with agent-based methodology.

3.2 Relationship Between Holons and the Source Code of a Program

In the earlier sections, the problem of robot's program modularity has been discussed. This approach is desired particularly in the cases of use of advanced programming techniques, like programming by demonstration or task-level programming. The holonic approach naturally harmonizes with the modular organization of a program, because of the following properties:

- any subroutine or function could be regarded as a holon, and the program as a holarchy,
- subroutines or functions could be separated from the program code and reused in the other program,
- on the higher level of hierarchy, which is the program level, all of the holons constitute the unit,
- the particular holons were triggered by procedure call or function call.

The programming language keywords, mathematical expressions etc. could be regarded as holons too, because they are triggered by language interpreter or – after compilation – by processor. This approach allows the programmer to analyze the program code on the different levels of generality. It also gives the possibility to generate universal code snippets that could be reused in different kind of tasks. The database of the snippets could be also used during the synthesis of the robot's program in the PbD or task-level programming.

4 The Petri Nets

4.1 Introduction

Petri nets are a kind of modelling tool, which has both the graphical and mathematical aspect. The methodology was presented by Carl Adam Petri in his dissertation, submitted in 1962. From the mathematical point of view, the Petri Net is a 4-tuple [16]

$$N = (P, T, F, V) \tag{1}$$

where P (places) and T (transitions) are finite sets and:

$$P \cap T = \emptyset,\ P \cup T \neq \emptyset \tag{2}$$

F is a relation of arity 2 with

$$F \subseteq (P \times T) \cup (T \times P) \tag{3}$$

and

$$V : F \rightarrow \mathbb{N}^{+} \tag{4}$$

where F is the flow relation of N and elements of F are called arcs, V are the weights of arcs. According to the mentioned definition, the Petri net is a two-colored, weighted, directed and finite graph [16].

The graphical representation of Petri net is very simple. The places are represented by circles, while the transitions by rectangles or bars. If the Petri net is marked, then inside the circles – which represent places – small dots could be located. Every dot represents a marker and is called token. The token could be moved to the other place after firing the transition. The example of marked Petri net is shown in Fig. 2.

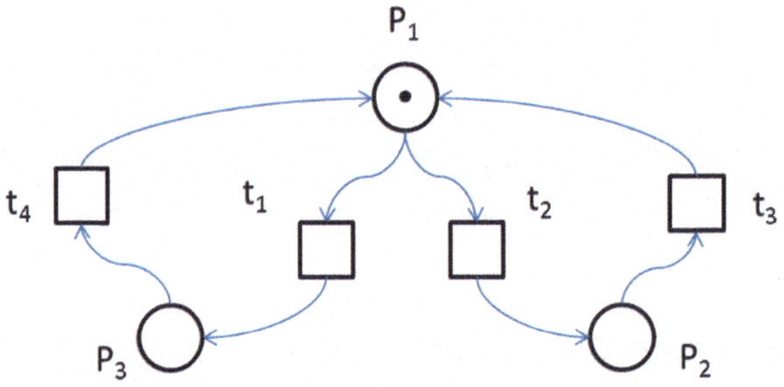

Fig. 2. Example of marked Petri net.

4.2 The Petri Nets as a Model of a Program of the Robot

Considering the Petri net graphical model as a "place-transition-place" sequence there is visible similarity to the "state-action-state" model of the programming by demonstration. By using this fact, it is possible to graphically show the subsequent actions performed by the robot. The model in a form of marked Petri net may also represent the states of the external equipment (machines, buffers, etc.). Besides showing the precise picture of the process, such model gives also the possibility to handle the information that come from the sensors or the other machines.

5 The Example of Using the Modular Approach to the Robot's Task Planning by Means of Holons and Petri Nets

Let's consider the simple workcell, shown in Fig. 3. The robot takes the object from Conveyor 1 and transports it to the lathe. After the machining, robot takes the object from the lathe and put it on the Conveyor 2.

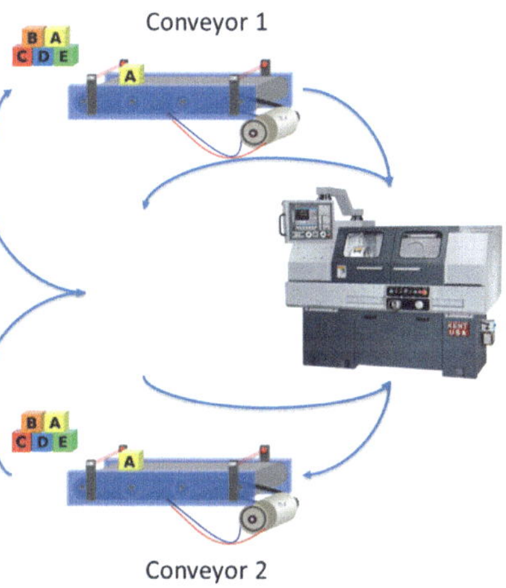

Fig. 3. The considered workcell and the route of the object of manipulation

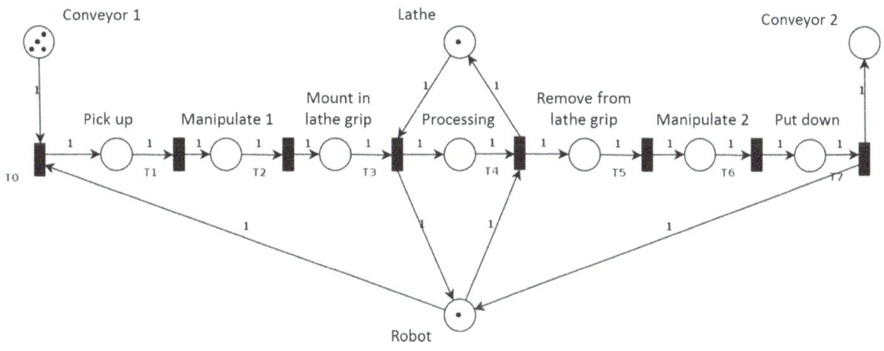

Fig. 4. The Petri net that illustrates the process shown in Fig. 3

The net, shown in Fig. 4 illustrates the subsequent steps of the process. According to the Petri nets theory, firing the transition causes the token to "move" to the neighboring place (in fact, every transition "consumes" and "produces" the tokens, according to the value of arcs' weights). This behavior is very similar to the holons that are triggered by the signal and acts according to the rules (program). Concomitantly, this approach is consistent with the manner, in which the actions are decomposed into modules during the programming by demonstration. The module of the program could be defined by the actions that are described by the name of the place, while both of transitions specify the start and the end of block respectively (Fig. 5).

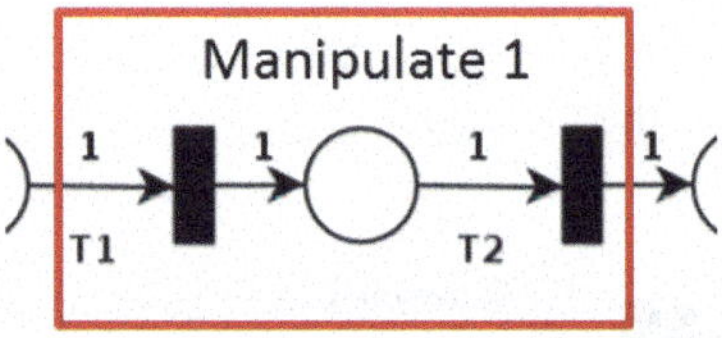

Fig. 5. The example of a program's block represented by Petri net

Going further, the block of the program could be decomposed into primary components. By analyzing these components and their mutual relations, the database of the code snippets could be created. In this manner the robots' task planning could be performed using the "top-down" approach, where the operator starts with the most general description of the task (like for task-level programming) – that may has the form of graph (e.g. Petri net) – and then consequently goes to the lower levels, using increased number of details.

6 Conclusions

The modern manufacturing follows the "Industry 4.0" concepts and philosophy. As machines become "smarter", the techniques of programming become more user-friendly. On the other hand, such approach requires more computing power and more resources, because it uses advanced numeric methods. The application of these techniques is also difficult, because the robot's programming languages are not standardized.

The method presented in this paper allows the operator to plan the tasks for robots on the higher level of generality without the necessity to use a programming language. In turn, he may use the Petri net (that are quite simple in use and understandable for non-programmers) or the other graph methods to generally describe the actions performed by the robot. Later, the graph may be decomposed in blocks. Increasing the detailedness, the operator could program the actions, writing the minimum of required code and using snippets for known, repetitive actions.

The described method is functional, but the theoretical one. It is planned that further development of the presented approach will be focused on the user-assisted, semi-automatic development of the robot program in the context of task-level programming of the robots.

References

1. Lasi, H., Fettke, P., Kemper, H.-G., Feld, T., Hoffmann, M.: Industry 4.0. Bus. Inf. Syst. Eng. **6**, 239–242 (2014)
2. Norman, D.A., Draper, S.W.: User Centered System Design: New Perspectives on Human-Computer Interaction. L. Erlbaum Associates, Mahwah (1986)
3. Smith, D.C., Cypher, A., Tesler, L.: Programming by example: novice programming comes of age. Commun. ACM **43**, 75–81 (2000)
4. Lieberman, H.: Your Wish is My Command: Programming by Example. Morgan Kaufmann Publishers, Burlington (2001)
5. Koenig, N., Matarić, M.J.: Robot life-long task learning from human demonstrations: a Bayesian approach. Auton. Robots. **41**, 1173–1188 (2017)
6. Lozano-Perez, T.: Robot programming. Proc. IEEE **71**, 821–841 (1983)
7. Hörmann, K., Werling, G., Frommherz, B., Hornberger, J., Pezzinga, A., Gallerini, R., Bison, P., Mirolo, C., Pagello, E., Stocchiero, L.: Task-level programming. In: Integration of Robots into CIM, pp. 122–167. Springer Netherlands, Dordrecht (1992)
8. Pedersen, M.R., Nalpantidis, L., Andersen, R.S., Schou, C., Bøgh, S., Krüger, V., Madsen, O.: Robot skills for manufacturing: from concept to industrial deployment. Robot. Comput. Integr. Manuf. **37**, 282–291 (2016)
9. Csiszar, A., Hoppe, M., Khader, S.A., Verl, A.: Behavior trees for task-level programming of industrial robots. In: Tagungsband des 2. Kongresses Montage Handhabung Industrieroboter, pp. 175–186. Springer, Heidelberg (2017)
10. Zieliński, C.: The robots' programming languages. (in Polish). AutomatykaOnline.pl. http://automatykaonline.pl/Artykuly/Robotyka/jzyki-programowania-robotow-przemysowych. Accessed 09 Jan 2017
11. Muench, S., Kreuziger, J., Kaiser, M., Dillmann, R.: Robot programming by demonstration (RPD)-using machine learning and user interaction methods for the development of easy and comfortable robot programming systems. In: Proceedings of the International Symposium on Industrial Robots, p. 685 (1994)
12. Billard, A., Calinon, S., Dillmann, R., Schaal, S.: Robot programming by demonstration. In: Springer Handbook of Robotics, pp. 1371–1394. Springer, Heidelberg (2008)
13. Simmons, R., Apfelbaum, D.: A task description language for robot control. In: Proceedings of 1998 IEEE/RSJ International Conference on Intelligent Robots and Systems (IROS), Innovations in Theory, Practice and Applications (Cat. No. 98CH36190), pp. 1931–1937. IEEE (1998)
14. Somani, N., Rickert, M., Gaschler, A., Cai, C., Perzylo, A., Knoll, A.: Task level robot programming using prioritized non-linear inequality constraints. In: 2016 IEEE/RSJ International Conference on Intelligent Robots and Systems (IROS), pp. 430–437. IEEE (2016)
15. Koestler, A.: Beyond atomism and holism - the concept of the Holon. In: Shanin, T. (ed.) The Rules of The Game: Cross-disciplinary Essays on Models in Scholarly Thought, pp. 233–247. Routledge, Abingdon (2013)
16. Popova-Zeugmann, L.: Timed Petri nets. In: Time and Petri Nets. Springer, Heidelberg (2013)

Detection and Recording of Acoustic Emission in Discrete IGBT Transistors

Radosław Gordon(✉)

Faculty of Machine Engineering, Maritime University of Szczecin,
1–2 Wały Chrobrego St., 70-500 Szczecin, Poland
r.gordon@am.szczecin.pl

Abstract. The article presents the results of experimental research, which is to show a correlation between the change of operating status of single IGBT transistor and its acoustic emission. Acoustic emission signals were obtained using Vallen sensor. Received signal was obtained with oscilloscope in order to further process it digitally and determine possibility of the damage to the element based on registered acoustic signal.

Keywords: IGBT modules · IGBT transistor · Acoustic emission

1 Introduction

Power industrial equipment and electrotechnics and electronic systems cannot exist without discrete semiconductor circuits as well as blocks connected into monolithic. Particularly rapid development of this kind of systems in recent years has been noticed in electronics and signal control systems with high output power. Control of those systems is theoretically relatively simple, but by switching high currents and voltages in semiconductors that often require many switching per unit time, it makes it really difficult in practical applications. The occurrences inside the semiconductors structure are often at borderline between classical and quantum physics. Therefore, watching what is happening inside such structures is a quite big challenge for designers, producers and users.

There are many methods of non-invasive observation of occurrences in semiconductors elements and their description, such as observation of currents and voltages, both input and output, temperature changes etc. and their profound analysis [1, 2].

Another very interesting solution is to test the systems with forbidden status in normal operations of IGBT bridges, made in the idle state of motor [3].

The knowledge of possible micro damages to the individual IGBT transistors or whole modules increasingly affects the safety of equipment and people. Among the applications of IGBT technology the following can be mentioned: control and operation of wind turbine generators [4, 5], urban transport control of subway motors, trains, trolleybuses and trams [3], control of tracer motors in jet controls and electric network on ships or offshore units [6].

J. Świder et al. (Eds.): MECHATRONICS 2017, AISC 934, pp. 144–150, 2019.
https://doi.org/10.1007/978-3-030-15857-6_15

This article presents an analysis of the acoustic emission of the popular IGBT transistor using the Vallen acoustic emission sensor. The research has been conducted since 2016 at the Institute of Electrical Engineering and Automation of the Maritime University and is constantly confronted with the current state of the art on this subject in the world. This type of research is conducted also by the team of Finnish scientists with ABB Group [1, 7]. The research description conducted by this center as well as work [7] do not answer the questions that arise during the analysis of this problem. Because of the specifics of this research, the proper equipment and measuring methods were used according to the needs of this system for the acquisition of data and results analysis.

The results of this research are intended to show that received data might be useful for further processing as well as to identify failures or signals indicating significant degradation of semiconductors structure (impending damage) of an IGBT transistor.

2 Measuring Method

A popular oscilloscope for filtering and recording digital data was used for measurements. The measuring element was Vallen's VS600-Z1 acoustic converter, which is characterized by transmission of spectrum within the range of 550–730 kHz (600 kHz) [8].

The scheme of measurement system is shown in Fig. 1.

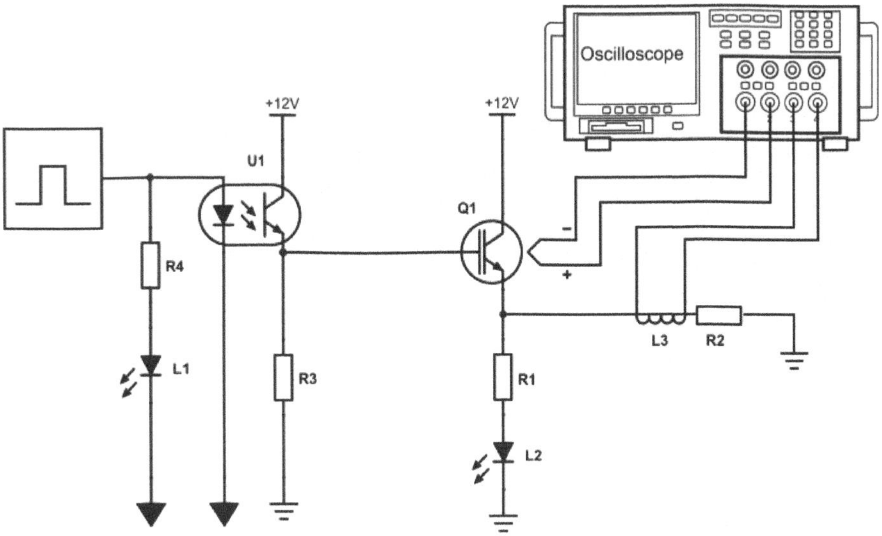

Fig. 1. Scheme of measurement system used in experiment (own study).

3 Operation

Transistor is controlled by generator which is separated from the gate with optocoupler. The L1 diode informs about the state at the generator output ("slow-changing" rectangular signals). The optocoupler transistor controls the gate of tested IGBT transistor. Transistor operates in two conditions: closed or open. The charge of the transistor is provided by the R2 resistor and system composed of resistor R1 and a diode denoted as L2 (small), which signals the switching of the transistor. The clamp ammeter was placed over the wire connecting the output of the emitter and R2 resistor. The ammeter gives waveforms of the current to the oscilloscope input. Signal from acoustic emission sensor is given directly to second oscilloscope channel.

Figure 2 shows the example of reading from the oscilloscope.

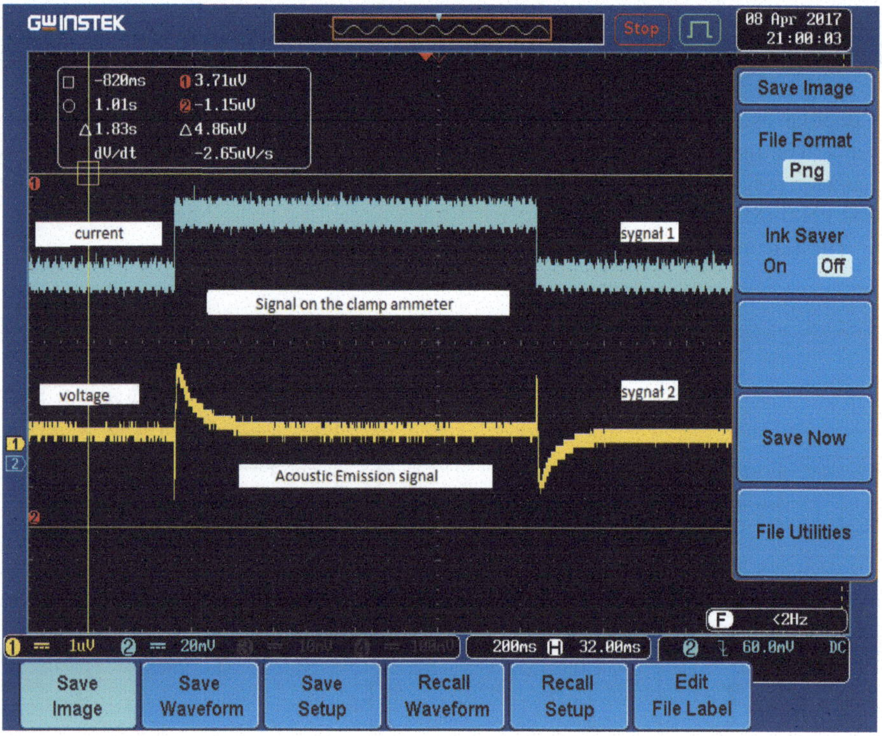

Fig. 2. Waveforms of acoustic emission and corresponding transistor current acquired on the digital oscilloscope (own study).

4 The Analysis of Obtained Results

It can be seen that signal 1 shows the current on the transistor output from clamp ammeter while signal 2 is obtained from acoustic emission sensor, which is placed directly on the surface of the IGBT transistor.

One of the most important factors during the measurement was proper fitting of acoustic emission sensor, which had to be placed directly above the silicon structure. Special treat was provided to preparation of contact point between silicon surface of power element and an AE sensor. Appropriate contact between sensor and tested element was assured through the special grease recommended by the manufacturer of the sensor, and without it the results may vary from expected [9].

The measurements were carried at different values of switched current and voltage. As each parameter increased the signal recorded on the oscilloscope was also proportionally larger.

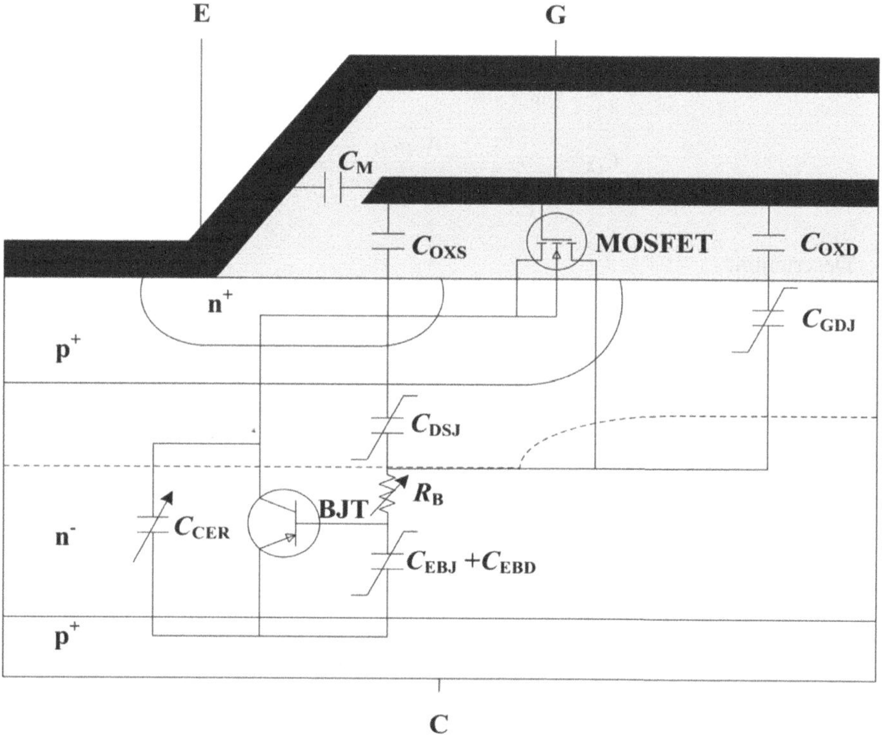

Fig. 3. Internal structure of IGBT transistor based on [2]

Figure 3 shows the structure of IGBT transistor and it explains the form of signal obtained in time. By analyzing this structure, capacitances can be seen which are weak link in the transistor inner structure, and they affect the signal at the time of switching.

Typical charging and discharging processes of parasitical capacitances in Fig. 2. Due to small currents, the signal is not smooth, but it clearly shows the capacitive response to the transistor operating when the state change. The response is similar when switching from open status to closed status and in reverse, and that means that sensor has recorded the transistor switching.

It proves the chance of non-invasive measurements and proper detection of the transistor operating status. Moreover, the presented method can confirm the correctness of IGBT operation. With proper tools it is possible to define the parasitical capacitances of transistor, which gives further possibility of identification the inner silicone structure damages and the effects caused by the "aging" of the element. Such diagnostic allows to define when the damage to the given element would occur.

Based on Fig. 3 and [2] correlation between capacitance changes and voltage on individual elements of internal structure can be defined:

$$C_{GC} = \begin{cases} C_{OXD} & V_{CE} < V_{GE} - V_{GE(th)} \\ \frac{C_{OXD} C_{GDJ}}{C_{OXD} + C_{GDJ}} & V_{CE} \geqslant V_{GE} - V_{GE(th)} \end{cases} \tag{1}$$

$$C_{GC} = \frac{A_{GD} \varepsilon_{Si}}{\sqrt{2\varepsilon_{Si}\left(V_{CE} - V_{(th)}\right)/q/N_B}} \tag{2}$$

Description:

C_{GC} - capacitance collector - emitter, consisting of capacitance in the point of contact of silicon oxide gate - drain C_{OXD} and capacitance of saturation gate - gate drain

C_{GDJ} - depending on V_{CE} voltage (collector - emitter), where A_{GD} is the area of the point of surface contact

ε_{Si} - constant of silicon dielectric and liminal voltage gate emitter $V_{GE(th)}$

N_B - area of concentration of majority carriers.

As it turns out, aforementioned formulas have the biggest influence on acoustic emission waveforms obtained on oscilloscope (Fig. 2).

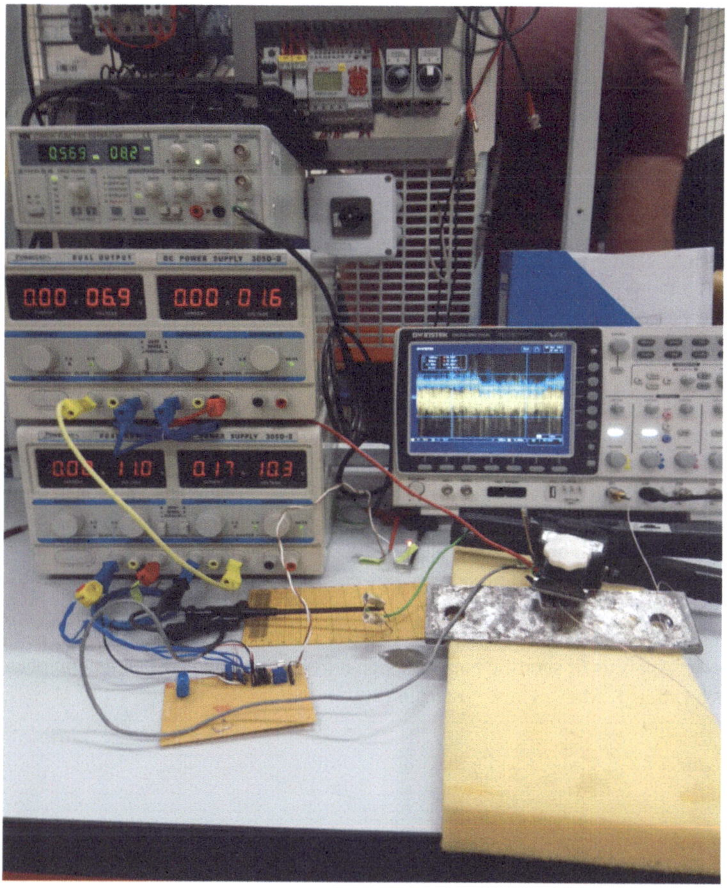

Fig. 4. The picture of measurement system and transistor with acoustic emission sensor (own study).

5 Conclusion

The possibility of defining the correlation between switching IGBT transistor and generated signals of acoustic emission was the main goal of the presented research. Acquired AE signals can be post-processed and after some analysis the status of transistor can be specified (good, damaged or nearly damaged). This will be the object of the further research.

The tests showed, that switched IGBT transistor is characterized by measurable acoustic emission signals and such signals correspond to the theoretical assumptions.

Measurements were carried for relatively small voltages values and switching currents in order to proper values recording. The tested element was subjected to small current changes condition. Increasing of currents should result in improved readings of acoustic emission sensor signals.

References

1. Kärkkäinen, T.: Observations of Acoustic Emission in Power Semiconductors. Lappeenrannan teknillinen yliopisto Yliopistopaino 2015 (2015). ISBN 978-952-265-912-5 (PDF)
2. Luowei Zhou, S., Sun, P.: Monitoring potential defects in an IGBT module based on dynamic changes of the gate current. IEEE Trans. Pow. Electron. **28**(3), 1479–1487 (2013)
3. Yeke, L., Dawei X., Yifan, F.: Prognosis of chip-loss failure in high-power IGBT module by self-testing. In: 2016 IEEE IECON 2016 - 42nd Annual Conference, Department of Electrical Engineering, Tongji University Shanghai, China (2016)
4. Wang, B., Wang, X., Wang, X., Shao, C., Liu, S.: Reliability evaluation of voltage-source converter-based multi-terminal direct current integrated offshore wind plants. IET Renew. Pow. Gener. **10**(6), 761–766 (2016)
5. Zhu, P., Liu, Y., Roesner, R., Hao, X.: Offshore wind converter reliability evaluation. In: 8th International Conference on Power Electronics - ECCE Asia 30 May–3 June 2011, The Shilla Jeju, Korea (2011)
6. Kozak, M., Gordon, R., Bejger, A.: Control of squirrel-cage electric generators in a parallel intermediate DC circuit connection. Zeszyty Naukowe, Akademia Morska w Szczecinie **45** (117), 17–22 (2016). ISSN 1733-8670
7. Kärkkäinen, T., Talvitie, J.P., Kuisma, M., Hannonen, J., Strom, U., Mengotti, E., Silventoinen, P.: Acoustic emission in power semiconductor modules—first observations. IEEE Trans. Pow. Electron. **29**(11), 6081–6086 (2014)
8. AE-Sensor Overview, Vallen-Systeme GmbH Data Sheet (2015)
9. Bejger, A.: Analysis of acoustic emission signal frequency bands dependent on phenomena in an injector. Sci. J. Marit. Univ. Szczecin (2008)

Zero-Sum Differential Game in Wheeled Mobile Robot Control

Zenon Hendzel and Paweł Penar(✉)

Rzeszow University of Technology, al. Powstańców Warszawy 12,
Rzeszów, Poland
{zenhen, ppenar}@prz.edu.pl

Abstract. Zero-sum differential game is a combination optimum control method and solution H_∞ control problem. Its solutions are based on Bellman's principle of optimality, for which the solution for nonlinear dynamic object is not available. In this case, approximation method based on actor-critic algorithms are used. One of the approximation method – SPIA [1] was applied in wheeled mobile robot tracking control problem and presented in the article. Numerical tests for the solution of the zero-sum differential game approximating algorithm were compared with the classical PD algorithm.

The obtained results confirm theoretical assumptions concerning the relationship between zero-sum differential game and H_∞ control problem.

Keywords: Differential game · Optimal control · Approximation dynamic programming

1 Introduction

Searching for optimal control algorithms is a trend of modern control theory [2, 3]. Such algorithms allow to define solution H_∞ control problem in the presence of the worst case disturbance. Determination of optimal control in the presence of the worst case disturbances is related to the zero-sum differential game solution [4, 5, 15]. In the zero-sum differential games, one of the players minimises, and the other maximises the defined value function. For a description of a dynamic systems in the state space, this is a task of dynamic optimization of the mini-max type [6]. Solution of zero-sum differential game is related to solution of nonlinear Hamilton-Jacobi-Issac (HJI) equation [4, 5, 7]. In order to solve HJI approximative equation, actor-critic algorithms are used. In this approach actor approximates the control signal, while critic influences actor behavior by approximation of value function. These are solutions that belong to the approximate dynamic programming [7]. Actor-critic algorithms are based on reinforcement learning methods [8]. The authors of many paper, i.e. [1, 9] based on Bellman approximate dynamic programming, propose many implementation of HJI equation approximative solution. Approximation structure of the presented method is executed by using neural network. In the paper [1], an SPIA (Synchronous Policy Iteration Algorithm) algorithms were proposed, which use the modified Levenberg-Marquardt method to teach neural network weights. In this case, critic weights adaptation rule of required knowledge on the internal dynamic of the object. Therefore, the

J. Świder et al. (Eds.): MECHATRONICS 2017, AISC 934, pp. 151–161, 2019.
https://doi.org/10.1007/978-3-030-15857-6_16

SPIA algorithm is a model-based method. In the paper [9], an algorithm is proposed which does not require complete knowledge on dynamics of the object. In this approach, critic weights adaptation rule which is more complex, because the algorithm refers to the data of previous state of a controlled dynamic object.

In this paper approximation algorithm was used for tracking control of the Wheeled Mobile Robot (WMR). In the theoretical part problems of zero-sum differential game, H_∞ control problem and the solution of the HJI equation approximating algorithm are presented. In the simulation part, numerical solution of the zero-sum differential game and H_∞ control problem were performed.

2 H_∞ Optimal Control. Zero-Sum Differential Game

A dynamic control object, in particular a mechanical object, can be described in the state space in the form [1, 5, 9]

$$\dot{x} = f(\boldsymbol{x}) + g(\boldsymbol{x})\boldsymbol{u} + \mathrm{k}(\boldsymbol{x})\mathbf{d} \tag{1}$$

where $\boldsymbol{x} = \boldsymbol{x}(t) = [x_1, x_2, \ldots, x_n]^T \in \boldsymbol{R}^n$ is a state space vector, $f(\boldsymbol{x}), g(\boldsymbol{x}), \mathrm{k}(\boldsymbol{x})$ are functions the state vector $\boldsymbol{x}$, $\boldsymbol{u} \in \boldsymbol{R}^{n_u}$ is a control and $\boldsymbol{d} \in \boldsymbol{R}^{n_d}$ is a disturbance. From Eq. (1) linked to the output

$$y = \boldsymbol{C}\boldsymbol{x} \tag{2}$$

where $\boldsymbol{C} \in \boldsymbol{R}^{nxn}$ is a constant project matrix.

The expression (1) can be reduced to a description of mechatronic object such as manipulation robot or wheeled mobile robots [10, 11].

2.1 L_2 - Gain. H_∞ Control Problem

For nonlinear system defines L_2- gain [4, 5]. For a dynamic object for (1) with output (2) L_2 - gain is less than or equal to γ if the inequality is satisfied

$$\frac{\int_0^\infty \boldsymbol{x}^T \boldsymbol{Q}\boldsymbol{x} + \boldsymbol{u}^T \boldsymbol{R}\boldsymbol{u}}{\int_0^\infty \boldsymbol{d}^T \boldsymbol{d}} \le \gamma^2 \tag{3}$$

where $\boldsymbol{Q} = \boldsymbol{C}^T\boldsymbol{C}$.

Using the L_2-gain concept, H_∞ stability problem can be defined, which is input-output stability. The solution of H_∞ control problem is reduced to the determination of such a course of control $\boldsymbol{u}^*$, disturbance $\boldsymbol{d}^*$ and gain $\gamma^* > 0$, for which the inequality $\gamma > \gamma^*$ is satisfied [4].

The H_∞ stability and control problem is connected in dissipative system theory [5, 12], which analyses the energy flow in the dynamic system. Energy in this theory is expressed by the storage function $V(\boldsymbol{x})$, which can be represented by

$$V(\boldsymbol{x}) = \int_0^\infty \boldsymbol{x}^T \boldsymbol{Q}\boldsymbol{x} + \boldsymbol{u}^T \boldsymbol{R}\boldsymbol{u} - \gamma^2 \boldsymbol{d}^T \boldsymbol{d} \tag{4}$$

The function $V(\boldsymbol{x})$ of the form (4) in the zero-sum differential game is called the value function. The value function (4) combines H_∞ control problem with the optimal control in the presence of the worst case disturbance [4].

2.2 Zero-Sum Differential Game

In a zero-sum differential game the control signal $\boldsymbol{u}$ plays a minimising role and disturbance signal $\boldsymbol{d}$ plays maximising role. The goal of the player is to optimize the value function $V(\boldsymbol{x})$ of the form (4). In the paper i.e. [6], the solution of zero-sum differential game are signals $\boldsymbol{u}^*$ and $\boldsymbol{d}^*$, for which the inequality

$$V(\boldsymbol{x}, \boldsymbol{u}^*, \boldsymbol{d}) \leq V(\boldsymbol{x}, \boldsymbol{u}^*, \boldsymbol{d}^*) \leq V(\boldsymbol{x}, \boldsymbol{u}, \boldsymbol{d}^*) \tag{5}$$

is satisfied. The control of the object (1) by the optimal pair of signals $(\boldsymbol{u}^*, \boldsymbol{d}^*)$ is optimal for the value function (4) when

$$0 = \min_u \max_d H(\boldsymbol{x}, \boldsymbol{u}, \boldsymbol{d}) = \min_d \max_u H(\boldsymbol{x}, \boldsymbol{u}, \boldsymbol{d}) \tag{6}$$

where $H(\boldsymbol{x}, \boldsymbol{u}, \boldsymbol{d})$ is Hamiltonian of the form

$$H(\boldsymbol{x}, \boldsymbol{u}, \boldsymbol{d}, V) = \boldsymbol{x}^T \boldsymbol{Q}\boldsymbol{x} + \boldsymbol{u}^T \boldsymbol{R}\boldsymbol{u} - \gamma^2 \boldsymbol{d}^T \boldsymbol{d} + \frac{\boldsymbol{dV}}{\boldsymbol{dx}} [f(\boldsymbol{x}) + g(\boldsymbol{x})\boldsymbol{u} + \mathrm{k}(\boldsymbol{x})\mathbf{d}] \tag{7}$$

In consequence of Eqs. (5–7) it follows that, optimal control of dynamic object of the form (1) with respect to the function (4), is defined as

$$\begin{cases} \boldsymbol{u}^* = -\frac{1}{2}\boldsymbol{R}^{-1} g^T(\boldsymbol{x}) \frac{dV^*}{dx} \\ \boldsymbol{d}^* = \frac{1}{2\gamma^2} k^T(\boldsymbol{x}) \frac{dV^*}{dx} \end{cases} \tag{8}$$

where V^* is a value function calculated by optimal pair of signals $(\boldsymbol{u}^*, \boldsymbol{d}^*)$. By inserting Eq. (8) into the Eq. (7) the following form is received [1, 9]

$$\boldsymbol{x}^T \boldsymbol{Q}\boldsymbol{x} + \frac{dV^*}{dx} f(\boldsymbol{x}) - \frac{1}{4}\left[\frac{dV^*}{dx}\right]^T g(\boldsymbol{x})\boldsymbol{R}^{-1} g^T(\boldsymbol{x}) \frac{dV^*}{dx} + \frac{1}{4\gamma^2}\left[\frac{dV^*}{dx}\right]^T k(\boldsymbol{x})k^T(\boldsymbol{x})\frac{dV^*}{dx} = 0 \tag{9}$$

Equation (9) is the HJI equation. Its solution is equivalent to the solution zero-sum differential game, which is correspondent to the solution of H_∞ control problem.

3 Approximation Solution

In the non-linear case, the analytical solution of the HJI equation is difficult or impossible, which determines the use of approximation algorithms. The authors of many paper, i.e. [1, 9, 13, 14] use adaptive dynamic programming method. These solutions are based on PI (Policy iteration) algorithm [7] and Eq. (7), which allows approximation solution of zero-sum differential game. It has disadvantage in the form of internal loop, which prevents its use online [1]. Therefore, in this area actor-critic algorithms [7] are used, which belong to the reinforcement learning algorithms. The task of the actor is to approximate pairs of signals $(\boldsymbol{u}^*, \boldsymbol{d}^*)$, and the task of critic is to approximate the value function and to modify the actor.

3.1 Approximation a Zero-Sum Differential Game

In the paper [1] an online solution zero-sum differential game has been proposed. The solution is to use the actor-critic structure in synchronously generating of approximation of the value function and the approximation of the minimizing and maximizing player signals approximation.

Approximation value function $V(\mathbf{x})$ can be implemented by neural network (NN). Then, approximation $\widehat{V}$ is in the form [1, 7]

$$\widehat{V}(\boldsymbol{x}) = \widehat{\boldsymbol{W}}^T \psi(\boldsymbol{x}) \tag{10}$$

where $\widehat{\boldsymbol{W}}$ is an approximation of the unknown weights of NN critic and $\psi(\boldsymbol{x}) = [\psi_1, \psi_2, \ldots, \psi_N]^{\boldsymbol{T}}$ is a basic vector for function connected with neurons.

Assuming knowledge of the dynamics of the control object (1) and considering Eq. (10), approximations pairs of signals $(\boldsymbol{u}^*, \boldsymbol{d}^*)$, Eq. (8) can be expressed as

$$\begin{cases} \widehat{\boldsymbol{u}} = -\frac{1}{2}\boldsymbol{R}^{-1} g^T(\boldsymbol{x}) \left[\frac{d\psi(x)}{dx}\right]^T \widehat{\boldsymbol{W}}_1 \\ \widehat{\boldsymbol{d}} = \frac{1}{2\gamma^2} k^T(\boldsymbol{x}) \left[\frac{d\psi(x)}{dx}\right]^T \widehat{\boldsymbol{W}}_1 \end{cases} \tag{11}$$

From Sect. 2.2 it is known that the solution of the zero-sum differential game satisfies Eq. (7). The value function reaches the lowest value when

$$\widehat{H}\left(\boldsymbol{x}, \widehat{\boldsymbol{u}}, \widehat{\boldsymbol{d}}, \widehat{V}\right) = \boldsymbol{x}^T \boldsymbol{Q} \boldsymbol{x} + \widehat{\boldsymbol{u}}^T \boldsymbol{R} \widehat{\boldsymbol{u}} - \gamma^2 \widehat{\boldsymbol{d}}^T \widehat{\boldsymbol{d}} + \widehat{\boldsymbol{W}}_1 \frac{d\psi(\boldsymbol{x})}{d\boldsymbol{x}} \left[f(\boldsymbol{x}) + g(\boldsymbol{x})\widehat{\boldsymbol{u}} + \mathrm{k}(\boldsymbol{x})\widehat{\mathbf{d}}\right] \tag{12}$$

equals zero. Therefore the task of the NN approximating value function is to minimise an error defined as

$$E = \frac{1}{2}\widehat{H}^2 \tag{13}$$

Using modified Levenberg-Marquard method, the authors of the paper [1] have proposed weight adaptation rule

$$\dot{\widehat{W}}_1 = -\eta \frac{\widehat{\sigma}_1}{\left(1+\widehat{\sigma}_1^T\widehat{\sigma}_1\right)^T}\left[\sigma_1^T\widehat{W}_1 + x^TQx + \widehat{u}^TR\widehat{u} - \gamma^2\widehat{d}^T\widehat{d}\right] \tag{14}$$

where $\widehat{\sigma}_1 = \left[\frac{d\psi(x)}{dx}\right]\left(f(x) + g(x)\widehat{u} + \mathrm{k}(x)\widehat{\mathbf{d}}\right)$ and η is learning rate.

4 Wheeled Mobile Robot

Figure 1 shows kinematics scheme of the wheeled mobile robot.

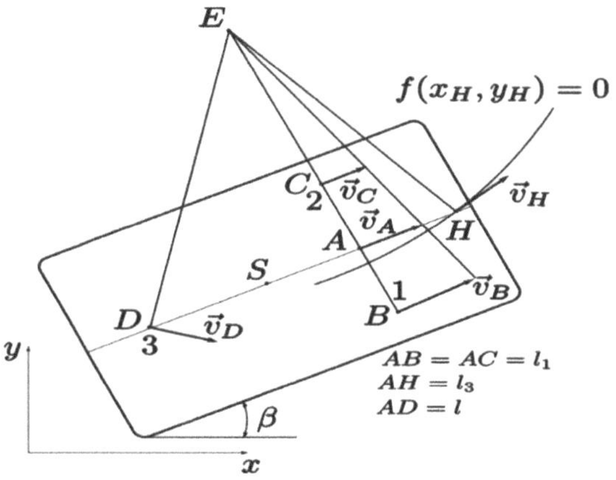

Fig. 1. WMR kinematics scheme.

Driving wheels were marked by 1 and 2. The angle of a self-turn the drive wheels form a vector generalized coordinates $\boldsymbol{q} = [q_1, q_2]^T$. In addition, Fig. 1 indicates: 3 – which is a free rolling wheel, S – which is a centre of mass and BC i AH which are lengths respectively $2l_1$ and l_3.

4.1 WMR Kinematics

In the paper [10] WMR kinematics is described. WMR is nonholonomic system, which means that

$$\dot{y}_A = \dot{x}_A tg\beta \tag{15}$$

where $\dot{x}_A, \dot{y}_A$ are velocity vector $\vec{v}_A$ projections.

Figure 1 shows that

$$\begin{aligned} x_H &= x_A + l_3 cos\beta \\ y_H &= y_A + l_3 sin\beta \end{aligned} \tag{16}$$

hence

$$\begin{aligned} \dot{x}_H &= v_A cos\beta - l_3\dot{\beta} sin\beta \\ \dot{y}_H &= v_A sin\beta + l_3\dot{\beta} cos\beta \end{aligned} \tag{17}$$

where $\dot{x}_A = v_A cos\beta, \dot{y}_A = v_A sin\beta$.

The characteristic point H, belonging to the WMR frame, along a path defined by function f, which after differentiation follows a form

$$\dot{f}(x_H, y_H) = 0 \tag{18}$$

Assuming the path of movement the course of function x_H, y_H, β can be determined during the time which allows to determine the angle of a self-turn of drive wheels

$$\dot{q}_1 = \frac{v_A}{r} + \dot{\beta}\frac{l_1}{r}, \dot{q}_2 = \frac{v_A}{r} - \dot{\beta}\frac{l_1}{r} \tag{19}$$

where r is the radius of the wheels. Finally, the angular parameters $q = [q_1, q_2]^T$ and their derivatives will form the desired trajectory of movement for the control system.

4.2 WMR Dynamics

The dynamical equation of the WMR [10] can be expressed as

$$\boldsymbol{M}\ddot{\mathbf{q}} + \boldsymbol{C}(\dot{\boldsymbol{q}})\dot{\boldsymbol{q}} + \boldsymbol{F}(\dot{\boldsymbol{q}}) + \boldsymbol{\tau} = \boldsymbol{U} - \boldsymbol{D} \tag{20}$$

where

$$\mathbf{M} = \begin{bmatrix} a_1 + a_2 + a_3 & a_1 - a_2 \\ a_1 - a_2 & a_1 + a_2 + a_3 \end{bmatrix}, \boldsymbol{C}(\dot{\boldsymbol{q}}) = \begin{bmatrix} 0 & 2a_4(\dot{q}_2 - \dot{q}_1) \\ -2a_4(\dot{q}_2 - \dot{q}_1) & 0 \end{bmatrix} \tag{21}$$

where $a_1 - a_6$ is a parameter based on geometry, masses and mass moments of inertia WMR. Control vector $\boldsymbol{U} = [M_1, M_2]^T$ presents torques that drive the robot wheels. Vector $\boldsymbol{F}(\dot{\boldsymbol{q}}) = [a_5 sgn(\dot{q}_1), a_6 sgn(\dot{q}_1)]^T$ is the friction of roller-wheel rolling (in numerical test, function sgn is approximation by sigmoid function), $\boldsymbol{\tau}$ presents external disturbances and $\boldsymbol{D}$ donates the vector of parameter inaccurancies. Disturbance $\boldsymbol{D}$ is associated with the description of motor resistance, in particular the friction of roller-wheel rolling.

Control error is defined as $\boldsymbol{e} = \boldsymbol{q}_d - \boldsymbol{q}$ where $\boldsymbol{q}_d$ is desired trajectory and generalised error

$$\boldsymbol{s} = \dot{\boldsymbol{e}} + \boldsymbol{\lambda e} \tag{22}$$

where $\boldsymbol{\lambda} > 0$ is the design matrix, the tracking control problem can be transformed into a problem stabilization $\boldsymbol{s}$ around zero.

Using the Eq. (22) the dynamics of WMR can be presented in the generalised error space. Hence, the following was obtained

$$\boldsymbol{M}\underline{\boldsymbol{s}} = -\boldsymbol{C}(\dot{\boldsymbol{q}})\boldsymbol{s} + f(\boldsymbol{x}) - \boldsymbol{U} + \boldsymbol{D} \tag{23}$$

where $f(\boldsymbol{x}) = \boldsymbol{M}[\ddot{\boldsymbol{q}}_d + \boldsymbol{\lambda}\dot{\boldsymbol{e}}] + \boldsymbol{C}(\dot{\boldsymbol{q}})[\dot{\boldsymbol{q}}_d + \boldsymbol{\lambda}\boldsymbol{e}] + \boldsymbol{F}(\dot{\boldsymbol{q}})$ contains the nonlinearity of a robot.

5 Numerical Simulation

In this point using zero-sum differential game, numerical simulations of WMR tracking control with assumed disturbance will be presented, for which the solution will be approximated. Obtained results will be compared to results obtained from control of the PD controller.

5.1 Desired Trajectory

Assuming the velocity profile $\mathrm{v_A(t)}$ [10] as

$$v_A(t) = v_A^*\left(\frac{1}{1+e^{-c(t-b_1)}} + \frac{1}{1+e^{-c(t-b_2)}}\right) \tag{24}$$

where v_A^* is a point A velocity in steady state, $\mathrm{b_1, b_2, c}$ are a project coefficient. Assuming path of the motion in a form of function

$$f(x_H, y_H) = \left(\frac{x_H - a}{a}\right)^4 + \left(\frac{y_H}{b}\right)^4 - 1 = 0 \tag{25}$$

and assuming $v_A^* = 0.425\left[\frac{\mathrm{m}}{\mathrm{s}}\right]$, $\mathrm{b_1} = 5$, $\mathrm{b_2} = 38$, $\mathrm{l_1} = 0.163[\mathrm{m}]$, $\mathrm{l_2} = 0.2[\mathrm{m}]$, $\mathrm{r} = 0.825[\mathrm{m}], \mathrm{a} = \mathrm{b} = 4, \mathrm{c} = 5, \mathrm{t} \in (0, 45)$, WMR kinematic parameter was determined.

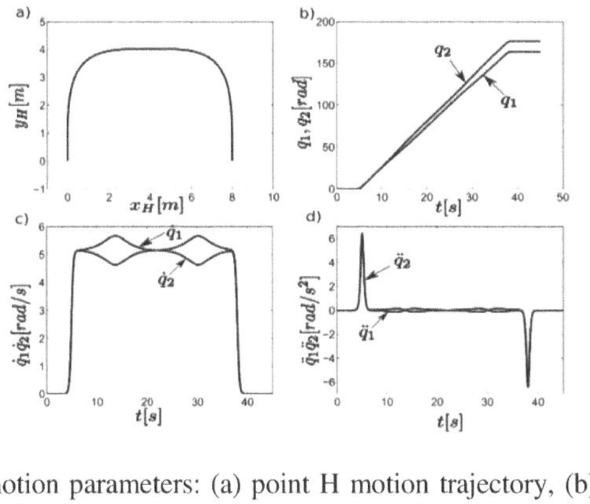

Fig. 2. Determined motion parameters: (a) point H motion trajectory, (b) course of revolution angle (b), angular velocity (c) and angular accelerations (d) of driving wheels of WMR.

Figure 2b–d shows graphs of kinematic parameters of drive wheels of WMR, which characteristic point H accomplishes the trajectory shown in Fig. 2a. Such define kinematic parameters of driving wheels will state the desired motion trajectory for the control system.

5.2 Simulation of Differential Game

At this point the approximation algorithm described in point 3.1 is used. Knowing the dynamic equations of the WMR motion in the generalized error space, we used the SPIA algorithm to solve the approximation of the zero-sum differential game.

Selecting desired trajectory from point 5.1, values of parameters $a_1 = 0.0265$, $a_2 = 0.0666$, $a_3 = 0.007$, $a_4 = 0.002$, $a_5 = a_6 = 2.411$ oraz $\boldsymbol{q}(0) = \dot{\boldsymbol{q}}(0) = [0,0]^T$, $\boldsymbol{R} = 0.5\boldsymbol{I}$, $\boldsymbol{Q} = 270\boldsymbol{I}$, $\widehat{\boldsymbol{W}}(0) = 3.5\boldsymbol{I}_{3x1}$, $\eta = 0.03$, $\gamma = 4.25$ $\psi(\boldsymbol{x}) = \left[s_1^2, s_1 s_2, s_2^2\right]^T$ performed numerical test of approximation algorithms.

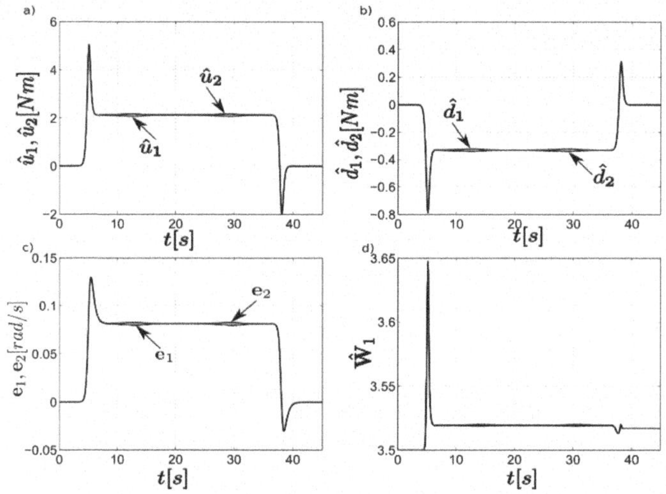

Fig. 3. The simulation results for the SPIA algorithm: (a) Behaviour of $\widehat{\boldsymbol{u}}$ which is the minimising player (b) behaviour of maximising player signal $\widehat{\boldsymbol{d}}$, (c) behaviour of the generalized error s; (d) behaviour critic weights $\widehat{\boldsymbol{W}}$.

Figure 3 shows the result of numerical simulation, which uses an algorithm SPIA in the tracking control. Figure 3a shows the control signal $\widehat{\boldsymbol{u}}$, which behaviour is the response to the worst case disturbance, i.e. the signal $\widehat{\boldsymbol{d}}$ shown in Fig. 4b. During the simulation, the weight $\widehat{\boldsymbol{W}}$ is adopted using Eq. (14), as shown in Fig. 3d. During the numerical experiment, the error $\boldsymbol{e}$ is bounded (Fig. 3c).

The final values of the weights $\widehat{\boldsymbol{W}}$ obtained from the zero-sum differential simulation will be used to solve the H_∞ control problem.

5.3 Simulation H_∞ Control Problem

By using NN weight values determined for approximation algorithm, the H_∞ control problem was solved. In order to verify it, disturbances of motion resistance in the form of changing the rolling friction of wheels by adding parametric disturbances in the form $D = [\Delta a_5, \Delta a_6]^T$ where $\Delta a_5 = \Delta a_6 = \begin{cases} 0, t < 15 \\ 1, t \geq 15 \end{cases}$.

At $t \in [15, 45]$, there has been an increased amplitude of the control signal $\hat{\boldsymbol{u}}$ (Fig. 4a), which is a response to added disturbance. The error $\boldsymbol{e}$ and $\dot{\boldsymbol{e}}$ obtained in the test is shown in Fig. 4b and c respectively. As shown in Fig. 4d, the condition $H_\infty(\gamma > \gamma^*)$ is satisfied. This means that the control system is not sensitive to disturbance.

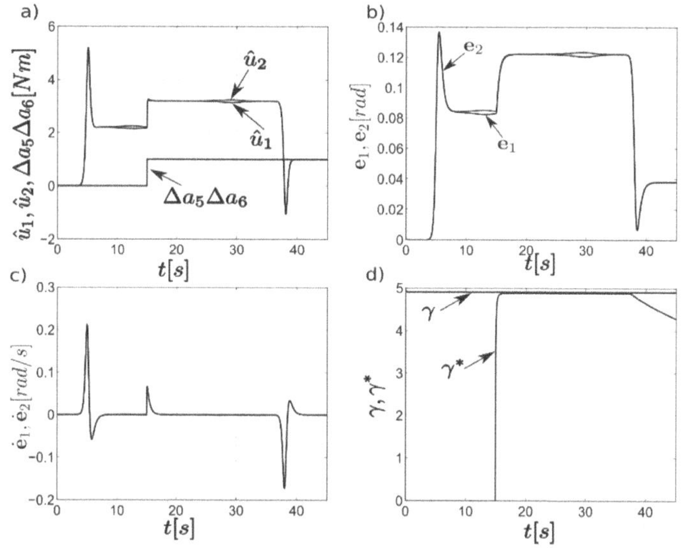

Fig. 4. The simulation results for H_∞ control problem (SPIA algorithm): (a) behaviour of control problem; behaviour of error $\boldsymbol{e}$ (b) end $\dot{\boldsymbol{e}}$ (c), behaviour of γ, γ^* gain coefficient (d)

PD control simulation was used to assess quality control procedure generated by the approximation algorithm. The following quality indicators were used to compare the result:

- module of the maximum errors: $|e_{max,k}|, |\dot{e}_{max,k}|, |s_{max,k}|, k = 1, 2$
- mean square errors: $\varepsilon_k = \sqrt{\frac{1}{n}\sum_{i=1}^{n}(e_{i,k})^2}, \dot{\varepsilon}_k = \sqrt{\frac{1}{n}\sum_{i=1}^{n}(\dot{e}_{i,k})^2}$, $\sigma_k = \sqrt{\frac{1}{n}\sum_{i=1}^{n}(s_{i,k})^2}, k = 1, 2.$

where k is the number drive wheels and n is the sample number. The values of the selected quality indicators for PD ($\boldsymbol{u}_{PD} = \boldsymbol{K}_d \boldsymbol{s}, \boldsymbol{K}_d = \boldsymbol{I}$) and SPIA algorithms summarised in Table 1.

Table 1. Values of the selected quality indicators for SPIA algorithm and PD regulator

		$\lvert e_{max,k} \rvert$	$\lvert \dot{e}_{max,k} \rvert$	$\lvert s_{max,k} \rvert$	ε_k	$\dot{\varepsilon}_k$	σ_k
PD	$k = 1$	1.7731	1.9848	4.3154	1.2271	0.3375	2.4768
	$k = 2$	1.7731	1.9849	4.3161	1.2451	0.3377	2.5126
SPIA	$k = 1$	0.1367	0.2125	0.3963	0.0969	0.0292	0.1961
	$k = 2$	0.1367	0.2125	0.3963	0.0976	0.0292	0.1380

Based on the value of the quality indicators shown in Table 1, it can be concluded that the control quality is better for the SPIA procedure. This is due to the fact that the PD algorithm is dedicated to linear objects and has a non-zero tracking error. The second advantage of using the SPIA procedure is that it is suboptimal to the value function of the form (4).

6 Summary

The paper presents the example zero-sum differential game application in wheel mobile robot control. This approach is equivalent to solving the H_∞ control problem. The satisfied condition $\gamma > \gamma^*$, is equivalent to the fact that the designed control system is stable in the L_2-gain sense [4]. The article compares results obtained by the approximation of the solution of the equation for HJI and the controller PD. The verification H_∞ control problem confirms the condition $\gamma > \gamma^*$. The obtained results confirm the theoretical considerations concerning the stability of the HJI equation solution.

References

1. Vamvoudakis, K.G., Lewis, F.L.: Online solution of nonlinear two-player zero-sum games using synchronous policy iteration. Int. J. Robust Nonlinear Control **22**, 1460–1483 (2012)
2. Vamvoudakis, K.G., Lewis, F.L.: Online actor-critic algorithm to solve the continuous-time infinite horizon optimal control problem. Automatica **46**(5), 878–888 (2010)
3. Zhu, Y., et al.: Iterative adaptive dynamic programming for solving unknown nonlinear zero-sum game based on online data, pp. 1–12 (2016)
4. Van der Schaft, A.J.: L2-gain analysis of nonlinear systems and nonlinear state feedback H-infinity Control. IEEE Trans. Autom. Control **37**(6), 770–784 (1992)
5. Abu-Khalaf, M., et al.: Nonlinear H2/H-Infinity Constrained Feedback Control. Springer, London (2006)
6. Starr, A.W., Ho, Y.C.: Nonzero-sum differential games. J. Optim. Theory Appl. **3**(3), 184–206 (1969)
7. Wang, F.-Y., Zhang, H., Liu, D.: Adaptive dynamic programming: an introduction. IEEE Comput. Intell. Mag. **4**(May), 39–47 (2009)
8. Sutton, R.S., Barto, A.G.: Reinforcement Learning: An Introduction. The MIT Press, Cambridge (1998)
9. Yasini, S., et al.: Online concurrent reinforcement learning algorithm to solve two-player zero-sum games for partially unknown nonlinear continuous-time systems. Int. J. Adapt. Control Signal Process. **22**(4), 325–343 (2008)

10. Żylski, W.: Kinematic and Dynamic of a Wheeled Mobile Robot (in Polish). Rzeszow University of Technology, Rzeszow (1996)
11. Gierlak, P., Hendzel, Z.: Control of a Wheel and Manipulation Robot (in Polish) (2011)
12. Brogliato, B., et al.: Dissipative Systems Analysis and Control. Springer, London (2007)
13. Wang, D., et al.: Event-based input-constrained nonlinear H-infinity state feedback with adaptive critic and neural implementation. Neurocomputing **214**, 848–856 (2016)
14. Yasini, S., et al.: Approximate dynamic programming for two-player zero-sum game related to H-infinity control of unknown nonlinear continuous-time systems. Int. J. Control Autom. Syst. **13**(1), 99–109 (2014)
15. Hendzel, Z., Penar, P.: Application of differential games in mechatronic control system. Int. J. Appl. Mech. Eng. **21**(4), 867–878 (2016)

Numerical Analysis of the Dynamic Impact of a Gun Barrel During Firing

Krzysztof Jamroziak[1(✉)], Miroslaw Bocian[1], Dariusz Pyka[1], and Maciej Kulisiewicz[2]

[1] Department of Mechanics, Materials Science and Engineering, Wroclaw University of Science and Technology, Smoluchowskiego 25, 50370 Wroclaw, Poland
{krzysztof.jamroziak,miroslaw.bocian, dariusz.pyka}@pwr.edu.pl

[2] Faculty of Technology and Engineering, Wroclaw University of Science and Technology, Armii Krajowej 78, 58302 Walbrzych, Poland
maciej.kulisiewicz@pwr.edu.pl

Abstract. The phenomena found in the UKM-2000P machine gun barrel during the firing were investigated. The aim was to evaluate how the change in the center of the mass of the projectile being fired affects the occurring barrel vibrations. Numerical simulations were performed in the ABAQUS program, and then the results were compared with the experimental results. The relationship between the vibrations caused by the mass movement in the barrel and the direction of the projectile was determined.

Keywords: Dynamic impact · Gun barrel · Barrel vibration

1 Introduction

The phenomenon of gun barrel vibrations is an important issue that must be emphasized in order to optimize the construction of a barrel. The accuracy of a gun barrel affects several matters, such as pressure changes occurring in terms of the firing phenomenon behind a moving projectile, causing vibrations and transferring them also on the mass of the projectile by impacting its rear surface [1]. In this way, the resulting vibrations are one of the factors that determine the natural gun dispersion and the nutational movement of the projectile axis, which translates into the accuracy of the weapon. Hence, the rigidity of the barrel is the overall rigidity of the whole launching system since unsuitable barrel stiffness will result in the increase in the vibrations manifesting themselves in bending the barrel end due to wave stresses generating the bore's displacement because of the point of impact.

Therefore, in setting out to design new models of weapons, the analysis of the influence of pressure changes on gun barrel vibrations is becoming crucial when it comes to issues related to the adoption of such a mathematical model that is further numerically and experimentally verified [2, 3].

J. Świder et al. (Eds.): MECHATRONICS 2017, AISC 934, pp. 162–174, 2019.
https://doi.org/10.1007/978-3-030-15857-6_17

Complex stress states (longitudinal, radial, tangential and lateral) occurring in the barrel directly affect the accuracy of fire - dispersion of shots at a given target. When the stresses resulting from the expansion of gases and the shift of the in-bore projectile are transferred to the thread roots they cause the variable deformation distribution in the material. The remaining part of the present work focuses on the above issue by performing the overall analyses followed by experimental and numerical studies to determine the relevant characteristics of the newly developed machine gun.

2 Analytical Approach to the Issue

The purpose of the analytical calculations is to provide a preliminary estimate of the intrinsic frequency of transverse barrel vibrations. The transverse vibrations of the barrel determine the projectile dispersion zone size since their basic form is characterized by the node at the cartridge chamber and the deflection arrow at the end of the barrel [4]. Therefore, the smaller the amplitude of the arrow at the end of the barrel, the smaller the dispersion. In addition, the accuracy of the projectile is also influenced by the time of projectile movement in the barrel that should be selected in such a way that, when the projectile is leaving the barrel the end of the barrel is in the extreme maximum (bottom or top) position. In these positions, the vertical components of the initial projectile velocity are close to zero. Dependencies on the basic frequency of the barrel vibrations were developed in the middle of the 20th century [5–7]. These patterns have still been used in the design of new small arms. They were derived based on the equation for stiffness of the cantilever beam (the beam is rigidly attached on one side), which is known as [2, 4, 8]:

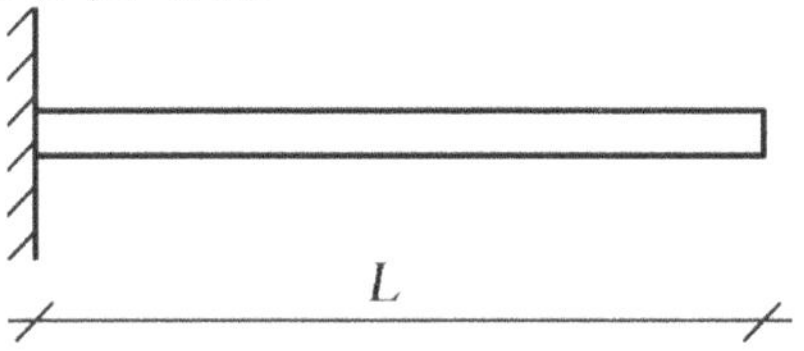

$$\Rightarrow k = \frac{3EJ_x}{L^3} \tag{1}$$

where: E - Young's modulus of beam material, J_x - moment of inertia of beam section, L - beam length.

If we assume that for the first form of vibrations the mass of the moving beam is about 0.25 of the whole beam mass m, the pattern for the fundamental vibration frequency takes the shape:

$$\omega^2 \cong \frac{k}{m_{zr}} = \frac{k}{0.25m} = \frac{4k}{m} = \frac{12EJ_x}{mL^3} = \frac{12EJ_x}{\rho \cdot \Omega L^4} \tag{2}$$

where: ρ - material density, Ω - cross-sectional area.

Then, the so-called A.A. Blagonavrov's formula is obtained in the form:

$$\omega = \frac{\beta^2}{L^2}\sqrt{\frac{EJ_x}{\rho \cdot \Omega}} \tag{3}$$

where: factor (exactly $\beta = 1.875$ is generally assumed as the effect of experimental studies).

When the beam is considered as the mass of variable external radius, the following formula is applied:

$$\omega \cong \omega_1 = \left(0.81\sqrt{R_1^2 + r_1^2} - 0.5302\sqrt{R_0^2 + r_1^2}\right)\frac{2\pi \cdot c_z}{l^2} \tag{4}$$

where: R_1 - outer radius of the barrel muzzle, R_0 - outer radius of the muzzle, r_1 - radius of inside diameter of the barrel, c_z - velocity of propagation of the transverse vibrations, l - height of the barrel taper.

Dependencies (2) and (4) are approximate and are used to estimate the basic form of vibrations. Since the analytical approach to the analysis of the phenomenon can be based on certain models of mechanics and the theory of elasticity of homogeneous bodies, it is necessary to use numerical technique grounded in the finite element method, which is the subject of further part of this work.

3 Subject and Purpose of Research

The subject of research is the barrel of the UKM-2000P machine gun. This gun was designed by the Polish armaments industry for the standard 7.62 × 51 mm cartridge used in the NATO countries (Fig. 1).

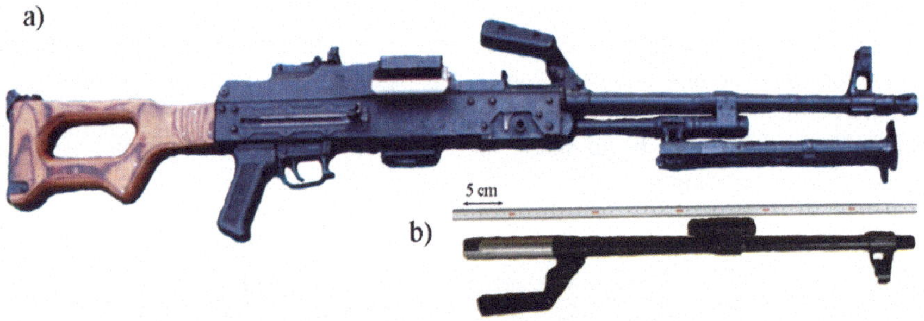

Fig. 1. General view: (a) UKM-2000P machine gun, (b) gun barrel.

The purpose of the work was to analyze the influence of changes in the position of the Ball type projectile on the transverse vibrations of the barrel by means of the developed numerical model using the ABAQUS program.

4 Experimental Studies

The scope of the experimental work was carried out in two stages. In the first stage, numerous measurements were made on the Prototype ballistic barrel using the EPVAT test barrel for caliber 7.62 × 51 mm with a length of 562 mm and a thread pitch of 304.8 mm. The ballistic barrel was equipped with two sensors, one of which was mounted in the chamber at the distance of 54 mm from the base (the beginning of the cartridge chamber) of the barrel (Sensor A) and the other was mounted at the distance of 381 mm (Sensor B). Kistler 6215 sensors were used in the pressure measurement and the signal was recorded using the Kistler Transient Recorder Type 2519A with appropriate software. The ammunition produced by "MESKO" was used for the test. The 7.62 × 51 mm cartridge has the following specifications: the cartridge weight of 23.2–24.5 g, the FMJ type projectile weight of 9.35–9.55 g, where the projectile is composed of the following elements: a brass jacket and a lead core with antimony added; the propellant charge weight is of 3 g and the average velocity of the projectile is of $v_{25} = 825 \pm 10$ m/s.

The test was conducted on the outdoor shooting range for external conditions: $ta = 18$ °C, $pa = 1050$ hPa and humidity of 75%. The recorded pressure measurements gave the following results, which are summarized in Fig. 2.

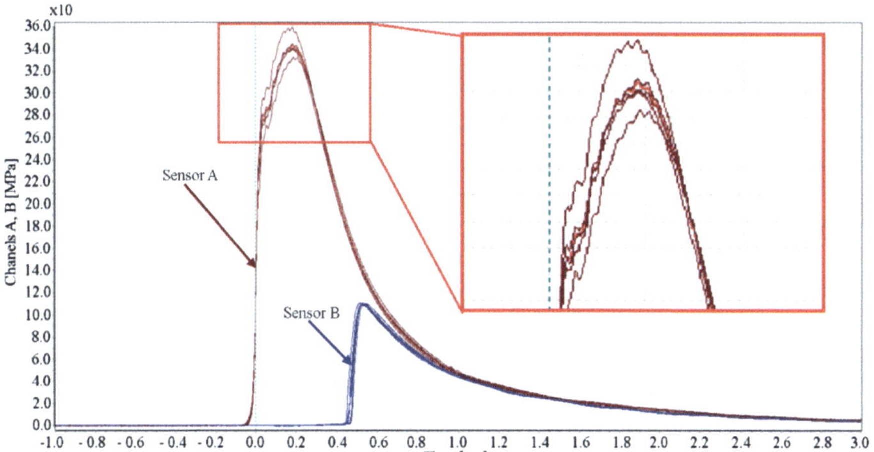

Fig. 2. Pressure distribution in the barrel for five shots with the 7.62 × 51 mm Ball cartridge of MESKO production.

The average result obtained for the sensor A is 343.5 MPa and for the sensor B – 110.1 MPa. According to the catalog data, the maximum average pressure of the MESKO 7.62 × 52 mm Ball cartridge is 365 MPa, i.e. the difference is of 6%.

The second stage of the study involved the measurement of pressure variations on the research facility that is the UKM-2000P machine gun with a length of 547 mm and a 328 mm thread pitch in similar ambient conditions. In this case, the research was limited to measuring the pressure at the barrel muzzle due to the impossibility of interfering with the structure. The MHP 050 piezoelectric muzzle sensor was used. The exemplary recorded result is depicted in Fig. 3. The estimated pressure of exhaust gases is approximately of 40.5 MPa. In this way, the parameters measured in the experiment constituted the basis for further work on the development of geometrical and numerical models of the analyzed barrel of the UKM-2000P machine gun.

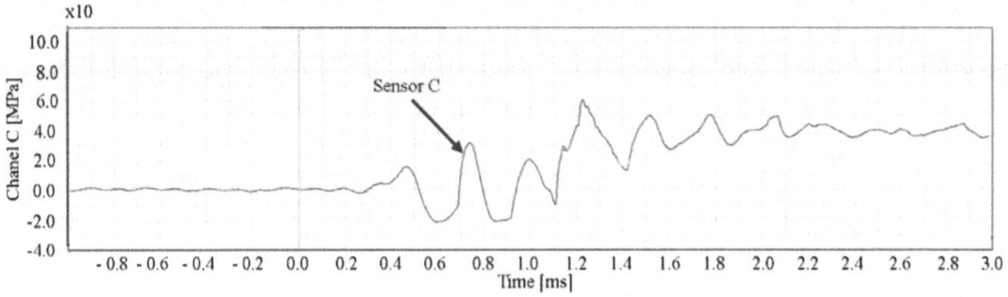

Fig. 3. Pressure distribution in the barrel of the UKM 2000P machine gun. The sensor C was mounted on the barrel muzzle.

5 Numerical Analysis FEM

5.1 Assumptions for Construction of Models

The geometric model of the UKM-2000P machine gun barrel was created for the purpose of this work. Two variants of the barrel were adopted: Model 1 was the bore without its components, Model 2 was the barrel with all its fixture, i.e. foresight, gas line, piston with piston rod and flash suppressor (Fig. 4).

For the analyses carried out, it was necessary to obtain a parametric spatial CAD model including data on the shape. To this end, the 3D scanning method was adopted using the linear RS3 laser scanner integrated with the measuring arm. The arm with the integrated RS3 scanner and Real Time Quality Meshing technology (real-time creation of triangle meshes from point clouds) allows up to 460,000 points per second to be taken for 4600 points on the line with the linear frequency of 100 Hz with declared accuracy at the level of 30 microns at 2-sigma. Based on the above, geometric models were developed relatively accurately, with particular regard to the cartridge chamber (Fig. 5).

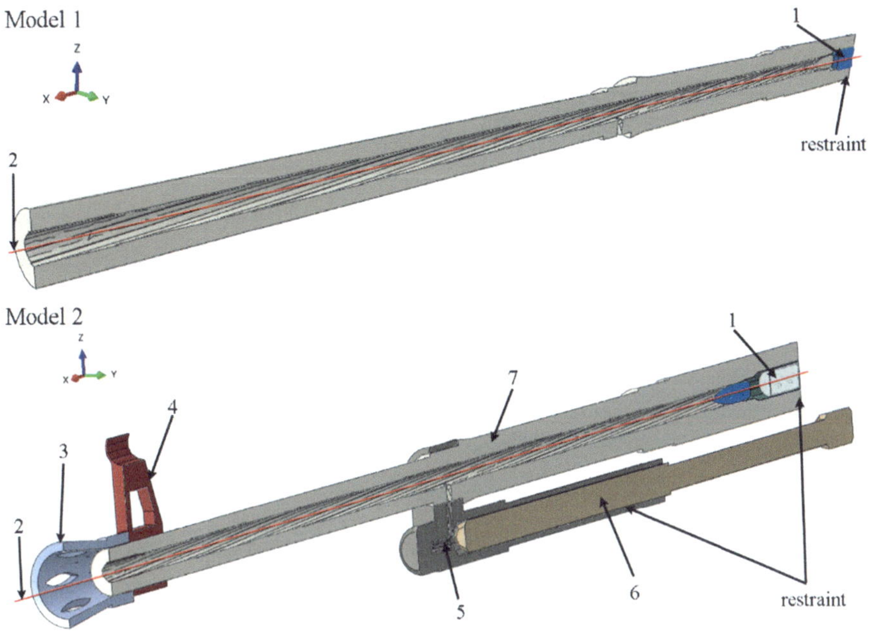

Fig. 4. Barrel geometric models: 1 - projectile, 2 - axis of the barrel line, 3 - silencer, 4 - rear sight, 5 - gas chamber, 6 - piston with piston rod, 7 - barrel.

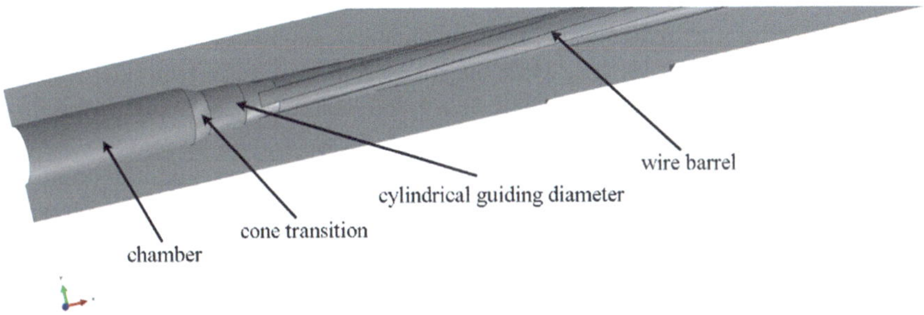

Fig. 5. The geometric model of the chamber.

Thereafter, the barrel line (bore) was modeled along with the projectile using the finite element method (FEM) as shown in Fig. 6.

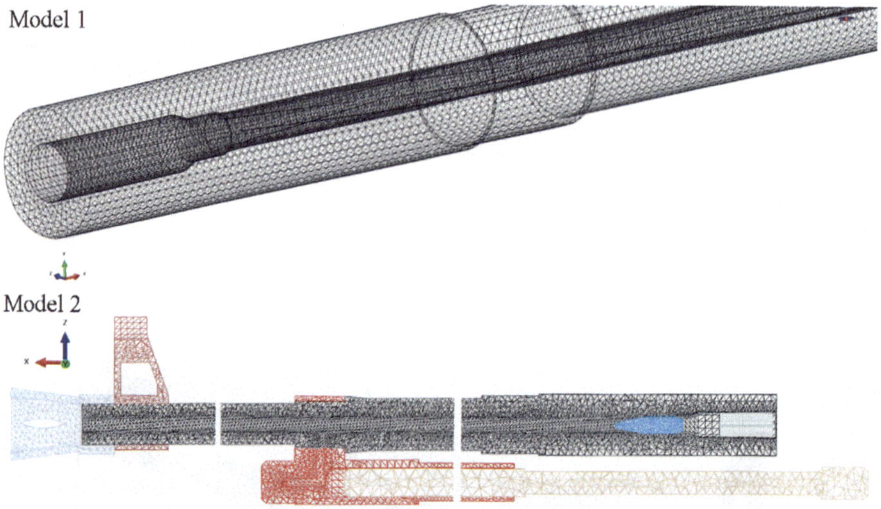

Fig. 6. The numerical model of the barrel and the barrel with its fixture.

The propellant was modeled as a solid element by means of the smoothed particle hydrodynamics (SPH) for time $t = 0$. At the initial moment, the examined body was divided into areas, each of which was replaced with a material particle.

5.2 Simulation and Results

The description of the material in the mathematical simulation was based on the Johnson-Cook (JC) model [9, 10], while the Jones-Wilkins-Le (JWL) model was used to describe the cartridge in the pyrostatic and pyrodynamic way [11, 12]. Appropriate parameters for the material of the barrel as well as the firing process were established according to the methodology employed in [10, 12–15]. The ignition point was determined by the coordinates in the generalized system $X = Y = Z = 0$, which was to simulate the ignition induced by a primer. The aim of the simulations was to obtain the following results. Figure 7 shows an example of the propagation of stress in the cartridge chamber caused by the combustion of the gunpowder spread over time. In the simulation, the air element in the bore was omitted.

The remaining residual stresses in the cartridge chamber for Model 1 were obtained at 363 MPa (Fig. 8). At the same time, the load on the protruding parts of the thread pitch is visible through the pressure on the projectile jacket.

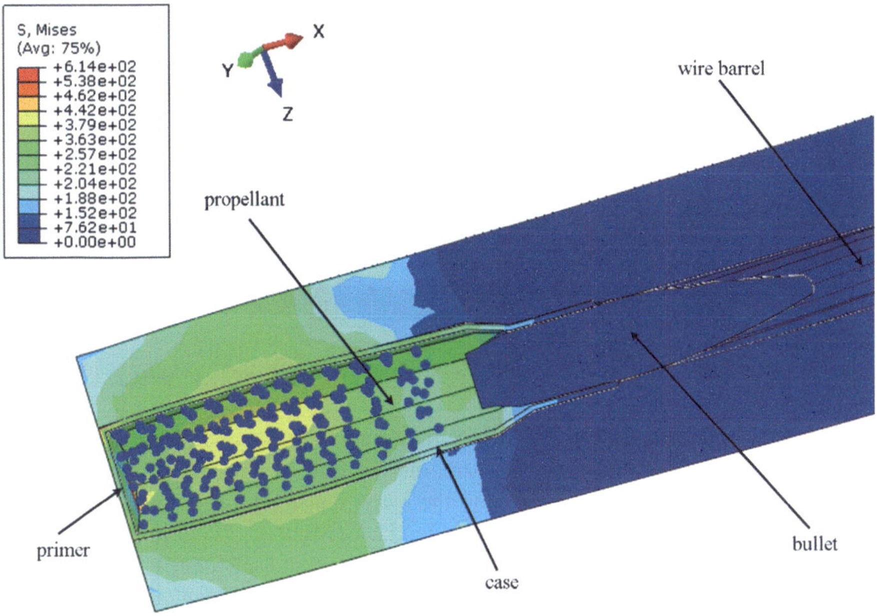

Fig. 7. Stress distribution in the chamber.

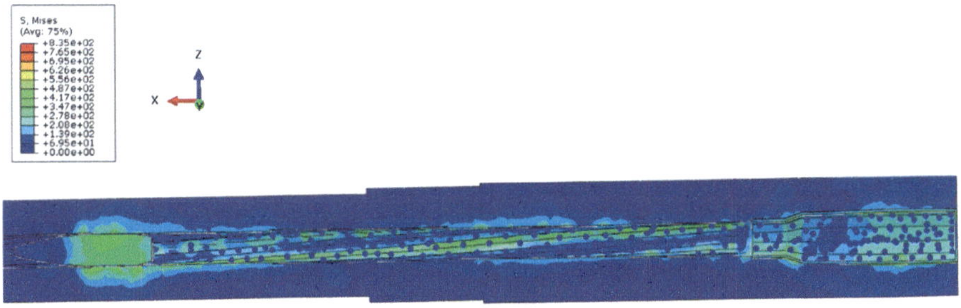

Fig. 8. Stress distribution in the barrel line for the time t = 0.005 s (Model 1).

For Model 2 the residual stresses were slightly higher, i.e. at the level of about 380 MPa (Fig. 9). This is understandable since the analyzed system in Model 2 is rigid due to the fact that the gas chamber together with the piston and piston rod, as part of the system ensuring the firearm action, stiffens the barrel.

Fig. 9. Stress distribution in the bar for the time $t = 0.005$ s (Model 2).

The range of vibrations transmitted by the changing pressure of expanding powder gases in the process of firing was analyzed on the basis of phase diagrams expressed through the energy that was emitted onto the barrel construction. The example results are summarized in Fig. 10 for Model 1 and Fig. 11 for Model 2. According to the results shown, the vibration range for Model 2 is twice lower than for Model 1. This indicates that the barrel must be modeled with all its assembly since a significant dispersion of test results can be obtained and this leads to erroneous conclusions as to the results in the process of numerical modeling of the impact of vibrations on the barrel.

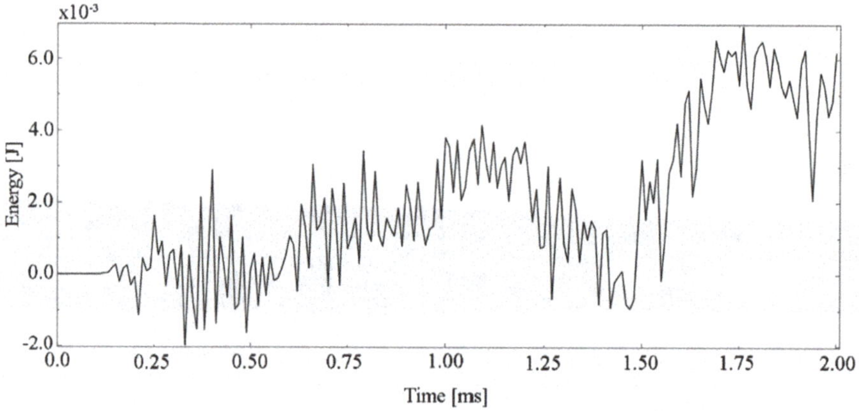

Fig. 10. The vibration energy for the barrel.

The effect of residual pressure on the bottom part of the projectile and the muzzle proved to be an important undertaking to illustrate the results (Fig. 12). During the simulation of the combustion process using the propellant with the SPH method applied, the influence of the number of free elements on the pressure in the barrel line was investigated. After tripling the amount of SPH that the propellant consisted of, the increase of stresses from 363 MPa to 365 MPa with more local nature was observed – a greater concentration of stressed appeared in the middle of the cartridge chamber and some of the stresses moved to the neck of the cartridge chamber (Fig. 9).

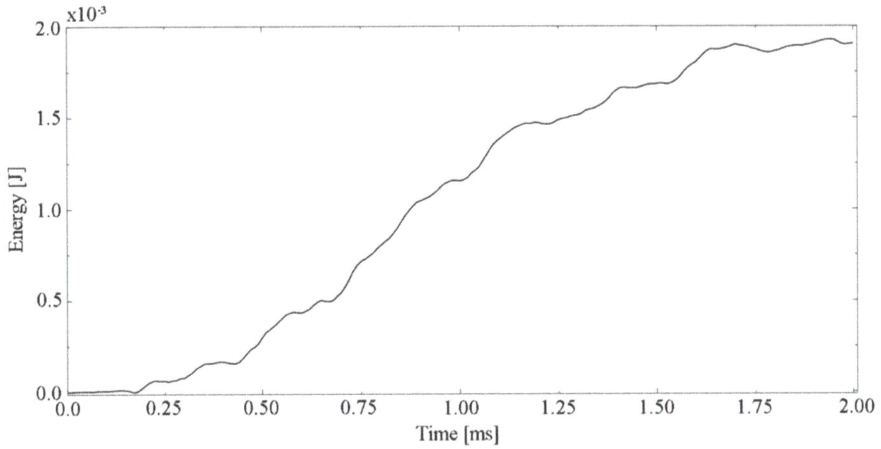

Fig. 11. The vibration energy for the barrel with the gas chamber.

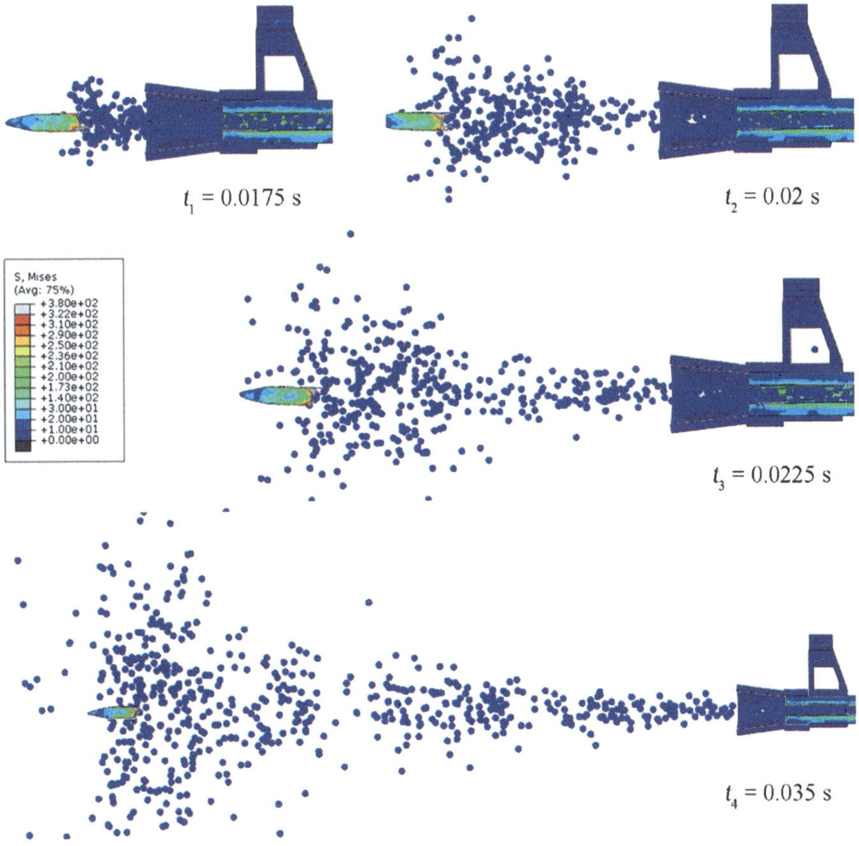

Fig. 12. Pressure acting in the barrel muzzle in specific times.

6 Comparison of Results

The findings summarized in Table 1 are based on calculations carried out with the use of the ABAQUS program and results recorded in the experiment. The overall result of pressure on the sensor A and the sensor B for the ballistic barrel and only on the C sensor for the gun barrel was averaged from 5 recorded measurements. These results were compiled with the average result of the numerical simulation. When modeling the cartridge chamber pressure, the result was of 5.4% higher than the average of the experimental results, and the pressure in the barrel fluctuated around 11.2%.

In addition, the analysis covered the increased amount of SPH components in the propellant to check to what extent the result could be disturbed, and what was observed was that it did not significantly affect the results, and the dispersion oscillated at around 0.5% with the elongation of calculations up to 3 times. Distribution of stresses in Model 2 reached higher values and their character was more universal. In the case of Model 1, the stress concentration was cylindrically centered on the axis of the bore, with the range not exceeding the outline of the thread. The stresses at the bottom part of the simulated projectile reached the value of 380 MPa. Figure 12 shows the deviation of the projectile axis from the main axis of the barrel line since the projectile is in the ascending phase.

Table 1. Comparison of experimental and numerical results.

Recorded pressure	Ballistic barrel	Average pressure	*Gun barrel	Average pressure	FEM
Pressure in the chamber [MPa]	340.4				
	332.4				
	340.4	343.5	–	–	363
	359.4				
	344.8				
Pressure in the barrel [MPa]	109.9		42.2		
	110.3		40.4		
	109.0	110.1	39.5	40.5	124
	110.7		42.3		
	110.5		38.1		

*pressure recorded on the muzzle

The disparity between the results (Table 1) mainly stems from the fact that the model construction and assumed boundary conditions were idealized.

7 Conclusions

In the article the authors presented the possibility of analyzing the vibrations of the bore using numerical FEM calculations and their influence on the accuracy of fire. Two models of the machine gun barrel were adopted for the analyses. As demonstrated above, modeling of the barrel without taking account of other fixture on it (sights, gas chambers, elements of a piston and piston rod) leads to erroneous conclusions. On the example of the diagrams (Figs. 10 and 11), the barrel's susceptibility to vibrations caused by the change in pressure of firing can be observed. The barrel of Model 2 is half the amplitude of vibrations. The restrictions imposed during experimental studies resulted from the lack of interference in the test specimen due to other law regulations. The research conducted on the ballistic barrel was further used to develop optimum input data for numerical modeling and, finally, to obtain relevant results that largely correspond to experimental results. The simulation of powder gases using the SPH method yields satisfactory results. In summary, the assumed purpose of the work fully confirms the correctness of the methodology adopted for works implemented on the optimization of the weapon accuracy resulting from the influence of the pressure in the bore on the barrel vibration.

Acknowledgements. Calculations have been carried out in Wroclaw Centre for Networking and Supercomputing (http://www.wcss.pl), grant No. 452.

References

1. Kathe, E., Gast, R., Vottis, P., Cipollo, M.: Analysis of launch induced motion of a hybrid electromagnetic/gas gun. IEEE Trans. Magn. **33**(1), 178–183 (1997)
2. Platek, P., Damaziak, K., Malachowski, J., Kupidura, P., Wozniak, R., Zahor, M.: Numerical study of modular 5.56 mm standard assault rifle referring to dynamic characteristics. Def. Sci. J. **65**(6), 431–437 (2015)
3. Lazowski, J., Malachowski, J., Niezgoda, T.: FE analysis of vibrations for barrel's tip at shot phase. Biul. WAT **57**(3), 235–245 (2008)
4. Budynas, R., Young, W.: Roark's Formula for Stress and Strain, 7th edn. McGraw-Hill, New York (2002)
5. Hunt, F.V.: Stress and strain limits on the attainable velocity in mechanical vibration. J. Acoust. Soc. Am. **32**(9), 1123–1128 (1960)
6. Ungar, E.E.: Maximum stresses in beams and plates vibrating at resonance. J. Eng. Ind. **84** (1), 149–155 (1962)
7. Gaberson, H.A., Chalmers, R.H.: Modal velocity as a criterion of shock severity. Shock Vib. Bull. **40**(2), 31–49 (1969)
8. Blagonravov, A.A.: Osnowanija projektowanija awtomaticzeskogo orużija. GIOP, Moscow (1940)
9. Johnson, G.R., Cook, W.H.: A constitutive model and data for metals subjected to large strains, high strain rates, and high temperatures. In: Proceedings 7th International Symposium on Ballistics, The Hague, pp. 541–547 (1993)
10. Sobolev, A.V., Radchenko, M.V.: Use of Johnson-Cook plasticity model for numerical simulations of the SNF shipping cask drop tests. Nucl. Energy Technol. **2**(4), 272–276 (2016)
11. Liu, M.B., Liu, G.R., Zong, Z., Lam, K.Y.: Computer simulation of high explosive explosion using smoothed particle hydrodynamics methodology. Comput. Fluids **23**, 305–322 (2003)

12. Sanchidriá, J.A., Castedo, R., López, L.M., Segarra, P., Santos, A.P.: Determination of the JWL constants for ANFO and emulsion explosives from cylinder test data. Cent. Eur. J. Energ. Mater. **12**(2), 177–194 (2015)
13. Malachowski, J., Mazurkiewicz, P.: Elastic-plastic half cylindrical surface response under blast loading. In: 19th International Conference on Computer Methods in Mechanics, pp. 341–342. Publ. House of the Warsaw University of Technology, Warsaw (2011)
14. Barnat, W.: Experimental and numerical study of influence of incidence angle of shock wave created by explosive charge on the steel plate. Bull. Pol. Acad. Sci. Tech. Sci. **62**(1), 151–163 (2014)
15. Bocian, M., Jamroziak, K., Kosobudzki, M.: Analysis of material punching including a rotational speed of the projectile. In: Solid State Phenomena, vols. 220–221, pp. 571–576 (2015)

Concept of Sensor for Mining Machines Powered by Pressure Changes

Dariusz Jasiulek(✉)

KOMAG Institute of Mining Technology, Gliwice, Poland
djasiulek@komag.eu

Abstract. In mining plants there are the machines and technological lines, in which use of additional sensors e.g. temperature or pressure ones, is recommended, but due to high temperature in workings it is difficult. Regulations resulting from the ATEX Directive limit a possibility of modernization of the existing systems for control of machines and use of additional measuring sensors with wires. A concept of electricity generator, designed to power the wireless sensors, which uses the pressure changes in a hydraulic system of machines, is presented. Presented generator is planned to be used in wireless temperature or pressure sensors intended to be used in mobile mining machines equipped with hydraulic systems.

Keywords: Energy harvesting · Piezoelectric transducer · Hydraulic systems · Pressure sensor · Self-powered sensor

1 Introduction

Development of electronic systems of low energy consumption, dedicated for energy harvesting, as well as development of radio data transmission standards cause that more and more solutions of wireless battery-powered or self-powered sensors appear on the market [5, 8, 10].

Self-powered sensors become more and more popular, among others in ventilation and air-conditioning systems, monitoring of migration of animals, automatics systems as well as systems for monitoring the technical parameters [2, 14–18]. Such sensors enable entering new information to Human Machine Interface (HMI) improving machine control and operator's safety.

Self-powered sensors are equipped with generators converting the energy from the sensor area into electricity. Depending on generator's design, the self-powered sensors can be powered from light, gas flow, mechanical vibrations, rotary movement or electromagnetic field. Electromagnetic generators or piezoelectric generators are most often used in the case of conversion of energy of mechanical vibrations. Regarding the use of piezoelectric materials for building the electricity generators, the vibrations of mechanical components with piezoelectric material are analysed [3, 6, 7, 11, 13, 19, 24–26].

In mining plants there are the machines and technological lines, in which use of additional sensors e.g. temperature or pressure ones, is recommended, but due to high temperature in workings it is difficult. Regulations resulting from the ATEX Directive

J. Świder et al. (Eds.): MECHATRONICS 2017, AISC 934, pp. 175–183, 2019.
https://doi.org/10.1007/978-3-030-15857-6_18

limit a possibility of modernization of the existing systems for control of machines and use of additional measuring sensors with wires [12]. In the case of transportation routes or complex mechanization systems it is recommended to use wireless communication. However, existing wireless sensors, measuring e.g. pressure in a hydraulic system of mining machines [5] are battery powered.

Considering the specific character of processes realized in hard coal mines, the sources of energy, which can be used to supply components of self-supplying, wireless network of sensors, were identified. Sources of energy available in undergrounds of mining plants are as follows:

- sources of mechanical energy: vibrations of machines, rotary movement, braking energy, human motion, pressure changes;
- sources of thermal energy: temperature of machines and equipment (energy of losses), temperature of human body, temperature of rock mass;
- sources of energy coming from electromagnetic radiation: electric motors, transformers, power cables;
- sources of energy coming from the air flow: flow of air in vetube fans, flow of air ventilating the mine workings.

Basing on a possibility of using the "energy harvesting" solutions in undergrounds of mining plants, research work on recovery of energy from heat [23] or from vibrations of machines to generate electricity with use of piezoelectric transducers [12, 20], as well as on communication between wireless sensors [21, 22] is carried out. Existing solutions of systems for conversion of energy in a result of pressure changes in a hydraulic system are presented in [4, 9].

Results of testing the pile of piezoelectric transducers of electricity generator, which is supplied by impact (changes in pressure), are given. It was assumed that there will be no deflection of pile of transducers. The amount of energy generated by the transducers in the case of mechanical force is determined. Identified demand for energy of the system for measurement, analysis and transmission of data in comparison with amount of energy generated by the pile of piezoelectric transducers will enable to develop a concept of generator.

2 Concept

The concept of piezoelectric generator assumes using the piezoelectric transducers of disk shape of diameter 25 mm. In [1] the authors present the examples of using the piezoelectric transducers for generation of electric energy from impact.

A concept of generator, built of transducers (3) using the simple piezoelectric effect that occurs in a result of mechanical impact on crystals, is presented in Fig. 1. In the generator the piezoelectric transducers are separated from working medium in a hydraulic system by a baffle (2), what makes the solution versatile. The chambers under (7) and over (6) piezoelectric transducers are filled with gas of proper pressure, correlated with pressure of working medium in a hydraulic installation. Gas will be fed to the generator's chambers through the check valve (5). The chambers under (7) and over (6) transducer are connected by adjustable throttle valve (4), which enables

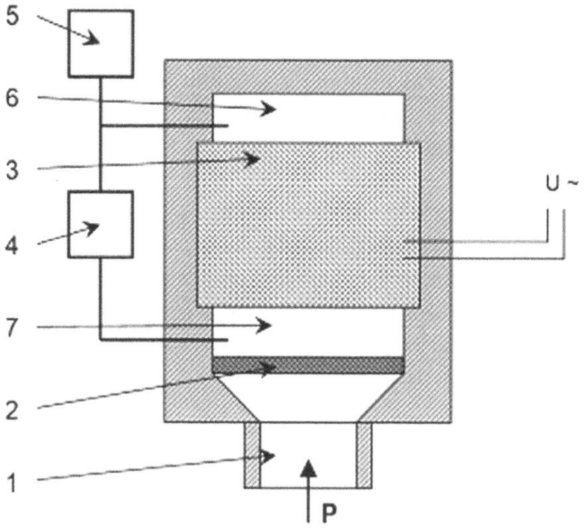

Fig. 1. Scheme of electricity generator powered by pressure changes of operational medium in a hydraulic system [12]

equalization of pressures in chambers, so the forces acting on a transducer are controlled automatically. Automatic control of forces ensures adjustment of transducer to frequency characteristics of pressure changes in a hydraulic system and protection of transducer against mechanical damage. The generator is connected to the hydraulic system by a process connection (1).

Electronic part of the generator will be designed basing on LTC3588 converter. Capacitors collecting the energy will be selected during the tests of the generator's model. It has been assumed that the sensor will operate in area threatened by explosion of methane and/or coal dust. To meet the requirements of the standard, the generator must be separated from surrounding environment and output systems (radio system) should be properly protected.

3 Demand of Energy for Wireless Sensor

Work associated with development of piezoelectric generator was preceded by the analysis of amount of electricity indispensable for operation of wireless sensor. Operations of wireless sensor during one measurement depend on its design and method of radio communication. The main stages of sensor operation are as follows: standby mode (sleep mode), measurement and data processing, radio transmission of data. Transmission of data is the most energy-consuming process. Due to minimization of electricity consumption, radio communication will be optimized during development of the sensor design. During analysis the amount of energy indispensable for the sensor to realize one operation of measurement and data transmission was determined. It is equal to 90 μJ. This amount was estimated on the basis of examples of energy consumption curves for the sensors of similar design [2, 15]. Consumption of energy by wireless sensor, which is a part of the system for detection of leakages in pipelines – water distribution networks, is presented in Fig. 2 [15]. The sensor measures the pressure indirectly. The power consumption was minimized to 2.2 μW (the measurement is taken every 6 h to extend operational time of the network supplied from batteries). It causes that the network of sensors can operate for 100 years when supplied from two AA batteries. Another example of energy consumption is presented in [2]. It is a concept of using the thermoelectric generator to supply the system for measurement of environmental parameters in the server systems. A prototype of device enables measurements of temperature, light intensity and supply voltage. Consumption of power by the device is equal to 2.86 μW (2.2 V and 1.3 μA). The energy necessary for one measurement and transmission is equal to 84.69 μJ. In a sleep mode the energy consumption is equal to 2.84 μJ.

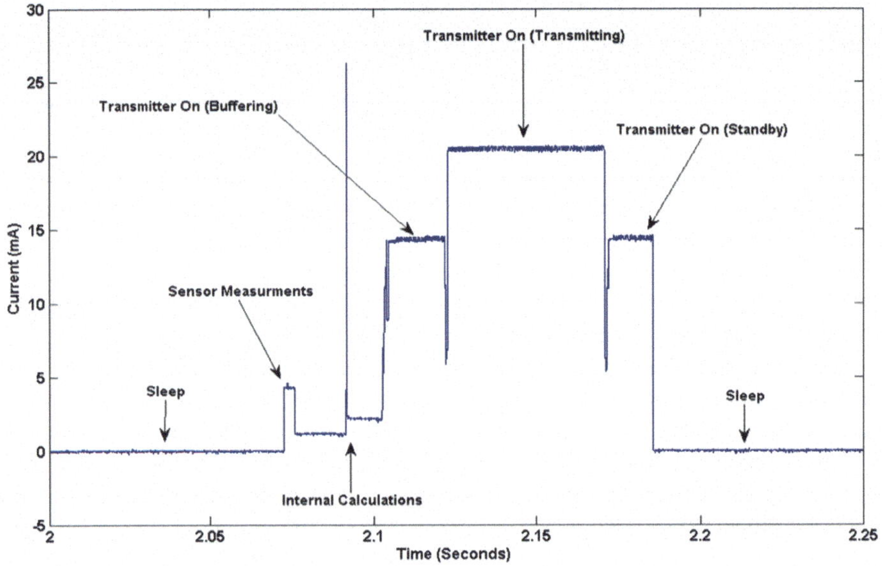

Fig. 2. Consumption of energy by the sensor of wireless network, battery-supplied with voltage 2.5 V [15].

4 Tests of Piezoelectric Generator

As previously mentioned, a piezoelectric pile, which is a component converting the energy resulting from pressure changes into electricity, was suggested to be used to generate the electricity. Piezoelectric piles are commonly used as the components, which process the electric current by using the effect that occurs in used piezoelectric materials and consists in changing the dimensions in a result of flow of electric current. Piezoelectric material generates the voltage proportionally to the force that changes its geometry (pressing, deflection). According to the results presented in [4] it is possible to use this feature and to build the electric generator with use of piezoelectric pile.

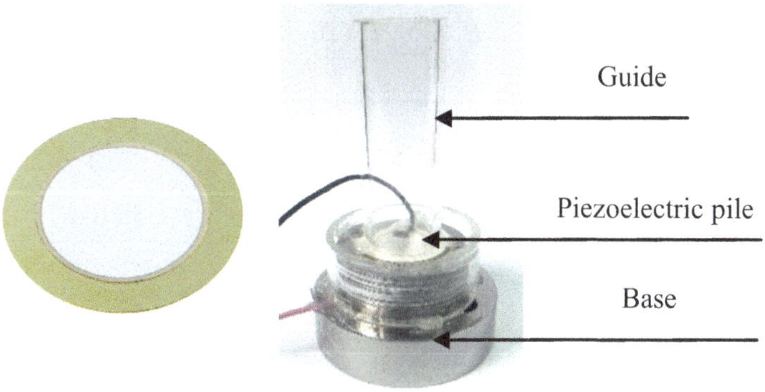

Fig. 3. Piezoelectric transducer and model of piezoelectric pile [12].

4.1 Model of Piezoelectric Pile

Piezoelectric transducers connected in series were used to build the model of pile (Fig. 3). The pile consisted of five transducers. The dimensions of a single transducer are as follows: external diameter of bronze disc $\phi = 35$ mm, diameter $\phi = 25$ mm and thickness h = 0.25 mm of piezoelectric ceramic. Electrical connections were led out of surface of upper transducer and metal base on which the pile was placed. The transducers were placed in a plastic tube to secure their position. A tube of a length 6 cm, which plays a role of guide for weights dropped on the piezoelectric pile, was glued to the upper transducer. The model of piezoelectric pile was connected to the measuring system.

The following weights were used to determine the impact energy: 2.4 g; 5.2 g; 9.2 g; 11.6 g. The weights were dropped from the height of 6 cm. It was assumed that in the case of free fall the kinetic energy EK at impact is equal to potential energy E in a field of gravity:

$$E_K = E = m * g * h \text{ [J]} \tag{1}$$

where:

g – gravity – 9.81 m/s^2
h – height of fall – 0.06 m

Kinetic energy of impact for the given weights was as follows:

- m = 2.4 g => E_K = 0.00141 J - m = 5.2 g => E_K = 0.00306 J
- m = 9.2 g => E_K = 0.00541 J - m = 11.6 g => E_K = 0.00682 J.

4.2 Energy Generated by the Pile of Piezoelectric Transducers

The electricity was calculated as a sum of momentary values of power in relation to Δt sampling time base of a signal.

$$E = \sum P * \Delta t \ \ [\mathrm{J}] \tag{2}$$

where:

E – energy generated during impact of weight [J]
P – power [W]
Δt = 0.000001 s.

4.3 Test Results

Time curves of voltage generated by a pile of five piezoelectric transducers are presented in Figs. 4, 5, 6 and 7.

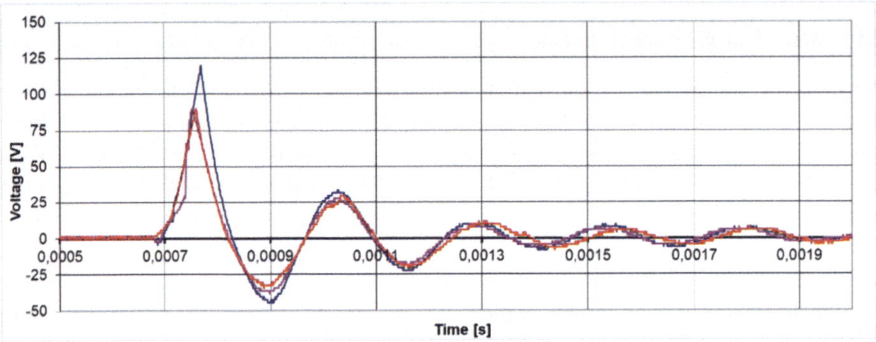

Fig. 4. Time curve of voltage generated during impact of energy 0.00141 J

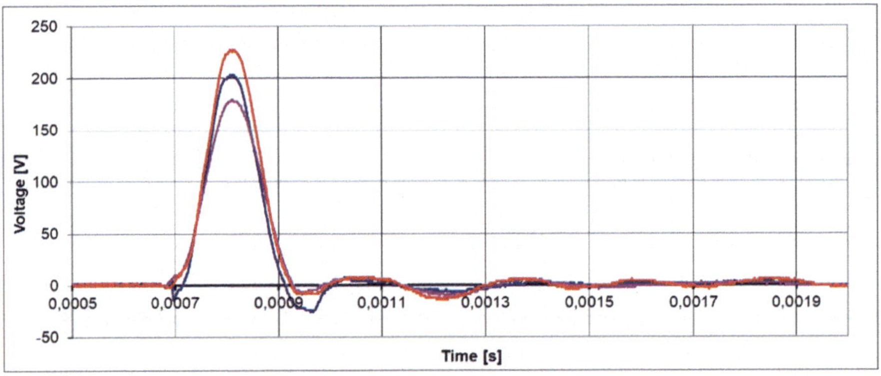

Fig. 5. Time curve of voltage generated during impact of energy 0.00306 J

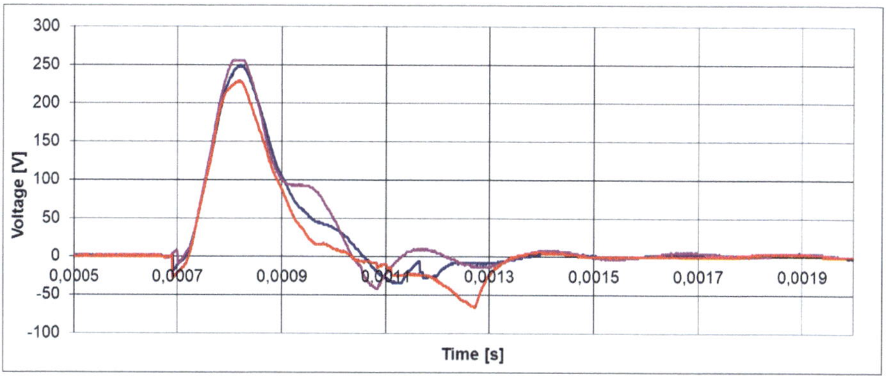

Fig. 6. Time curve of voltage generated during impact of energy 0.00541 J

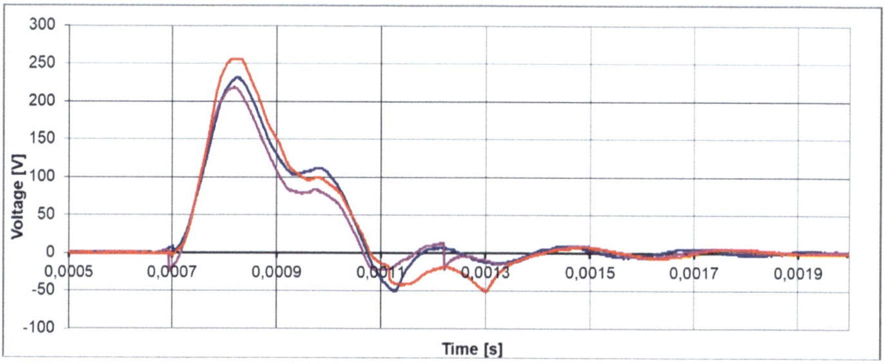

Fig. 7. Time curve of voltage generated during impact of energy 0.00682 J

Energy generated by a pile of five piezoelectric transducers during impact of different energy is listed in Table 1.

Table 1. Energy obtained from a pile of five piezoelectric transducers [J]

Load/measurement	2.4 g	5.2 g	9.2 g	11.6 g
Measurement 1	5.02068E−06	3.43931E−05	6.08408E−05	6.80042E−05
Measurement 2	2.58752E−06	2.92244E−05	6.72597E−05	5.58135E−05
Measurement 3	2.80928E−06	4.41884E−05	4.92622E−05	8.13848E−05
Mean value	3.4 μJ	35.9 μJ	59.1 μJ	68.4 μJ

5 Summary

A concept of the generator for self-powered pressure or temperature sensor is presented. The generator uses energy of pressure fluctuation in the hydraulic system of the machine. The results of initial stage of work on development of the generator/sensor aiming at verification of design assumptions and determination of amount of energy possible to be obtained from a pile of piezoelectric transducers are given.

Model of a pile consisting of five piezoelectric transducers was built and tested. A set of weights of the following kinetic energy was used for dynamic load: 0.00141 J, 0.00306 J, 0.00541 J, 0.00682 J. Voltage signal generated by the transducers during release of kinetic energy was recorded and then the energy generated by a pile of transducers was calculated.

It was assumed that the amount of energy indispensable for operation of wireless sensor is equal to 90 μJ. The energy obtained during the tests was equal to 68.4 μJ. Thus it will be necessary to re-compress the piezoelectric pile. Optimization of electronic system and reducing the energy demand of the sensor will be another method to solve the problem of insufficient amount of energy.

The results regarding the amount of generated energy should refer to the energy generated by pressure fluctuations in the hydraulic system. Construction of testing facility equipped with hydraulic receivers and switching system (split valves) is planned for that purpose. The manufactured prototype of generator will be tested to determine the amount of generated energy.

Development of details of the automatic system for pressure control in the generator and development work on using other, already made plates of piezoelectric material to obtain more energy, is also planned.

References

1. Beeby, S.P., Tudor, M.J., White, N.M.: Energy harvesting vibration sources for microsystems applications. Meas. Sci. Technol. **17**(12), 175–195 (2006)
2. Brunelli, D., Passerone, R., Rizzon, L., Rossi, M., Sartori, D.: Self-powered WSN for distributed data center monitoring. Sensors **16**(1), 57 (2016)
3. Buchacz, A., Płaczek, M., Wróbel, A.: Control of characteristics of mechatronic systems using piezoelectric materials. J. Theoret. Appl. Mech. **51**(1), 225–234 (2013)
4. Cunefare, K.A., Skow, E.A., Erturk, A., Savor, J., Verma, N., Cacan, M.R.: Energy harvesting from hydraulic pressure fluctuations. Smart Mater. Struct. **22**, 025036 (2013). (10 pp)
5. Catalogue card of Elgór-Hansen EH-PressCater system. www.elgorhansen.com
6. Erturk, A., Inman, D.J.: Piezoelectric Energy Harvesting. Wiley, Hoboken (2011)
7. Erturk, A., Hoffmann, J., Inman, D.J.: A piezomagnetoelastic structure for broadband vibration energy harvesting. Appl. Phys. Lett. **94**, 254102 (2009)
8. Gilbert, J.M., Balouchi, F.: Comparison of energy harvesting systems for wireless sensor networks. Int. J. Autom. Comput. **5**(4), 334–347 (2008)
9. Grzybek, D.: Piezoelectric generators: materials and structures. Pomiary Automatyka Robotyka, October 2013

10. Jasiulek, D.: The use of lost energy for supplying the dispersed network of sensors. Problemy Eksploatacji – Maint. Prob. **97**(2), 51–59 (2015)
11. Jasiulek, D.: Testing the piezoelectric energy harvester's deflection on the amount of generated energy. In: Awrejcewicz, J., Kaliński, K.J., Kaliczyńska, M., Szewczyk, R. (eds.) Mechatronics: Ideas, Challenges, Solutions and Applications, pp. 95–112 (2016)
12. Jasiulek, D.: Alternative power sources for sensors used in mining. Own work ITG KOMAG (not published)
13. Latalski, J.: Modelling of macro fibre composite piezoelectric active elements in ABAQUS system. Eksploatacja i Niezawodnosc – Maint. Reliab. **4**, 72–78 (2011)
14. Radkowski, S., Lubikowski, K., Piątak, A.: Vibration energy harvesting in the transportation system: a review. Diagnostyka – Appl. Struct. Health Usage Cond. Monit. **4**(64), 39–44 (2012)
15. Sadeghioon, A.M., Metje, N., Chapma, D.N., Anthony, C.J.: SmartPipes: smart wireless sensor networks for leak detection in water pipelines. J. Sens. Actuator Netw. **3**(1), 64–78 (2014)
16. Sardini, E., Serpelloni, M.: Passive and self-powered autonomous sensors for remote measurements. Sensors **9**, 943–960 (2009)
17. Sardini, E., Serpelloni, M.: Self-powered wireless sensor for air temperature and velocity measurements with energy harvesting capability. IEEE Trans. Instrum. Meas. **60**(5), 1838–1844 (2011)
18. Sazonov, E., Li, H., Curry, D., Pillay, P.: Self-powered sensors for monitoring of highway bridges. IEEE Sens. J. **9**(11), 1422–1429 (2009)
19. Shena, H., Qiu, J., Balsi, M.: Vibration damping as a result of piezoelectric energy harvesting. Sens. Actuators A **169**, 178–186 (2011)
20. Sodano, H.A., Inman, D.J.: A review of power harvesting from vibration using piezoelectric materials. Shock Vib. Dig. **36**, 197–206 (2004)
21. Stankiewicz, K.: The method of swarm self-organization in monitoring and control of equipment in underground mining. Maszyny Górnicze **4**, 10–13 (2011)
22. Stankiewicz, K.: The concept of the simulation environment to evaluate the self-organization of routing in the sensor network. Maszyny Górnicze **2**, 3–8 (2015)
23. Świder, J., Woszczyński, M.: Use of the system for energy recuperation and control in diesel machines. Mach. Dyn. Res. **38**(1), 73–79 (2014)
24. Zolkiewski, S.: Vibrations of beams with a variable cross-section fixed on rotational rigid disks. Latin Am. J. Solids Struct. **10**, 39–57 (2013)
25. Zolkiewski, S.: Damped vibrations problem of beams fixed on the rotational disk. Int. J. Bifurcat. Chaos **21**(10), 3033–3041 (2011)
26. Chen, X., Yang, T., Wang, W., Yao, X.: Vibration energy harvesting with a clamped piezoelectric circular diaphragm. Ceram. Int. **38**, S271–S274 (2012)

Model of Dynamics of the Three Wheeled Mobile Platform

Anna Jaskot(✉) and Bogdan Posiadała

Institute of Mechanics and Machine Design Fundamentals,
Czestochowa University of Technology,
ul. Dąbrowskiego 73, 42-201 Częstochowa, Poland
{jaskot, b.posiadala}@imipkm.pcz.pl

Abstract. In this work the dynamics model of the three wheeled mobile robot, with two rear wheels and one front castor wheel, was proposed. It has been assumed, that every wheel is equipped with the independent drive units. The motion parameters of the mobile platform have been determined by adopting classical approach of mechanics. The presented model is designed to analyze and to get the motion parameters, such as position, velocity and acceleration. In the work the differential equations were formulated and the results by using the proposed Runge-Kutta method of the fourth order can be obtained. In the model the relations between the active and passive forces with their influence to each other have been introduced. The problem of the forced motion and free motion of the platform with the possibility of modification of the drive modulus positions has been considered. The work is dedicated to the designing motion of the three wheeled mobile platform under the unsteady conditions.

Keywords: Dynamics of motion · Motion parameters · Mobile platform

1 Introduction

While solving problems of motion of the wheeled mobile platforms, the considerations of parameters and constraints during motion are an important matter. Investigations about the character of motion, trajectory, and the behavior of the platform under working cycles are presented in this paper.

This work is a result of studies of the three wheeled mobile platform. The authors, so far, have researched the model of the four wheeled mobile platform [1], and the proposed model of the three wheeled mobile platform in work [2]. In this work the model of the three wheeled mobile platform, but with different design of the construction of the chassis is presented.

This work contains description of a model of platform with two rear wheels and one front caster wheel, and it has been described in further sections. The platform prototype design has also been described. The kinematic analysis and the experimental results of the three wheeled driving machines have been published in [3]. The analysis of motion planning with consideration of the dynamic equations of the three wheeled mobile robot has been proposed among others in [4]. The analysis of the system stability and the internal dynamics have been presented in [5]. The dynamic and kinematic models

J. Świder et al. (Eds.): MECHATRONICS 2017, AISC 934, pp. 184–191, 2019.
https://doi.org/10.1007/978-3-030-15857-6_19

of the two-wheeled differential drive mobile robots with two fixed and in-line with each other electric motors have been gathered in [6].

The studies about platform with adaptive robust sliding mode control with four Mecanum wheels are widely described in [7]. Motion trajectory tracking with models of kinematic and dynamic models by adopting the Lagrange formulation has been researched and included in [8]. The formulation of tracking control of a group of mobile robots with guaranty of no collisions between robots is proposed and described in [9].

The main purpose of this work is to get to know the behavior of the designed prototype of the three wheeled mobile platform with its work conditions. The description of the dynamics of the platform is necessary in order to obtain the motion parameters in planar movement. The methodology gathered in this work is a base to provide the fundamental parameters such as position, velocity and acceleration of center of mass of the platform, by taking into account the causes and effects of motion. The research clue is to know the conditions of platforms operation, which has a significant meaning for both the optimization of the construction solutions and the selection of the operating conditions for the considered group of machines.

2 Dynamics of the Platform

2.1 Design of the Prototype of the Platform

The design work of the prototype solution system of the three wheeled mobile platform have been undertaken. The prototype is designed with the purpose of experimental verification of adopted models considering the planar motion. In order to build an universal model of dynamics, taking into account the active and passive forces, the prototype of the platform has been designed.

The actions aimed to adopt the proven research methods to analyze problems related to motion of wheeled mobile platforms have been undertaken.

During the design process the following assumptions need to be taken into account:

a. motion will be realized in any desired direction,
b. wheels will be in constant contact with the ground,
c. wheels will be able to rotate to set in any position,
d. platform will move on a flat surface.

The system provides an equipment of three electric drive units, with possibility to disconnect the selected units in order to study the following motion. The platform is designed from a rigid frame on the sides of which the two rear wheels were placed.

The platform was constructed in order to ensure motion in every desired direction and to study free and forced motion of the platform. The considerations of both, the forced motion and free motion of the platform with ability to modify the drive modulus positions have been included. It was assumed, that the wheels will be attached to rigid frame by main hitches, which in turn will be useful in changing the distance between them and along the longer side of the frame. The description of the dynamics with the geometry of the platform is included in the next subsection.

2.2 Model of Dynamics

The mobile platforms belong to a group of machines designed to overcome the distances, carry the load and drive in places inaccessible for a human.

The model of a simplified chassis of the mobile platform is presented graphically in Fig. 1.

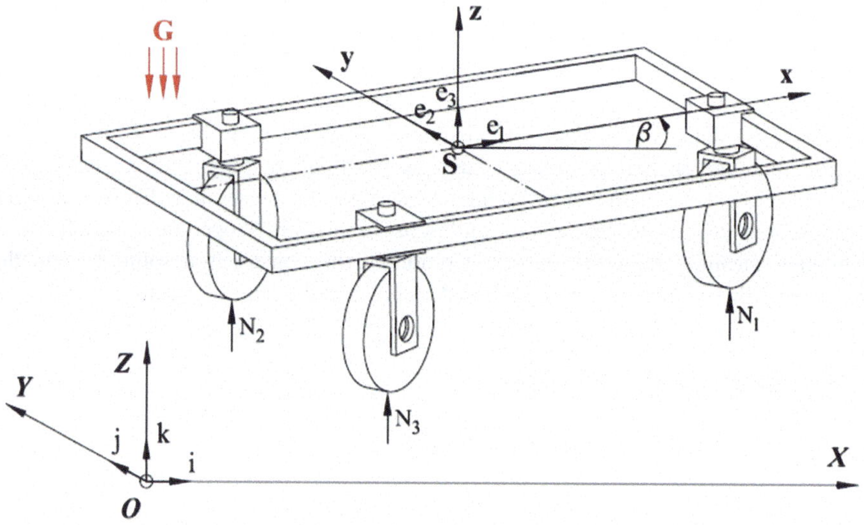

Fig. 1. Model of the three-wheeled mobile platform in the reference frame.

The geometrical scheme of the three wheeled mobile platform, according to Fig. 1, was a base in determining the description of the dynamics in global coordinate system *OXYZ*. **G** vectors represent the continuum load deriving from the mounted equipment, materials or tools. **N** vectors are the pressure forces and represent the reactions from the ground caused by the internal and external load. *S* point is a gravitational center as well as the center of mass of the platform.

The local coordinate system *Sxyz* is denoted to *S* point. The **i**, **j**, **k** vectors are unit vectors in global coordinate system, and the $\mathbf{e}_1$, $\mathbf{e}_2$, $\mathbf{e}_3$ respectively are unit vectors in local coordinate system. Active and passive forces are connected to wheels of the platform. Because of the dependencies between the forces there is a need to include those relations in dynamic description. The global and local coordinate systems, and the geometric relations between the positions of the respective drive modules are shown in the model scheme. Each wheel is associated with its local coordinate systems, as presented in Fig. 2. The consecutive axes of the local coordinate systems are parallel one to another. Parameter β is an initial inclination angle of the platform at the beginning of the analyzed motion, measured between *X*-axis of the global coordinate system and the longer side of the frame.

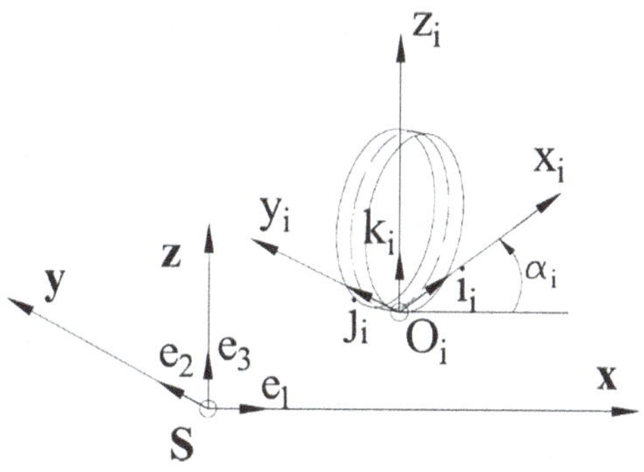

Fig. 2. Local coordinate system of the *i*-th wheel.

According to Fig. 2, α_i is an inclination angle of the platform's wheel, measured between *X*-axis of the global coordinate system and x_i-axis of the wheel local coordinate system.

The motion parameters are determined with respect to the global reference frame *OXYZ*. The motion is composed of the progressive motion with position, velocity and acceleration of the center of mass and the rotational motion with the angular parameters around the center of mass of the mobile platform. In establishing the parameters of motion, the motion equations have to be formulated.

In order to properly determine the values of the described parameters the scheme of the occurring forces is presented in Fig. 3.

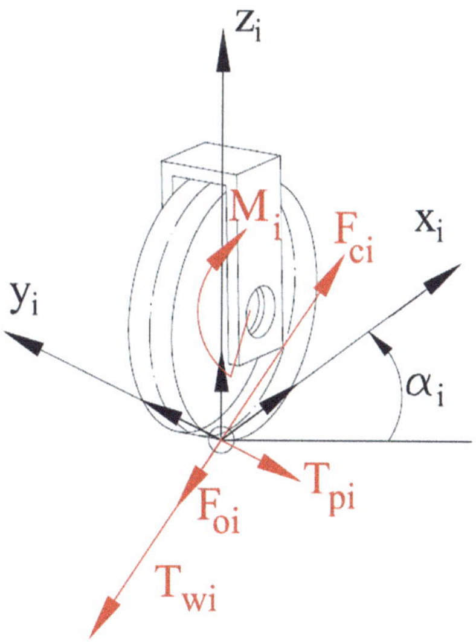

Fig. 3. Active and passive forces in the local coordinate systems of the *i*-th wheel.

$\mathbf{M}_i$ represents the drive torque, which is the initiator of motion. $\mathbf{F}_{ci}$ represents the active force. $\mathbf{F}_{oi}$ force represents the motion resistance force. **T** vectors represents the passive forces, which are the friction forces in longitudinal $\mathbf{T}_{wi}$ and transverse $\mathbf{T}_{pi}$ directions. The model of dynamics has been determined taking into account the forces for each of the three steering wheels and then brought back to the rigid structure of the mobile robot.

2.3 Description of the Dynamics of the Three-Wheeled Mobile Platform

The considerations about the dynamics of platform as the rigid structure resulted in description of three equations of the planar translations and of the rotation around Z–axis. It is important to pay special attention to the fact, that the same force can be an internal force for one system and, at the same time, for other system it can be an external force.

The internal forces are among others gravitational force, friction forces, trajectory reactions or motion resistance forces. All forces, which were taken into account in this work have been presented in previous subsection in Fig. 3.

In describing the dynamics of the three wheeled mobile platform the Coulomb friction model was adopted.

Forces which are consequent from the contact of the wheel with the ground, but most of all, the active forces deriving from the drive torque are described below.

$$\mathbf{F}_{ci} = \frac{M_i}{r} \cdot \mathbf{e}_1^i \tag{1}$$

where: r–radius of an i-th drive wheel.

The passive forces, especially the friction forces $\mathbf{T}_{wi}$ and $\mathbf{T}_{pi}$, are needed to be taken into account in the description. When the platform is starting to move, the values of the passive forces are changing from zero during the platform motion, to values obtained from Eqs. 2 and 3.

$$\mathbf{T}_{wi} = -\mu_w \cdot N_i \cdot sign(v_{wi}) \cdot \mathbf{e}_1^i \tag{2}$$

$$\mathbf{T}_{pi} = -\mu_p \cdot N_i \cdot sign(v_{pi}) \cdot \mathbf{e}_2^i \tag{3}$$

where: μ_w, μ_p are the coefficients of friction for the longitudinal and transverse directions and the v_{wi} and v_{pi} are the velocity components for the longitudinal and transverse directions. N_i is a reaction deriving from the wheel load on the ground. Reaction force N_i can be designated according to the following equation:

$$\mathbf{N}_i = \mathbf{G}_i \tag{4}$$

where: $\mathbf{G}_i$ is value of load force acting on each wheel ($\mathbf{G} = mg$).

The force, which does not allow the platform to move and is deriving from the static friction, is a resistance force and has been designated as $\mathbf{F}_{oi}$.

In this model it was assumed that motion of the platform is possible, when the limit values of the static friction forces are exceeded by values of active forces.

This dependence is denoted in formula in Eq. 5.

$$\mathbf{F}_{\mathrm{oi}} = -f \cdot N_{\mathrm{i}} \cdot sign(v_{\mathrm{wi}}) \cdot \mathbf{e}_1^i \tag{5}$$

where: f is a coefficient of static friction.

$$\mathbf{F}_{\mathrm{ci}} = \begin{cases} \frac{M_i}{r_i} & for \ \frac{M_i}{r_i} < T_i \\ T_i & for \ \frac{M_i}{r_i} \geq T_i \end{cases} \tag{6}$$

In the above formula, Eq. 6, both cases of the possible analysis have been included. To properly present the motion, when the wheels slip on the ground, it is necessary to compare it with the situation without slippage. During motion of the platform, and when the values of active forces are higher than friction forces, the platform will move safely, without slippage, but when the active forces will be exceeded by the developed friction forces, the platform will lose control, and the slippage occurs.

The values of the resultant forces $\mathbf{W}_{\mathrm{i}}$ have been calculated separately for each wheel as follows:

$$\mathbf{W}_{\mathrm{i}} = \mathbf{F}_{\mathrm{ci}} + \mathbf{F}_{\mathrm{oi}} \tag{7}$$

The procedure of determining is based on the methodology described in [2] which concerns the three wheeled mobile platform, but with different configuration of driven wheels. The translational motion equation is included below:

$$m\mathbf{a} = \sum_{i=1}^{3} \mathbf{W}_{\mathrm{i}} \tag{8}$$

where: m – mass, $\mathbf{a}$ – acceleration of the center of mass of the platform.

The equation of the rotational motion around Z-axis of the center of mass of the platform has been denoted in Eq. 9.

$$\frac{d\mathbf{K}}{dt} = \sum_{i=1}^{3} \mathbf{s}_{\mathrm{i}} \times \mathbf{W}_{\mathrm{i}} + \sum_{i=1}^{3} \mathbf{M}_{\mathrm{mi}} \tag{9}$$

where: $\mathbf{K}$ – angular momentum vector of the whole platform, $\mathbf{s}_{\mathrm{i}}$ – location vectors of each of the drive wheels, $\mathbf{M}_{\mathrm{mi}}$ – moment deriving from the motion around center of mass in the slippage circumstances.

By taking into account the initial conditions according to the starting values of the motion parameters the dynamics equations of motion can be formulated. The equations of motion, written in the form of differential equations of the center of mass in the reference frame, have been determined and are presented in Eqs. 10–12.

$$\ddot{X} = \frac{1}{m}\sum_{i=1}^{3} W_{\mathrm{ix}} \tag{10}$$

$$\ddot{Y} = \frac{1}{m}\sum_{i=1}^{3} W_{\mathrm{iy}} \tag{11}$$

$$\ddot{\beta} = \frac{1}{I_z}\sum_{i=1}^{3} \left(s_{\mathrm{ix}} \cdot W_{\mathrm{iy}} - s_{\mathrm{iy}} \cdot W_{\mathrm{ix}}\right) \tag{12}$$

where: $\ddot{X}$ –acceleration of center of mass on the *X*–axis in reference frame, $\ddot{Y}$ –acceleration of center of mass on the Y–axis in reference frame, $\ddot{\beta}$ –angular acceleration around center of mass of the platform.

In this work the model of the three wheeled mobile platform has been determined. Based on the differential equations 10–12 the task can be completed. By using numerical methods, in this work the Runge-Kutta method of the fourth order, the motion parameters will be obtained.

3 Conclusions

The model of the prototype of the three wheeled mobile platform with one front wheel and two rear wheels has been presented. The model of the dynamics of such a system is included. The dynamic model is useful in designating the various possible configurations of motion, with adjustment of the wheel behavior during motion. The model is also useful in performing the analysis, in which the rates of the motion parameters (among others position, velocity and acceleration) in planar translations, as well as the angle, angular velocity and angular acceleration in rotations around Z-axis, will be possible.

The model is accurate in studies about the conditions of work, which in consequence allows to determine the parameters affecting to the movement of both the platform and its components. The possibilities of both, the forced motion and free motion of the platform with ability to modify the drive modulus positions have been considered.

The model can also be developed by introducing other elements of the real object to the mathematical description.

The proposed design concept is intended to facilitate further research, both experimental and theoretical, the dynamic effects, which may occur during the movement of objects, in particular in the case of maximizing the motion velocities and to analyze the platform motion during the change of the motion direction.

The analysis of motion and the sample results will be the subject of further publications.

References

1. Jaskot, A., Posiadała, B., Śpiewak, S.: Motion modeling of the four-wheeled mobile platform under slippage conditions. Mach. Dyn. Res. **40**(2) (2016, in print)
2. Jaskot, A., Posiadała, B., Śpiewak, S.: Dynamics model of the three-wheeled mobile platform. In: 22nd International Conference Engineering Mechanics, Svratka, Czech Republic, pp. 250–253 (2016)
3. Abdelgabar, Y.E., Lee, J.H., Okamoto, S.: Motion control of a three active wheeled mobile robot and collision-free human following navigation in outdoor environment. In: Proceedings of the International Multi-conference of Engineers and Computer Scientists, Hong Kong, vol. I (2016)
4. Zohar, I., Ailon, A., Rabinovici, R.: Mobile robot characterized by dynamic and kinematic equations and actuator dynamics: trajectory tracking and related application. Robot. Auton. Syst. **59**, 343–353 (2011)
5. Eghtesad, M., Necsulescu, D.S.: Study of the internal dynamics of an autonomous mobile robot. Robot. Auton. Syst. **54**, 342–349 (2006)
6. Salem, F.A.: Dynamic and kinematic models and control for differential drive mobile robots. Int. J. Curr. Eng. Technol. **3**(2), 253–263 (2013)
7. Alakshendra, V., Chiddarwar, S.S.: Adaptive robust control of Mecanum-wheeled mobile robot with uncertainties. Nonlinear Dyn. **87**, 2147–2169 (2017)
8. Ali, Z.A., Wang, D., Safwan, M., Jiang, W., Shafiq, M.: Trajectory tracking of a nonholonomic wheeled mobile robot using hybrid controller. Int. J. Model. Optim. **6**(3), 136–141 (2016)
9. Do, K.D.: Output-feedback formation tracking control of unicycle-type mobile robots with limited sensing ranges. Robot. Auton. Syst. **57**, 34–47 (2009)

Control of Bucket Conveyor's Output

Sebastian Jendrysik[1(✉)] and Gabriel Kost[2]

[1] Komag, Institute of Mining Technology, Gliwice, Poland
sjendrysik@komag.eu
[2] Silesian University of Technology, Gliwice, Poland

Abstract. New method for controlling the bucket conveyor, which transports materials from pulsating water jigs is presented. Correlation between conveyor output and its load is the method essence. That results in reduction of electric energy consumption and can extend the conveyor life.

The suggested solution of controlling the conveyor speed is based on information about the degree of opening the discharging slot of the jig as well as about speed and load of the conveyor.

The analysis of the process of loading the product leaving the jig on the bucket conveyor is made to select those parameters, which have an impact on the conveyor load. On the basis of this analysis, the model of the process was suggested and its identification was carried out including laboratory test, preparation of measuring data, selection of the model structure and its approximation. The model was used in a prediction control system, which was implemented in PLC controller. The results of industrial verification tests are also presented.

Keywords: Jig · Elevator · Control system · Identification

1 Introduction

Increasing demand of mining industry and coal processing industry for higher production efficiency and effectiveness contributed to the development and improvement of minerals transportation technology with the use of bucket conveyors. After achieving the required reliability of the machines, attention of designers and users was drawn to reduction of operational costs by implementing advanced solutions of control systems and driving systems. First, the developed solutions have to reduce consumption of electric energy and also they have to extend life of the conveyors. The most effective use of electric energy takes place, when the conveyor operates with nominal load [5]. So far, majority of bucket conveyors in Polish mines operate with constant speed, so development of controlling methods was focused on adjustment of conveyor speed to its current load.

In KOMAG, research projects on automation of bucket conveyors are carried out in the context of automation of operation of entire jig node.

J. Świder et al. (Eds.): MECHATRONICS 2017, AISC 934, pp. 192–200, 2019.
https://doi.org/10.1007/978-3-030-15857-6_20

2 Characteristics of Bucket Conveyor Operation in the Jig's Mineral Processing Node

Bucket conveyors are used in hard coal processing plants for transportation and dewatering of materials obtained from the beneficiation process in pulsing water jigs [4, 8, 19]. When the raw coal feed has a small amount of waste rock and/or semi-product, the conveyors operate for a long time with low load. In such situation there is unjustified power consumption as well as wear of the conveyor is speeded up. We can eliminate this by the use of automatic control, which would change speed of bucket movement adjusting it to the changing load. Changeability of load is a characteristic feature of operation of bucket conveyors in jig's beneficiation nodes (see Fig. 1).

Fig. 1. Jig's hard coal beneficiation node [8].

Change in volume of the material fed on the conveyor results from the change in jig load, change in gravimetric composition of beneficiated material or intervention of the machine operator. In this case speed of the conveyor should be adapted to new load conditions to avoid possible concentration of fed material in the conveyor's loading zone. Such situation can cause emergency stop of conveyor and in consequence stop of the jig's beneficiation node.

As the beneficiation process in pulsing jigs takes place in water (buckets are half immersed in water) it is difficult to determine their degree of load at the loading station.

The following theoretical output of bucket conveyor was calculated for the purpose of its control [4, 19]:

$$Q_t = 3.6 \cdot \frac{V}{a} \cdot \gamma \cdot v \cdot \Psi [\mathrm{kg/s}] \quad (1)$$

where:

V – capacity of a single bucket [m^3],
a – bucket spacing [m],
γ – bulk density of transported material [kg/m^3],
v – linear speed of buckets [m/s],
Ψ – coefficient of the bucket filling $\langle 0 \div 1 \rangle$.

From the Eq. (1) it results that the bucket conveyor's output can be controlled through change of its speed. Change in flow intensity of bottom product from the jig resulting from action of the controller, which maintains constant height of waste product layer by opening/closing draining flap, decides significantly about materials volume of the bucket conveyor. However, it should be remembered that bottom product drained out form the jig settles in water medium and has to cover few meters to the conveyor's loading zone. This delay caused the need of using the prediction system for automatic control as well as demand of the process model.

3 Identification of Model

In the analysed process it has been assumed that the weight of the material transported on the conveyor depends on conveyor's speed and on the degree of opening the drainage flap in the jig, through which the bottom product, directed to the conveyor, is passing. Such assumption has sense only when the jig operates in stable conditions. The structure of the model is given in Fig. 2 [16, 22].

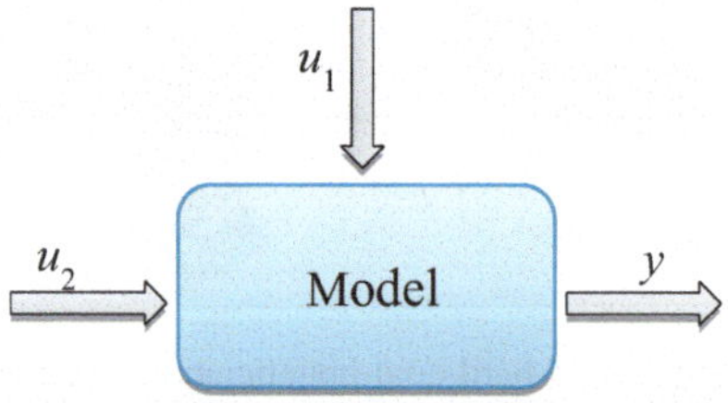

Fig. 2. Block structure of the process model [8].

where:

u_1 – degree of opening the drainage flap,
u_2 – linear speed of conveyor's buckets,
y – weight of the material on the conveyor.

Measurements of input/output signals, required for the identification process, were taken in Budryk Coal Mine on the bucket conveyor, which transported waste product from OS36D3E jig designed for three-product beneficiation of the material in the grain class 60(80) – 12(0) mm.

From identification of data recorded during measurements four sets of data of duration 3600 s were selected. One of these sets was used for estimation of parameters of the model, while remaining three sets were used in verification of the parameters. Model of the best average fit to the identification data was used in the system for automatic control. The following four parametric models were tested [1, 3, 7, 10, 15–17, 22]:

- ARX,
- ARMAX,
- Box-Jenkins (BJ),
- Outpu Error (OE).

Parameters of polynomials of each model were determined from optimization of the model structure regarding the Akaike information criterion [10, 16, 22]:

$$AIC(p) = 2p/N + \ln(V) \tag{2}$$

where:

p – number of model parameters,
N – number of signal samples,
V – noise variation.

Average fitting of the models, determined for three historical data is presented in Fig. 3.

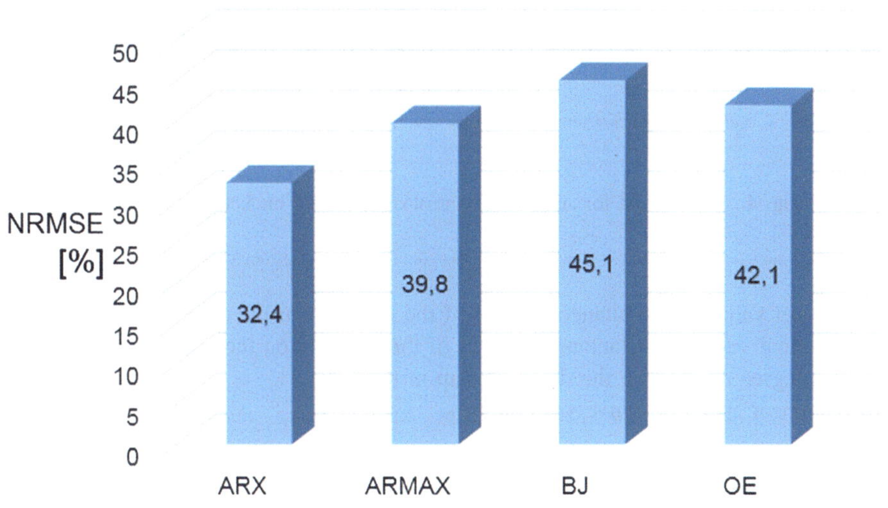

Fig. 3. Average fitting of the models, determined for three historical data [8].

Accuracy of the models was determined by NRMSE parameter. Value of this index expressed in % enables easy interpretation of the results. This parameter is defined by the following equation [1, 2, 10, 12, 13, 18]:

$$NRMSE = \left[1 - \frac{\|y - \hat{y}\|}{\|y - \bar{y}\|}\right] \cdot 100\% \tag{3}$$

where:

$\hat{y}$ – prediction of output,
$\bar{y}$ – average value from series of data.

Most accurate approximation was achieved for the Box-Jenkins (BJ) model and this model was accepted for creation of automatic control system.

4 System for Automatic Control of Bucket Conveyor Speed

Due to changeable character of carrying the material out of the jig we face constant change of load to the bucket conveyor. Thus, the algorithm controlling the conveyor operation assumes constant correction of bucket conveyor speed to keep total mass of the material on the conveyor constant (constant degree of the buckets fill in). Modified control system with Smith [14, 16] predictor was used in conveyor speed control. Block diagram of the system is presented in Fig. 4.

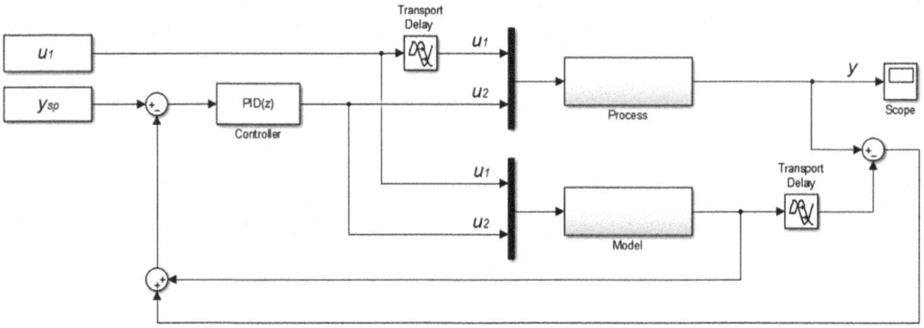

Fig. 4. The system for automatic control of speed for bucket conveyor [8]

where:

y_{sp} – set value of instantaneous mass of the material on the conveyor,
y – actual value of instantaneous mass of the material on the conveyor,
u_1 – degree of opening the drainage flap in the jig,
u_2 – linear speed of conveyor's buckets.

In the prediction system presented in Fig. 4 the real process is the object with delayed information and model of the process is an object without information delay. The model delivers information to the controller about load with some advance.

PI controller of the following structure was used in the given system [2, 11, 20, 21, 23]:

$$G_{PI} = K_P - \left(1 - \frac{1}{T_i s}\right) \tag{4}$$

where:

K_p – multiplication of proportional part,
T_i – integration time.

Setting of the controller was realized basing on the stability criterion as well as static and dynamic accuracy, determined from the selected features of frequency characteristics of a closed system.

5 Verification Tests of Impact of Automatic Control on Energy Consumption of the Conveyor's Motor

In the verification tests, consumption of electric energy by the motor of bucket conveyor was analysed in the following four variants [5, 6]:

- variant 1 – with automatic control system – industrial measurements,
- variant 2 – with constant control (80% fill up) – theoretical discussion,
- variant 3 – with constant control (100%) – theoretical discussion,
- variant 4 – with constant (nominal) rotational speed – theoretical discussion.

Verification tests were carried out on the same bucket conveyor (BUDRYK coal mine) as during identification tests. Hardware arrangement and software of PLC controller for regulation of pulsing jig's operation were modified. Controller's software was extended by new functional blocks to control conveyor speed. Operator's control panel was also modified by adding new synoptic screen which enabled monitoring of the control process (Fig. 5).

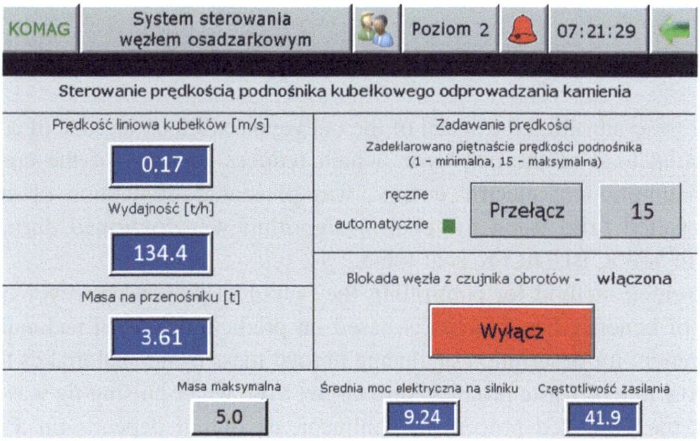

Fig. 5. Synoptic screen which enabled monitoring the operation of bucket conveyor [4, 9].

In Table 1, energy consumption ability of bucket conveyor in four variants of motor's rotational speed control is given.

Table 1. Energy consumption ability of bucket conveyor in four variants of its speed control [8].

Variant	Time of operation [min]	Mass of transported material [kg]	Energy consumed [kWh]	Percent of consumed energy [%]
1	600	832238	69.5	83.8
2			67.8	81.7
3			62.3	75.0
4			83.0	100

Ten groups of measurements were randomly selected, each covering a period of 60 min. Only those cases when the conveyor operated with load were analysed. Each measurements group is from a different day and different shifts. Total time of measurements taken into account in the verification tests was 600 min. During this time 32 238 kg of waste product was transported. In the case when the conveyor operated with automatic control system (variant 1), 69.5 kWh of energy was consumed. This amount was obtained with the control system set to 80% of filling in the conveyor's buckets with waste product. However, at the selected time intervals, 100% of filling in the conveyor's buckets was observed, so increasing the set value higher seemed to be not safe. We assumed the 80% of filling in the conveyor's buckets is the limit value for safe operation of the conveyor. Ability for energy consumption was compared with theoretically calculated consumption with the same filling in (variant 2), which was 67.8 kWh (difference 1.7 kWh). Theoretically calculated consumption with filling in equal to 100% (variant 3) was 62.3 kWh. Energy consumption by the motor with constant rotational speed (variant 4), which was also calculated and it was equal to 83 kWh. Treating these calculated value as 100% of energy consumption, use of the developed automatic control system enables saving the energy consumption by 16.2%. Maximal reduction of energy consumption at higher conveyor load can reach 25%.

6 Summary

The method for controlling the speed of the conveyor based on an idea of adjusting the speed to actual load of the conveyor, which reduces idle run of the conveyor and reduces consumption of electric energy, was presented. Reduction of energy consumption resulted from using the control algorithm was confirmed during the tests carried out in KWK BUDRYK coal mine.

The presented method for controlling the output of bucket conveyor operating in the jig nod of beneficiation process is based on prediction control technique. Simplified, linear model for description of change in total mass of the material as the result of change in flow rate of waste products discharged from water pulsing jig was used in the method. As the modelled process in nonlinear, its output depends on a part of the stream of waste product from pulsating jig and conveyor speed also on many other variables, thus, the model conformity is met only around the point for which model parameters were determined.

The results enable to conclude that implementation of the method to industry can significantly affect operational parameters of bucket conveyors, first of all resulting in reduction of energy consumption. Verification tests enabled analysis of new controlled method applied for industrial bucket conveyor.

During industrial tests, stable operation of the control system was reported and no excessive accumulation of waste material at the loading zone was observed.

Effectiveness of the presented solution is mainly limited by the ability of recreating the complex loading and transportation of material on the conveyor. This problem will be the subject of further tests.

References

1. Brandt, S.: Analiza danych, Wydawnictwa Naukowe PWN (1998)
2. Brzózka, J.: Regulatory i układy automatyki, MIKOM (2004)
3. Czemplik, A.: Modele dynamiki układów fizycznych dla inżynierów. Wydawnictwo Naukowo-Techniczne, Warszawa (2008)
4. Gawliński, A., Jendrysik, S., Kowol, D., Rogala-Rojek, J., Stankiewicz, K., Woszczyński, M.: Doświadczenia z badań i wdrożenia systemu sterowania osadzarką pulsacyjną OS36 w KWK Budryk. In: KOMEKO 2011, Innowacyjne i przyjazne dla środowiska techniki i technologie przeróbki surowców mineralnych. Bezpieczeństwo - Jakość - Efektywność, Instytut Techniki Górniczej KOMAG, Gliwice, pp. 137–144 (2011)
5. Glinka, T.: Zużycie energii elektrycznej w układach napędowych przenośników taśmowych. Zeszyty Problemowe – Maszyny Elektryczne **73**, 37–42 (2005)
6. Górecki, H.: Optymalizacja i sterowanie systemów dynamicznych. Uczelniane Wydawnictwa Naukowo – Dydaktyczne AGH, Krakowie (2006)
7. Janiszowski, K.: Identyfikacja modeli parametrycznych w przykładach. Akademicka Oficyna Wydawnicza EXIT (2002)
8. Jendrysik, S.: Metoda sterowania przenośnikiem kubełkowym w osadzarkowym węźle wzbogacania. Praca doktorska, Politechnika Śląska (2016)
9. Jendrysik, S., Woszczyński, M., Bartoszek, S., et al.: Zintegrowanie systemu sterowania osadzarki i podnośnika kubełkowego w kopalni KWK Budryk. Praca statutowa ITG KOMAG, Gliwice (2011). (materiały niepublikowane)
10. Kasprzyk, J., et al.: Identyfikacja procesów. Wydawnictwo Politechniki Śląskiej, Gliwice (2002)
11. Kosmol, J.: Napędy Mechatroniczne. Wydawnictwo Politechniki Śląskiej, Gliwice (2013)
12. Kost, G., Nierychlok, A.: Napęd hybrydowy. Koncepcja sterowania. Przegląd Mechaniczny, marzec (2011)
13. Kowal, J.: Podstawy automatyki. Uczelniane Wydawnictwo Naukowo-Dydaktyczne AGH, Kraków (2003)
14. Kost, G., Łebkowski, P., Węsierski, Ł.: Automatyzacja i robotyzacja procesów produkcyjnych, Polskie Wydawnictwo Ekonomiczne (2013)
15. Królikowski, A., Horla, D.: Identyfikacja obiektów sterowania, metody dyskretne parametryczne. Wydawnictwo Politechniki Poznańskiej, Poznań (2010)

16. Ljung, L.: System Identification: Theory for the User. Prentice-Hall, Englewood Cliffs (1987)
17. Maciejowski, J.M.: Predictive Control with Constraints. Prentice Hall, London (2002)
18. Mańczyk, K.: Metody identyfikacji wielowymiarowych obiektów sterowania. Wydawnictwa Naukowo - Techniczne (1971)
19. Morari, M., Lee, J.H.: Model predictive control: past, present and feature. Comput. Chem. Eng. **23**, 667–682 (1999)
20. Mukherjee, A.K., Mishra, B.K.: An integral assessment of the role of critical process parameters on jigging. Int. J. Miner. Process. **81**(2006), 187–200 (2006)
21. O'Dwyer, A.: Handbook of PI and PID Controller Tuning Rules. Imperial College Press, London (2006)
22. Soderstrom, T., Stoica, P.: System Identification. Prentice-Hall, Hemel Hempstead (1989)
23. Świder, J., Baier, A., Kost, G., Zdanowicz, R.: Sterowanie i automatyzacja procesów technologicznych i układów Mechatronicznych. Wydawnictwo Politechniki Śląskiej, Gliwice (2002)

Magnetorheological Suspension Based on Silicone Oil

Sławomir Kciuk[1(✉)], Monika Kciuk[2], Tomasz Machoczek[1], and Wojciech Klein[1]

[1] Institute of Theoretical and Applied Mechanics, Silesian University of Technology, Konarskiego Street 18a, 44-100 Gliwice, Poland
Slawomir.Kciuk@polsl.pl

[2] Institute of Engineering Materials and Biomaterials, Silesian University of Technology, Konarskiego Street 18a, 44-100 Gliwice, Poland
monika.kciuk@polsl.pl

Abstract. Magnetorheological materials (MR fluids) are a class of smart materials whose rheological properties (e.g. viscosity) may be rapidly varied by applying a magnetic field. Under influence of magnetic field the suspended magnetic particles interact to form a structure that resists shear deformation or flow. This process takes only a few (1–10 ms) and magnetorheological fluid rapidly changes its viscosity.

The main aim of this article was to present the investigation results of magnetorheological fluids (MR) composed of carbonyl iron (CI) particles and analyse their flow behaviour in terms of the internal structure formation by a control of applied external magnetic field. The paper presents also the results of the experimental studies of a prototype magnetorheological damper at various magnitudes of control current as well as the manner of modelling electromagnetic phenomena occurring in the damper are presented in this paper.

The researches were carried out on model MR fluid which was prepared using silicone oil OKS 1050 mixed with carbonyl iron powder CI.

Keywords: Computational material science · Finite element method · Smart materials · Magnetorheological materials · MR damper

1 Introduction

Magnetorheological fluids belong to a group of non-Newtonian fluids, i.e. such that do not meet Newton's law of fluid friction (the η kinetic viscosity coefficient is variable). They also represent a group of purely-viscous fluids with the liquid limit and the controllability of their properties by means of a magnetic field [1–5].

A magnetorheological fluid (MR fluid) is a type of smart fluid in a carrier fluid, usually a type of mineral oil, synthetic oil, water or ethylene glycol. Typical magnetorheological fluids are the suspensions of micron sized, magnetizable particles (iron, iron oxide, iron nitride, iron carbide, reduced carbonyl iron, unreduced carbonyl iron,

J. Świder et al. (Eds.): MECHATRONICS 2017, AISC 934, pp. 201–219, 2019.
https://doi.org/10.1007/978-3-030-15857-6_21

chromium dioxide, low-carbon steel, silicon steel, nickel, cobalt, and combinations thereof [4]) with diameter in range from 3 to 5 microns [6, 7].

A variety of additives (stabilizers and surfactants) are used to prevent gravitational settling and promote stable particles suspension, enhance lubricity and change initial viscosity of the MR fluids. The stabilizers serve to keep the particles suspended in the fluid, whilst the surfactants are adsorbed on the surface of the magnetic particles to enhance the polarization induced in the suspended particles upon the application of a magnetic field [8–10].

In the absence of an applied field, MR fluids are reasonably well approximated as Newtonian liquids. For most engineering applications a simple Bingham plastic model is effective at describing the essential, field-dependent fluid characteristics. A Bingham plastic is a non-Newtonian fluid whose yield stress must be exceeded before flow can begin [10]. Thereafter, the rate-of-shear vs. shear stress curve is linear. In this model, the total yield stress is given by (1):

$$\tau = \tau_o(H) + \eta\dot{\gamma} \tag{1}$$

where:

τ_o – yield stress caused by applied magnetic field, [Pa]
H – magnetic field strength, [A/m]
$\dot{\gamma}$ – shear rate, [s^{-1}]
η – plastic viscosity, [Pa s]

When MR fluid is subjected to a magnetic field, the fluid greatly increases its apparent viscosity, to the point of becoming a viscoelastic solid in the presence of a magnetic field, the particles magnetize and form chain-like structures that align in the direction of the applied field as depicted in Fig. 1.

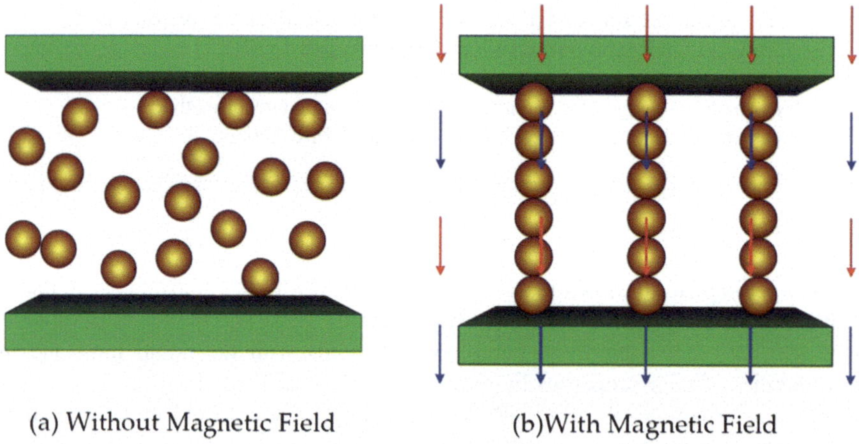

Fig. 1. Micro-structures of MR fluids without/with external magnetic field [11].

Advances in the application of MR materials are parallel to the development of new, more sophisticated MR materials with better properties and stability. Many smart systems and structures would benefit from the change in viscosity or other material properties of MR. Nowadays, these applications include brakes, dampers, clutches and shock absorbers systems.

In the recent several years magnetorheological dampers have been widely investigated both, due to their interesting phenomenological characteristics as well as potential possibilities to use them for controlling vibrations in pneumatic systems. They are especially important in the automotive industry, where improved driving conditions are inherent to a driver's safety [8–10].

MR dampers are semi-active devices that contain magnetorheological fluids. After application of a magnetic field the fluid changes from liquid to semi-solid state in few milliseconds, so the result is an infinitely variable, controllable damper capable of large damping forces. MR dampers offer an attractive solution to energy absorption in mechanical systems and structures and can be considered as "fail-safe" devices [11–16].

Figure 2 shows schematic example of MR fluid dampers. It can be seen that the operation of this device is fundamentally different from that of brakes and clutches, in that MR fluid is forced through annular orifices rather than being directly sheared [17].

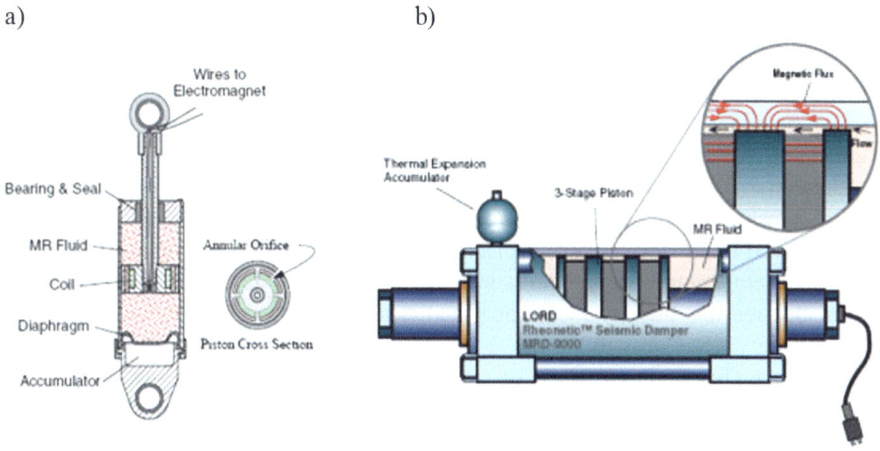

Fig. 2. Schematic diagram of the MR fluid dampers (a) small-scale SD-1000 MR fluid damper, (b) the prototype 20-ton large-scale MR fluid damper [17, 18].

In the SD-1000 MR fluid damper (Fig. 2a), magnetorheological fluid flows from a high pressure chamber to a low pressure chamber through an orifice in the piston head. The damper is 21.5 cm long in its extended position, and the main cylinder is 3.8 cm in diameter. The main cylinder houses the piston, the magnetic circuit, an accumulator, and 50 ml of MR fluid. The damper has a ±2.5 cm stroke. The magnetic field, which is perpendicular to the fluid flow, is generated by a small electromagnet in the piston head. Forces of up to 3,000 N can be generated with this device [18].

The damper shown in Fig. 2b, uses a particularly simple geometry in which the outer cylindrical housing is part of the magnetic circuit. The effective fluid orifice is the entire annular space between the piston outside diameter and the inside of the damper cylinder housing.

Movement of the piston causes fluids to flow through this entire annular region. The damper is double-ended. The advantage of this arrangement is that a rod-volume compensator does not need to be incorporated into the damper, although a small pressurized accumulator is provided to accommodate thermal expansion of the fluid. In this damper maximum damping force of 200,000 N can be generated.

2 Experimental

In order to prepare the prototype magnetorheological fluid the carbonyl iron powder was selected (CI, reduced penta-carbonyl iron, SiO_2 – coated, BASF, Germany) as for a model particle suspended system.

The carbonyl iron particles are coated with silica which provides an effective stabilization against aggregation. CI is magnetically soft material characterized by high saturation magnetisation. The average particle size and tap density was 7.0 μm and 4.3 g/cm^3, respectively. The concentration of CI was fixed at 20 and 40 wt%. In the aim of reduction of sedimentation process of the CI particles additional components (1 against CI amount) were added to the fluids. For the silicon oil the fumed silica (Aerosil 200, Degussa, Germany) was chosen.

The function of a carrier liquid is primarily to provide a liquid in which the magnetically active phase particulates are suspended.

The carrier liquid should also be largely non-reactive towards the magnetic particles. Similarly, the carrier liquid should be non-reactive toward the components/materials used in the device. When selecting a carrier liquid, it is important to consider the boiling temperature, vapour pressure at elevated temperatures and freezing point.

For that reasons magnetorheological particles were dispersed in colorless silicone oil (SO, type OKS, Germany).

Metallographic examinations of the material structure were made on the SUPRA^{TM}25 ZEISS scanning electron microscope with EDAX energy dispersion X-ray spectrometer with 20 kV accelerating voltage.

Magnetorheological characteristics were examined using laboratory test stand, which could work as viscometer or it could perform a function of laboratory clutch (Fig. 3).

The sedimentation was measured by visual observation of the position changes of boundary between clear and turbid part of carrier oil. Prepared samples were placed into cylindrical glass test tubes (length 1 m, diameter 30 mm) for a few days. As a

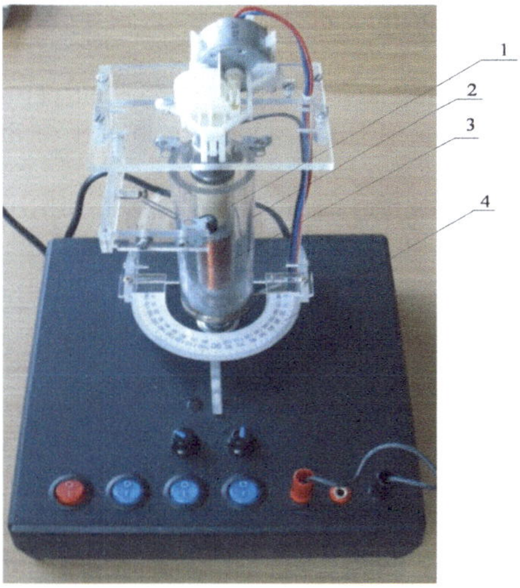

Fig. 3. Apparatus for MR effect measurement. 1 – shaft (drive) and inner cylinder, 2 – outer cylinder, 3 – coil, 4 – needle and angular displacement scale

result sedimentation ratio (R) was calculated. Sedimentation ratio is simply defined as a proportion between length of clear and turbid part of MR fluid (2):

$$R = \frac{a}{a+b} \cdot 100\% \tag{2}$$

where: R [%] – sedimentation ratio, a – length of the clear part, b – length of the turbid part.

A prototype of the MR damper was designed and produced as part of own work (Fig. 4).

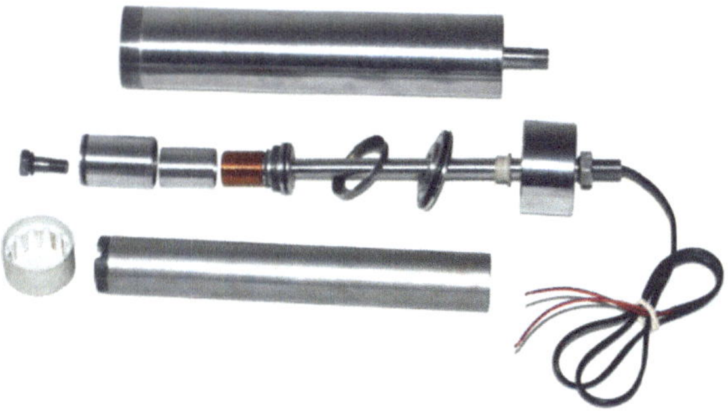

Fig. 4. Prototype MR damper

3 Results and Discussion

Metallographic microscopic research that was carried out permitted to observe morphologies of the carbonyl iron (CI) and applied stabiliser of the magnetorheological fluid – fumed silica (Aerosil 200).

The carbonyl iron particles possess in general spherical shape (Fig. 5).

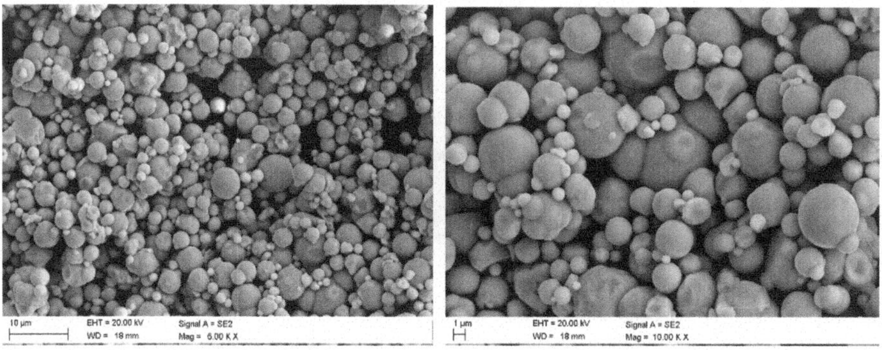

Fig. 5. Scanning electron micrograph of carbonyl iron particles used to preparation of model MR fluids.

Initially constructed chain structure of these particles can be transformed into an assembled structure of individual chains through the chain rupture and reformation process by shear deformation under an applied magnetic field.

The surface of CI is coated by silicon dioxide layer with the purpose of protection carbonyl iron particles against agglomeration.

On the base of metallographic observations it was found that analysed silica (Aerosil 200) was characterized by irregular shape (Fig. 6). The silica surrounds the spherical CI particles, ensuring better protection against their agglomerations in suspension.

Fig. 6. Scanning electron micrograph of silicon dioxide (Aerosil 200).

The effect of additives on the stability of the magnetorheological fluid has been investigated experimentally. The filled MR fluids having different compositions were set in a static condition, until it reached asymptotic values. Analyzes the liquid MR about viscosity 50 mPa s and concentration of carbonyl iron 40%, investigation showed slow decrease of the sedimentation coefficient in the case of the fluid with and without the silica (Fig. 7). It was also noticed that stability of investigated fluids improves after adding the stabilizer. This is the effect of the silica gel arising net, which prevents agglomeration of the carbonyl iron particles and the same descent of the agglomerates on the bottom of the reservoir.

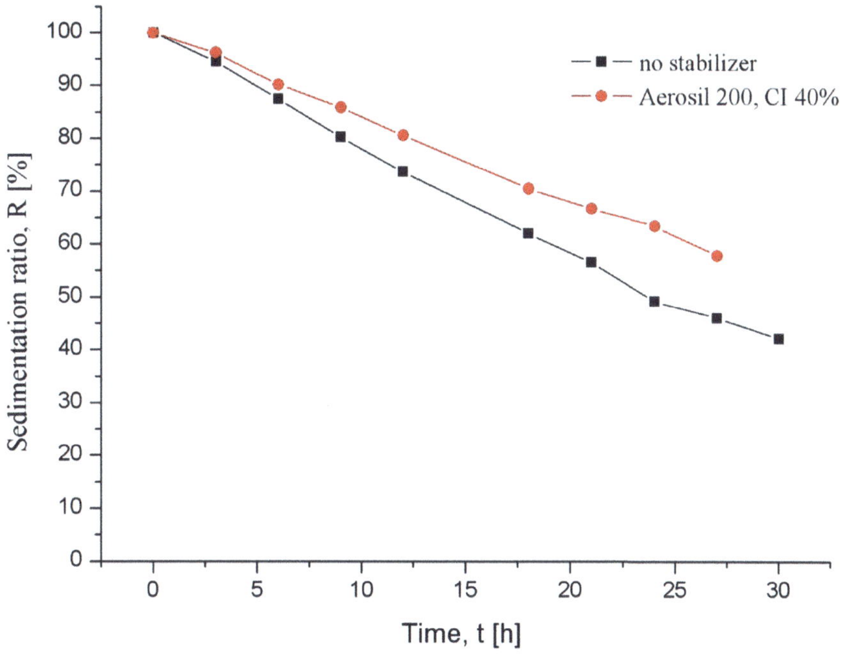

Fig. 7. Sedimentation ratio R of MR fluid (OKS 1050, 50 mPa s).

The main aim of the investigations was the qualification of the influence of external magnetic field on the dynamic viscosity of the prototype magnetorheological fluid. For this reason under consideration were taken following factors: concentration of magnetic particles (carbonyl iron) and influence of applied stabilizer (Aerosil 200).

The most important for the magnetorheological properties is presence of the stabilizing additives which further increase the MR effect. Besides that in the presence of stabilizers MR fluid are more homogeneous and faster respond to the external magnetic field at lower magnetic field intensity. It was affirmed on the basis of experimental results that the dynamic viscosity of studied liquids increase with enlargement of the intensity of the external magnetic field after the addition of silica considerably (Fig. 8). This is caused that added silica prevents concentrating of the CI particles as a result of

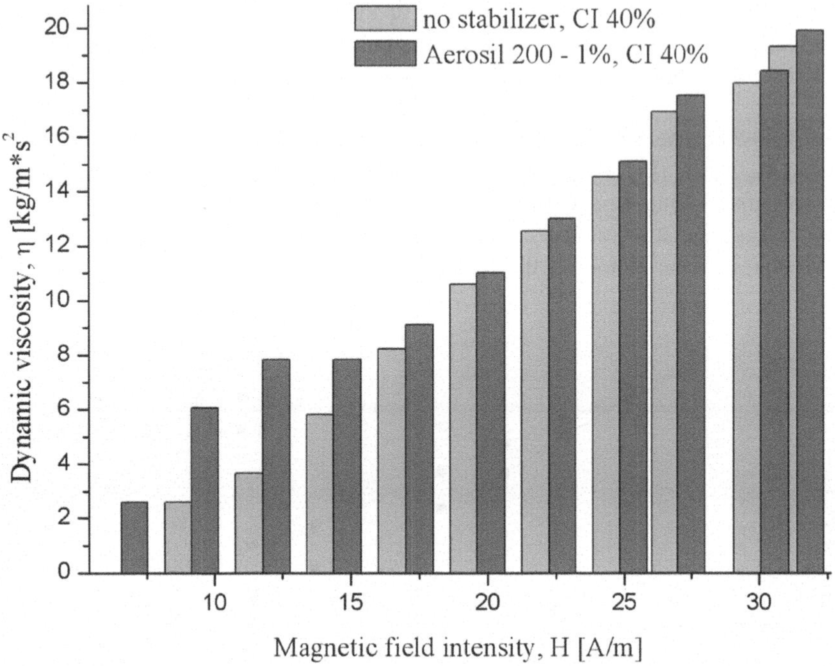

Fig. 8. The influence of the stabilizer on the dynamic viscosity of MR fluid (OKS 1050, 50 mPa s).

the gel mesh formation. For this reason the particles of the iron slowly fall on the bottom of the drum and larger quantity of magnetic particles has been still subjected to the applied external magnetic field.

It was affirmed on the basis of conducted investigations that with the enlargement of the concentration of magnetic particles (CI) increase the dynamic viscosity of analysed liquids together with the extension of the intensity of the external magnetic field. Higher content of CI increase also the stability of the investigated fluids. The values of the dynamic viscosity increase up to 21 kg/m s^2 in the investigated MR fluid (Fig. 9).

A piston rod of the prototype magnetorheological damper was the subject of a numerical analysis concerning electromagnetic effects. The following calculation modules were chosen for the process of electromagnetic phenomena modelling in the prototype MR damper:

- a mechanical module responsible for strength properties,
- a magnetic module that allows to accomplish a field distribution characteristic based on a colour flat model or a spatial model created by revolving a 2-D model around a vector located on the left-hand side,
- a dynamic module showing fluid behaviour relative to the displacing piston.

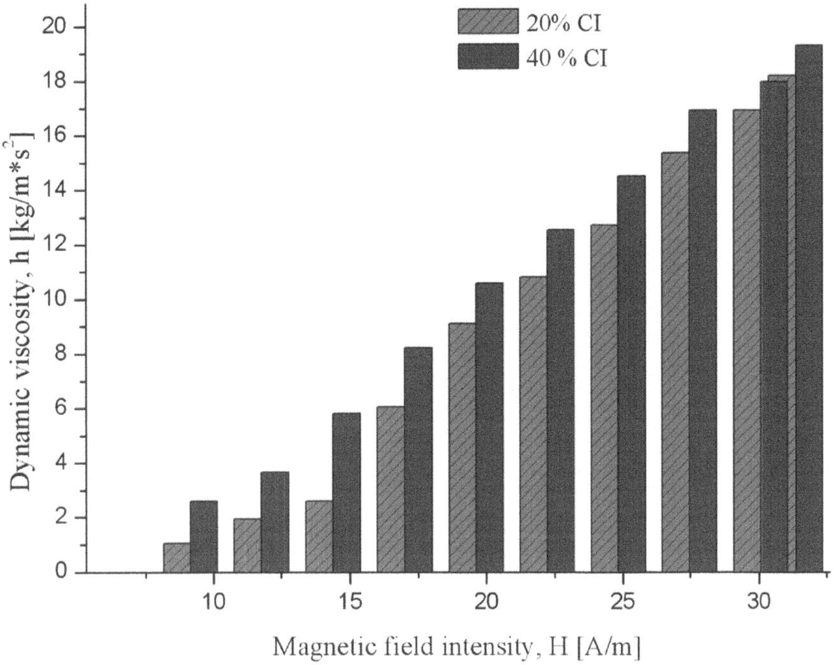

Fig. 9. Dependence of MR fluid (OKS 1050, 50 mPa s) dynamic viscosity on the magnetic field intensity.

An axisymmetric damper model (Fig. 10) was created with a special focus on the elements where the distribution of the magnetic field has the largest effect. The coordinate values defining the location of the individual constituent parts and their geometry were introduced for this purpose.

All the surfaces were divided into finite elements using quadrilateral, octanodal elements with the four degrees of freedom in the PLANE 53 type node described with a low order polynomial. The used option allows us to divide the grid in a coarse manner, but allowing to achieve the required field distribution at the expense of shortening the calculation time.

Materials were chosen for the adopted damper areas and their material properties were defined, in particular magnetic permeability (Table 1).

Table 1. The relative magnetic permeability of the materials making up the piston rod.

Diamagnetic materials	μ_r
Copper	0.9999999
Ferromagnetic materials	μ_r
Carbon steel (E295)	50–220
Prototype magnetorheological fluid	5
Paramagnetic materials	μ_r
Air	1.0000004

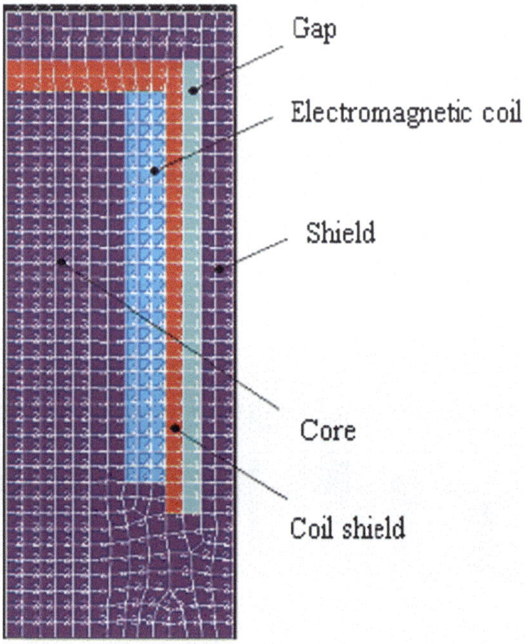

Fig. 10. Piston division into finite elements – the calculation areas

The magnetic permeability μ_r is a value characterising the behaviour of a given material body as a result of the magnetic field's activity. It depends on the magnetic induction vector $\vec{B}$ the magnetic field intensity vector $\vec{H}$ (3):

$$\vec{B} = \mu_0 \cdot \mu_w \cdot \vec{H} \tag{3}$$

where:

$\mu_0 = 4 \cdot \pi \cdot 10^{-7} \left[\frac{H}{m}\right]$ denotes the magnetic permeability of vacuum,
$\mu_w = \frac{\mu_r}{\mu_0}$ is relative magnetic permeability.

One can distinguish in dependence of magnetic permeability values:

- ferromagnetic materials ($\mu w \gg 1$),
- paramagnetic materials ($\mu w > 1$),
- diamagnetic materials ($\mu w < 1$).

All the piston's surfaces were limited with a line beyond which the magnetic field's distribution is of negligible importance. This process once again permits to simplify the calculation stage and to observe the line of forces of the piston port.

The geometric and electromotoric parameters of electromagnetic coil were defined:

- space underneath the electromagnetic coil,
- copper wire for electromagnetic winding was chosen,
- the active coil field was identified considering the completing coefficient,
- current's value was defined based on the system's resistance,
- current's density necessary to determine the magnetic fields' intensity was defined.

The images of magnetic field line distribution were produced when selecting the relative magnetic permeability (acc. to Table 1) for the elements.

The actual characteristics of magnetisation identified experimentally were used as the values of relative magnetic permeability of the piston housing material (E295) – Fig. 11.

An analysis of the non-linear model and the distribution of magnetic field line reveals that the lines along which the forces of this field are active change their inclination angle insignificantly thus being closer to the required straight angle improving the working conditions.

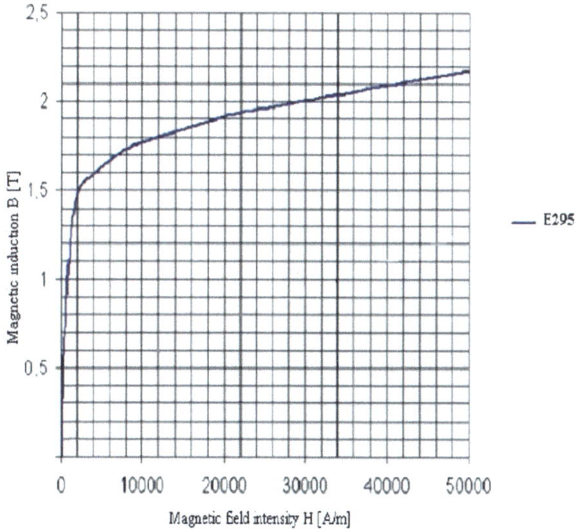

Fig. 11. The actual characteristic of magnetisation for the selected constituent parts of the magnetorheological damper.

The activity of magnetic forces occurs to the largest extent at the beginning and at the end of the active part of the port.

It was stated that the largest effect of concentration occurs at the height of the start and the end of the coil winding, and it is falling when nearing its centre (Fig. 12).

Distribution of the magnetic field lines, values of magnetic flux density and corresponding magnetic field strengths have been obtained in the form of maps and in the form of a vector field, which enable determination of dynamic damping forces that occur in the model of magnetorheological fluid (Figs. 13, 14, 15, 16 and 17).

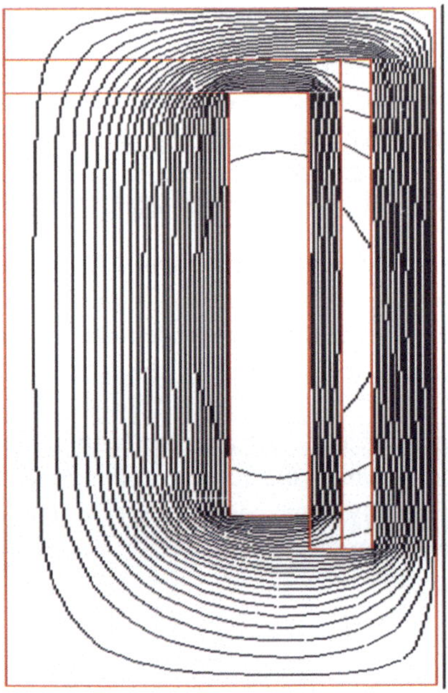

Fig. 12. Distribution of magnetic field line.

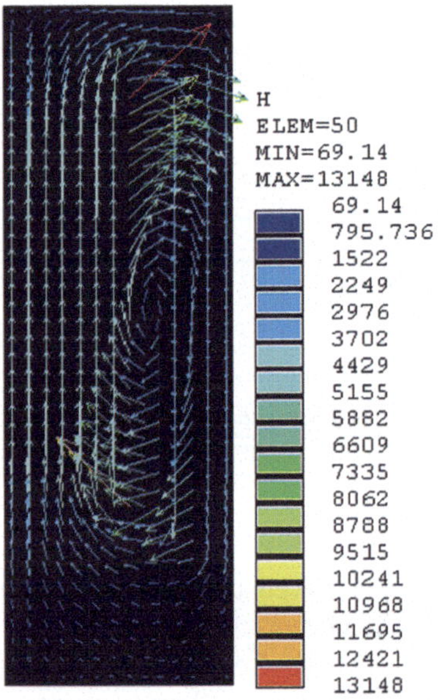

Fig. 13. The concentration of the magnetic field represented with vectors.

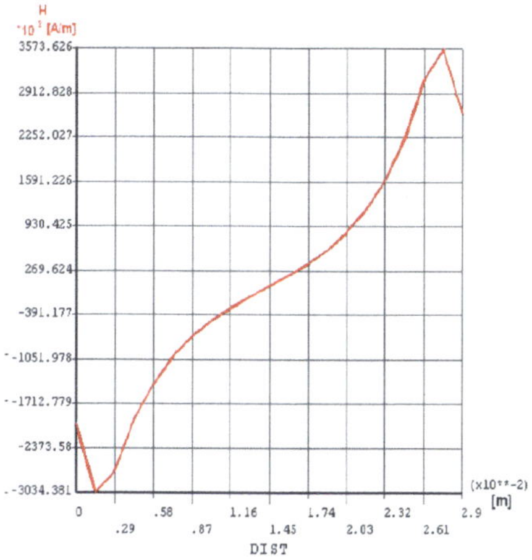

Fig. 14. The diagram of magnetic concentration relative to the port length.

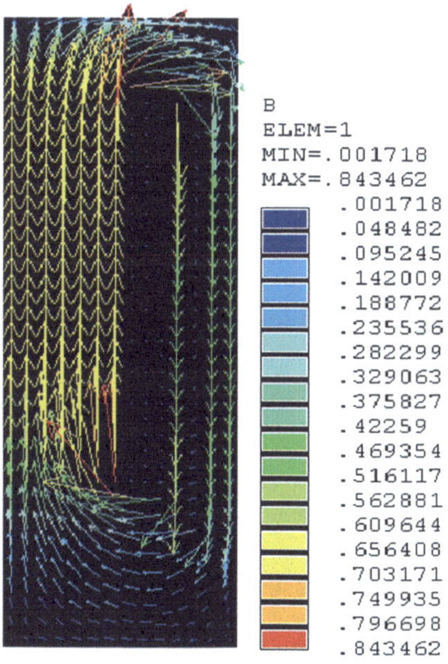

Fig. 15. The vectoral distribution of magnetic induction for non-linear analysis.

Fig. 16. A colour map of magnetic induction distribution – non-linear model.

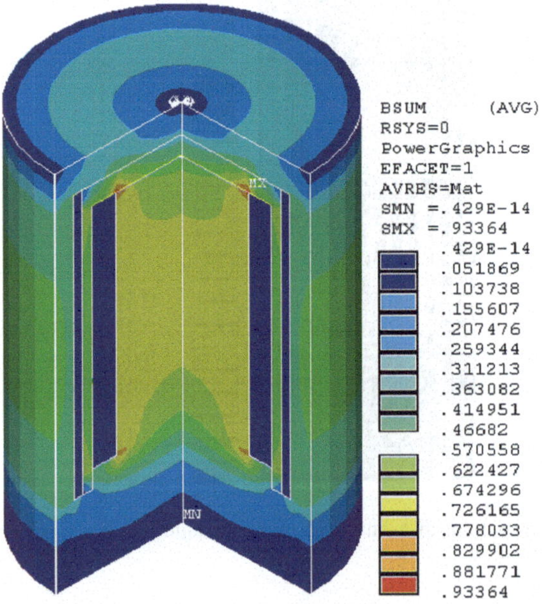

Fig. 17. A spatial view of magnetic induction distribution.

On the base of the investigations it was revealed that the flux density increases linearly from the start of gap length to the point where coil winding starts. It then decreases nonlinearly down to zero in the middle of the total gap length (Fig. 18).

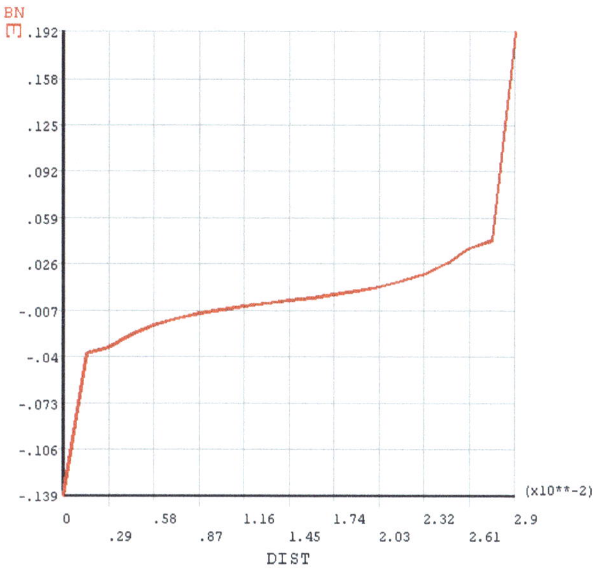

Fig. 18. Magnetic induction chart according to port length – non-linear model.

An identification experiment was carried out in the next stage of the modelling process for a semi-active suspension system where the dynamic characteristic of the prototype MR damper was produced.

The obtained damping force curves were dependent upon the speed of the piston force and the fluid flowing in the coil of the prototype MR damper.

The identification experiment was made on the MTS testing machine (Fig. 19).

Dynamic characteristics were identified in the tests by assuming:

- coil current of 0 A to 5 A,
- piston rod speed of 0 m/s to 1 m/s.

The characteristics produced for the selected parameters of the adopted range are shown in the Figs. 20, 21 and 22.

The damper's dynamic model was established in Simulink software on the basis of the characteristics describing the prototype MR damper's dynamics recorded in the experiment including their geometric form (piston rod's stroke).

The block identified as F(v, I) represents the damper dynamics (Fig. 23) identified in the experimental studies and recorded in Simulink.

The established dynamic MR damper model was used for performing numerical simulations to revise the adopted assumptions were correct. The simulations were made for the system with two degrees of freedom.

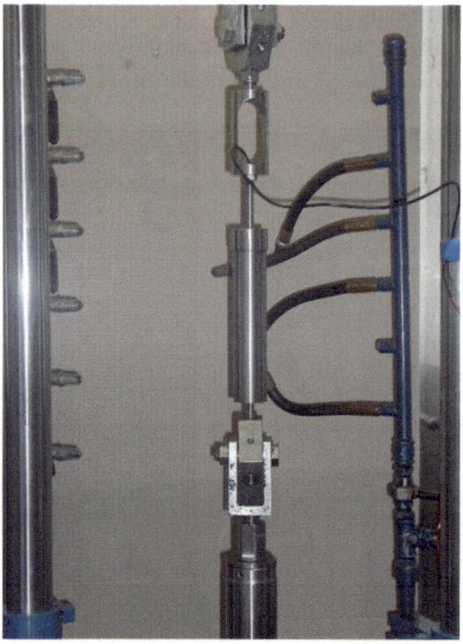

Fig. 19. MTS test station for the evaluation of dynamic characteristics of the prototype MR damper.

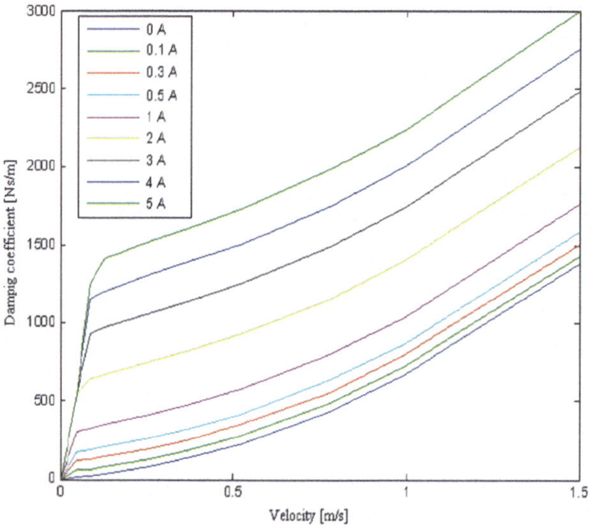

Fig. 20. A group of damping force characteristics according to speed and piston force for different coil current values of the prototype MR damper.

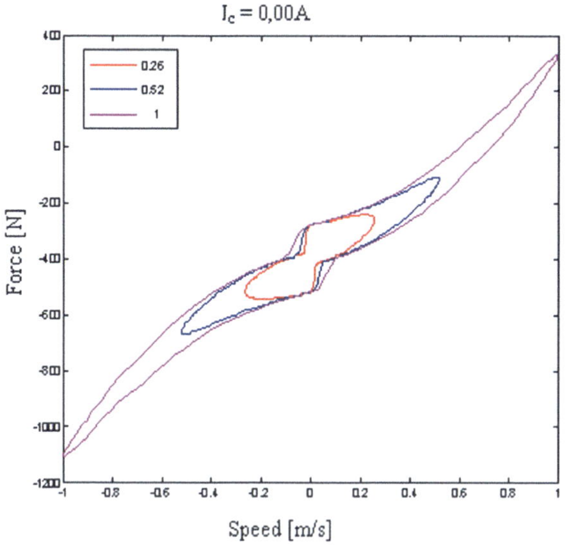

Fig. 21. A group of damping force characteristics according to three piston force speeds for the coil current of 0 A.

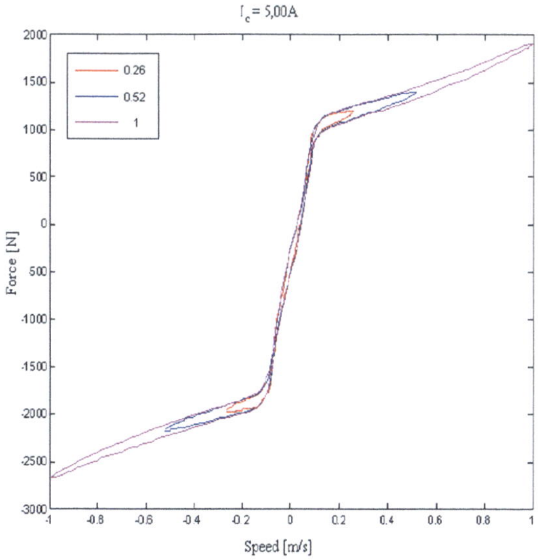

Fig. 22. A group of damping force characteristics according to three piston force speeds for the coil current of 5 A.

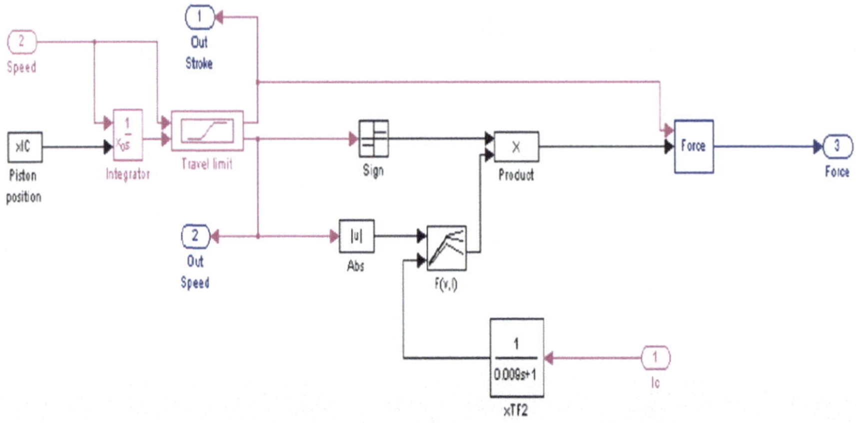

Fig. 23. The prototype MR damper model.

4 Conclusions

On the basis of conducted investigations it was stated that with the increase of carbonyl iron particles concentration the analysed magnetorheological fluids exhibit high stability in the course of time.

Added submicron-sized particles (fumed silica) in 1% with relation to CI amount further inhibited sedimentation of the carbonyl iron particles especially effective for high loaded (40%) MR fluid.

The dynamic viscosity of investigated magnetorheological fluids increase in the response to presence of an external magnetic field with the increase of CI amount and In the presence of the stabilizer additive (Aerosil 200).

The results presented in this paper show the good efficiency of vibration damping within the full range of excitation frequencies occurring while driving a car. The sprung mass' displacements for the Skyhook type control used and for all the studied excitation frequencies were lower than for the system with passive suppression.

One can observe a tendency of the active system's higher vibration isolation efficiency in relation to the passive system along with the growing excitation frequency.

Another characteristic of the established system is its much lower sensitivity within the range of resonance frequency.

Furthermore, active damping helps the vibrating object to reach its fixed state more quickly.

Acknowledgment. Authors would like to thank BASF AG for the kindly provided carbonyl iron samples.

References

1. Dorfmanna, A., Ogden, R.W., Wineman, A.S.: A three-dimensional non-linear constitutive law for magnetorheological fluids, with applications. Int. J. Non-Linear Mech. **42**, 381–390 (2007)
2. Bossis, G., Lacis, S., Meunier, A., Volkova, O.: Magnetorheological fluids. J. Magn. Magn. Mater. **252**, 224–228 (2002)
3. Premalatha, S.E., Chokkalingam, R., Mahendran, M.: Magneto mechanical properties of iron based MR fluids. Am. J. Polym. Sci. **2**(4), 50–55 (2012)
4. Kciuk, M., Kciuk, S., Turczyn, R.: Magnetorheological characterization of carbonyl iron based suspension. J. Achiev. Mater. Manuf. Eng. **33**(2), 135–141 (2009)
5. Turczyn, R., Kciuk, M.: Preparation and study of model magnetorheological fluids. J. Achiev. Mater. Manuf. Eng. **27**(2), 131–134 (2008)
6. Sapiński, B.: Real-time Control of Magnetorheological Dampers in Mechanics System. AGH University of Science and Technology Press, Cracow (2008)
7. Jolly, M., Bender, J.W., Carlson, J.D.: Properties and applications of commercial magnetorheological fluids. In: SPIE 5th Annual International Symposium on Smart Structures and Materials, San Diego, CA (1998)
8. Klein, W., Otorowski, M.: Review of magnetorheological fluids applications in mechanical systems. In: 9th International Seminar of Applied Mechanics, Wisła 2005, pp. 75–82 (2005). (in Polish)
9. Klein, W., Mężyk, A., Otorowski, M.: The application overhaul of magnetorheological fluids in mechanical engineering. In: XIII International Scientific Conference TEMAG, Gliwice-Ustroń 2005, pp. 95–105 (2005)
10. Pranoto, T., Nagaya, K.: Development on 2DOF-type and Rotary-type shock absorber damper using MRF and their efficiencies. J. Mater. Process. Technol. **161**, 146–150 (2005)
11. Ashfak, A., Saheed, A., Rasheed, K.A., Jaleel, J.A.: Design, fabrication and evaluation of MR damper. Int. J. Aerosp. Mech. Eng. **53**, 358–363 (2009)
12. Yang, G., Spencer Jr., B.F., Carlson, J.D., Sain, M.K.: Large-scale MR fluid dampers: modeling and dynamic performance considerations. Eng. Struct. **24**, 309–323 (2002)
13. Sapiński, B., Rosół, M.: MR damper performance for shock isolation. J. Theoret. Appl. Mech. **45**(1), 133–145 (2007)
14. Ahmadian, M.: On the isolation properties of semi-active dampers. J. Vib. Control **5**(2), 217–232 (1999)
15. Muhammad, A., Yao, X., Deng, Z.: Review of magnetorheological (MR) fluids and its applications in vibration control. J. Marine Sci. Appl. **5**(3), 17–29 (2006)
16. Chooi, W.W., Oyadiji, S.O.: Design, modelling and testing of magnetorheological (MR) dampers using analytical flow solutions. Comput. Struct. **86**, 473–482 (2008)
17. Carlson, J.D., Jolly, M.R.: MR fluid, foam and elastomer devices. Mechatronics **10**, 555–569 (2000)
18. Kciuk, M., Turczyn, R.: Properties and application of magnetorheological fluids. J. Achiev. Mater. Manuf. Eng. **18**(1–2), 127–130 (2006)

Experimental Verification of the Filtration Phenomena in Hydraulic Systems

Klaudiusz Klarecki(✉) and Dominik Rabsztyn

Institute of Technological Processes, Automation and Integrated Manufacturing System, Silesian University of Technology, Konarskiego Str. 18 A, 44-100 Gliwice, Poland
klaudiusz.klarecki@polsl.pl

Abstract. Article presents comparing of the authors' mathematical model of the filtration phenomena with the results of the experimental tests. Experimental studies have shown that the concentration of solid contaminant particles in the hydraulic fluid is affected by: the filter efficiency in removing solid particles of specific size x (in microns) also named as separation capacity β_x, relation of the pump flowrate to tank capacity Q/V, the filtration time. The obtained results indicate that for the cleaning of the hydraulic fluid is important the flushing time. During flushing of the hydraulic system the concentration of the contaminant particles decreases asymptotically to a certain value, which is dependent upon the particular hydraulic system, used filter and the environmental factors. The contamination levels of hydraulic fluids for tested hydraulic systems were measured using the Parker icount Oil Sampler. Samples of fluid were taken using the sampling valve, which was mounted in pressure line. Obtained experimental results show that efficiency of filter elements with the filtration ratings 10 μ and 20 μ dramatically decreasing for lower sizes of particles.

Keywords: Hydraulic fluid filtration · Filter B ratio · Flushing

1 Introduction

The hydrotronics is the present and future of hydraulic drives. But advanced control systems not mean that hydrotronic systems are more dirt-resistant then typical hydraulic units. If the designers of hydrotronic systems want to increase their reliability, they must thoroughly analyze factors that affect the contamination degree of the hydraulic fluid. Excessive contamination of hydraulic fluids is a major cause of damage and malfunction of hydraulic systems [1, 5]. It follows that the selection of a suitable hydraulic filter and its location into circuit is a critical moment in the design process of the hydraulic system [3, 4]. It is generally assumed that when choosing a filter element the most important is its beta ratio β_x. The results of this article suggest that besides beta ratio are other factors that affect the cleanliness of the hydraulic fluids.

J. Świder et al. (Eds.): MECHATRONICS 2017, AISC 934, pp. 220–230, 2019.
https://doi.org/10.1007/978-3-030-15857-6_22

2 Model of the Filtration Phenomena for Hydraulic Systems

Very important feature of hydraulic filters is Beta ratio β_x. Beta ratio β_x is a formula used to calculate the filtration efficiency of a particular fluid filter using base data obtained from multi-pass testing (Fig. 1). The formula used to calculate β_x is [2]:

$$\beta_x = \frac{N_u}{N_d} \tag{1}$$

where:

- N_u – number of solid particles in upstream fluid,
- N_d – number of solid particles in downstream fluid,
- x – minimum size of counted solid particles [μm].

Particulate removal efficiency n_f can be seen as the probability of capturing particles flowing through the filter (Fig. 2). The relationship between beta ratio and n_f is given by the formula:

$$n_f = \left(1 - \frac{1}{\beta_x}\right) \cdot 100\% \tag{2}$$

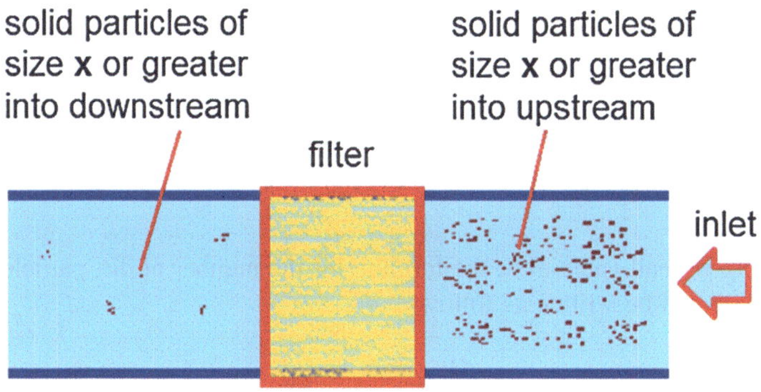

Fig. 1. Multi-pass testing

Model of filtration phenomena was based on the following assumptions:

- filter efficiency is equal the probability of capturing particles with size x or greater,
- particulates are evenly distributed in the liquid,
- the inner and outer particulates with total flux Θ enters the hydraulic liquid, particulate flux is accepted as constant,
- particulates appear uniformly throughout the entire volume of hydraulic fluid.

The probability of capture contaminant particle over unit time λ can be accept as [3]:

$$\lambda = \frac{n_f}{100\%} \cdot \frac{Q}{V} \tag{3}$$

where:

- Q – flow rate through filter [dm^3/min],
- V – total volume of hydraulic fluid into circuit [dm^3].

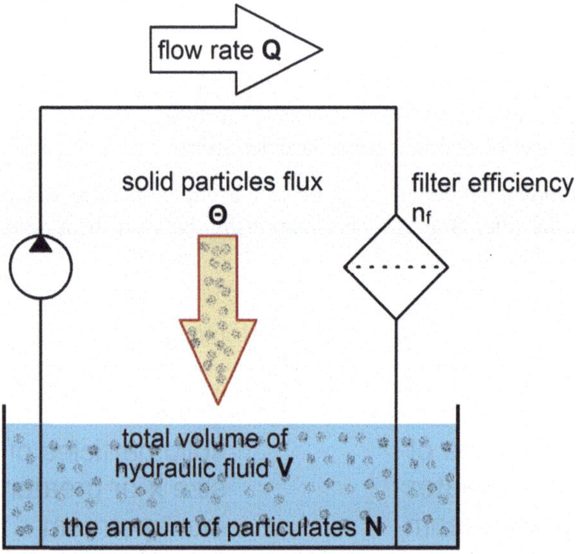

Fig. 2. Sketch of the analysed hydraulic system.

Assuming a small value of λ, the formula for the number of dirt particles N in the hydraulic fluid can be written as follows:

$$\frac{dN}{dt} = -\lambda N + \Theta \tag{4}$$

where:

- Θ – solid particulates flux

It can be seen that formula (4) is same to describing the phenomenon of radioactive decay.

Combining (3) and (4) were obtained:

$$\frac{dN}{dt} = -\frac{n_f}{100\%} \cdot \frac{Q}{V} N + \Theta \tag{5}$$

In general, the factors that determine the number of particles of pollutants according to (5), are not fixed. For example, filtration efficiency may depend on the degree of filling of the filter. Also, the liquid volume in the system may be varied, e.g. due to leakages.

Determination of the number of pollutants, taking into account the variability of factors n_f, Q, V and Θ most likely require the use of numerical methods to solve the Eq. (5). Further analysis was accepted a simplified model of filtration phenomenon in which factors have fixed values.

According to the accepted simplification, the solution of Eq. (5) is particular integral in a particular as a:

$$N = N_0 e^{-\frac{n_f}{100\%}\frac{Q}{V}t} + \Theta \frac{V \cdot 100\%}{n_f \cdot Q}\left(1 - e^{-\frac{n_f}{100\%}\frac{Q}{V}t}\right) \tag{6}$$

where:

- N_0 – initial number of contamination particles.

Determination of total contamination particle number N is not convenient. It is better to transform the formula (6) on the form, which will be determined the concentration of particulates ρ_c in the adopted unit volume. As the volume unit may be set to any value of the volume, for example 100 cm^3. This allows user the result relate directly to the ISO-codes according ISO 4406:1999. It should also be the calculated flow rate and fluid volume, including a new unit of volume according ISO4406, namely one volume unit is equal 100 cm^3.

$$\rho_c = \rho_{c0} e^{-\frac{n_f}{100\%}\frac{Q}{V}t} + \frac{\Theta \cdot 100\%}{n_f \cdot Q}\left(1 - e^{-\frac{n_f}{100\%}\frac{Q}{V}t}\right) \tag{7}$$

where:

- ρ_{c0} – initial concentration of particulates.

For formulas (6) and (7) can enter the time constant filtration T_f as:

$$T_f = \frac{V \cdot 100\%}{n_f \cdot Q} \tag{8}$$

It should be noted that for the filtration times $t \gg T_f$, the concentration of particulates approaching the values:

$$\rho_c = \frac{\Theta \cdot 100\%}{n_f \cdot Q} \tag{9}$$

From (9) it follows that for a given flux of particulates Θ, the particle concentration can be reduced by increasing filter efficiency n_f or increasing the flow through the filter Q.

3 Experimental Tests of Cleanliness of the Hydraulic Fluid During the Flushing of the Hydraulic Power Packs

Experimental tests were realized using two hydraulic power packs type HYBOX Flex Mini with 30 dm^3 volume tanks. Hydraulic power pack type Hybox flex mini contains return filter mounted on its valve basic block. Tested hydraulic units were equipped with gear positive displacement pump size 5 cm^3 powered by asynchronous motor 1420 rpm. Test stand is shown as the diagram of hydraulic circuit on Fig. 3.

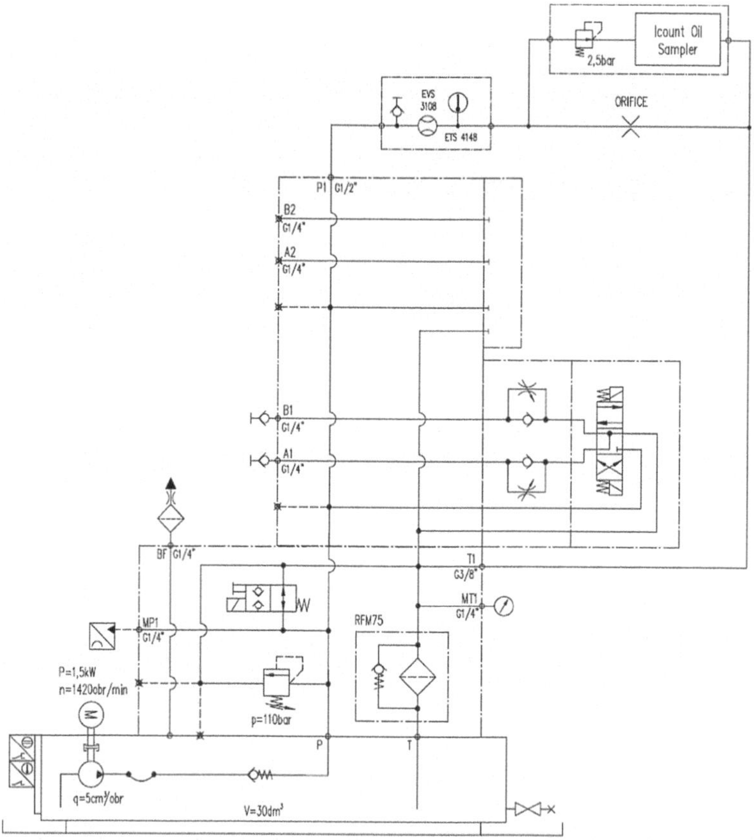

Fig. 3. Hydraulic diagram of the test stand.

The cleanliness of hydraulic fluid was measured using the icountOS portable unit for condition monitoring of hydrocarbons produced by Parker Hannifin. The icountOS can report the measuring results to the user in either ISO4406:1999 or NAS1638.

The tests were realized using in sequence four types of return filter elements made by HYDAC:

- 0075 R 020 ON (filter element for particles size 20 μm),

- 0075 R 010 ON (filter element for particles size 10 µm),
- 0075 R 005 BN3HC (filter element for particles size 5 µm),
- 0075 R 003 BN3HC (filter element for particles size 3 µm).

For technical reasons, the tests with second hydraulic unit were limited to the use of 20 µm and 10 µm filter elements.

Sample results of cleanliness measurements during flushing of hydraulic system using filter element type 0075 R 020 ON are shown on Fig. 4 and for filter element type 0075 R 010 ON are shown on Fig. 5.

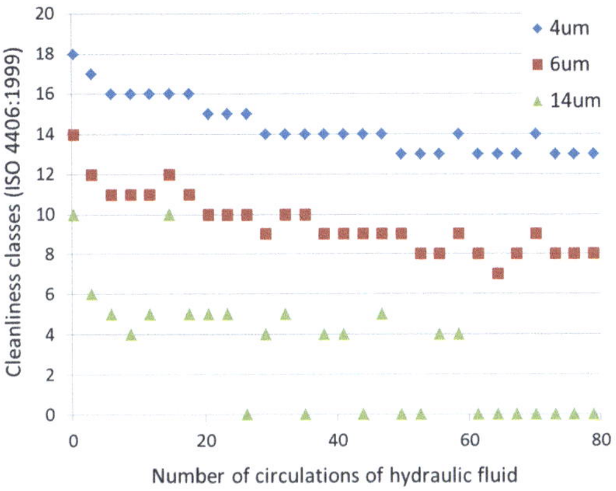

Fig. 4. Classes of contamination during the flushing of hydraulic system (according ISO4406:1999) using the filter element type 0075 R 020 ON.

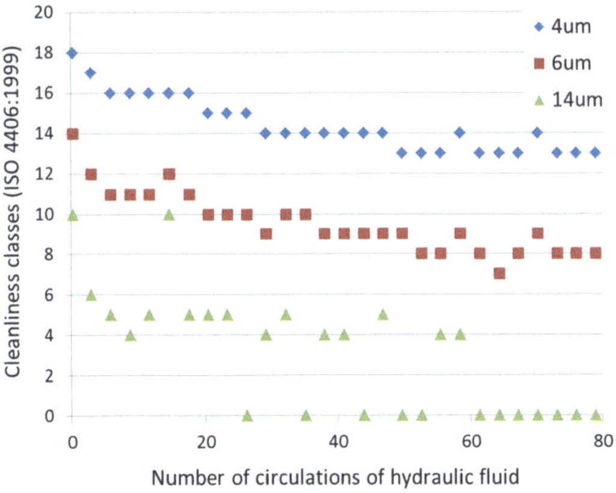

Fig. 5. Classes of contamination during the flushing of hydraulic system (according ISO4406:1999) using the filter element type 0075 R 010 ON.

For abscissa axes on Figs. 4 and 5 was accepted dimensionless unit [number of circulation of hydraulic fluid]. One circulation of hydraulic fluid is the time (for example called $\boldsymbol{T_{CoHF}}$) it takes the pump of hydraulic system for pumping through the filter all the liquid contained in the tank:

$$T_{CoHF} = \frac{V}{Q} \quad (10)$$

The adoption of such a unit of time allows for generalization of the results of research. For tested hydraulic power packs one full circulation of hydraulic fluid through filter lasted about 3.5 min. The $\boldsymbol{T_{CoHF}}$ value is the result of volume of oil in system Q = 26 dm^3 and pump flowrate Q = 7.4 dm^3/min.

The obtained results of cleanliness classes in steady state for flushing using different filter elements are presented in Table 1.

Table 1. Total results of cleanliness tests according ISO4406:1999.

	Filter element (0075 R ***)			
Range of particles size	020	010	005	003
	1 st hydraulic unit type HYBOX Flex Mini			
≥4 μm	**12**	**10**	**8**	**8**
≥6 μm	**8**	**6**	**4**	**4**
≥14 μm	**0**	**0**	**0**	**0**
	2 nd hydraulic unit type HYBOX Flex Mini			
≥4 μm	**13**	**11**		
≥6 μm	**8**	**7**		
≥14 μm	**0**	**0**		

Due to the unknown values of fluxes of particulates $\boldsymbol{\Theta}$ for each range of particles sizes (according Formula 9), it can only be carried out a comparative analysis of the filter efficiencies $\boldsymbol{n_f}$ of the tested filter elements.

Due to the constant ambient conditions assumed that for each tested filter element at the specific hydraulic unit occured the same value of flux particulates.

Please notice that one contamination class more is equal the double concentration of particles.

To comparing two filter elements (A and B) can be used Formula 9.

$$\rho_{cA} = \frac{\Theta \cdot 100\%}{n_{fA} \cdot Q} \qquad \rho_{cB} = \frac{\Theta \cdot 100\%}{n_{fB} \cdot Q}$$

where:

n_{fA} – filter A efficiency,

n_{fA} – filter B efficiency.

After transformation was eliminated the flux of particulates $\boldsymbol{\Theta}$.

$$\frac{\rho_{cA}}{\rho_{cB}} = \frac{n_{fB}}{n_{fA}} \tag{11}$$

where:

ρ_{cA} – value of concentration of solid particles into hydraulic liquid achieved using filter A,

ρ_{cB} – value of concentration of solid particles into hydraulic liquid achieved using filter B.

According ISO4406:1999 the concentration of particles is approximately:

$$\rho_c = 2^{cont\_class} \tag{12}$$

where:

- *cont_class* – measured cleanliness class according ISO4406:1999.

Combining (11) and (12) were obtained:

$$\frac{2^{count\_classA}}{2^{count\_classB}} = \frac{n_{fB}}{n_{FA}} \tag{13}$$

According formula (13) the filter elements have been compared each other for each range of particle size. Results are shown at Tables 2, 3, 4 and 5.

Table 2. Relationship of efficiencies of filter elements for particles size ≥4 μm (hydraulic unit No. 1).

		Code of filter element (equal to filtration rating [μm])			
		020	010	005	003
Code of filter element	020	1	$2^{(12-10)}$	$2^{(12-8)}$	$2^{(12-8)}$
	010	$2^{(10-12)}$	1	$2^{(10-8)}$	$2^{(10-8)}$
	005	$2^{(8-12)}$	$2^{(8-10)}$	1	$2^{(8-8)}$
	003	$2^{(8-12)}$	$2^{(8-10)}$	$2^{(8-8)}$	1
		020	010	005	003
	020	1	4	16	16
	010	0.25	1	4	4
	005	0.0625	0.25	1	1
	003	0.0625	0.25	1	1

Table 3. Relationship of efficiencies of filter elements for particles size ≥ 6 μm (hydraulic unit No. 1).

		Code of filter element (equal to filtration rating [μm])			
		020	010	005	003
Code of filter element	020	1	$2^{(8-6)}$	$2^{(8-4)}$	$2^{(8-4)}$
	010	$2^{(6-8)}$	1	$2^{(6-4)}$	$2^{(6-4)}$
	005	$2^{(4-8)}$	$2^{(4-6)}$	1	$2^{(4-4)}$
	003	$2^{(4-8)}$	$2^{(4-6)}$	$2^{(4-4)}$	1
		020	010	005	003
	020	1	4	16	16
	010	0.25	1	4	4
	005	0.0625	0.25	1	1
	003	0.0625	0.25	1	1

Table 4. Relationship of efficiencies of filter elements for particles size ≥ 4 μm (hydraulic unit No. 2).

		Code of filter element (equal to filtration rating [μm])				
		020	010		020	010
Code of filter element	020	1	$2^{(13-11)}$	020	1	4
	010	$2^{(11-13)}$	1	010	0.25	1

Table 5. Relationship of efficiencies of filter elements for particles size ≥ 6 μm (hydraulic unit No. 2).

		Code of filter element (equal to filtration rating [μm])				
		020	010		020	010
Code of filter element	020	1	$2^{(8-7)}$	020	1	2
	010	$2^{(7-8)}$	1	010	0.5	1

The results in the tables should be understood that the efficiency of filter in table column is compared with the filter in the table row.

For particles equal to or greater than 14 μm all filter elements are equally very high effective. According to [1] the beta ratio for particles size 14 μm of filter element with filtration ratings 20 μ is approximately 25. For example in same range of the particles size the filter element with filtration ratings 5 μ has beta ratio approximately 400000 times higher. However, according to formula (2), the filtration efficiencies n_f of these two filter elements for 14 μ particles do not differ much, for element with filtration rating 20 μ its n_f efficiency is equal 0.96 and for element with filtration rating 5 was obtained $\boldsymbol{n_f} = 1$. For particles equal to or greater than 14 μm all filter elements are equally effective. For particles equal to or greater than 4 μm and 6 μm filter elements with the filtration ratings 3 μm and 5 μm (003 and 005) are equally effective. Additionally, their efficiencies are significant higher than filter elements with the filtration ratings 10 μm and 20 μm (010 and 020). Minimum typical value of filter beta ratio for nominal size of particles is 75. It follows that filter efficiency for nominal size of particles is greater than 0.987, so can be assumed $\boldsymbol{n_f = 1}$. For filter elements with the filtration ratings 3 μm and 5 μm can be assumed the filtration efficiency $\boldsymbol{n_f}$ equals 1 for each range of particles size. The filter element with the filtration ratings 10 μm (010) is characterised by the filtration efficiency $\boldsymbol{n_f}$ equals 0.25 for particles equal to or greater than 4 μm and for particles equal to or greater than 6 μm. The filter element with the filtration ratings 20 μm (020) is characterised by the filtration efficiency $\boldsymbol{n_f}$ equals 0.0625 for particles equal to or greater than 4 μm and for particles equal to or greater than 6 μm. This means that the filter element (020) is four times less efficient than the filter element (010) to capture the small particles (sizes 4 μm and 6 μm). This is confirmed by the results presented in Table 4, although the results of Table 5 shows that the difference in efficacy compared filter elements (020 vs 010) is not so strong.

4 Estimation the Total Flux Value Θ of Solid Particulates

Assuming the value of filtration efficiency n_f as 1 for filter elements (005) and (003) can be estimated the total flux value Θ of solid particles for each range of particles size. After transformation formula (9) was obtained:

$$\Theta = \frac{n_f \cdot Q \cdot \rho_c}{100\%} \tag{14}$$

The results of solid particles flux Θ calculations are shown at Table 6, as concentration of particulates were accepted upper limits of the measured cleanliness classes according ISO4406:1999.

Table 6. The total flux of particles for tested hydraulic unit.

Range of particles size	Filter element 0075 R		Max concentration of particles [1/100 cm^3]	Total flux of solid particles [1/min]
	005	003		
≥ 4 μm	8	8	250	18500
≥ 6 μm	4	4	16	1184
≥ 14 μm	0	0	1	74

Should be noted that shown at Table 6 total flux value of solid particles for the laboratory environment was obtained.

5 Summary

It is generally accepted that the presence of the solid contaminant particles in the hydraulic liquid is main cause of the damages of hydraulic units. For the improve of the reliability and durability of the hydraulic components the extent of the hydraulic liquid contamination should be controlled and kept low.

The most important result of this research is that it is possible to determine the total flux value Θ of solid particles. The measurements of Θ can be carried out both in the laboratory and industry environments. Knowledge of the real total flux value of solid particles is necessary for optimal design of filtration system for hydraulic systems.

By the way it has been shown how the cleanliness of hydraulic fluid depends on the filtration rating of the filter element, although no significant difference between the filter elements with the filtration ratings 3 μm and 5 μm.

References

1. Eissa, A.S., Mohammed, N.I., Abd-Allah, R.I., El-Sheltawy, S.T.: Characterization of used and fresh hydraulic oils of heavy machinery plants. Pet. Sci. Technol. **31**(6), 625–632 (2013)
2. HYDAC Filtration Handbook No. E70.000.0/02.08. www.hydac.com.au
3. Klarecki, K.: Filtration and cleanliness of hydraulic fluids. Driv. Controls **1**(19), 72–77 (2017)
4. Michael, P.W., Wanke, T.S., McCambridge, M.A.: Additive and base oil effects in automatic particle counters. In: Automotive Lubricant Testing and Additive Development. Book Series, vol. 1501, pp. 109–115. American Society for Testing and Materials Special Technical Publication
5. O'Donnell, A., Porter, B., Steiger, G., Brinkman, D., Ventura, M.: Effects of cross-flow filtration and tramp oil removal on the components and performance properties of metalworking fluids. Lubri. Eng. **58**(11), 23–28 (2002)

Role of Didactical Stations in Education Process of Industrial Automatics Technical Staff

Janusz Kobiałka and Wojciech Zawiejski(✉)

Department of Mechanised Construction,
Automatization and Assembling Technologies,
Institute of Mechanised Construction and Rock Mining, ul. Mrówcza 243,
04-697 Warsaw, Poland
w.zawiejski@imbigs.pl

Abstract. Educational stations made at IMBiGS were described in the paper and their structure and functionalities were presented. Those stands were dedicated for developing practical skills during training process of technical staff. Wide range of stations can be used in all levels of schools.

Keywords: Educational stations · Didactical stations · Industrial network systems · Manufacturing control systems · Industrial automation · PLC controllers · HMI operator panels

1 Introduction

Mechatronics is an interdisciplinary field of science which requires from nowadays engineers or technicians sound knowledge within the scope of mechanics, electronics, software and control systems. Education and training programs dedicated for technical staff have to guarantee not only theoretical knowledge but also practical skills. The Industrial Automatics laboratories which are located in many of technical vocational schools, high schools and universities enable the realization of such training. The range of didactical stations which were designed and manufactured by Warsaw Institute of Mechanised Construction and Rock Mining, was to fulfil the needs of modern education units. Those didactical stations can be assigned to the following groups: desktops, objects and industrial networks. Selected examples of each group of educational stations were described in this paper and their structure and functionalities were presented.

2 Desktop Didactical Stations

Desktop stations are a sub group of didactical stations which are designed to create environment for practising operations such as: programming, data readout and acquisition, visualization of industrial processes for variety of modules and devices used in industrial automatic systems. Characteristic feature of desktop station is the focus on one single or a pair of modules or devices.

J. Świder et al. (Eds.): MECHATRONICS 2017, AISC 934, pp. 231–239, 2019.
https://doi.org/10.1007/978-3-030-15857-6_23

2.1 Desktop Didactical Stations Consisting of a PLC Controller and HMI Panel Are Designed to Allow Learning the Following Skills

- PLC controllers and HMI panels programming language,
- industrial communication protocols of serial network between PLC controller and HMI panel,
- creating visualisations of cooperation between controller and panel,
- methods of configuration of devices working in serial network and Ethernet network,
- the use of analog signals and discreet signals in control circuits,
- creating applications for the controller where output signals are indicated by dedicated LEDs,
- safety assurance in industrial automatic systems,
- control algorithms for high frequency and low frequency variable processes (Fig. 1).

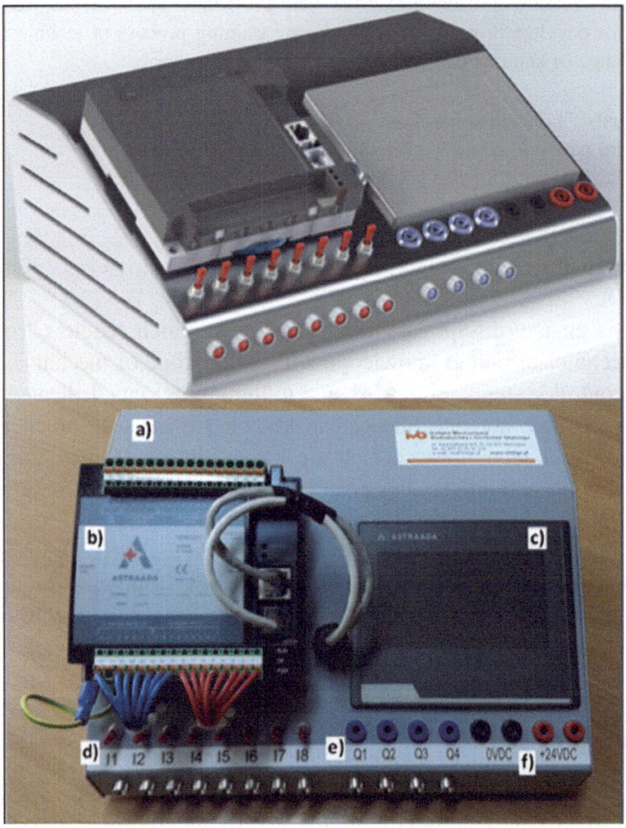

Fig. 1. View of the desktop with PLC controller and HMI operator's panel: (a) desktop case, (b) PLC controller, (c) HMI panel, (d) switches with LEDs, No I1 to I18, connected with PLC inputs, (e) sockets with LEDs, marked as Q1 to Q4, connected with PLC outputs, (f) socket 0 V DC and 24 V DC connected with 24 V DC power supply outputs.

The software of the PLC controller allows functions such as: hardware configuration, complex algorithms building according to IEC61131, creating operator interface applications. Creating control software can be carried out in one of 5 following programming languages:

- LD – graphic, ladder diagrams, analogical to electric diagrams,
- FBD – graphic, block diagrams, applications are created as functional blocks which assign the relationship between input and output signals,
- IL – text, list of instructions, simple language, similar to Assembler,
- ST – text, structural text record, hi level language, based on Pascal,
- SFC – text/graphic, graph of sequential functions SFC, flowing graph FC, program is similar to status diagrams, describes the sequence and parallel algorithms.

The desktop station should be equipped with USB cable, Ethernet RJ45 cable, 1 mm^2 LGY type cables with laboratory type single male plugs, RJ45 connector – 2 × female plug 8p8c. The station can work with a PC or another didactical station (Fig. 2).

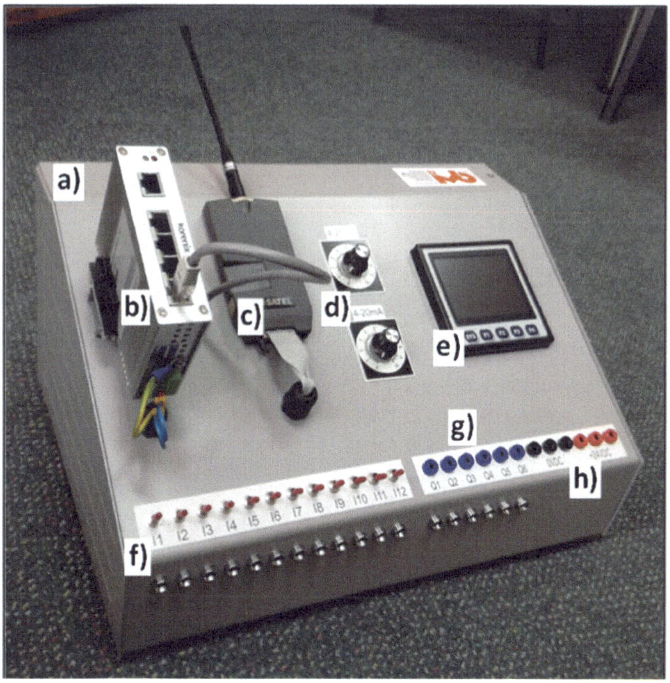

Fig. 2. View of the desktop with PLC and UHF modem: (a) desktop case, (b) Ethernet switch, (c) UHF modem, (d) two potentiometer referencing units 4–20 mA, (e) PLC unit, (f) switches with LEDs, marked as I1 to I12, connected with PLC inputs, (g) sockets with LEDs, marked as Q1 to Q6, connected with PLC outputs, (h) sockets 0VDC and 24VDC, connected with 24 V DC power supply.

2.2 Desktop Didactical Station Consisting of a PLC Controller and UHF Modem (Fig. 2) Can Be Used to Carry out Following Tasks

- learning programming languages of PLC controllers equipped with an LCD,
- learning industrial network communication protocols, including wireless network,
- learning hardware configuration methods of network equipment,
- using analog and discreet signals in control circuits,
- safety assurance in industrial automatics systems,
- control algorithms for high frequency and low frequency variable processes,
- learning industrial communication protocols used to connect distributed modules and devices by the wireless data acquisition (two stations required),
- learning configuration methods of measuring equipment,
- identification of the configured object,
- recognition of variety of sensors used in technological lines,
- making visualizations of process data of a given technological line.

2.3 Desktop Didactical Station Consisting of an Industrial Computer

This desktop station is designed to allow operation in environmental software dedicated to create applications for PLC controllers. The station is equipped with a PC and USB, Ethernet and RS-232 sockets. It can work with other didactical stations. When using special software it is possible to set the parameters of other connected devices of industrial automatic systems (Fig. 3).

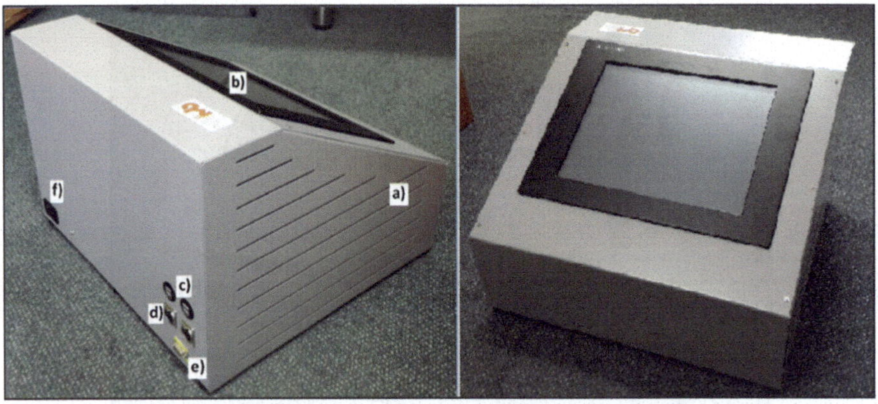

Fig. 3. Desktop with industrial computer: (a) desktop case, (b) industrial PC, (c) two USB ports, (d) two double Ethernet ports, (e) RS-232 port, (f) Power supply socket, 230 V AC, 50 Hz, with on/off switch and fuse case

2.4 Desktop Didactical Station Equipped with Frequency Inverter

The station main unit is the case equipped with power supply and frequency inverter which is controlled from operator panel placed on the front wall. The complement of the station comprises the lower part where an electric motor with a rotating disc are placed. They simulate work of a circular saw.

Inverter didactical station creates possibility to learn the following:

- practise electric motor control methods with use of frequency inverter unit,
- understand the idea of preserving constant torque while various rotation speeds are set,
- understand the idea of scalar control method (U/f) and vector control method,
- get to know the aspects of safety behaving in automatic environment with rotating units,
- learn basic design rules for increased safety industrial installations (Fig. 4).

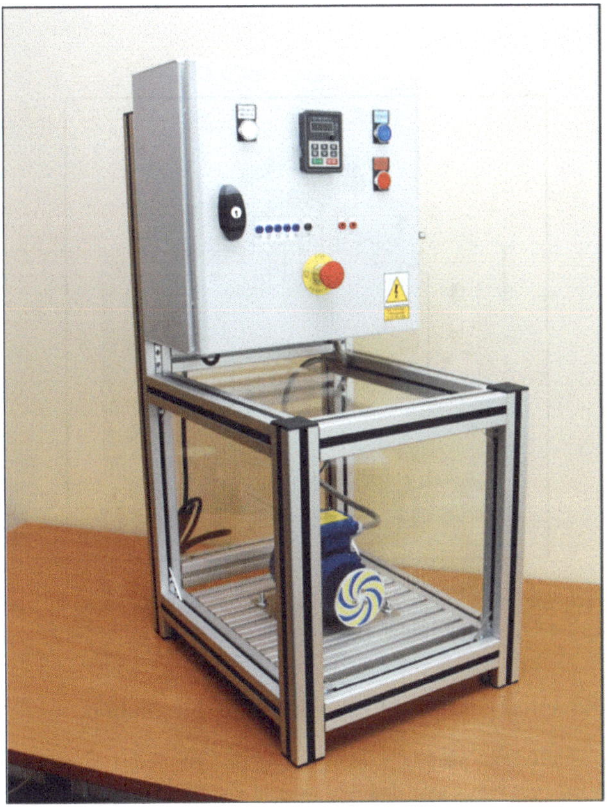

Fig. 4. Station with frequency inverter.

3 Object Didactical Stations and Industrial Network Didactical Stations

Object stations are intended to be used for learning selected topics from the field of automation in industrial processes using a set of devices built in a small demonstrative circuit. For example the following processes can be studied:

- transport and selection of details,
- servo drives shaft synchronization,
- fluid pumping unit management,
- redundant control circuits.

Industrial network stations are for the recognition of industrial network problematics and the tasks which have to be overcome while building, starting up and operation. Similarly to object stations they are based on a small demonstrative circuit which reflects the real industrial automatic environment.

The structure idea of such stands is shown in Fig. 5.

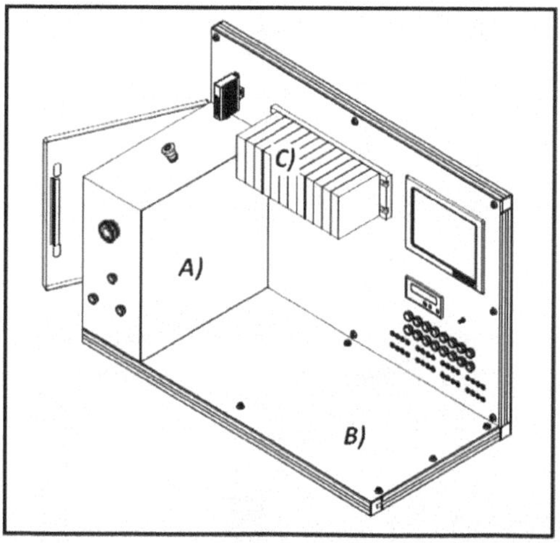

Fig. 5. General scheme of modular construction station. (A) Power supply and control units case. (B) Workstation supporting frame. (C) Control devices and other auxiliaries.

3.1 Object Didactical Station - Processing Line Simulation

Processing line station is used for learning about the control processes of details segregation. Those processes are discreet and are realized using a section of sensors, traps and servo motors placed along a belt conveyor. Various material and colour pucks are used to simulate transported details. Optical and induction sensors are used to determine the material dimensions and colour. Following the sensors a block of traps can realize segregation as described in the algorithm programmed by the operator (Fig. 6).

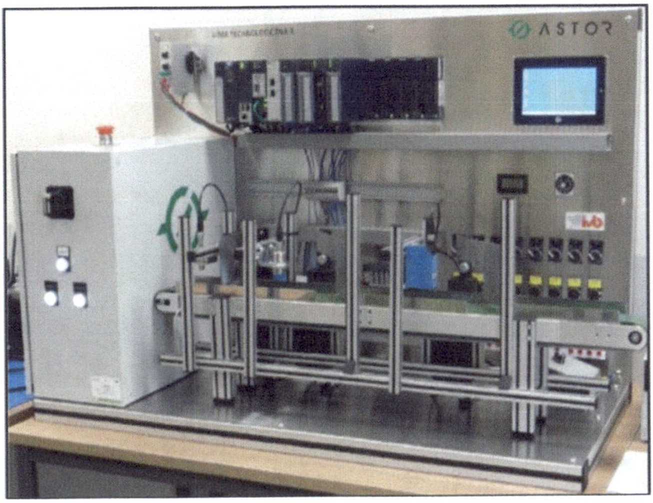

Fig. 6. Object station – processing line.

3.2 Plotter Didactical Station

Station is based on PAC controller combined with axis controlling devices. The executive unit consist of a two axial plotter with a servo motor driving each axle. The connection between controller and servo is based on fiber optic transmission for faster response and increased accuracy. Axes are equipped with positioning sensors which can function as limiters according to programmed function. Additional sensor is used for the zero - reference position placing.

The station is excellent tool for learning programming methods for industrial installation with multiple servo motors which work synchronically (Fig. 7).

3.3 Serial I/O – Industrial Network Station

Serial I/O station is equipped with a PAC controller and an electronic weight. Controller has two separate serial ports supporting Modbus RTU (master/slave) protocol. Weight is connected via RS-232 port. The station is for learning the rules of serial communication. It enables writing own network protocols for specific needs in symbol communication mode. Tasks such as reading weight indication, processing them, modifying weight working parameters etc. can be realized using Serial I/O station (Fig. 8).

Fig. 7. Object station – plotter.

Fig. 8. Industrial network station – Serial I/O.

4 Conclusions

Didactical stations presented in this article offer a very wide range of practical taska which are indispensable in education process of an automation engineer. Those stations should be used in industrial automatic laboratories to facilitate understanding of the theory and enabling putting the theory in practice. Various automatic devices which are widely used in industrial infrastructure are placed in the didactical stations offered by IMBiGS Institute. Those modules and devices are placed in environment which enables profound understanding of their role and functional principle in automatic systems. The main advantage of learning on didactical stations is the possibility to work on a single module or a separated sub system which is not connected to the whole system. Hence, students are not influenced by needles data while practising. In effect well equipped laboratories can improve the level of professional qualifications of future technical staff graduating from various automatics profiled schools and universities.

References

1. Łunarski, J.: Projektowanie procesów technicznych, produkcyjnych i gospodarczych. Oficyna Wydawnicza Politechniki Rzeszowskiej (2012)
2. Czarnul, J., Pienkowski, J.: Określanie właściwej strategii dla układów automatyki i sterowania. Utrzymanie Ruchu, October 2005
3. Łunarski, J., Szabajkowicz, W.: Automatyzacja procesów technologicznych montażu maszyn. WN-T (1993)
4. http://zap-robotyka.com.pl
5. Dokumentacja techniczno-ruchowa i serwisowa użytych do budowy stanowisk podzespołów oraz elementów

Image-Based Method for Knee Ligament Injuries Detection

Piotr Kohut[1(✉)] and Rafał Obuchowicz[2]

[1] Department of Robotics and Mechatronics, AGH University of Science and Technology, Mickiewicza Av. 30, 30-059 Krakow, Poland
pko@agh.edu.pl

[2] Department of Radiology, Jagiellonian University Medical College, Street św. Anny 12, 31-008 Krakow, Poland
gechrzas@cyf-kr.edu.pl

Abstract. Anterior cruciate ligament tear is one of the most frequent knee injuries, with reported incidence of 120,000 per year in the US alone. Different forms of everyday activities and contact and noncontact sports associated with pivot shift, dashboard, clip and hyperextension injury may lead to ACL rupture. Various forms of rupture of this important ligament are reported. Among these, the most difficult form to diagnose with imaging methods is reported to be the fan-shaped deformation with fiber discontinuity and no stump formation. Magnetic resonance has proven to be the contemporary gold standard in visualization of the ACL, with specificity and sensitivity reaching 86 and 95%, respectively, in comparison to arthroscopic evaluation. Although high field MRI is very accurate in the evaluation of ACL tears, fatigue and routine of medical professionals is potentially hazardous, and creates the possible risk of overlooking the lesion. Therefore, a semi-automatic diagnostic system is very much awaited by the medical community.

In this paper a new image-based method for the detection of fan-like deformation of acute ACL injury is presented. The developed algorithm is based on digital image processing and analysis. The demonstrated solutions are implemented in the MATLAB environment. The proposed method is investigated and examined on sets of MRI scans with cases of healthy and damaged knee joints. The efficacy of the method is analyzed based on obtained findings, which offered some important conclusions as well.

Keywords: Digital image processing · Watershed segmentation · Image analysis · Knee ligament injuries

1 Introduction

Software which supports decision-making for injury diagnosis offers a new area for experimental science, at the interface of computer science and medicine.

The anterior cruciate ligament (ACL) is statistically one of the most commonly damaged ligament structures in the knee joint cavity, which is a due to the mechanical role it plays in limiting multidirectional motion of the tibia bone in relation to the femur

J. Świder et al. (Eds.): MECHATRONICS 2017, AISC 934, pp. 240–247, 2019.
https://doi.org/10.1007/978-3-030-15857-6_24

bone, so that they remain in a coaxial position during flexion, extension and accompanying rotation [1, 2].

The ACL is a connective tissue structure composed of collagen strands that form two bundles. The majority of injuries causing damage to the ACL take place during the axial counter rotation of the femur bone in relation to the tibia bone. This is the pivot-shift injury typical of the sprained knee joint [3, 4]. This type of rotation causes overload, which leads to rupture of the ligament fibers when it exceeds 4% of the initial length of the ACL [5, 6]. Among available contemporary diagnostic techniques, nuclear magnetic resonance (NMR) techniques have been verified intraoperatively in many randomized studies and are viewed as the gold standard in diagnostic imaging of this type of injury. NMR renders visibility of injury to the ligament to the same degree as intraoperative [7, 8]. Fat-saturated proton density (PDFS) is the most commonly used sequence; it guarantees both high tissue contrast (image of the differences between structures of various chemical composition) and excellent visual anatomical rendition. Sagittal plane imaging is most often used; it displays the ligament in its anatomical position [9]. Therefore PDFS sequences taken in the sagittal plane were used for analysis. Ligament damage visible in imaging using the above procedures may involve rupture of the ligament with stump formation - when fluid is visible between the ligament elements. A less typical form of injury is fan-shaped damage, when the ligament fibers are discontinuous but keep their contour, and are only broadened and irregular internally – due to the atypical picture of this type of injury, it may be overlooked in analysis scans, which results in an improper diagnosis and potential medical error. A characteristic feature of this type of ligament damage in imaging scans is swelling – related to the increased amount of water in the tissue. It is identifiable in the NMR scan in the proposed sequence as a high signal (light) area with broadening of the ligament tissues related to the altered anatomy [10].

In addition, programming should effectively identify characteristics such as: ligament position, maintenance of continuity and presence of swelling in the ligament.

MRI segmentation is well known in the literature and dates to the previous century [11]. Watershed segmentation has been widely used in many fields of image processing including segmentation of medical images [12, 13]. It is characterized as an intuitive and fast method, the result of which is a segmentation of images even in the case of low contrast images and touching objects. In the scope of extracting complex anatomical structures based on the images obtained from magnetic resonance imaging (MRI), computer tomography (CT), mammography, or ultrasound (US), watershed segmentation it has become very popular and attractive technique [14–16].

Numerous papers have investigated knee meniscus diagnosis and injuries [17–20]. The paper by Kohut et al. [21] concerns an analysis of medial meniscal tears and demonstrates the potential of image processing methods. These studies have confirmed the possibility of qualitative and quantitative detection of meniscus injuries, even in the case of minor damage characterized by complex element shapes. Furthermore, two image-based methods were tested and compared: watershed image segmentation and classical image processing techniques.

In the literature, there is little work on the fully computer-aided diagnosis of the ACL knee lesion. An automatic segmentation of ACL is challenging due to the properties of MRI images: adjacent soft tissues with similar intensity distribution, inhomogeneous intensity of structures inside the ACL, in particular the part of the structure attached to the tibia.

Unfortunately, none of the studies reported in the literature concern the problem of semi-automated detection of fan-like ACL injury. This is the main reason for this paper, which demonstrates a new image-based method for the detection of fan-shape deformation in acute ACL injury. The developed algorithm employs digital image processing techniques and encompasses the following main operations: image pre-processing; segmentation involving watershed segmentation and image analysis phase. The demonstrated solutions are implemented in the MATLAB environment with application of the Image Processing Toolbox. The proposed method is investigated and examined on sets of MRI images with cases of healthy and damaged knee joints. The efficacy of the method is verified, based on obtained findings.

The goal of the conducted study was to propose the development of a method and algorithm to serve injury diagnosis and initial classification of injury type, regarding fan-shaped ACL damage.

2 Methodology

120 sagittal images of the knee were taken on a high-field MRI system (Siemens Sommatom Essenza) equipped with 4-channel knee coil; they were selected by a diagnostic imaging specialist with special interest in the musculoskeletal system. Sets of sagittal images presenting different types of ACL ruptures were taken, including images from the control group. Midline sagittal plane images of the knee were included in the study. Patients' age was 25–46 years. Patients with osteoarthritis of the knee and those presenting signs of systemic diseases or local knee lesions were excluded from the study. The following parameters were used to create a knee image in the sagittal plane: field of view 160, plane thickness 3 mm, relaxation time 3500, echo time 35, averages 1, concentrations 1, flip angle 150, matrix 256×256, voxel size $0.6 \times 0.6 \times 0.3$, obtained signal noise ratio (SNR) was 1. No image filters in the postprocessing stage of the raw data were used. Data was collected on the Siemens Syngo 12.1 PACS system and was uploaded to the local workstation equipped with OsriX MD Pixmeo 8.5.2 on a Linux platform (Apple) and analyzed on 4K resolution monitor.

MATLAB with Image Processing Toolbox was used for the development of a programming tool to identify the state of ligament tears. Verification of the developed algorithm was conducted on an existing database of scans of both injured and healthy knees. These scans underwent successive transformation based on the developed algorithm (Fig. 1).

2.1 Developed Algorithm

The developed method encompasses the following main digital image processing operations (Fig. 1).

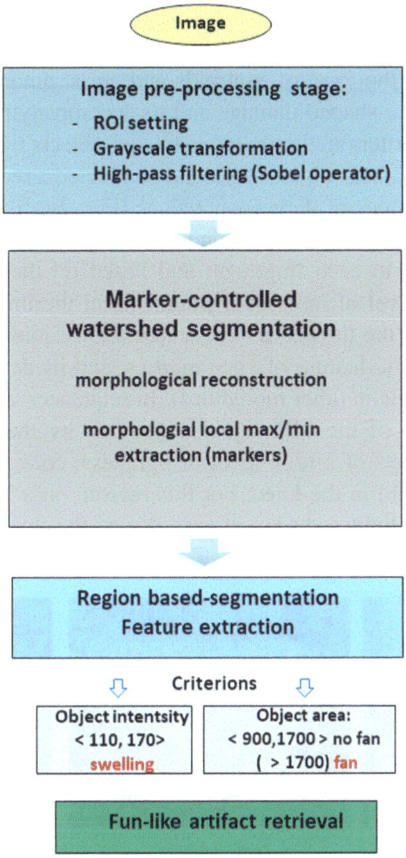

Fig. 1. The developed algorithm for detection of fan-shaped ACL knee tear.

- image pre-processing including high pass filtering, set of morphological operations;
- segmentation involving watershed segmentation;
- image analysis phase with feature extraction.

In the image preprocessing stage, after ROI selection and RGB image to grayscale transformation (Fig. 2b), an image gradient was calculated using the Sobel operator (Fig. 2c).

Before conducting marker-controlled watershed segmentation a series of morphological transformations were applied [22]. In order to discover and enhance the foreground markers, the opening-by-reconstruction and then closing-by-reconstruction operations were utilized. The opening and closing-by-reconstruction operations are characterized by greater effectiveness than basic opening or closing functions because they effectively reduce minor image artifacts without impacting overall object shape. Next, regional maxima of individual regions were marked and very small objects were removed (Fig. 2e).

After image gradient modification, which caused regional minima to be found only in marker pixel locations, watershed segmentation was carried out (Fig. 2f).

In order to enhance individual watershed segmented objects, the image was binarized and then object characteristics of the analyzed objects was extracted using the *regionprops* function in the form of centroids and areas, among others.

In order to expose fan-shaped damage and its accompanying swelling, two criteria were applied. The first criterion eliminated extracted objects by conducting tests based on their lightness level values, while the second conducted a reduction of the remaining objects by testing the values of their areas values (Fig. 2g, h).

The most important step in the detection of fan-shaped damage is the loop in which a histogram is assigned to each fragment, and based on the histogram, the average value of the lightness level of individual fragments in the image is determined. This allows for the setting of the threshold over which a knee joint demonstrates swelling. Swelling is a characteristic feature of knee injuries, and its detection is made possible by MRI (and is impossible in other modalities). In sequences with a long T2 relaxation time, the reconstruction of the MR signal conducted by the computer presents the maximal signal for objects of a high level of lightness, corresponding to high proton density areas that are fluid in the knee. For this reason, on a T2 sequence image they can be distinguished by lighter shades of grey due to the fluid they contain.

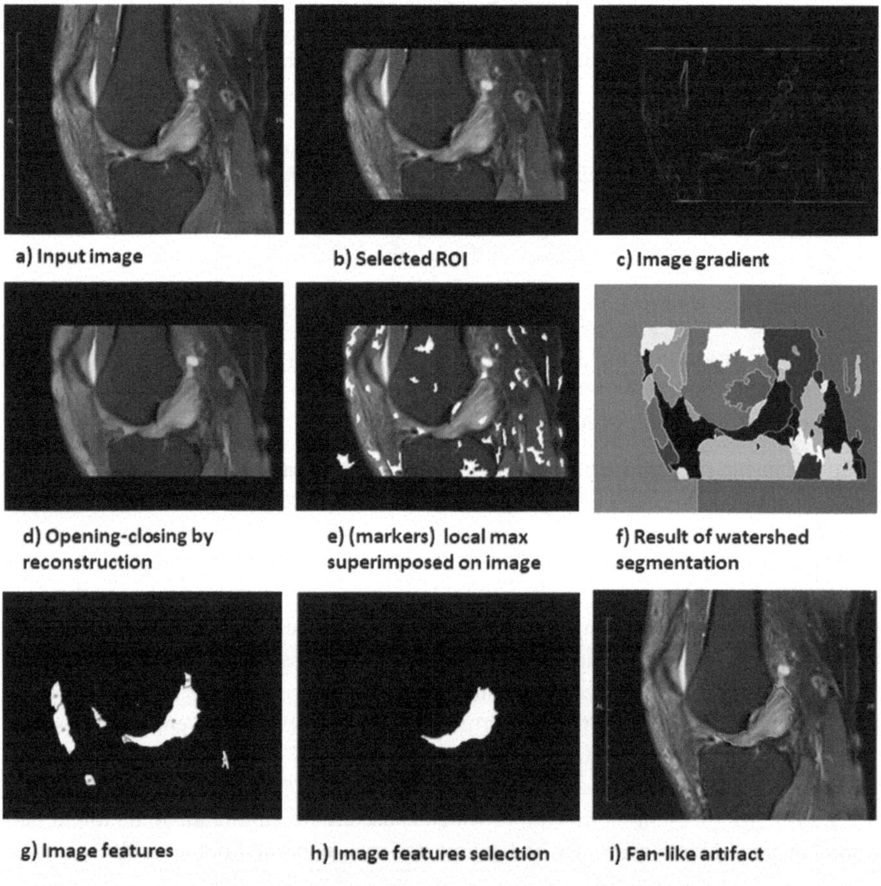

Fig. 2. Results of image processing and analysis

For the studied series of knee images the threshold values ranges indicating swelling were set to be: 110 < swelling < 170. The range chosen offers satisfactory results in swelling detection due to the fact that the studied images show varied distribution of signal maximas for the proton nuclei and that fluid is present. After marking the average values of grey levels of the extracted fragments of the knee structures, and defining the black frame of the image, logical operations were applied to differentiate the analyzed fragment of the object (Fig. 2g).

The structures obtained that passed the first criterion selection (Fig. 2g) again underwent a region-based image analysis using the *bwlabel* and *regionprops* functions, and their features were marked (areas and centroids, among others). The conducted second criterion tests of the remaining objects were based on size of their areas (Fig. 2h). This criterion eliminates both those regions that do not meet the acceptable ranges for the areas and also the structures found to the left of the center of the image.

As a result, the regions obtained had surface areas located in the given first range <900, 1700> [pixels] and marked areas indicating swelling, whereas the extracted fragment found to the right of the center of the image and with a surface area exceeding 1700 pixels indicates a fan-shaped type injury.

The presented ranges for surface area were chosen experimentally on the basis of an image bank (adult patients) to enable the automatic rejection of irrelevant knee joint structures.

Figure 2(i) presents the identified fan-shaped injury superimposed on the analyzed initial image.

3 Results

The implemented method was investigated and tested on a number of images of the knee acquired from healthy subjects and patients with a diagnosed pathology.

Regarding the set of images (10 images) with fan-shaped knee joint damage, 80% were correctly classified as containing a fan-shaped injury. Of the set of healthy images (10 images), 80% were classified as healthy (Table 1).

Table 1. Effectiveness of the developed method

Type of images\Knee joint state	Fan like injury detection	Healthy state
Healthy Knee join		80%
Damaged Knee joint	80%	

The obtained results confirmed the effectiveness of the developed method.

4 Conclusions

Due to the steady growth in the number of images created in various radiological modalities, namely magnetic resonance (MR), which is excessively used for knee lesion detection, semi-automatic analysis is expected increasingly. This is particularly true for detection of the most common ACL lesion. Characteristic features of the MR image in sagittal view are being used by the medical professional community (namely radiologists) to analyze ligament shape and width, although apparent continuity of the ligament and regularity of its outlines may be misleading. In the presence of routine fatigue or even lack of experience, abnormalities may be overlooked. Thus, image segmentation is a very promising tool when implemented with medical software to evaluate different types of lesions visualized in magnetic resonance both in musculoskeletal system structures as well as in other (parenchymal) organs.

The software tool developed to detect the state of knee joint injuries could facilitate the evaluation of diagnostic images for the presence of specific characteristics of a torn ligament (fan-shaped injury) and thus minimize the probability of error during the diagnostic process.

The algorithms implemented were investigated and tested on a number of images of the knee acquired from healthy subjects and patients with a diagnosed pathology. The obtained findings confirmed efficiency and effectiveness of the developed method.

A novelty of the proposed solutions addresses the issues related to semi-automatic segmentation of selected elements of the knee joint. Although the demonstrated technique is used for detection of the state of the most important structures of the knee such as the ACL, it still poses a challenge in the complex task of diagnosing fan-shaped injuries. The presented technique, developed and tested on examples of the sagittal view of the knee, may also be implemented on axial and coronal images, which would possibly increase the diagnostic confidence of the proposed technique – and thus this issue will be addressed in upcoming studies.

Acknowledgements. The research reported in this paper has been financed from the state budget for science. The authors wish to acknowledge Sebastian Turek for participation in the reported research.

References

1. Kaeding, C.C., Léger-St-Jean, B., Magnussen, R.A.: Epidemiology and diagnosis of anterior cruciate ligament injuries. Clin Sports Med. **36**(1), 1–8 (2017)
2. Anderson, M.J., Browning, W.M., Urband, C.E., Kluczynski, M.A., Bisson, L.J.: A systematic summary of systematic reviews on the topic of the anterior cruciate ligament. Orthop J. Sports Med. **4**(3), 2325967116634074 (2016)
3. Balazs, G.C., Pavey, G.J., Brelin, A.M., Pickett, A., Keblish, D.J., Rue, J.P.: Risk of anterior cruciate ligament injury in athletes on synthetic playing surfaces: a systematic review. Am. J. Sports Med. **43**(7), 1798–1804 (2015)
4. Posthumus, M., Collins, M., September, A.V., Schwellnus, M.P.: The intrinsic risk factors for ACL ruptures: an evidence-based review. Phys Sportsmed. **39**(1), 62–73 (2011)
5. Shimokochi, Y., Shultz, S.J.: Mechanisms of noncontact anterior cruciate ligament injury. J. Athl. Train. **43**(4), 396–408 (2008)
6. Louw, Q., Gillion, N., van Niekerk, S.M., Morris, L., Baumeister, J.: The effect of vision on knee biomechanics during functional activities - a systematic review. J. Sci. Med. Sport **18** (4), 469–474 (2015)

7. Van Dyck, P., Vanhoenacker, F.M., Lambrecht, V., et al.: Prospective comparison of 1.5 and 3.0-T MRI for evaluating the knee menisci and ACL. J. Bone Joint Surg. Am. **95**, 916–924 (2013)
8. Crawford, R., Walley, G., Bridgman, S., Maffulli, N.: Magnetic resonance imaging versus arthroscopy in the diagnosis of knee pathology, concentrating on meniscal lesions and ACL tears: a systematic review. Br. Med. Bull. **84**, 5–23 (2007)
9. Kosaka, M., Nakase, J., Toratani, T., et al.: Oblique coronal and oblique sagittal MRI for diagnosis of anterior cruciate ligament tears and evaluation of anterior cruciate ligament remnant tissue. Knee **21**, 54–57 (2014)
10. Boks, S.S., Vroegindeweij, D., Koes, B.W., Hunink, M.G., Bierma-Zeinstra, S.M.: Follow-up of posttraumatic ligamentous and meniscal knee lesions detected at MR imaging: systematic review. Radiology **238**, 863–871 (2006)
11. Sijbers, J., Scheunders, P., Verhoye, M., van der Linden, A., van Dyck, D., Raman, E.: Watershed-based segmentation of 3D MR data for volume quantization. Magn. Reson. Imaging **15**(6), 679–688 (1997)
12. Roerdink, J., Meijster, A.: The watershed transform: definitions, algorithms and parallelization strategies. J. Fundamenta Informaticae **41**(1), 187–228 (2001)
13. Beucher, S.: The watershed transformation applied to image segmentation. In: 10th Pfefferkorn Conference on Signal and Image Processing in Microscopy and Microanalysis Cambridge, UK, 16–19 September 1991. Scanning Microscopy International, suppl. 6, pp. 299–314 (1992)
14. Włodarczyk, J., Wojciechowski, W., Czaplicka, K., Urbanik, A., Tabor, Z.: Fast automated segmentation of wrist bones in magnetic resonance images. Comput. Biol. Med. **65**, 44–53 (2015)
15. Iordanescu, G., Venkatasubramanian, P.N., Wyrwicz, A.M.: Automatic segmentation of amyloid plaques in MR images using unsupervised support vector machines. Magn. Reson. Med. **67**(6), 1794–1802 (2012)
16. Cui, Y., et al.: Malignant lesion segmentation in contrast-enhanced breast MR images based on the marker-controlled watershed. Med. Phys. **36**(10), 4359–4369 (2009)
17. Saygılı, A., Albayrak, S.: A new computer-based approach for fully automated segmentation of knee meniscus from magnetic resonance images. Biocybern. Biomed. Eng. **37**(3), 432–442 (2017)
18. Boniatis, I, Panayiotakis, G, Panagiotopoulos, E.: A computer-based system for the discrimination between normal and degenerated menisci from Magnetic Resonance Images. In: 2008 IEEE International Workshop on Imaging Systems and Techniques, pp. 335–339. (2008)
19. Swamy, M.S.M., Holi, M.S.: Knee joint menisci visualization and detection of tears by image processing. In: 2012 International Conference on Computing Communication and Applications, pp. 1–5 (2012)
20. Patel, A.J., Modi, H., Patel, H.: Measurement of cartilage thickness in osteoarthritis and visualization of meniscus tear of knee MRI image processing. Int. J. Comput. Sci. Mob. Comput. **5**, 39–52 (2016)
21. Kohut, P., Holak, K., Obuchowicz, R.: Image processing in detection of knee joints injuries based on MRI images. J. Vibro Eng. **19**(5), 3822–3831 (2017)
22. Woods, R., Gonzales, R.: Digital Image Processing. Prentice Hall, Upper Saddle River (2008)

Application of Surface Electromyographic Signals for Electric Rotor Control

Agnieszka Konopelska(✉) and Mariola Jureczko

Faculty of Mechanical Engineering, Institute of Applied and Theoretical Mechanics, Silesian University of Technology, 44-100 Gliwice, Poland
Agnieszka.Konopelska@polsl.pl

Abstract. The aim of this study was to develop an algorithm to control an electric rotor using surface electromyographic (sEMG) signals. The paper presents a design of an automated mechatronic robot for rehabilitation of both upper and lower limbs. Due to the implemented controller it is possible to program exercises in full spectrum and with various loads.

Keywords: Surface electromyography (sEMG) · Rehabilitation · Mechatronics system

1 Introduction

Kinezotherapy is one of the most commonly used methods of mobility rehabilitation. Its main purpose is to rule out physical unfitness of a given patient [1, 2]. Nowadays rehabilitation is carried by means of various equipment and accessories, which aim at simplifying and accelerating the process, simultaneously releasing other body parts. When analyzing the development of rehabilitation equipment its greater and greater mechatronization can be seen [3]. It allows practicing exercises without the supervision of a physiotherapist, who thanks to that can supervise several patients at the same time. As a result it would be possible for hospitals or rehabilitation centers to serve more patients simultaneously.

We can interfere to a human body by means of internal energy sources. At the same time, based on energy emitted by a human body, we can evaluate the condition of an individual. Thus, information collected in this way can be read, processed and analyzed with e.g. computer diagnostics, which makes it possible to eliminate subjective evaluation by a physiotherapist. The disadvantages of this methods are resolution of the used converters, frequency of data collection and system sensitivity to noises. The basic methods for detecting changes in a researched system are: statistic, correlation and limitation control methods or researching equilibrium points of a researched signal. Detected characteristic features of a signal can be presented in a form of a bistate passive diagnostic matrix or given to fussy logics controllers in order to evaluate variables after being converted to linguistic values [4–6]. Next, thanks to the application of numerical environments for laboratory analysis e.g. LabVIEW it is possible to

J. Świder et al. (Eds.): MECHATRONICS 2017, AISC 934, pp. 248–257, 2019.
https://doi.org/10.1007/978-3-030-15857-6_25

elaborate programs and operator panels for visualization of phenomena which occur in a researched object e.g. human body and for monitoring diagnostic system operation. Simultaneous control of many different physical values is necessary for automatic monitoring of the operation of mechatronic rehabilitation device. The choice of proper parameters for observation requires the knowledge of practices used in traditional rehabilitation [7–11].

The use of simple machines with advanced control systems requires determination of a group of signals which form a premise for taking decisions by a controller. One of the basic measurable signals is the activity of a locomotor system presented as voltage measured by sEMG (surface electromyography). The most important signals for the concept of the authors are EMG signals from human muscles measured on the skin. The voltage of controlling EMG signals is huge including, among others, detection of muscle fatigue or steering the robot arms which are much bigger than the arms of a controlling person [12].

Many researchers in their publications reported that mean values of EMG signals increased linearly with an increase in force of muscle contraction under either isometric or constant velocity isotonic conditions [13–15]. Other investigators described a nonlinear relationship between mean values of EMG signals and values in force of muscle [15, 16]. According to [17] the relationship between mean values of EMG signals and mean force of muscle is linear, but not at all conditions. When the entire range of swing phase and stance phase forces are considered together, the relationship between mean values of EMG signals and force of muscle is nonlinear.

The paper presents the design of a mechatronic stationary rotor for rehabilitation of both upper and lower limbs. Its purpose is to accelerate healing and alternative processes in locomotor system and inner organs of a human. The designed mechatronic device uses EMG signals read from sensors placed on the patient's body, which come from skeleton muscles in order to put a signal on a winch drive, which is an actuator of this mechatronic system. The data together with advanced diagnostic system, which uses various algorithms allow to make use a wide range of exercises. The usage of automatic system of control with feedback enables programming exercises in full range and of different types of load, from passive exercises through loaded, up to loads with constant or variable value.

2 Construction of Stationary Mechatronic Rotor

Elaboration of the concept of stationary mechatronic rotor which enables carrying rehabilitation exercises in non-weight bearing in patient's bed was based on a general scheme of automatic regulation system. Figure 1 shows the block diagram of a control system of mechatronic stationary rotor.

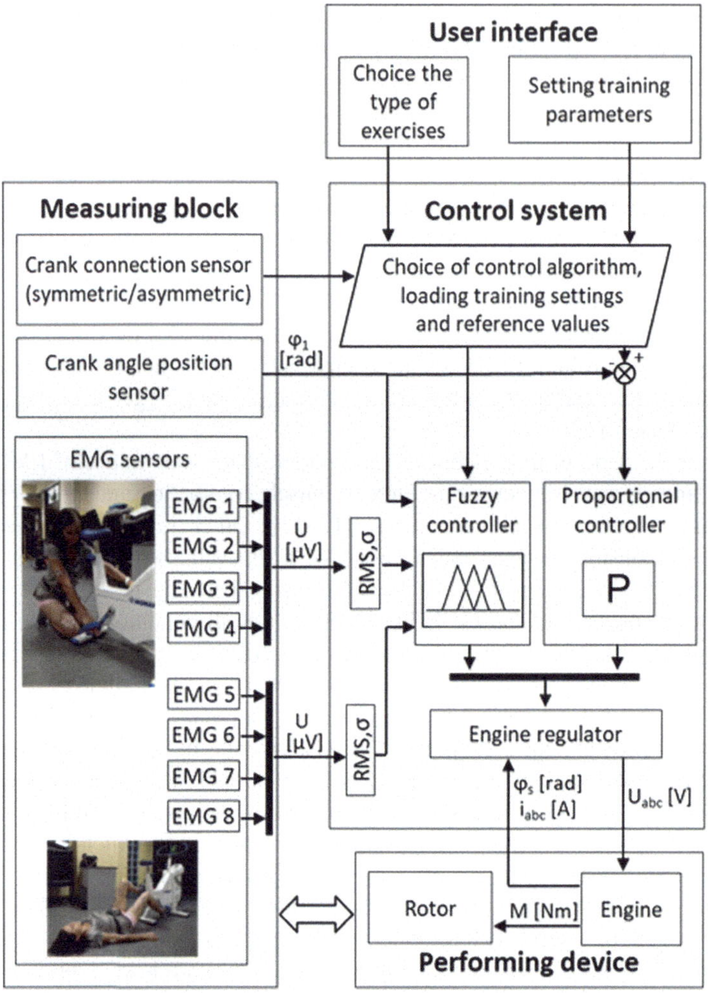

Fig. 1. Block diagram of a control system of a mechatronic stationary rotor. The symbols used: ϕ_1 [rad] – crank rotation angle, U [μV] – voltage (EMG signal), ϕ_s [rad] – rotation angle of the motor shaft, i_{abc} [A] – stator current, U_{abc} [V] – stator phase voltages, M [Nm] – motor torque, RMS, σ – EMG signal processing (RMS – root-mean-square, σ – standard deviation)

Measuring block consists of three kinds of sensors which carry three different tasks, which are:

- sensor for crank angle position which is the sensor of absolute position,
- sensors determining the way of crank assembly i.e. sensors which allow determination of rotor operation (symmetric or asymmetric) as well as correctness of crank assembly,
- EMG sensors which enable evaluation of the work of antagonistic muscle of left and right limb with minimum of 4 sensors per each limb.

The use of so many sensors allows to determine both the type of a rotor as well as the correctness of crank assembly. At the same time, it enables carrying rehabilitation exercises in four modes with two crank settings, namely:

- Symmetric passive exercises,
- Asymmetric passive exercises,
- Symmetric active-passive exercises,
- Asymmetric active-passive exercises,
- Symmetric non-weight bearing active exercises,
- Asymmetric non-weight bearing active exercises,
- Symmetric non-weight bearing passive exercises,
- Asymmetric non-weight bearing passive exercises.

The structure of **stationary rotor** gives a possibility of mounting it at the front or the rear part of a bed frame by means of belts. The arm of a truss can be suspended over patient's head or by the legs. In order to facilitate the performance of both symmetric and asymmetric exercises the solution in a form a crank with one arm removable and multi inlets has been used. A pin made of soft polymer enables quick assembly and disassembly of a crank. In order to put a crank on, everything one has to do is to put in on a shaft and align with resistance surface. Next, the pin is led in, pressing in the crank. The rotation of an arm around the shaft is blocked by a lug, which when pressed inside the shaft switches the sensor of crank location. The position sensors, placed on the shaft enable detection of kinds of exercises. When there is no signal from any of them the control system doesn't allow the rotor to work.

The BLDC motor as **an actuator** allows to use drive system of relatively small size. This motor enables operation in passive mode and with resistance. Additionally, the control over the map of BLDC motor has been implemented directly in the driver. The motor and the crank have been mounted on a support placed at the rear of the truss and have been equipped with a belt and a gear train which in turn result in significant reduction of both the engine speed and vibrations occurring at its changes (Fig. 2).

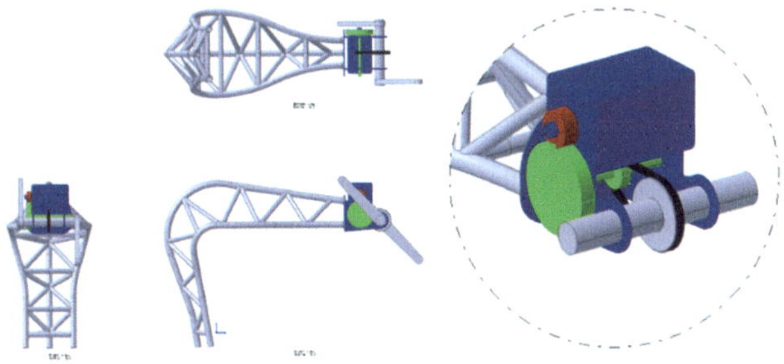

Fig. 2. The concept of mechatronic stationary rotor

The task of the **control system** is first of all to monitor a few input values. Both modes of a rotor – symmetric and non-symmetric are based on the same algorithms, the difference being the volume of input values. The driver is Arduino Mega, which ensures a possibility of using digital-analogue modules for monitoring the BLDC motor. The control over modes which do not require checking the EMG sensors is done by a simple algorithm, where the signal output to the motor is proportional to the set parameters at the input. However, control over modes with measurements uses fuzzy logics based on changes in standard deviation values of displaceable register including subsequent EMG measurements. As a result, the increase in values is interpreted as the increase in muscle activity, whereas the decrease in values as transition to quiescent output voltage. The program code for control of the engine is presented in Fig. 3.

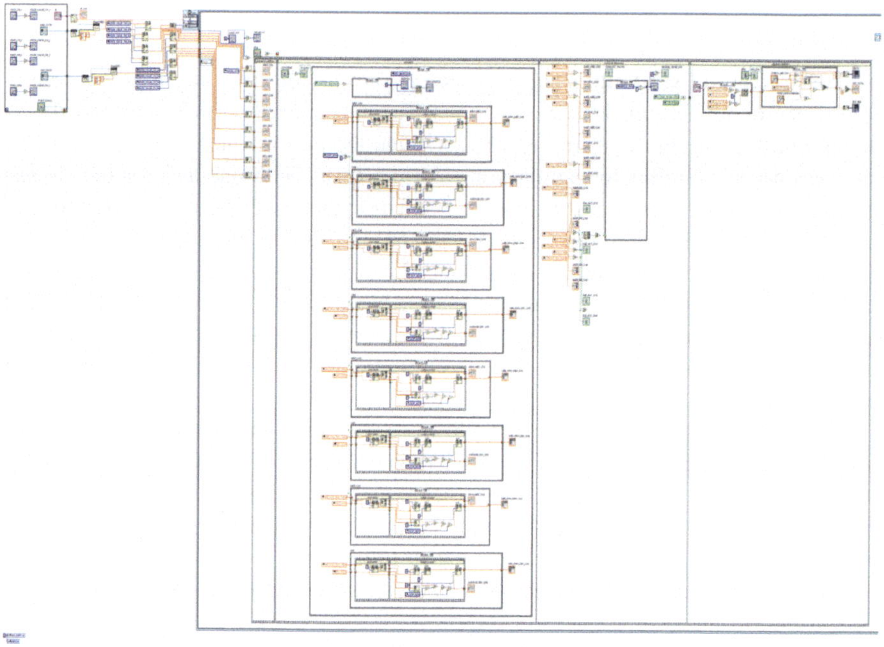

Fig. 3. The program code for control of the engine

The program which controls motor operation aims at detecting muscles activity and based on that it gives a signal to the crank drive. The values which are decisive here are: RMS – root-mean-square and σ – standard deviation of EMG signals. Standard deviation of EMG signals shows the smallest change in value at the moment of muscle activation and that is why this very value has been considered as the most useful for detecting movement initiation. Thus, based on that value a decision is made on motor activity. The rotary speed of movement, however, is regulated on the basis of root-mean-square of EMG signals. When there is a decrease in standard deviation value below given set value then the program terminates. This value is set individually for each patient while setting the system.

Regulator follows the set value if it goes beyond the insensitivity sphere. In case of too low levels of muscles activity while exercising, the application deducts the number of cycles left for switching off the motor.

3 Simulation of Operation of Stationary Mechatronic Rotor

EMG signal measurements, which served as input data for original automatic system of control, were carried on Noraxon TELEMYO 2400T G2 device. The electrodes were mounted on patient's body according to SENIAM project [18]. For upper limb signal measurements 6 electrodes were used, 3 for each side which can be seen in Fig. 4. The electrodes were placed on medial head of triceps brachii muscle (MED. TRICEPS), lateral head of triceps brachii muscle (LAT. TRICEPS) and biceps brachii muscle (BICEPS BR.). For upper limb measurements 8 electrodes were used, 4 for each side. The following muscles were taped: rectus femoris muscle (RECTUS FEM.), biceps femoris muscle (BICEPS FEM.), vastus lateralis obliquus muscle (VLO) and vastus medialis obliquus muscle (VMO). The measurements were recorded with various loads 1 N and 10 N or using only abductor muscles force with 10 N load:

a. upper and lower limbs:
   - crank rotations,
   - pulse, alternately increasing and decreasing muscles activity with minimal force that does not cause the crank to rotate,
b. crank rotation with only right upper limb,
c. crank rotation with only right lower limb.

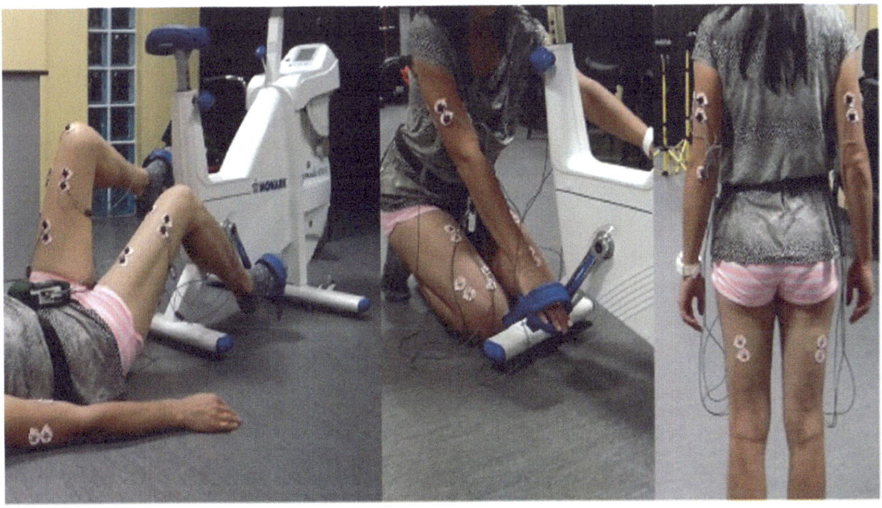

Fig. 4. Exercises carried in order to verify the operation of stationary mechatronic rotor

The results were analyzed in order to select a group of muscles for steering the program which collaborates with four channels giving values of measured sEMG signal. Exemplary graphs of activities of particular muscles of lower limbs (a) and upper limbs (b) while exercising with 10 N load or 1 N load respectively, were presented in Figs. 5 and 6. The symbols used to mark the muscles have been explained in the text, where the abbreviation LT refers to the left limb and the abbreviation RT to the right limb. The differences in the value of muscle tension result from amplification set in the original program for data acquisition from a sensor. Alternating increase in muscles activity can be clearly seen. Additionally, slight time shift can be seen at the time of momentary activation of muscles within the collaborating group.

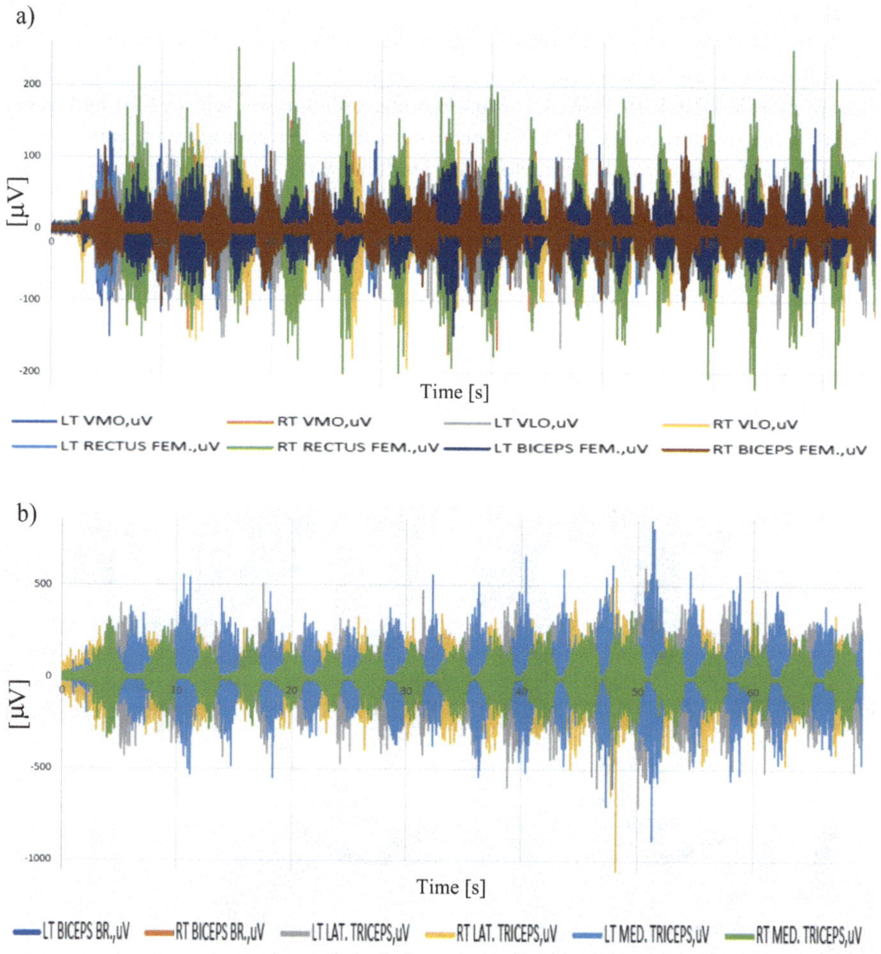

Fig. 5. Activity diagram of particular muscles of lower limb (a) and upper limbs (b) generated at rotations on ergometer with 10 N load

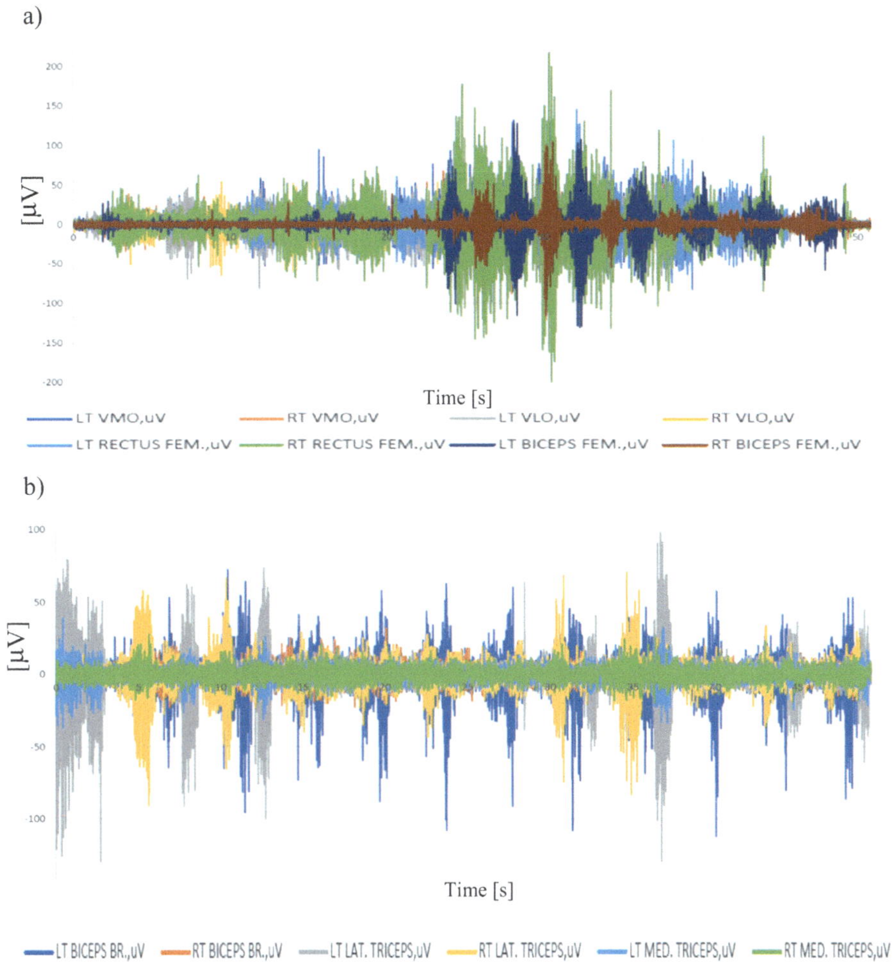

Fig. 6. Activity diagram of particular muscles of lower limb (a) and upper limbs (b) generated at rotations on ergometer with 1 N load

While increasing muscle activity either by load or motion, the amplitude of the muscle tension on the skin increases. By finding sections of greater value of EMG signals and comparing them to the value at lower activity we can determine the degree of muscle activity, which in turn allows to determine which muscles in a given phase of motion are more active and how it changes while exercising [19, 20].

4 Conclusions

Rehabilitation equipment available on the market needs mechatronization of current solutions. The use of measuring devices, drivers and actuators does not change a function or principle of operation of them. However, it enables significant expansion of methods or range of mechanisms operation. Thanks to the solution put forward by the authors it will be possible to perform rehabilitation exercises not only with or without loads but also in non-load bearing mode which up till now has been carried either with complicated pulley mechanisms or the power of muscles of physiotherapist.

The use of EMG signals coming from human muscles and read by sensors stuck to the body allowed to program control over crank drive.

References

1. Zembaty, A.: Kinezyterapia. Tom 1. Zarys podstaw teoretycznych I diagnostyka kinezyterapii. Wydawnictwo KASPER (2015)
2. Kiwerski, J. (ed.): Rehabilitacja Medyczna. Wydawnictwo Lekarskie PZWL, Warszawa (2005)
3. Paśniczek, R.: Bioinżynieria w rehabilitacji narządu ruchu. Oficyna Wydawnicza Politechniki Warszawskiej, Warszawa (2014)
4. Kruse, R., Gebhardt, J., Klawonn, F.: Foundations of Fuzzy Systems. Wiley, Hoboken (1994)
5. del Boca, A., Park, D.C.: Myoelectric signal recognition using fuzzy clustering and artificial neural networks in real time. In: IEEE International Conference on Neural Networks and IEEE World Congress on Computational Intelligence, vol. 5, pp. 3098–3103 (1994)
6. Popkorn, S.: First Steps in Modal Logic. Cambridge University Press, Cambridge (1994)
7. Erdmann, W.S.: Inżynieria rehabilitacji ruchowej. Wydawnictwo Politechniki Gdańskiej (2016)
8. Kazuo, K., Yoshiaki, H.: An EMG-based control for an upper-limb power-assist exoskeleton robot. IEEE Trans. Syst. Man Cybern. – Part B: Cybern. **42**(4), 1064–1071 (2012)
9. Ahmad, I., Ansari, F., Dey, U.K.: A review of EMG recording technique. Int. J. Eng. Sci. Technol. **4**(2), 530–539 (2012)
10. Wege, A., Zimmermann, A.: Electromyography (EMG) sensor based control for a hand exoskeleton. In: Proceedings of the 2007 IEEE International Conference on Robotics and Bio-mimetics, 15–18 December 2007, Sanya, China, pp. 1470–1475 (2007)
11. Roy, S.H., Luca, G.D., Cheng, M.S., Johansson, A., Gilmore, L.D., Luca, C.J.D.: Electromechanical stability of surface EMG sensors. J. Med. Bio. Eng. Comput. **45**, 447–457 (2007)
12. Jain, R.K., Datta, S., Majumder, S.: Design and control of an EMG driven IPMC based artificial muscle finger. In: Naik, G.N. (ed.) Computational Intelligence in Electromyography Analysis - A Perspective on Current Applications and Future Challenges. InTech. https://doi.org/10.5772/48814
13. Lippold, O.C.J.: The relation between integrated action potentials in a human muscle and its isometric tension. J. Physiol. **117**, 492–499 (1952)
14. Milner-Brown, S.R.B.: The relation between the surface electromyogram and force. J. Physiol. (London) **246**, 549–569 (1975)

15. Woods, J.J., Bigland-Ritchie, B.: Linear and non-linear surface EMG/force relationships in human muscle. J. Phys. Med. **6**, 287–299 (1982)
16. Komi, P.V., Buskirk, E.R.: Reproducibility of electromyographic measurements with inserted wire electrodes and surface electrodes. Electromyogr. Clin. Neurophysiol. **10**, 357–367 (1970)
17. Roberts, T.J., Gabaldón, A.M.: Interpreting muscle function from EMG: lessons learned from direct measurements of muscle force. Integr. Comp. Biol. **48**(2), 312–320 (2008). https://doi.org/10.1093/icb/icn056
18. Hermens, H.J., et al.: European Recommendations for Surface ElectroMyoGraphy. Results of the SENIAM Project. ISBN 90-75452-15-2. http://seniam.org/sensor_location.htm
19. Raez, M.B.I., Hussain, M.S., Mohd-Yasin, F.: Techniques of EMG signal analysis: detection, processing, classification and applications. J. Biol. Proc. Online **8**, 11–35 (2006)
20. Merlo, A., Farina, D.A.: Fast and reliable technique for muscle activity detection from surface EMG signals. IEEE Trans. Biomed. Eng. **50**(3), 316–323 (2003). https://doi.org/10.1109/TBME.2003.808829

Voltage Source Inverter Synchronization with the Use of FFT Algorithm

Maciej Kozak(✉)

Maritime University of Szczecin, Waly Chrobrego St. 1-2, Szczecin, Poland
m.kozak@am.szczecin.pl

Abstract. One of the very important questions in the new concept of distributed sources of electrical energy is how to find out frequency of rotating generators to match it to inverter synchronization requirements. The information about generator angular speed is crucial for synchronization process that is needed to avoid flow of high values of equalizing currents. In the paper VisualDSP++ 5.0 simulation results are presented, followed by experimental tests of the fast Fourier transform software function with VSI inverter connected to the self-excited synchronous generator.

Keywords: VSI inverter · FFT · Real-time · DSP · Synchronous generator · Synchronization

1 Introduction

In the most common types of generators used onboard of seagoing ships there are self-excited, brushless synchronous generators. Mostly the combustion engines drive electrical generators with constant rotational speed [1]. Because of the environmental and economic reasons there is a need for optimization of diesel engine and electrical system cooperation. One of the methods is creation of a control systems working in parallel with generating units in the real time. In the modern diesel engines control units there are maps of optimal operation regions embedded in their control software. These maps allow to change rotational speed in case of need of fuel consumption and emissions reduction [2]. In the commonly used generating sets constant rpm's problem occurs to maintain constant grid frequency. With the use of so-called grid side inverters it is possible to keep constant output frequency while changing revolutions of prime mover [3]. The grid side inverters can produce alternating voltage of desired parameters at the same time maintaining high enough value of intermediate circuit voltage. There are many control algorithms for accurate driving synchronous generator to achieve constant voltage on DC bus but before applying them both inverter and machine should be put in the parallel work. This process is called synchronization and is a standard procedure in ship power plants. Currently used synchronizers are designed mostly for operation with 50 or 60 Hz electrical power grid with little content of voltage harmonics [4]. Fundamental grid frequency can be chosen by operator and is set to one value with up to few Hertz deviations needed for control purposes.

J. Świder et al. (Eds.): MECHATRONICS 2017, AISC 934, pp. 258–268, 2019.
https://doi.org/10.1007/978-3-030-15857-6_26

2 Structure of Investigated System

The rotating synchronous generator produces voltages on its terminals, which are connected to inverter arms. In the case of self-excited generator resulting frequency depends on the rotational speed of generator (Fig. 1).

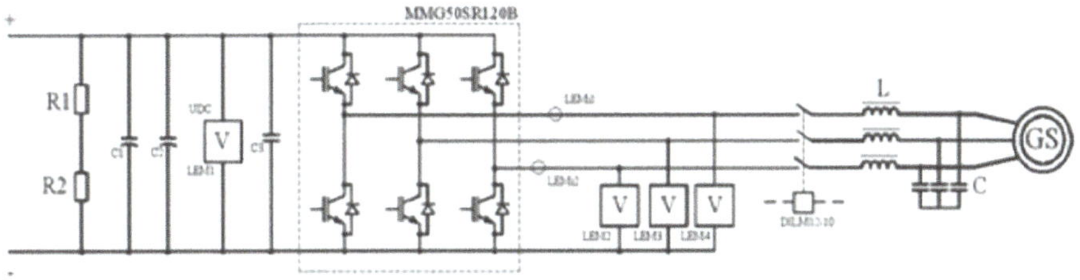

Fig. 1. Simplified scheme of investigated system [5].

The direct current circuit capacitors of machine side inverter are feed by power diodes in parallel connection with IGBT collectors. Common method used widely in synchronizers is calculation of frequency based on the zero crossing detection.

Because of the diodes and capacitors action there are some additional frequencies (harmonics) in measured voltages so problem of the multiple zero crossings resulting in improper voltage frequency calculation arises (Fig. 2).

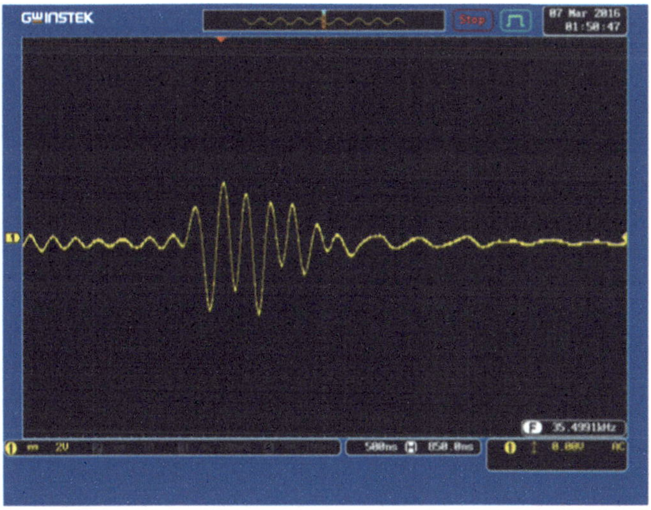

Fig. 2. Magnified waveforms of phase voltage with multiple zero value crossings for sample voltage frequency of 10 Hz.

Some kind of filtering can be used but it will significantly extend synchronization time. Additional problem in the proposed solution is a wide range of initial generator rotational speed, which is unknown in the sensorless control system. Under such conditions some kinds of more sophisticated methods of frequency detection have to be applied. One of the most popular is fast Fourier transform in its fastest Radix-2 decimation-in-time (DIT) form.

3 Basics of Radix-2 DIT

Cooley and Tukey [6] have popularized the algorithms which are based on decomposing the transformation into smaller transforms hierarchically and combining them to get the total transform.

The Radix-2 decimation-in-time fast Fourier transforms is one of the simplest FFT algorithms. Speed of operation is gained by reusing the results of smaller, intermediate computations to compute multiple DFT frequency outputs. As understood from the methodology of creating Fast Fourier Transform (FFT), it can be considered as a data parallelization algorithm [7].

The Radix-2 DIT algorithm divides the discrete Fourier transform equation into two parts: a sum over the even-numbered discrete-time indices n_e = [0, 2, 4, …, N − 2] and a sum over the odd-numbered indices n_o = [1, 3, 5, …, N − 1] and it can be presented as the following equations:

$$X(k) = \sum_{n=0}^{N-1} x(n)e^{-(i2\pi nk/N)} \tag{1}$$

where n is the sum of n_e and n_o.

$$X(k) = \sum_{n=0}^{(N/2)-1} x(2n)e^{-(i2\pi nk/(N/2))} + e^{-(i2\pi k/N)} \sum_{n=0}^{(N/2)-1} x(2n+1)e^{-(i2\pi nk/(N/2))} \tag{2}$$

$$X(k) = DFT_{N/2}[[x(0), x(2), \ldots, x(N-2)]] + W_N^k DFT_{N/2}[[x(1), x(3), \ldots, x(N-1)]] \tag{3}$$

The simplifications in (1) show that frequency outputs X(k) can be computed as the sum of the outputs of two length N/2 transforms [8], of the even-indexed and odd-indexed discrete-time samples, where the odd-indexed short discrete Fourier transform is multiplied by a so-called twiddle factor term:

$$W_N^k = e^{-(i2\pi k/N)} \tag{4}$$

Because of arranging samples in two independent groups this algorithm is called Radix-2 decimation-in-time (Radix-2 DIT) [8].

Frequency outputs of N/2 even numbered time samples, are denoted as G(k) (see Fig. 3) while odd numbered samples are depicted as H(k). With the use of two different twiddle factors G(k) and H(k) can be used to calculate two N length frequencies: X(k) and X(k + N/2).

The use of DIT algorithm makes FFT calculations in real time much faster. Direct computation of DFT requires N^2 multiplies and N^2–N additions and notably both of them are complex operations.

Calculations on two shorter DFTs are much more effective because complex multiplications are reduced to $N + N^2/2$ and a number of additions is reduced to $N^2/2$.

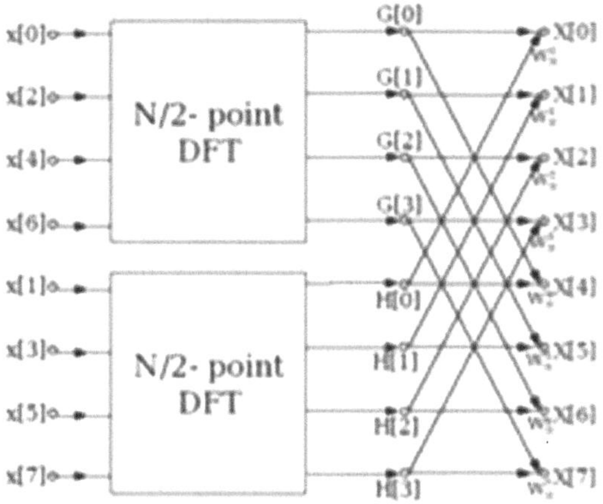

Fig. 3. DIT of *N* samples into two DFT's of length *N/2* [9].

To gain additional speed of calculations Radix-2 DIT can be applied to N/2 length DFTs which will divide them to the next two DFTs.

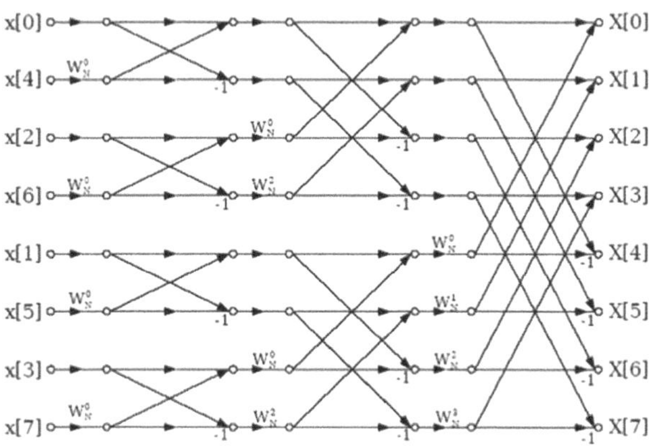

Fig. 4. Radix-2 DIT of 8 samples length signal [9].

In the case of DFT operation depicted in the Fig. 4. Radix-2 DIT of 8 samples length signal [9] calculations take $M = \log_2 N$ stages. In every stage each so-called butterfly operation requires one multiplication and two additions using complex numbers. Overall gain in mathematical operations from this method gives $(N/2) \log_2 N$ multiplies and $N \log_2 N$ additions. As an example, it can be easy calculated use of classic DFT algorithm for 512 samples takes 262144 multiplications and 261632 additions. For the same number of points DIT Radix-2 requires 2304 multiplications and 4608 adds. It is worth noticing that the presented algorithm gets relatively faster for longer vectors of incoming data. Additional increment can be achieved by the use of special type of twiddle factors with reduced number of multiplications [9].

4 Relationship Between FFT Data Vector Length and Frequency Resolution

The frequency resolution depends on the relationship between the FFT vector length and the sampling rate of the input signal. For calculations needed in specific hardware configuration of inverter and synchronous generator, an assumption must be made about the need for synchronization frequency value. For practical reasons, the frequency range covers the values contained between 10 and 50 Hz. Sufficiently accurate for most commercial synchronizers resolution of calculated frequency 0.2 Hz/bin was assumed. Because of Nyquist frequency theorem measurement of voltage signal must be done at minimal value of 100 Hz (twice the maximal frequency). To obtain 0.2 Hz/bin resolution for 100 Hz we have to get at least 500 points. Because of its nature, Radix-2 algorithm works properly when the number of sampled points is a power of two. The closest one is 512 and in modern real-time programming FFT function is embedded with appropriate suffixes.

For example in VisualDSP++ 5.0 function calculating FFT for 512 points is called rfft512. In the simple way bin resolution can be calculated as F_S/N where F_S is the input signal sampling rate and N is the number of DFT points. From the above it can be noticed that to obtain smaller FFT bins it has to either take more sample points or decrease sampling rate of measurement. In a case of using DFT algorithm to detect wide range frequency there is always a need to find compromise between frequency and time domain resolution. With attempts to reduce the size of FFT bins the measurements will take more time. Along with rapidly changing generator's output voltage frequency this property of DFT may cause problem. The frequency measurement will be too slow, which prevents synchronization process. On the other hand short time of acquiring samples will cause low FFT bin resolution. Despite of great capabilities of the real-time DSP systems at some point of software development the run time problem will occur. Taking a huge number of samples in quite short time can also cause extensive use of processor time and lock program execution. These parameters are heavily hardware dependent and are crucial for software creation and further system operation.

5 Investigated Control System Properties and Computer Simulations Results

Real-time system consists of ADSP-21363 processor and Altera FPGA (field-programmable gate array) circuitry. Digital outputs of FPGA are tied to inverter drivers. The FPGA printed circuit board is equipped with analog to digital converters, which are converting analog signal from current and voltage LEM transducers. The FPGA software is called in DSP's program subroutines and time provided for handling this event is set to 150 μs. Any attempt of prolonging the time of software subroutine, will end up with processor internal error and abort the DSP operation.

When interrupt time communication between FPGA and DSP is established, inverter gets control signals for IGBT's and DSP obtains set of acquired data to work out new control signals (Fig. 5).

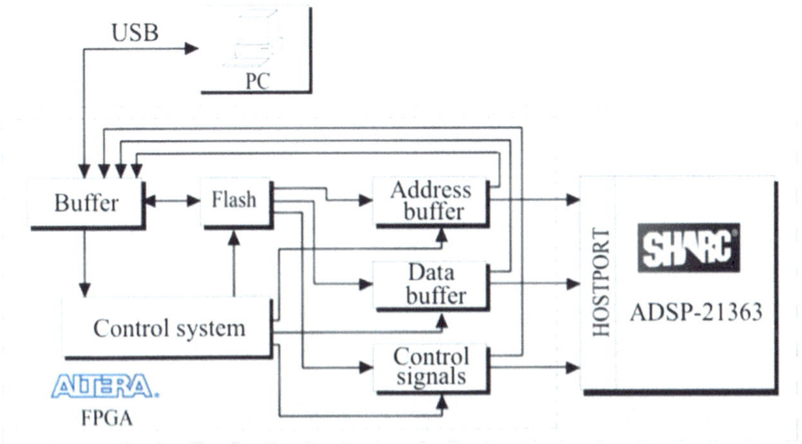

Fig. 5. Scheme of a laboratory control system [10].

The software used for programming DSP was generic Analog Devices Visual DSP ++ 5.0 (VDSP) while Altera was programmed in Quartus II. The VDSP is integrated development environment, which brings C++ dedicated editor, compiler, debugger and processor simulator just to mention the most important. The simulator of DSP is very interesting feature of IDE VDSP, because it allows to simulate software run-time and it all can be done in the absence of the real hardware.

For the simulation input vector data of the real voltages measurements from synchronous generator was taken. Few different data series for chosen one-phase voltage frequencies were written into the text file by the means of digital oscilloscope. As an example the waveforms of 10 Hz frequency was chosen for simulations tests presented in this paper. Obtained waveforms were embedded into VisualDSP++ code as vector filled with data. In the execution time, program takes data samples one by one and puts them into the vector that is an input for rfftN function. Amongst many interesting features of VDSP IDE one can find it very useful that every data vector after compilation of simulated program can be displayed as a waveform. In such simple way one can always check if embedded data is recognized properly. Next feature is a possibility of doing some basic math operations on embedded and vectorized data. It is accessible under View, Debug windows and Plot menu. Available Data Processing options are: converted to dB, FFT Magnitude and 2D FFT Magnitude. FFT Magnitude operation on sample vector of data was performed just to prove the validity of oscilloscope initial measurements. Additionally, sample rate value is needed to calculate values of FFT magnitude. Because of known oscilloscope measurement sample rate 5 kHz this value was inserted right into appropriate place in Plot Settings window.

New window with FFT magnitude pops up after completing calculations. As it turned out in simulation tests developed software recognized well the frequency of distorted voltages throughout all the assumed range with fully represent real world hardware application with its interruptions and subroutines supporting FPGA (Fig. 6).

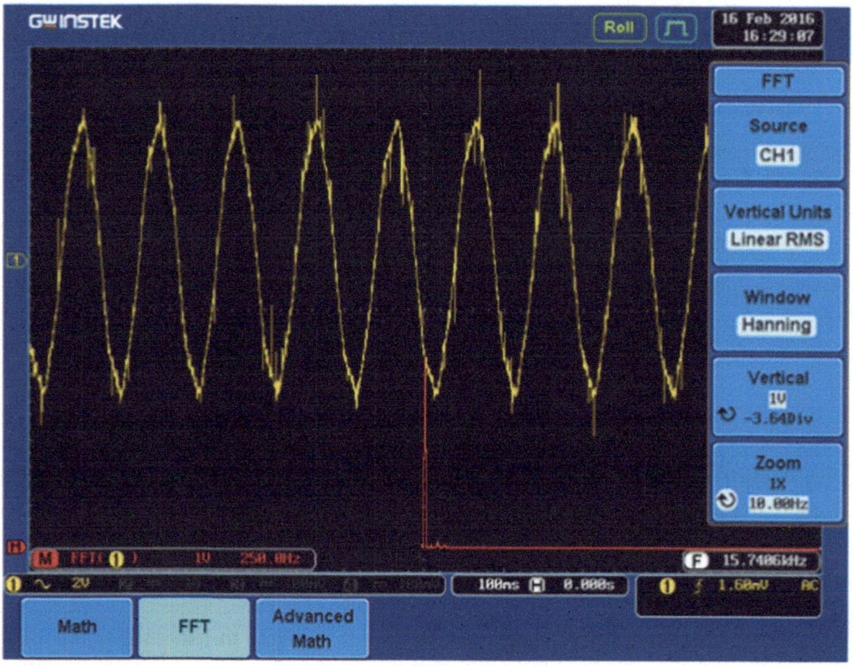

Fig. 6. Sample voltage waveforms and FFT results acquired by digital oscilloscope (voltage frequency of 10 Hz).

There are also no restrictions of input data vector length in simulation software thus maximal value of input FFT vector provided by IDE is 65536 points [10]. This software fits perfectly for testing and debugging the prepared code that will be used in real-time application.

6 Real-Time Experimental Test Bench and Chosen Results

Experimental setup consists of voltage source inverter in back-to-back connection of total power 11 kW. Each module was half of power i.e. 5.5 kW. One side of VSI inverter was connected to terminals of 6 kVA self-excited synchronous generator. Because of specific construction excitation system generator could not be connected directly in parallel with other synchronous generator or power system. This property of generator excitation system causes voltage amplitude variations and makes synchronization process even more complicated. The DSP along with Altera starts up with voltage supply, and after initialization program runs in infinite loop. Within this software loop there are measured voltages and two currents resulting from rotational movement of generator. In this time capacitors are charged until reaching certain value of DC voltage. After initial charging, the system closes main contactor and from this time is ready for operation. To make synchronization process possible there must be rotational speed calculated first and so for this purpose DFT was used. It is notable for this solution that angular speed can change during the operation of generator. Therefore some software means must be provided to detect frequency with high enough precision in reasonable time. Real time system must measure voltage within software subroutine execution time. It was called every 150 μs and there were performed other time consuming calculations during this time. In the presented system synchronization is one of many functions executed in the software interrupt time so with quite complex functions as DFT hardware resources can be easily exceeded resulting in freezing of processor operation. While performing experimental tests it turned out that maximal value of FFT points may be as high as 512. Because of this reducing resolution of FFT bins will occur. Having sampling frequency of 6666 Hz to measure 50 Hz we have to save every 66 sample to get desired 101 Hz. Incoming data can be saved with much higher resolution but there is no need to do this as long as we are not interested in other harmonics. Huge amount of incoming data would eventually slow down hardware performance. As it was aforementioned maximal number of points that can be used is 512 so this leads to conclusion that frequency detection of 50 Hz with 512 FFT points and bin resolution of 0.2 Hz will take around 5 s. It is possible to decrease data acquiring time with smaller number of data samples collected but maintaining maximal number of FFT points will worsen bin resolution. Attempts made to detect frequency with resolution of 1 Hz per bin soon will turn out that for 512 points input vector sampling frequency should be equal to 512 Hz. That is because 512 Hz divided by 512 FFT points gives resolution of 1 Hz/bin. In order to prove aforementioned assumptions, a series of experimental tests were undertaken. The tests included changes of generator rotational speed during real-time software frequency detection (Fig. 7).

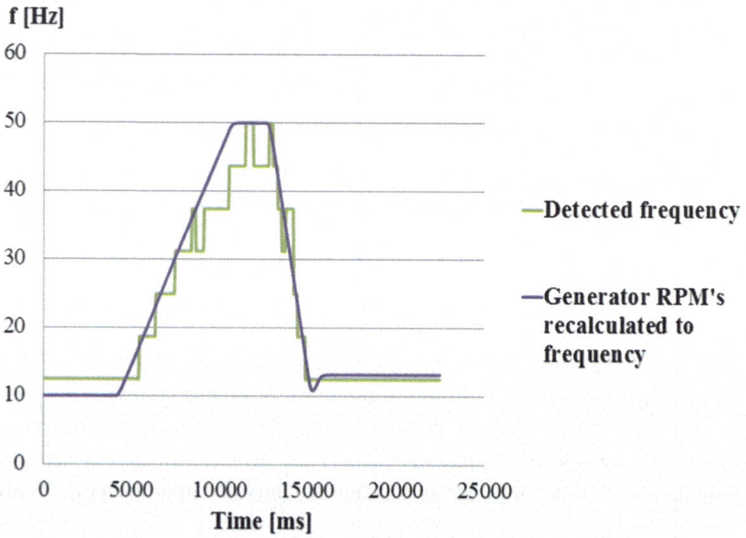

Fig. 7. Experimental results of frequency detection n real-time setup. Data consists of 512 points taken from 1024 samples.

Angular speeds were changed from 10 Hz to 50 Hz and then back to around 10 Hz. In the tests there were 1024 points acquired for every calculation period. This value is consistent with expected FFT function execution time. Such low FFT bin resolution is of course not satisfying so further tests with altered parameters were conducted (Fig. 8).

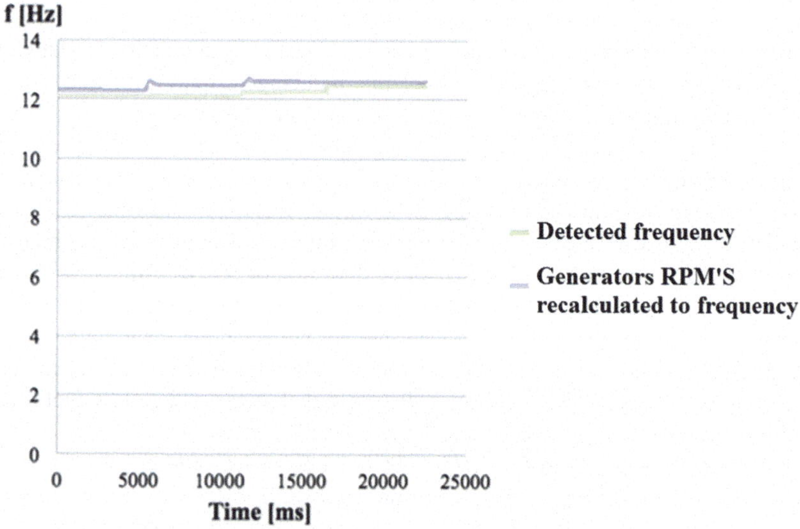

Fig. 8. Experimental results of frequency detecion in real-time setup. Data consists of 512 points taken from 66105 samples.

Every second sample was put into input FFT vector 512 points long. Analysis of obtained data shows that FFT bin resolution equals 6.5 Hz/bin. According to bin resolution formula for every second point acquired, the sampling frequency drops to 3333 Hz. Total number of points is 512 so after completing F_S/N we get 6.5 Hz per FFT bin. Minimal time needed for frequency calculation equals around 154 ms.

This value is consistent with expected FFT function execution time. Such low FFT bin resolution is of course not satisfying so further tests with altered parameters were conducted.

Taking measurements in longer period resolution of FFT bin has been improved to 0.2 Hz per bin but acquiring data thus calculation time significantly increased.

7 Conclusions

Completed real-time experiments clearly showed that there is always a trade-off between temporal and frequency resolution. The more accurate measurement is in the frequency domain, the rougher results are obtained in the time domain. Long time measurements create many points which need to be put in long data vectors. The calculations on so huge amount of data are time consuming which is inadvisable in rotational speed quick changing systems. Another disadvantage of the presented method is the lack of phase detection with some prediction needed in case of circuit breaker lag which occurs in every application. This experiment proved that for variable speed generator, FFT fundamental frequency detection is not optimal choice so some other computational means should be used. The experimental results were obtained with use of the test bench present in Fig. 9.

Fig. 9. View of the laboratory test stand.

It should be noted that other algorithms such Radix-4 DIT FFT are under development. These algorithms require less space and reduce computational power compared to the existing methods [11].

The research project is funded within a 1/S/IEiAO/2016 program.

References

1. Watson, G.O.: Marine Electrical Practice. Marine Engineering Series. Butterworth-Heinemann Ltd., Oxford (1990)
2. Choi, I.C., Jeung, Y.C., Lee, D.C.: Variable speed of diesel engine-generator using sliding mode control. In: 2017 IEEE Transportation Electrification Conference and Expo, Asia-Pacific (ITEC Asia-Pacific) (2017)
3. Benesch, N., Namyslo, L.: Grid-side control of inverters for power generation in microgrids. In: 2104 4th International Electric Drives Production Conference (EDPC) (2014)
4. IEEE Standard for Cylindrical-Rotor 50 Hz and 60 Hz Synchronous Generators Rated 10 MVA and Above, IEEE Standard C50.13-2005 (2005)
5. Przemiennik częstotliwości MMB005 IM, instrukcja użytkownika. MMB Drives Sp. z o.o. (2014)
6. Cooley, J.W., Tukey, J.W.: An algorithm for the machine calculation of complex Fourier series. Math. Comput. **19**(90), 297–301 (1965)
7. Prabhu, E., Mangalam, H., Karthick, S.: Design of area and power efficient Radix-4 DIT FFT butterfly unit using floating point fused arithmetic. J. Cent. S. Univ. **23**(7), 1669–1681 (2016)
8. Arndt, J.: Matters Computational. Ideas, Algorithms, Source Code. Springer, Berlin (2011)
9. Jones, D.L.: Decimation-in-Time (DIT) Radix-2 FFT Algorithms. OpenStax CNX homepage. https://cnx.org/contents/XaYDVUAS@6/Decimation-in-Frequency-DIF-Ra. Accessed 20 Jun 2017
10. VisualDSP++ 5.0 User's Guide. Analog Devices, Inc. (2007)
11. Cerna, M., Harvey, A.F.: The fundamentals of FFT-based signal analysis and measurement. National Instruments, Application Note 041, p. 4, National Instruments Corporation (2000)

Experimental Research Assessing Threat of EOD Technicians from Explosive Blast

Edyta Krzystała(✉), Krzysztof Kawlewski, Sławomir Kciuk, Tomasz Machoczek, and Grzegorz Bienioszek

Institute of Theoretical and Applied Mechanic, Silesian University of Technology, Gliwice, Poland
Edyta.Krzystala@polsl.pl

Abstract. The EOD technicians work in an environment with a higher threat. The job of the EOD technicians is to assist in identification of explosive ordnance and its destruction at the place of finding. Their working environment is influenced by physical, chemical, physiological and psychological factors. The main threats are injuries, illnesses, and work traumas resulting from explosions. To reduce the risk of blast injuries the protective bomb suits are used by the technicians. In the article results of experimental research assessing threat of EOD technicians from explosive blast are presented. Upper limbs are the most exposed to injuries in the work of pyrotechnic technicians. In the article the acceleration test results of upper limb are presented. During the research an acceleration of the particular body parts as well as pressure acting on the thoracic of the dummy surrogate were measured. The test was conducted in open air condition in Krupski Młyn area. Essential elements of the research were to develop and use proper devices for data acquisition in a test condition of blast. During the test the prototype autonomous measurement system developed in the Institute of Theoretical and Applied Mechanics was used. The results of the experimental research will be used to design and improve better personnel blast protection suits as well as educational materials for the future EOD technicians.

Keywords: Blast injury · Explosive ordnance disposal · Injury risk · Explosive technicians training

1 Introduction

Increasingly common events translate into a greater need for explosive ordnance disposal (EOD) technicians and a greater risk incurred by many EOD technicians employed by the military and law enforcement agencies. The job of the EOD technician is to assist in identifying explosive ordnance and its destruction at the place of finding. Their working environment is influenced by physical, chemical, physiological and psychological factors. The main threats are injuries, illnesses, and work traumas resulting from explosions. To reduce the risk of blast injuries the protective bomb suits are used by the technicians. Bomb suits are designed to protect against the characteristics of an explosive event that could result in human injury. The most commonly used are EOD bomb suit (Fig. 1) [2–7, 9]. Due to the extent of injuries that may result from an

J. Świder et al. (Eds.): MECHATRONICS 2017, AISC 934, pp. 269–276, 2019.
https://doi.org/10.1007/978-3-030-15857-6_27

explosion, bomb suits are designed to provide comprehensive, head-to-toe protection against a blast. Bomb suits should provide protection against four main factors: fragments, overpressure, heat wave as well as shock wave. In order to achieve a sufficient level of protection, the produced suits are invariably heavy and bulky, causing restricted body movement, fatigue, heat stress, and potentially reduced operational performance of the user [8]. Most suits are designed without protection for the user's hands because the technicians need them to have maximum dexterity and maneuverability in order to defuse a bomb [8]. The example of EOD 9 bomb suit was presented in Fig. 1.

Fig. 1. EOD bomb suit and helmet ensemble for CBRN threat [11]

In the case of injuries, the upper limbs are most exposed because they are in the closest position to the detonation point. In the article, results of experimental research assessing threat of EOD technicians from explosive blast, especially upper limb are presented. The results of the experimental research will be used to improve better personal blast protection gloves as well as educational materials for the future explosive technicians.

2 Methodology

The aim of a research was identification of threat for EOD technicians working with explosives materials. The methodology of parameters measurement was developed allowing assessment of threat of EOD technicians. Preparation of experiment setup included arrangement of accelerometers and placing the system in initial configuration (Fig. 2). The markers were put on the dummy and beams, which were buried in the ground. The distances between all experiment setup components were measured, so the proper calibration of Phantom v9.1 camera was possible. Essential elements of the research was to develop and use proper devices for data acquisition in a test condition of blast. During the test the prototype autonomous measurement system developed in the Institute of Theoretical and Applied Mechanics was used [10]. Thanks to this device acceleration measurements acting on a human body were conducted (Fig. 3c).

Fig. 2. Initial configuration showing the location of dummy, explosive charge and ground markers.

The research involved detonation of TNT explosive charges of different masses at a distance of 3.5 m from the dummy. The charge was placed at a height of approximately 1.5 m in the air, so the explosion occurred at the level of the human chest. During the tests, controlled explosions of the 75, 200, 400 and 1000 g of TNT charges were made. The dummy was in a standing position with arms outstretched (Fig. 2). The aim of selecting such a configuration was to reflect the situation, in which a sapper carries an explosive charge using a special boom. Preparation of experiment setup included allocation of accelerometers (Fig. 3) and placing the system in initial configuration (Fig. 3). The test was conducted in open air condition in Krupski Młyn area. An equivalent of 1 kg of TNT charge was repeated next day and the comparison of the test results was presented in the paper as well.

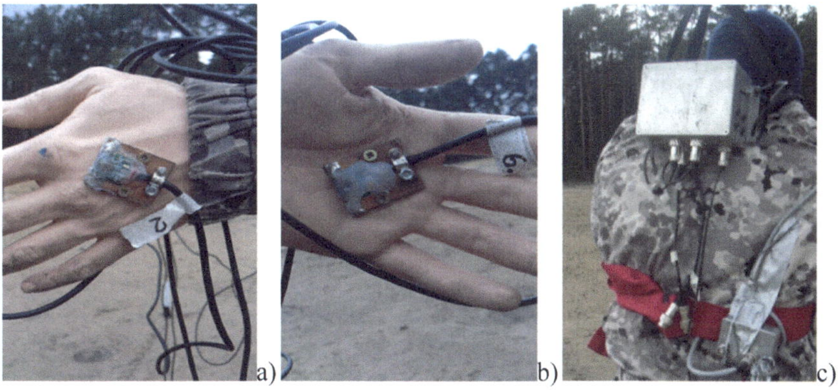

Fig. 3. Experiment setup: (a) channel 2 accelerometer (X axis), (b) channel 6 accelerometer (Z axis), (c) automatic measurement system

By using high speed cameras it was possible to make an identification of overall characteristics of explosions, registration of a shock wave propagation for the different types of explosion and different materials. The distances between all experiment setup components were measured, so the proper calibration of Phantom v9.1 high speed camera was possible (Figs. 4 and 5). The camera was located 17.7 m from the dummy, while recording explosions of the charge mass between 75 g to 400 g. For the safety reasons, during explosion of 1 kg of TNT it was moved further to the distance of 19 m. The speed of recording frames of a Phantom camera during explosion was 3600 frames/s.

Fig. 4. High speed cameras used during research

Fig. 5. Example of image frame recorded with the use of high speed camera

3 Results

Acceleration levels measured in two axes of dummy hand were shown in Figs. 6, 7 and 8. Hand acceleration was measured in two directions and the results were calculated. The reason was the huge rotation of arm around shoulder, which made it impossible to measure the acceleration just in one axis accurately.

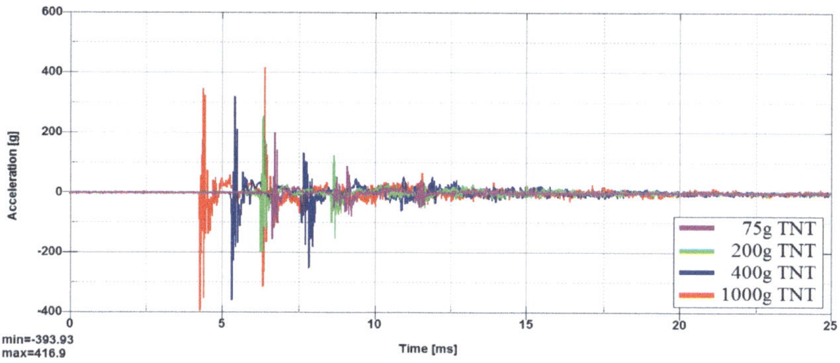

Fig. 6. Accelerations recorded for hand in X axis, during explosions of different masses of TNT charge (raw signals).

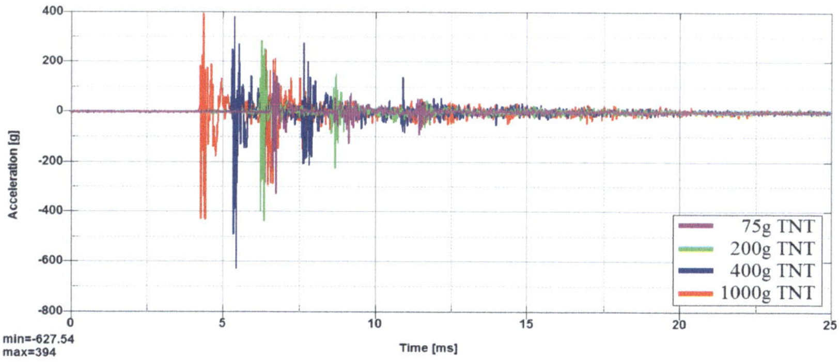

Fig. 7. Accelerations recorded for hand in Z axis, during explosions of different masses of TNT charge (raw signals).

Repeatability of records was checked by performing one additional explosion of 1 kg of TNT charge on another day. The results were shown in Figs. 9, 10 and 11.

The obtained results proved good repeatability of records for two explosions of 1 kg mass of TNT charge. It should be highlighted that, although experiment configuration was the same, the environmental conditions were not. Nevertheless, this fact did not have a relevant influence on the final results.

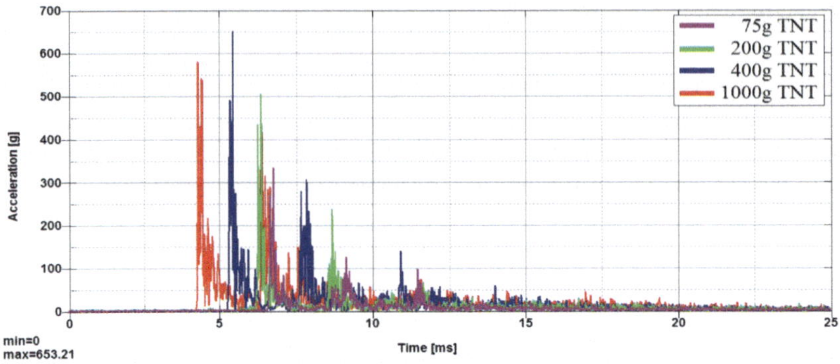

Fig. 8. Calculated resultant accelerations for hand during explosions of different masses of TNT charge (raw signals).

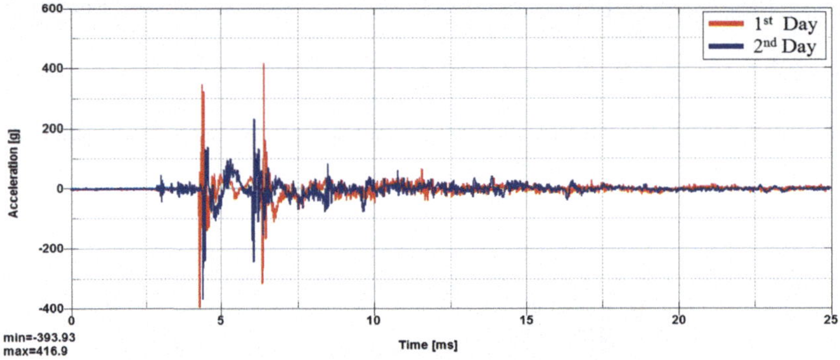

Fig. 9. Comparison of accelerations recorded for hand in X axis during two different explosions of 1 kg of TNT charge (raw signals).

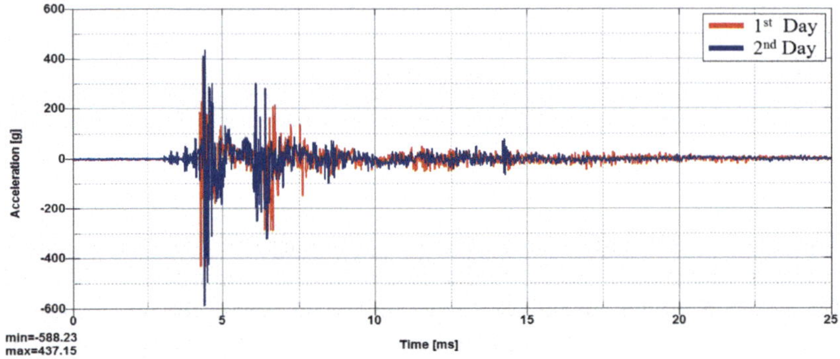

Fig. 10. Comparison of accelerations recorded for hand in Z axis during two different explosions of 1 kg of TNT charge (raw signals).

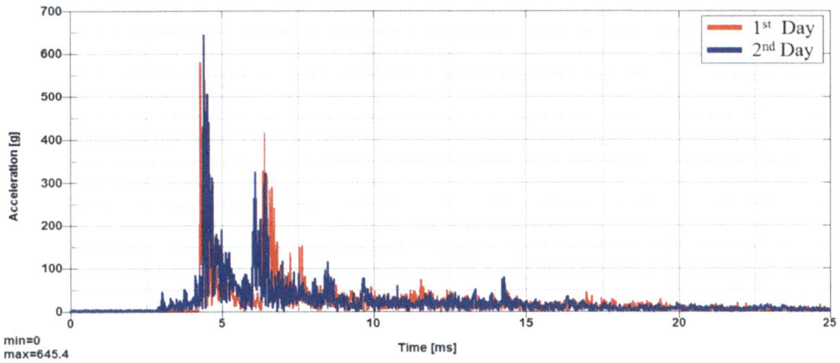

Fig. 11. Comparison of resultant accelerations calculated for hand for two different explosions of 1 kg of TNT charge (raw signals).

Furthermore, information about characteristic values describing the resultant acceleration signals was summarized in Table 1. Besides maximum acceleration level (amax), the average acceleration in two different time periods (t = 10 ms and t = 40 ms) were calculated according to following equation:

$$|a|_{avg\_10\,\mathrm{ms}} = max\left(\frac{\int_{t_1}^{t_2}|a|dt}{\Delta t}\right) \quad where\ \Delta t = t_2 - t_1 = 10\,\mathrm{ms} \tag{1}$$

$$|a|_{avg\_40\,\mathrm{ms}} = max\left(\frac{\int_{t_1}^{t_2}|a|dt}{\Delta t}\right) \quad where\ \Delta t = t_2 - t_1 = 40\,\mathrm{ms} \tag{2}$$

Table 1. Comparison of resultant acceleration characteristic values for different masses of TNT charge.

TNT charge mass		75 g	200 g	400 g	1000 g (1st test)	1000 g (2nd test)
Raw signal	a_{max}	336 G	507 G	653 G	583 G	645 G
	$\|a\|_{avg\_10ms}$	18 G	23 G	36 G	44 G	47 G
	$\|a\|_{avg\_40ms}$	7.1 G	9.1 G	12.5 G	15.6 G	16.0 G

4 Conclusions

The primary role of EOD technicians is to protect life. The results of the experimental research will be used to design and improve better personal blast protection gloves as well as educational materials for the future explosive technicians. The tests results show repeatable acceleration time histories for different trial as well as for repeated test of 1 kg TNT charge. For the last few decades, the need of implementation of blast resistant design in major construction projects, especially for government buildings, has been recognized by observing numerous terrorists' attacks on civilian structures [1].

The recorded acceleration signals proved that during explosion of higher amounts of TNT charge mass the hand injury risk grows rapidly. The good repeatability of measurements enhanced the reliability of the conducted research.

The performed work contributed to better understanding of blast wave impact onto human body, in particular hands which are exposed the most in case of EOD technicians.

References

1. Ullah, A., Ahmad, F., Jang, H.-W., Kim, S.-W., Hong, J.-W.: Review of analytical and empirical estimations for incident blast pressure. KSCE J. Civ. Eng. (2016). https://doi.org/10.1007/s12205-016-1386-4
2. Bowen, I.G., Fletcher, E.R., Richmond, D.R.: Estimate of man's tolerance to the direct effects of air blast. Technical Progress Report, DASA-2113, Defense Atomic Support Agency, Department of Defense, Washington, DC, October 1968
3. "Dale" Bass, C., Davis, M., Rafaels, K., "Steve" Rountree, M., Harris, R.M., Sanderson, E., Andrefsky, W., DiMarco, G., Zielinski, M.: A methodology for assessing blast protection in explosive ordnance disposal bomb suits. Int. J. Occup. Saf. Ergon. **11**(4), 347–361 (2005). https://doi.org/10.1080/10803548.2005.11076655
4. Kirkman, E., Watts, S., Cooper, G.: Blast injury research models. Philos. Trans. R. Soc. B **366**, 144–159 (2011). https://doi.org/10.1098/rstb.2010.0240
5. Explosive Ordnance Disposal operations, Headquarters Department of the Army, December 2011, this publication is available at Army Knowledge Online (www.us.army.mil) and General Dennis J. Reimer Training and Doctrine Digital Library at (www.train.army.mil)
6. MIL-STD-662F 18 December 1997 Superseding, Department of Defence Test Method Standard, V50 Ballistic Test for Armour
7. NASA memorandum human tolerance to rapidly applied accelerations: a summary of the literature By A. Martin Eiband Lewis Research Center Cleveland, Ohio (1959)
8. Public Safety Personal Protective Equipment for Disposal of Explosive Devices System Assessment and Validation for Emergency Responders (SAVER) TechNote, April 2014
9. Test Methodologies for Personal Protective Equipment Against Anti-Personnel Mine Blast RTO Technical report TR-HFM-089, Final Report of the RTO Human Factors and Medicine Panel (HFM) Task Group TG-024 (2004)
10. Szmidt, P., Kciuk, S., Mężyk, A., Krzystała, E.: An autonomous measuring system for the acquisition of quick change voltage signals, Patent 219 525
11. https://www.med-eng.com/Products/PersonalProtectiveEquipment/MedEngEODIEDD/EOD9SuitHelmet.aspx

Theoretical Analysis of Piezoelectric Transformers in Different Configurations

Paweł Łabędzki[1](✉), Rafał Pawlikowski[2], and Andrzej Radowicz[2]

[1] Faculty of Management and Computer Modelling, Kielce University of Technology, al. Tysiąclecia Państwa Polskiego 7, 25-314 Kielce, Poland
pawlab@tu.kielce.pl

[2] Faculty of Mechatronics and Mechanical Engineering, Kielce University of Technology, al. Tysiąclecia Państwa Polskiego 7, 25-314 Kielce, Poland
{rpawlikowski,radowicz}@tu.kielce.pl

Abstract. Piezoelectric transformer (PT) is an electromechanical transducer that employs the piezoelectric effect. The behavior of PTs is generally analyzed by solving piezoelectric equations or using equivalent electrical circuits. In this study, an implicit Lagrangian method was applied to predict the behavior of different types of PTs from the natural frequency of vibration and some geometric parameters. The calculation results secured good agreement with the data found in the literature on the subject.

Keywords: Piezoelectric ceramic resonator · Rosen–type transformer · Lagrange equations

1 Introduction

This article deals with models of piezoelectric transformers (PTs) which are based on direct and reverse piezoelectric effect [1–4]. Usually, the behavior of PTs is theoretically described by applying piezoelectric equations [4, 5] or by studying equivalent electrical circuits [4–7]. The approach proposed in this article involves using the Lagrangian formalism to derive equations of motion for piezoelectric devices [8]. With the assumption that the electrical load at the PT output is zero, these equations can be reduced to one equation of a driven harmonic oscillator (damped mechanically and driven by the term proportional to the externally applied voltage). The fundamental question arises how much the transformation ratio of a PT depends on such parameters as dimensions, polarizations of the segments, frequencies of the input voltage, quality factor, etc. To answer this question analytical considerations were performed. The calculation results secured agreement with the data found in the literature [4, 5, 8].

J. Świder et al. (Eds.): MECHATRONICS 2017, AISC 934, pp. 277–289, 2019.
https://doi.org/10.1007/978-3-030-15857-6_28

2 Fundamentals

2.1 Piezoelectric Materials

The basic coupled fields in piezoelectric materials could be described by constitutive equations [3, 4, 7, 9, 10]:

$$\boldsymbol{T} = \boldsymbol{c} \cdot \boldsymbol{S} - \boldsymbol{e} \cdot \boldsymbol{E}, \quad \boldsymbol{D} = \boldsymbol{e}^T \cdot \boldsymbol{S} + \boldsymbol{\varepsilon} \cdot \boldsymbol{E} \tag{1}$$

where, in Voigt notation: T = [T1, T2, …, T6]^T – stress tensor, S = [S1, S2, …, S6]^T – strain tensor, D = [D1, D2, D3]^T – electric displacement vector, E = [E1, E2, E3]^T – electric intensity field.

The structures of the stiffness, piezoelectric and permittivity matrices (c, e and ε, respectively) and values of their elements represent the internal structure and the properties of piezoelectric materials. The calculations discussed in this article were performed for PZT ceramic:

$$\boldsymbol{c} = \begin{bmatrix} c_{11} & c_{12} & c_{13} & 0 & 0 & 0 \\ c_{21} & c_{22} & c_{23} & 0 & 0 & 0 \\ c_{31} & c_{32} & c_{33} & 0 & 0 & 0 \\ 0 & 0 & 0 & c_{44} & 0 & 0 \\ 0 & 0 & 0 & 0 & c_{55} & 0 \\ 0 & 0 & 0 & 0 & 0 & c_{66} \end{bmatrix}, \boldsymbol{e} = \begin{bmatrix} 0 & 0 & e_{31} \\ 0 & 0 & e_{32} \\ 0 & 0 & e_{33} \\ 0 & e_{24} & 0 \\ e_{15} & 0 & 0 \\ 0 & 0 & 0 \end{bmatrix}, \boldsymbol{\varepsilon} = \begin{bmatrix} {}_{11} & 0 & 0 \\ 0 & {}_{22} & 0 \\ 0 & 0 & {}_{33} \end{bmatrix} \tag{2}$$

2.2 Lagrangian Formalism for Piezoelectric Devices

The phenomena associated with the operation of a piezoelectric transformer were described using the Lagrangian formalism [8, 11]. In the case considered here, the Lagrangian can be written as:

$$L = \frac{1}{2}\int_V \left[\rho \dot{\boldsymbol{u}}^T \dot{\boldsymbol{u}} + \boldsymbol{E}^T \boldsymbol{\varepsilon} \boldsymbol{E} + 2\boldsymbol{S}^T \boldsymbol{e} \boldsymbol{E} - \boldsymbol{S}^T \boldsymbol{c} \boldsymbol{S}\right] dV \tag{3}$$

where the first term represents kinetic energy density for the deforming PTZ body, the second term is electric energy density, the third term is the density of energy responsible for the interaction of the coupled fields and the last term is the elastic–potential energy density. Here $\boldsymbol{u}$ is displacement vector, $\boldsymbol{V}$ is the volume occupied by piezoelectric body and $\boldsymbol{\rho}$ is the material density. In processes with no dissipation present, the equations of motion have the following form:

$$\frac{d}{dt}\left(\frac{\partial L}{\partial \dot{q}_i}\right) - \frac{\partial L}{\partial q_i} = 0, \quad \frac{d}{dt}\left(\frac{\partial L}{\partial \dot{\lambda}_j}\right) - \frac{\partial L}{\partial \lambda_j} = 0, \quad i = 1, 2, \ldots, n_1, \; j = 1, 2, \ldots, n_2 \tag{4}$$

where q_i and λ_j are generalized coordinates (generalized displacements, flux linkages, [8, 11]). The influence of dissipation will be introduced into these equations in the form of the so–called mechanical quality factor [7, 12, 13].

3 Equations of Motion

3.1 PT Configurations

This article considers the following types of PTs:

1. a two–segments e31–e33 type PT – a classical Rosen–type transformer [1, 2] (Fig. 1a);
2. a two–segmented e33–e33 type PT (Fig. 1b);
3. a two–segmented e31–e31 type PT (Fig. 1c);
4. a three–segmented e31–e31 type PT with the driving segment in the middle ("middle–driving", "md"), (Fig. 1d);
5. a three–segmented e31–e31 type PT with the voltage generating segment in the middle ("middle–generating", "mg"), (Fig. 1e).

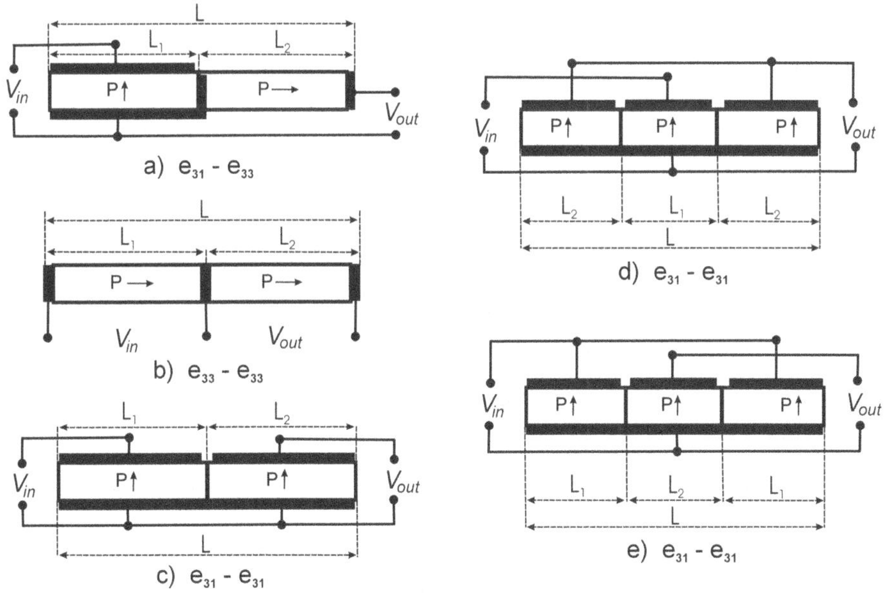

Fig. 1. PTs with different shapes and piezoelectric properties (capital P indicating the direction of the polarization vector).

Terminology and chosen coordinate system are shown in Fig. 2.

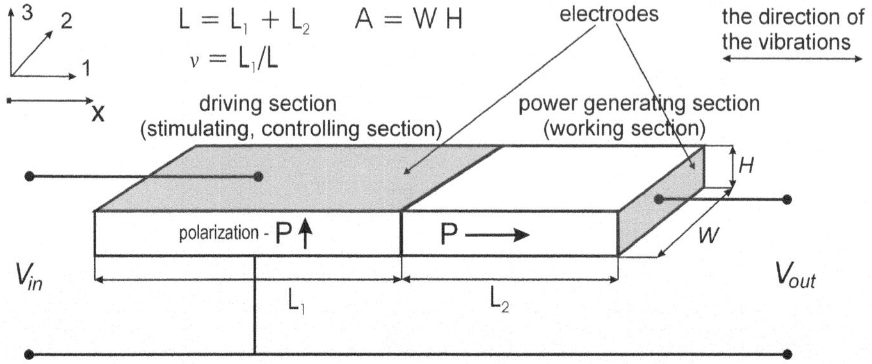

Fig. 2. Classical Rosen–type PT (type 1 PT) in assumed coordinate system.

3.2 Assumptions

The following are the basic assumptions made in this study:

- the transformer is a bar; only single–axis stress and single–axis strain are taken into account;
- the boundary condition is that the bar has both ends free;
- all physical fields are homogeneous; the elastic field on boundary between the segments is continuous;
- electric field vector is parallel to the polarization vector;
- the device is in the state of resonance in the *n-th* natural vibration mode $\phi_n(x)$; thus, the displacement $u(x, t)$ can be formulated as:

$$u(x,t) = Lz(t)\phi_n(x) = Lz(t)\cos\left(\frac{n\pi x}{L}\right) \tag{5}$$

where $z(t)$ is dimensionless parameter representing the time–dependent displacement vector.

The following generalized coordinates were selected to formulate Lagrange's equations of motion:

1. $z(t)$ – a dimensionless parameter representing displacement vector (5);
2. $\lambda(t)$ – a flux linkage, where $d\lambda/dt = V_{out}$ is an output voltage (Fig. 2).

3.3 Derivation of the Equations of Motion for the Two-Segments E31–E33 Type PT (Classical Rosen–Type PT [1, 2])

This PT consists of two piezoelectric bars whose polarizations are as in Fig. 1a. The Lagrange function (3) can be written as a sum of three terms:

$$L = T + W_d + W_g \tag{6}$$

$$T = \frac{1}{2}\int_o^L \rho A \dot{u}^2 dx \tag{7}$$

$$W_d = \frac{1}{2}\int_0^{L_1} A\left(\varepsilon_{33}E_3^2 + 2S_1 e_{31}E_3 - c_{11}S_1^2\right)dx \tag{8}$$

$$W_g = \frac{1}{2}\int_{L_1}^{L} A\left(\varepsilon_{33}E_1^2 + 2S_1 e_{33}E_1 - c_{11}S_1^2\right)dx \tag{9}$$

Here lowercase *d* refers to the driving segment of the bar, *g* indicates the generating segment and *A* is the cross–sectional area of the bar (Fig. 2). The electric field in the driving segment is $E_3 = V_{in}/H$, whereas strain $S_1 = \partial u/\partial x$ is the same in both the driving and generating segments. Now it is necessary to express the electric field E_1 in the generating segment of the device using the generalized coordinates λ and z. Applying Eq. (1):

$$E_1 = \frac{1}{\varepsilon_{33}}D_1 - \frac{e_{33}}{\varepsilon_{33}}S_1 \tag{10}$$

and Gauss' law $\partial D_1/\partial x = 0$ (with D_1 does not depend on x in driving segment), we get:

$$\dot{\lambda} = V_{out} = \int_{L_1}^{L} (-E_1)dx = -\frac{L - L_1}{\varepsilon_{33}}D_1 + \frac{e_{33}}{\varepsilon_{33}}\int_{L_1}^{L} S_1 dx \tag{11}$$

After performing standard calculations, from Eqs. (10) and (11) we obtain E_1 as a function of generalized coordinates.

Thus, the system of two coupled Lagrange's equations (4) is as follows:

$$\ddot{z} + \alpha z + \beta\dot{\lambda} = \gamma V_{in} \tag{12}$$

$$\ddot{\lambda} = \delta\dot{z} \tag{13}$$

where α, β, γ, δ are constants dependent on material and geometrical parameters [11].

The above system of Eqs. (12) and (13) was considered for the most practical case, i.e. for zero initial conditions ($z(0) = 0$, $\lambda(0) = 0$, $\partial z(0)/\partial t = 0$, $\partial \lambda(0)/\partial t = 0$).

By integrating (13):

$$\dot{\lambda} = \delta z \tag{14}$$

and substituting to (12), these two coupled Lagrange's equations can be transformed into one equation:

$$\ddot{z} + \omega_0^2 z = \gamma V_{in} \tag{15}$$

where $\omega_0^2 = \alpha + \beta\delta$. Equation (15) describes undamped harmonic oscillator driven by excitation proportional to the input voltage. The mechanical quality factor Q_m is arbitrarily introduced to take account of dissipation, which occurs in all dynamic processes [7, 12, 13]:

$$\ddot{z} + \frac{\omega_0}{Q_m}\dot{z} + \omega_0^2 z = \gamma V_{in} \tag{16}$$

giving a well-known equation of a driven damped harmonic oscillator.

The parameters in this equation are represented by the following formulas

$$\omega_0^2 = \frac{e_{33}^2}{2L^2\rho\varepsilon_{33}}\left[2n^2\pi^2 - 2n^2\pi^2 v + n\pi\,\sin(2\pi n v)\right] + \frac{c_{11}}{L^2\rho}n^2\pi^2 \tag{17}$$

$$\gamma = \frac{2e_{31}}{HL^2\rho}\left[\cos(n\pi v) - 1\right] \tag{18}$$

$$\delta = \frac{Le_{33}}{\varepsilon_{33}}\left[(-1)^n - \cos(\pi n v)\right] \tag{19}$$

v, H, L – geometrical parameters of the PT (Fig. 2).

The equations of motion for all the other PT configurations are derived in an analogous manner. In all those cases, the form of Lagrangian and the equations of motion have the same form, to within coefficients. The differences between the model of a classical Rosen–type PT and the other PT models will be discussed below.

3.4 Derivation of the Equations of Motion for the Two-Segments PT of the e33–e33 Type

This PT model consists of two piezoelectric bars whose polarizations are as in the Fig. 1b. As the differences appear in the polarization of the driving segment, the formula for W_d in this segment is:

$$W_d = \frac{1}{2}\int_0^{L_1} A\left(\varepsilon_{33}E_3^2 + 2S_1 e_{33}E_3 - c_{11}S_1^2\right)dx \tag{20}$$

and $E_3 = V_{in}/L_1$. Then, the suitable coefficients in this equation of motion have a following form:

$$\omega_0^2 = \frac{e_{33}^2}{2L^2\rho\varepsilon_{33}}\left[2n^2\pi^2 - 2n^2\pi^2 v + n\pi\sin(2\pi n v)\right] + \frac{c_{11}}{L^2\rho}n^2\pi^2 \tag{21}$$

$$\gamma = \frac{2e_{33}}{vL^3\rho}\left[\cos(n\pi v) - 1\right] \tag{22}$$

$$\delta = \frac{Le_{33}}{\varepsilon_{33}}\left[(-1)^n - \cos(\pi n v)\right] \tag{23}$$

3.5 Derivation of the Equations of Motion for the Two-Segments e31–e31 Type PT

This PT model consists of two piezoelectric bars whose polarizations are as in the Fig. 1c. As the differences appear in the polarization of the generating segment, the formula for W_g in this segment is:

$$W_g = \frac{1}{2}\int_{L_1}^{L} A\left(\varepsilon_{33}E_3^2 + 2S_1 e_{31}E_3 - c_{11}S_1^2\right)dx \tag{24}$$

where E_3 is expressed by generalized coordinates in a manner analogous to that of case 1 (classical Rosen-type PT). Thus, the coefficients in the equation of motion have the following form:

$$\omega_0^2 = \frac{e_{31}^2}{2L^2\rho\varepsilon_{33}}\left[2n^2\pi^2 - 2n^2\pi^2 v + n\pi\sin(2\pi n v)\right] + \frac{c_{11}}{L^2\rho}n^2\pi^2 \tag{25}$$

$$\gamma = \frac{2e_{31}}{HL^2\rho}\left[\cos(n\pi v) - 1\right] \tag{26}$$

$$\delta = \frac{Le_{31}}{\varepsilon_{33}}\left[(-1)^n - \cos(\pi n v)\right] \tag{27}$$

3.6 Derivation of the Equations of Motion for the Three-Segments E31–E31 Type PT with the Driving Segment in the Middle ("Middle–Driving" – "Md")

This model of PT consists of three piezoelectric bars whose polarizations are as in the Fig. 1d. The central segment performs a driving role whereas the left and right segments play a generating role. The generating segments have identical lengths (Fig. 1d). For this PT we write:

$$W_d = \frac{1}{2}\int_{L_1}^{L-L_1} A\left(\varepsilon_{33}E_3^2 + 2S_1 e_{31}E_3 - c_{11}S_1^2\right)dx \tag{28}$$

$$W_g = \frac{A}{2}\left[\int_0^{L_1}\left(\varepsilon_{33}E_1^2 + 2S_1 e_{31}E_1 - c_{11}S_1^2\right)dx + \int_{L-L_1}^{L}\left(\varepsilon_{33}E_1^2 + 2S_1 e_{31}E_1 - c_{11}S_1^2\right)dx\right] \tag{29}$$

where E_3 is expressed by generalized coordinates in a manner analogous to that of case 1. The coefficients in the equation of motion have the following form:

$$\omega_0^2 = \frac{e_{31}^2}{L^2\rho\varepsilon_{33}}\left[2n^2\pi^2 v - n\pi\sin(2\pi n v)\right] + \frac{c_{11}}{L^2\rho}n^2\pi^2 \tag{30}$$

$$\gamma = \frac{2e_{31}}{HL^2\rho}[-1+(-1)^n]\cos(\pi n v) \tag{31}$$

$$\delta = \frac{Le_{31}}{\varepsilon_{33}}[-1+(-1)^n][1-\cos(\pi n v)] \tag{32}$$

3.7 Derivation of the Equations of Motion for the Three-Segments E31–E31 Type PT with the Generating Segment in the Middle ("Middle–Generating" – "Mg")

This model of PT consists of three piezoelectric bars whose polarizations are as in the Fig. 1e. The segment in the middle performs a generating role while the segments on the left and right have a driving role. The driving segments have identical lengths (Fig. 1e). For this PT:

$$W_d = \frac{A}{2}\left[\int_0^{L_1}\left(\varepsilon_{33}E_3^2 + 2S_1e_{31}E_3 - c_{11}S_1^2\right)dx + \int_{L-L_1}^{L}\left(\varepsilon_{33}E_3^2 + 2S_1e_{31}E_3 - c_{11}S_1^2\right)dx\right] \tag{33}$$

$$W_g = \frac{1}{2}\int_{L_1}^{L-L_1} A\left(\varepsilon_{33}E_1^2 + 2S_1e_{31}E_1 - c_{11}S_1^2\right)dx \tag{34}$$

where E_3 is expressed by generalized coordinates in a manner analogous to that of case 1. The coefficients in the equation of motion have a following form:

$$\omega_0^2 = \frac{e_{31}^2}{L^2\rho\varepsilon_{33}}\left[n^2\pi^2(1-2v) + n\pi\sin(2\pi n v)\right] + \frac{c_{11}}{L^2\rho}n^2\pi^2 \tag{35}$$

$$\gamma = \frac{2e_{31}}{HL^2\rho}[-1+(-1)^n][1-\cos(\pi n v)] \tag{36}$$

$$\delta = \frac{Le_{31}}{\varepsilon_{33}}[-1+(-1)^n]\cos(\pi n v) \tag{37}$$

4 Results

4.1 Formulas

When a PT is driven by a sinusoidal input voltage $V_{in} = V_0\sin(\omega t)$, which is the most common practical case, the analytical formula for the output voltage of a transformer in a steady state is given by:

$$V_{out}(t) = \delta \cdot z(t) = \frac{\gamma\delta V_0}{\sqrt{\left(\omega_0^2 - \omega^2\right)^2 + \frac{\omega_0^2\omega^2}{Q_m^2}}} \cdot \sin\left(\omega t - arctg\left(\frac{\omega_0\omega}{Q_m\left(\omega_0^2 - \omega^2\right)}\right)\right) \quad (38)$$

From the above we determine the transformation ratio M and its maximum value M_{max} in the frequency domain:

$$M = \left|\frac{ampV_{out}}{ampV_{in}}\right| = \left|\gamma\delta/\sqrt{\left(\omega_0^2 - \omega^2\right)^2 + \frac{\omega_0^2\omega^2}{Q_m^2}}\right|, \quad M_{max} = M(\omega_0) = \frac{\gamma\delta Q_m}{\omega_0^2} \quad (39)$$

where $ampV_{in}$, $ampV_{out}$ are amplitudes of the input and output voltages, respectively.

4.2 Input Data

In this study, the calculations were performed for PTs made of PZT ceramic with the following dimensions L = *20 mm*, W = *4 mm* and H = *1* mm and the following material constants ρ = *7750[kg/m³]*, c_{11} = *77 [GPa]*, ε_{33} = *1.11* 10^{-8} *[F/m]*, e_{33} = *24.61 [C/m²]*, e_{31} = *−10.77 [C/m²]* and Q_m = *100* (the values being the same as in [5]).

4.3 Calculation Results

The calculations were performed for all the PT configurations and four natural vibration modes (n =1, 2, 3, 4).

Frequency Characteristics

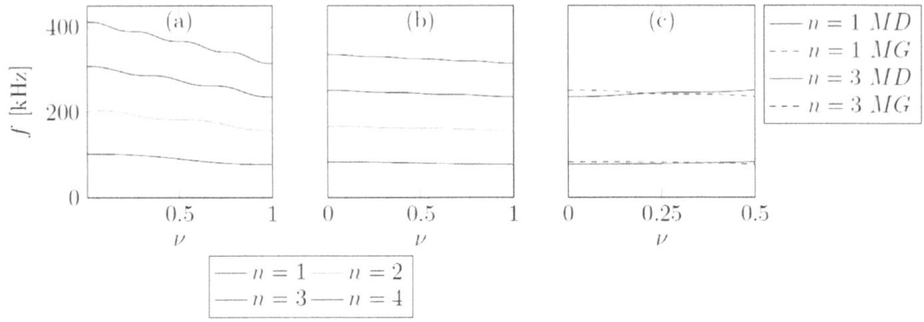

Fig. 3. Resonant frequencies vs. the geometric parameter ν in the n-th natural vibration mode for: (a) the two-segmented e31–e33 and e33–e33 type PTs, (b) the two-segmented e31–e31 type PT and (c) the three-segmented md and mg PTs.

Transformation Ratio

The transformation ratios for the analyzed PTs are illustrated in Figs. 4, 5, 6, 7 and 8 with the details being as follows:

a. maximal transformation ratio M_{max} in function of the geometrical parameter ν;

b. the transformation ratio M plotted against frequency for the selected values of ν with the transformer in the first natural vibration mode (n = 1);
c. the transformation ratio M plotted against frequency for the selected values of ν with the transformer in the second natural vibration mode (n = 2) (in the case of the two-segmented PTs) and the third natural vibration mode (n = 3) (in the case of the three-segmented PTs).

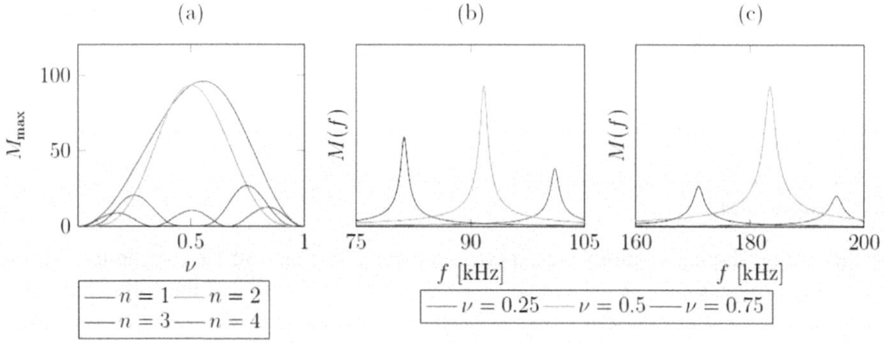

Fig. 4. Transformation ratios for the two-segmented e31–e33 type PT (classical Rosen-type).

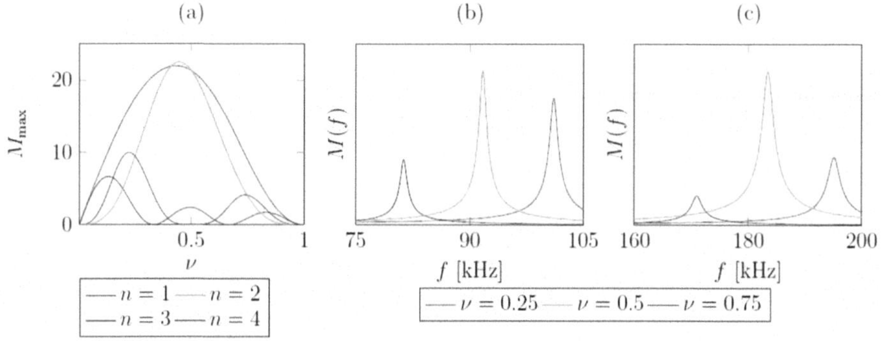

Fig. 5. Transformation ratios for the two-segmented e33–e33 type PT.

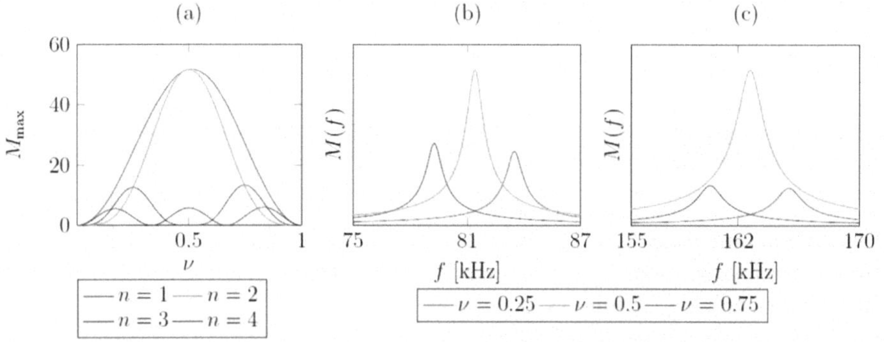

Fig. 6. Transformation ratios for the two-segmented e31–e31 type PT.

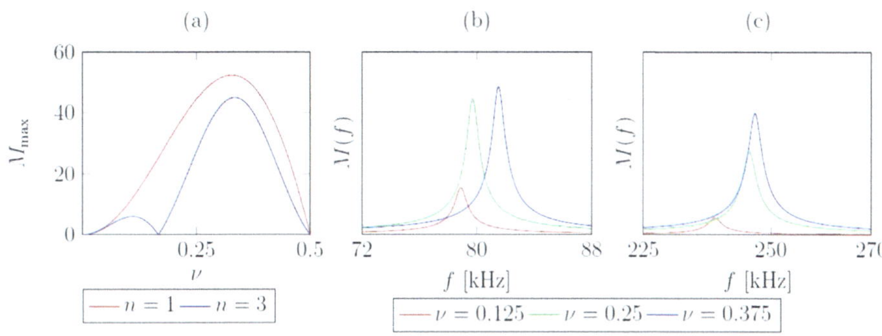

Fig. 7. Transformation ratios for the three-segmented e31–e31 type PT with the driving segment in the middle (md).

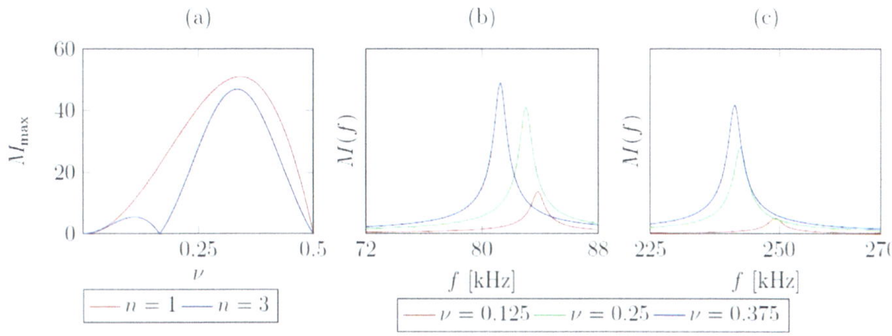

Fig. 8. Transformation ratios for the three-segmented e31–e31 type PT with the generating segment in the middle (mg).

5 Discussion

1. The formulas for the resonant frequency ω_0 obtained for the two-segmented PTs are similar. For the e31–e33 and e33–e33 type PTs, they are the same (Eqs. (17) and (21)) while for the e31–e31 type PT (Eq. (25)) the only difference is in the coefficient in the first term (e_{31} instead of e_{33}). In the case of the three-segmented PTs, the formulas for ω_0 (Eqs. (30) and (35)) have different forms but their values are similar (Fig. 3c).
2. The transformation ratios *M* for the three-segmented PTs are zero when the numbers of the *n*-th natural vibration mode are even (Eqs. (31), (32), (36), (37) and (39)).
3. When the transformation ratio is considered for the two-segmented PTs, only the first and second modes have practical meaning; for the three-segmented PTs only the first and third are of significance (Figs. 4, 5, 6, 7 and 8).
4. The most important data from Figs. 4, 5, 6, 7 and 8a are presented in the table below.

Table 1. Maximum values of the transformation ratio M_{max} and the values of v obtained at $n = 1$ and $n = 2$ for the two–segmented PTs and at $n = 1$ and $n = 3$ for the three-segmented PTs.

PT type	e31–e33		e33–e33		e31–e31		e31–e31 (mg)		e31–e31 (md)	
n	1	2	1	2	1	2	1	3	1	3
v	0.55	0.5	0.43	0.45	0.51	0.5	0.33	0.33	0.34	0.33
M_{max}	95.78	93.07	21.97	22.46	51.7	51.69	52.47	44.96	51.04	46.91

5. The maximum values of the transformation ratio M_{max} were observed for the classical Rosen-type (or e31–e33 type) PT, while the minimum values were reported for the e33–e33 type PT (Figs. 4 and 5, Table 1); the differences between the maximum values M_{max} of the two-segmented e31–e31 type PT and those of the three-segmented e31–e31 type PTs were small (Figs. 6, 7 and 8, Table 1).
6. From Figs. 4, 5, 6, 7 and 8b, c it is evident that by changing the value of the parameter v it is possible to adjust a PT to operate over a wide input voltage frequency range, despite the fact that the value of M is not maximum.
7. The working parameters of PTs (frequency and transformation ratio) can be controlled by selecting a suitable geometric parameter v (Figs. 4, 5, 6, 7 and 8b, c).
8. The working mode of the transformer is fixed by selecting a suitable input voltage frequency (Fig. 3).

6 Conclusions

The Lagrangian formalism was used to derive the equations of motion that describe the behavior of two- and three-segmented PTs. These were transformed into one equation of a driven damped harmonic oscillator. The analytical solutions, i.e. the formulas for the output voltage and the transformation ratio, were found assuming zero initial conditions, no load at the output and a sinusoidal input voltage. The calculations were performed for specific geometric and material parameters of the piezoelectric transformers to determine their resonant frequencies and select the optimal geometric parameters at the maximum value of the transformation ratio.

References

1. Rosen, C.A., Fish, K.A., Rothenberg, H.C.: U.S. Patent No 2,830,374, April 1954
2. Rosen, C.A.: Ceramic transformers and filters. In: Proceedings of the Electronic Component Symposium, pp. 205–211 (1956)
3. Auld, B.A.: Acoustic Fields and Waves in Solids. R. E. Krieger Publishing Company, Malabar (1990)
4. Arnau, A., et al.: Piezoelectric Transducers and Applications. Springer, Heidelberg (2008)
5. Erhart, J., Pulpan, P., Pustka, M.: Piezoelectric Ceramic Resonators: Topics in Maning. Metallurgy and Materials Engineering. Springer, Heidelberg (2017)

6. Lin, R.-L.: Piezoelectric transformer characterization and application of electronic ballast, Ph.D. Dissertation, Virginia Polytechnic Institute and State University (2001)
7. IEEE Standard on Piezoelectricity. American National Standards Institute (1987)
8. Preumont, A.: Mechatronics: Dynamics of Electromechanical and Piezoelectric Systems, vol. 136. Springer Science & Business Media, Dordrecht (2006)
9. Radowicz, A.: Fale powierzchniowe, Monograph. Kielce University of Technology, Kielce (1994). (in Polish)
10. Awrejcewicz, J., Koruba, Z.: Classical Mechanics: Applied Mechanics and Mechatronics. Springer Science + Business Media, New York (2012). LCC
11. Labedzki, P., Pawlikowski, R., Radowicz, A.: Analiza teoretyczno-symulacyjna transformatora piezoelektrycznego (Theoretical and simulation-based analysis of a piezoelectric transformer), Mechanik (2016). (in Polish)
12. Mezheritsky, A.V.: Elastic, dielectric, and piezoelectric losses in piezoceramics: how it works all together. IEEE Trans. Ultrason. Ferroelectr. Freq. Control **51**(6), 695–707 (2004)
13. Rajapurkar, A.M.: Loss mechanisms in piezoelectric PZT ceramics and single crystals. The Pennsylvania State University (2008)

Fracture Surface Analysis of the EN AW-2017A-T4 Specimens with Rectangular Section

Wojciech Macek(✉), Sebastian Faszynka, and Adam Deptuła

Opole University of Technology, Sosnkowskiego str. 31, 45-272 Opole, Poland
wojciech.macek@yahoo.com, a.deptula@po.opole.pl

Abstract. This paper presents an analysis of the state of fatigue fracture surface in specimens with rectangular cross-section. The specimens were made of EN AW-2017A-T4 aluminium alloy and had an external, unilateral, sharp and blunt one-sided notches, with the of radius $\rho = 0.2$ mm, 5 mm, 10 mm and 22.5 mm. Fatigue test subjected to bending, torsion, and proportional bending with torsion for stress ratio R = −1, −0.5, 0 were carried.

Fracture surfaces were observed using the focus variation microscope, which allows the acquisition of data sets with large depth of focus.

Influence of notches radius ρ, loading ratio r, stress ratio R and form of fatigue (LCF or HCF) on the fracture surface of EN AW-2017A-T4 was investigated. The authors especially focused on presenting the differences of roughness parameters of fracture propagation and rupture areas, caused by bending, torsion, and bending with torsion.

Keywords: Surface metrology · Focus variation method · Fractography · Fatigue of materials · Aluminium alloy

1 Introduction

Fatigue life tests of structural materials in miscellaneous loading conditions have been performed by scientists for nearly two centuries [1]. Methods of testing materials are endeavoring not only to extensively characterize their mechanical properties, but also to accurately depict real operational loads occurring in constructional elements and machine parts. Additionally to fatigue life testing under various load conditions, researchers observe the profile, surface and morphology of materials. Observations are carried out on a macroscopic and microscopic scale, also after the specimens have been destroyed. Fracture mechanics mainly concentrates on crack growth during the test. One of the elements in which, in some measure, the history of the loading is saved, is the surface of the fatigue fracture.

This development of the field of fatigue of materials allows the evaluation of advanced materials. Also more accurate methods for measurement, inter alia, surface topography [2], allow for a better understanding of the structure of materials.

It has been confirmed, for notched specimens, that the change from torsion to proportional bending with torsion fatigue life increases significantly, the same as changing load from bending with torsion to bending [3, 4]. For similar studies, it has been confirmed [5] that the change in stress ratio from R = 0 to R = −1 results in an

J. Świder et al. (Eds.): MECHATRONICS 2017, AISC 934, pp. 290–297, 2019.
https://doi.org/10.1007/978-3-030-15857-6_29

increase in fatigue life and decrease in fatigue crack growth rate (da/dN). It has also been found that the change in notch radius from $\rho = 0.2$ mm to $\rho = 22.5$ mm results in a significant increase in fatigue life and a decrease in the (da/dN). To find the impact of these test parameters on surface topography, fracture surface was further investigated. The results of these studies, and in particular the roughness parameter Ra, were analyzed in this paper.

2 Materials and Methods

2.1 Studied Material

The specimens were made of the EN AW-2017A-T4 aluminium alloy. This material was characterized by the mechanical properties and chemical composition [5] shown respectively in Tables 1 and 2.

Table 1. Mechanical properties of the EN AW-2017A-T4 aluminium alloy.

σ_y (MPa)	σ_u (MPa)	E (GPa)	ν (-)
382	480	72	0.32

Table 2. Chemical composition of the EN AW-2017A-T4 aluminium alloy [in wt%].

Cu	Mn	Zn	Mg	Fe	Cr	Si	Ti	Al
4.15	0.65	0.50	0.69	0.70	0.10	0.45	0.20	Balance

The specimens with rectangular cross-section dimensions 8 × 10 mm, shown in Fig. 1, were used in fatigue tests.

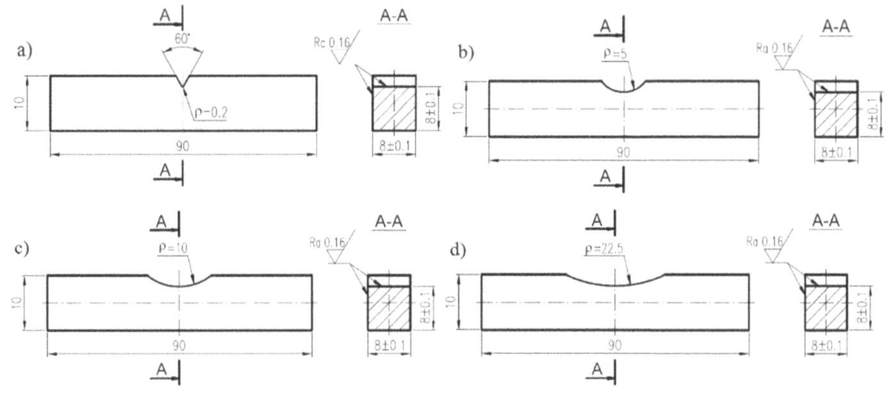

Fig. 1. Shape and dimensions of specimen for: (a) $\rho = 0.2$ mm, (b) $\rho = 5$ mm, (c) $\rho = 10$ mm, and (d) $\rho = 22.5$ mm.

The specimens were made of an extruded bar 16 mm in diameter. The specimens had an external, unilateral, sharp and blunt one-sided notches, which radius ρ = 0.2 mm, 5 mm, 10 mm and 22.5 mm. The notches in the specimens were cut with a milling cutter and their surfaces were polished with finer and finer emery papers.

2.2 Fatigue Experiments

Fatigue tests were performed on a MZGS-100 test stand [3–7] where specimens were loaded as shown in Fig. 2. Bending with torsion loadings was obtained by rotation of the head by angle α. Pure bending takes place when α angle is 0°, and for the α angle 90° we have torsion.

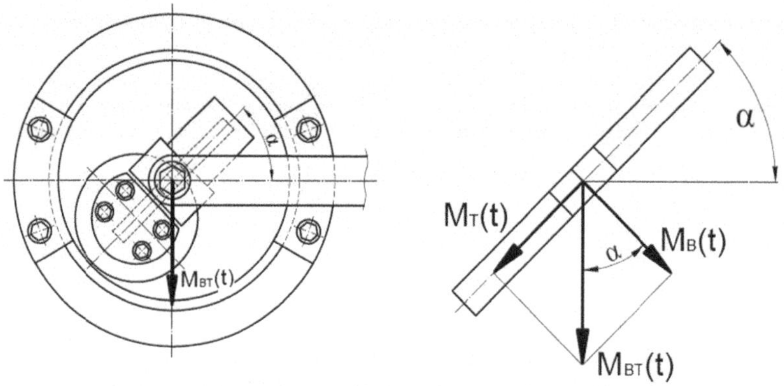

Fig. 2. Loading of the specimen under proportional bending with torsion.

In the specimen under torsion, cross-sections do not remain plane but are subjected to deformation or the so-called deplanation. This means that the individual point's cross sections can be rotated parallel to the y-axis of the specimen (Fig. 3).

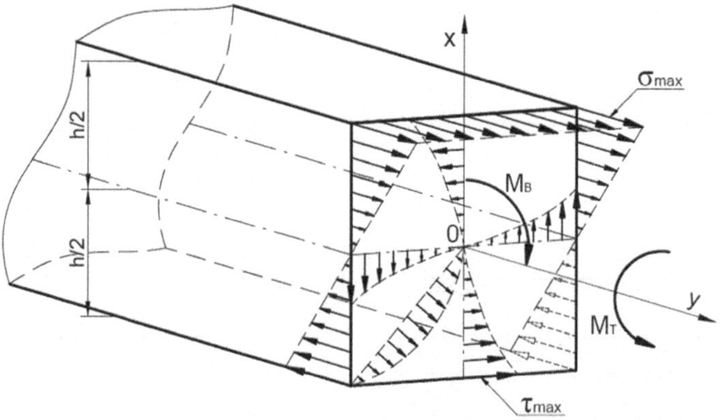

Fig. 3. Distribution of normal and shear stresses in the specimen of a rectangular cross-section under proportional bending with torsion.

In order to distinguish the type of loading stress ratio $r = \tau_{max}/(\sigma_{max} + \tau_{max})$ was employed.

2.3 Fracture Form and Roughness Measurements

Surface topography was examined on the test stand, shown in Fig. 4. The focus variation instrument is the main element of this set from the Infinite Focus produced by Alicona Imaging. The measurement device was equipped with a motorized nosepiece using five dedicated microscopic objective lenses from 2.5× to 100× magnification.

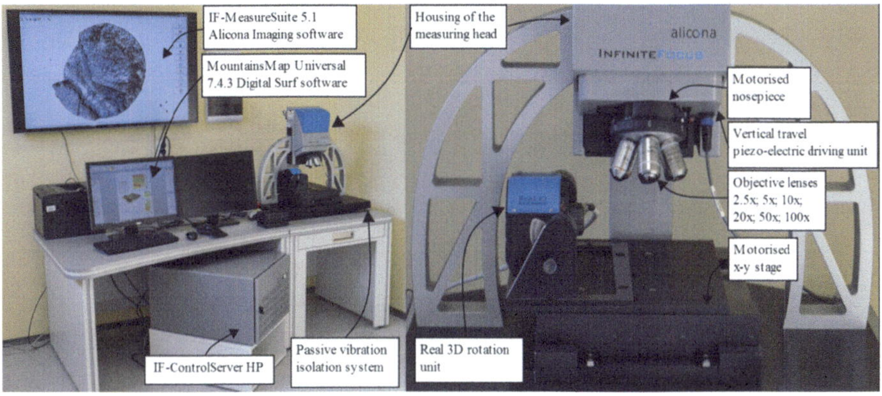

Fig. 4. Focus variation microscope InfiniteFocus IF G4 produced by Alicona Imaging.

The InfiniteFocus IF G4 focus variation microscope was supported by IF-MeasureSuit 5.1 software produced by Alicona Imaging and by advanced MountainsMap Universal 7.4 software provided by Digital Surf.

3 Results and Discussion

3.1 Fracture Surface Geometry

Figures 5 and 6 show exemplary 3D views of the surface topography of propagation and rupture area, respectively, for the 10× and 100× magnifications.

Figure 5a shows the surface area of the fracture for bending loadings, at R = 0 and −1, for notch radius ρ = 0.2 mm. There was a uniform development of fatigue cracks on both sides of the specimen.

Figure 5b shows exemplary fatigue fracture surfaces of specimens after torsion loading for stress ratio R = −1 and two different notch radiuses ρ = 0.2 mm, ρ = 22.5 mm. It can be seen that the crack propagation occurred in the form of non-linear fatigue strips moving to the center of the specimen.

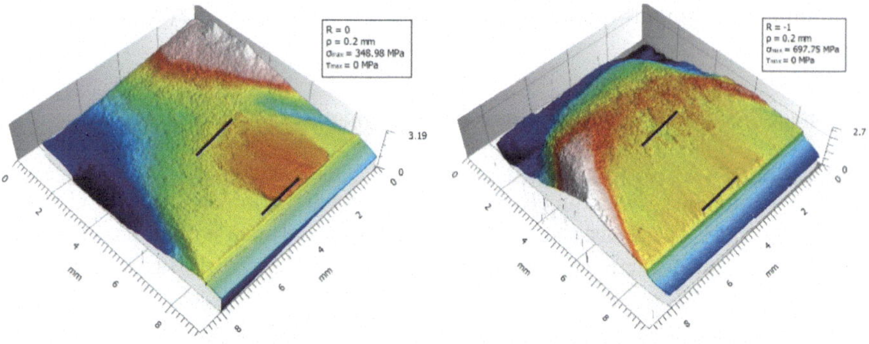

a)

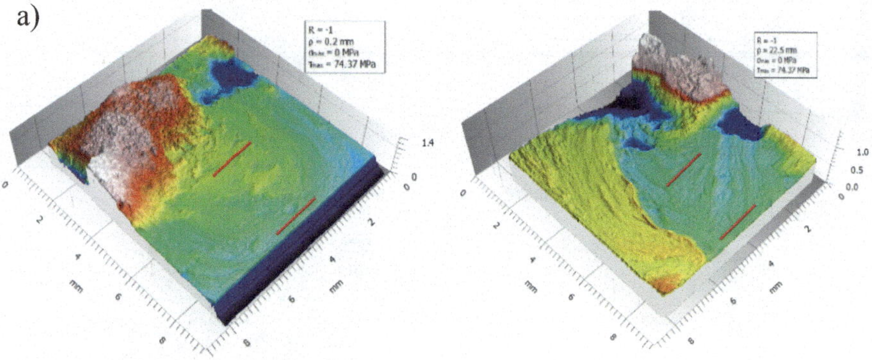

b)

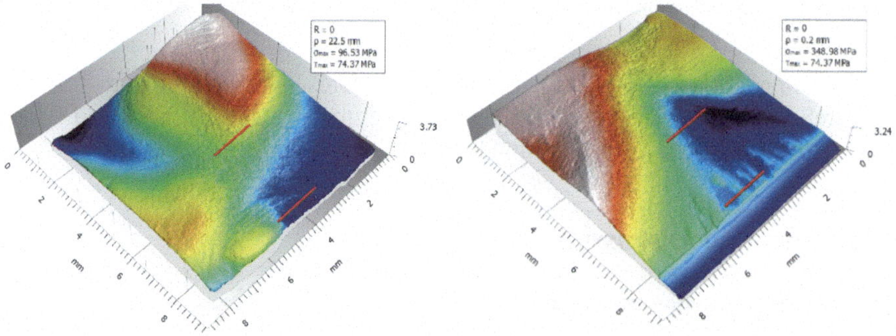

c)

Fig. 5. EN AW-2017A-T4 specimens with marked crack propagation and rupture roughness profiles, for exemplary specimens loadings subjected to (a) bending, (b) torsion, (c) bending with torsion.

The fracture surfaces, shown in Fig. 5c, was formed by bending with torsion, for stress ratio R = 0 and two different notch radius ρ = 22.5 mm and ρ = 0.2 mm. The crack propagation occurred in the form of fatigue (arc-shaped) lines moving to the center of the specimen.

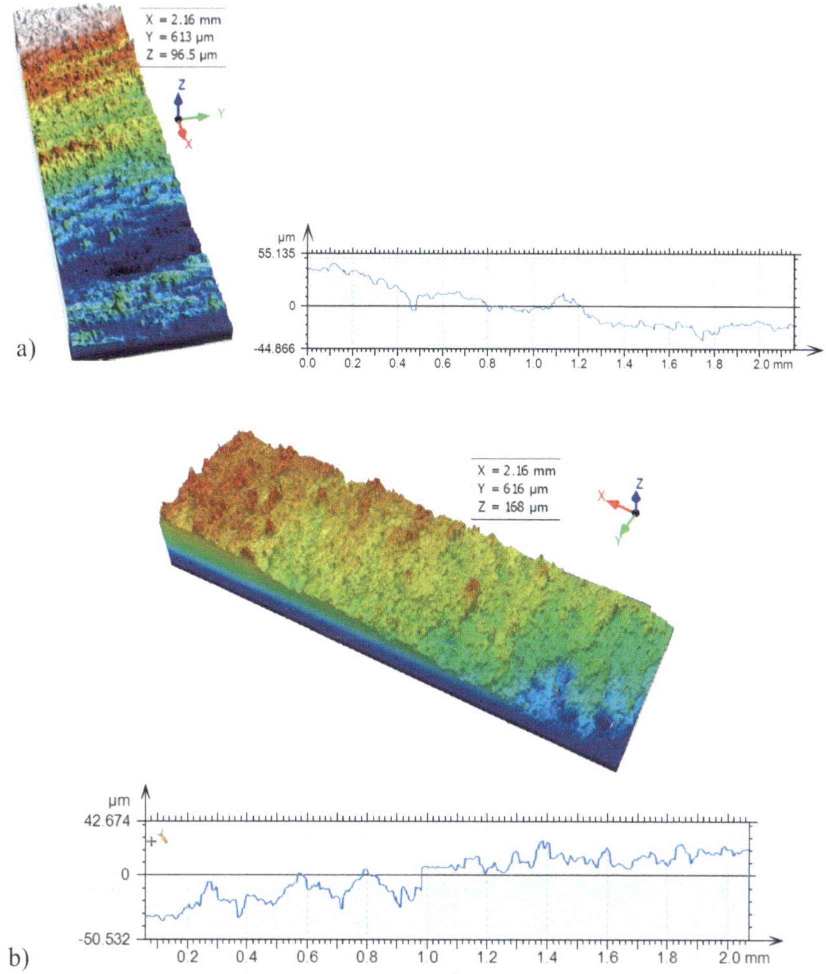

Fig. 6. The 3D view with magnitude 100× and profile diagram of EN AW-2017A-T4 specimen subjected to combined bending loading, r = 0, for (a) the propagation area, (b) the rupture area.

3.2 Propagation and Rupture Roughness Analysis

Figure 7 shows the relationship of propagation area Ra to the rupture area Ra, depending on (a) stress ratio R, (b) notch radius ρ, (c) loading ratio r and (d) form of fatigue.

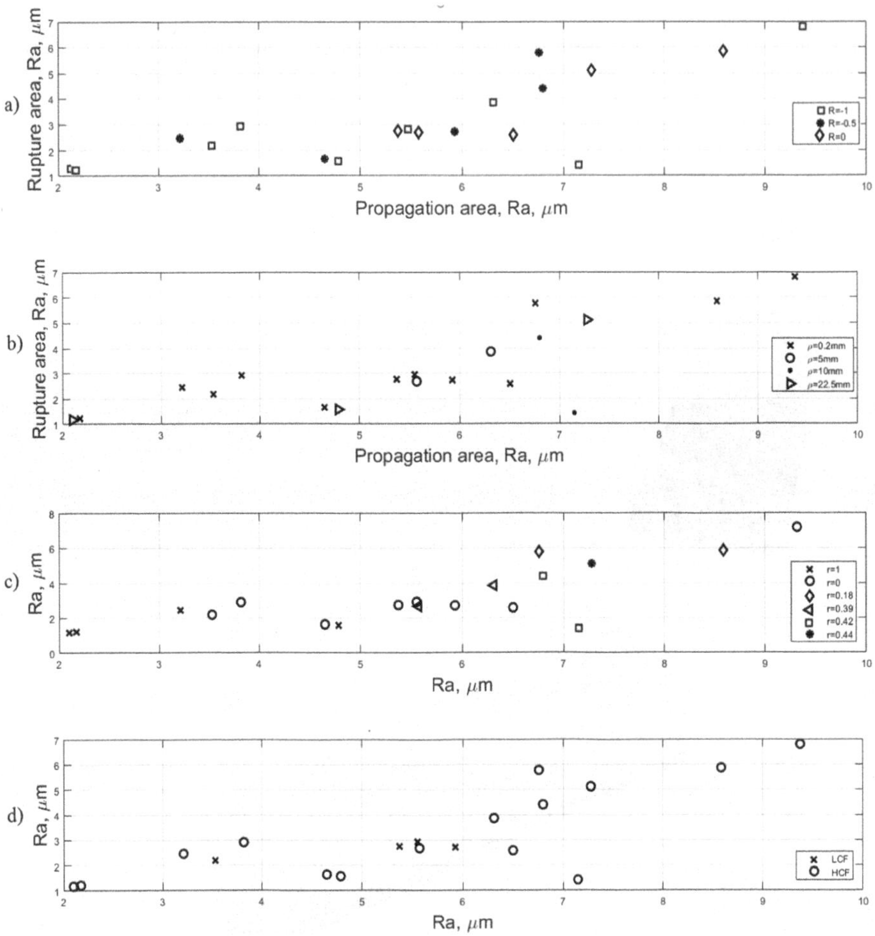

Fig. 7. Relation of propagation area Ra to the rupture area Ra, depending on (a) stress ratio R, (b) notch radius ρ, (c) loading ratio r and (d) form of fatigue.

The obtained results show (Fig. 7c) an increase in the roughness of Ra parameter as the share of shear stress τmax increases. This trend is preserved both in the propagation area as well as in rapture area.

Comparing the roughness Ra parameters of rupture to propagation areas, on average for all tested specimens 2 times increase is observed, with the minimum value of 1.2 times and the maximum one equal to 5 times.

4 Conclusions

Based on the present result, the biaxial fatigue fracture topography of EN AW-2017A-T4 aluminium alloy specimens with rectangular section and notches research, could be summarised as follows:

- arithmetical mean deviation of the roughness profile Ra, for the rupture area, is 2 times higher, in arithmetical mean value, than for the propagation area,
- the increase in the share of torsional loading components increases the roughness of the fracture surface.

References

1. Schütz, W.: A history of fatigue. Eng. Fract. Mech. **2**(54), 263–300 (1996). https://doi.org/10.1016/0013-7944(95)00178-6
2. Leach, R. (ed.): Optical Measurement of Surface Topography. Springer, Heidelberg (2011). https://doi.org/10.1007/978-3-642-12012-1
3. Faszynka, S., Rozumek, D.: Analysis of crack growth path under simple and mixed-mode loading in specimens with rectangular section. Solid State Phenom. **251**, 692–695 (2016). https://doi.org/10.4028/www.scientific.net/SSP.251.194
4. Rozumek, D., Faszynka, S.: Influence of the notch radius on fatigue crack propagation in beam specimens of 2017A-T4 alloy. Mater. Sci. **3**(53), 1–7 (2017). https://doi.org/10.1007/s11003-017-0090-y
5. Faszynka, S.: Influence of stress ratio and stress concentrator on the fatigue crack growth under bending with torsion. Doctoral dissertation (2016). (in Polish)
6. Macek, W., Macha, E.: The control system based on FPGA technology for fatigue test stand MZGS-100 PL. Arch. Mech. Eng. **62**(1), 85–100 (2015). https://doi.org/10.1515/meceng-2015-0006
7. Macek, W., Macha, E.: Energy-saving mechatronic system for fatigue tests of materials under variable-amplitude proportional bending and torsion. Solid State Phenom. **164**, 67–72 (2010) https://doi.org/10.4028/www.scientific.net/SSP.164.67

Autonomous Robot Control System for Automation of Manipulations

Julian Malaka(✉)

Silesian University of Technology,
Street Konarskiego 18 A, 44-100 Gliwice, Poland
julian.malaka@polsl.pl

Abstract. The aim of the research is to analyze the possibilities of creating a robot control system allowing automatic replication of manual actions. Functioning of the system would be based on the detection and identification of manipulations and environment changes related to a particular task. For the purpose, 3D machine vision is required in order to monitor people and objects around a robot. Acquired information would be used by an artificial intelligence system for creating relations between manipulation motions and events in a working area. This would allow the automation of a huge range of manual actions by showing the controller how a robot should perform them. In combination with motion path optimization and collision prevention, the presented solution could be an intelligent tool in the human-machine collaboration and machine learning. The article concerns the main concept of the autonomous robot control system, including techniques planned to be applied, and the plan of the research leading to the definition of principles allowing the creation of the system.

Keywords: Robotics · Automation · Artificial intelligence · Machine learning · Machine vision

1 Purpose of Research

Investigating and searching for the possibilities of automation lead to the development of technological processes. Thereby, new tools and systems are being developed and applied in the industry. The issues of workers' safety as well as the reliability, efficiency and simplicity of the operation of devices are considered to be of the utmost importance. The notion currently developed concerns human-machine cooperation in terms of manual activities. The spectrum of the applications of robots is being extended due to modern solutions, allowing a working area to be shared by a man and a robot in industrial conditions. It implies the increasing number of robots, as well as the shift in expectations in regard to their functionality. A machine, becoming a human's collaborator, needs to be characterized by its interactivity and adaptability in a given environment in order to react adequately in various situations. The aim to boost the possibilities of analyzing events and making decisions in control systems results in growing tendency to consider devices as intelligent objects, with which it is possible to establish a dialogue. The ways of conducting the dialogue themselves are undergoing

J. Świder et al. (Eds.): MECHATRONICS 2017, AISC 934, pp. 298–307, 2019.
https://doi.org/10.1007/978-3-030-15857-6_30

changes, as they are being rendered to be comprehensible for a man and adapted to his natural means of communication. The objective of the research presented in the article is to examine the aspect related to the autonomy of robots. Suggested solutions can prove themselves to be a step forward in the application of machine learning.

2 Research Concept and Plan

In the first place, an extensive review of the literature needs to be performed as a part of the research. The conception of an autonomous robot control system is innovative and includes elements which have not been described in scientific publications so far. It is planned to develop a synthetic set of knowledge in the fields related to those which constitute the subject matter of the research. This shall make it possible to choose the right course of searching new solutions. At this point, a hypothesis and an idea for proving it are determined. They are to undergo an analysis in theoretical considerations and be tested in laboratory conditions as a part of the research.

It is planned that computer tools will be used, allowing programming, integration and swift information exchange between separate components of the whole system. It needs to be verified whether operations can be performed in various environments; what is more, appropriate measures necessary for accomplishing intended objective need to be selected. Specific devices, such as an industrial manipulator, components constituting a vision systems, and sensors are indispensable for conducting the research. It is essential to ensure reliable and quick network communication within the system; this means developing the network of connections and a way to control it. In terms of artificial intelligence, it is necessary to analyze methods and select those which allow to create assumed learning mechanisms. Presumably, the application of artificial neural networks might prove to be useful in the given case. Their structure needs to be appropriately modified on the basis of knowledge acquired from the literature and experimental research. It is also vital to develop the model of input and training data.

Theoretical considerations are subject to the risk of a certain error due to the assumed simplification of methods and data inaccuracy. It may be transpired that developed solutions do not ensure proper precision of projected system operation. Since the research focuses on the conception being an innovation, conducting experiments arises from, among others, the need to take measurements and gather information concerning events which occur during specified operations in the system.

Below, a framework plan of the research is presented.

1. The literature review in the fields constituting the issue under consideration and the approximate ones.
2. The inspection and verification of methods taken into account as those enabling the fulfilment of intended objectives.
3. The development of measurement apparatus configuration, adjusted to the requirements of measurements necessary for the expected operation of the autonomous robot control system.
4. The selection of computer tools for programming and operation of devices planned to be used during the performance of the research.

5. Ensuring reliable and quick network communication within the system.
6. The implementation of appropriately selected and adjusted methods of artificial intelligence in the development environment.
7. The integration of the operation of devices and created computer programs in one system user interface.
8. Conducting tests and tuning of the created system in order to verify the proposed hypothesis, and acquire data necessary to formulate the theoretical foundations for the considered issue.

2.1 Research Method

A laboratory model of the system, fulfilling the expected functions, is planned to be created. The works would be divided into stages related to: the creation of a computer robot control system; the development of a vision and measuring system for use in monitoring a robot's working area; artificial intelligence being applied in order to analyze recorded actions and teach robot to perform tasks; and finally the implementation of anti-collision and anti-damage protection. It is essential to prepare a suitably equipped station with an industrial manipulator and a series of sensors and video cameras, where the tests of solutions developed throughout the project realization would be performed. Thereby, theoretical assumptions could be verified under laboratory conditions, which would result in the precise formulation of conclusions. Experiments performed from the early stage of the research would allow to accurately canalize further actions, which would ultimately be concluded with the elaboration of the foundations for a new issue in the field of robotics.

In the assumed system operation a man would play the role of a teacher, whereas a robot – the one of a student. The system itself would be an intermediate, an "artificial brain" of a machine, with a certain learning method developed. The verification and amelioration of a given method entail the observation and analysis of the effects it produces, likewise in the case of man's learning process. It imposes the suggested way of conducting the research as essential for the assumed objectives to be achieved.

3 Design Assumptions of Projected System

The objective of the autonomous robot control system is to bring the methods of machine learning closer to human learning based on the observation of cases. Activities performed within the framework of a given operation has to be registered and analyzed meticulously in order to reconstruct a specific process using available tools. It is important that the essence of an issue does not consist in an absolute repetition of movements performed at a certain place in accordance with given chronology. It is the transformation of an object from an initial state through an intermediate phase to a final state that needs to be pursued. Examples of performed likewise operations are supposed to aid in determining the sequence of activities leading to obtaining an expected result.

Taking into account the manipulation of objects, which is the subject matter of the research, the considerations on the creation of the autonomous robot control system may be commenced with the analysis of a simple operation of carrying an object from one place to another. The presence of an object in a first location constitutes an initial state, whereas a final state is determined by its presence in a second location. Let us assume that an intermediate goal is to avoid collision and the tool we dispose of is an articulated manipulator equipped with a gripper. The task of the system is to analyze and recreate occurrences within the stage of a robot. First of all, a relation between the movements of a manipulator and the change in the position of an object needs to be established. This requires learning data, which could be, among others, location change over time in relation to various elements in an operating space. The operation of carrying an object, performed by a worker, is then monitored by the system. In this way information is gathered about which item and from which place is to be taken, where it can be transported and where to put it. On the basis of the data, the trajectory of the movement of a manipulator and the course of the opening and closing signals of the gripper are created.

The tasks of the projected system are not limited to the repetition of specified operations. The ongoing inspection of the robot environment is to be performed, using monitoring tools, which can be additionally aided through the use of, for example, distance sensors. The recognition of man's presence within the working area and detecting environmental changes which may cause disruptions to the performance of a process are of great importance. Reacting to such occurrences is indispensable so that, in case of a hazard, it is possible to cease a movement or correct its trajectory or speed. The safety of people working in the direct vicinity of a robot, ensured without the use of physical barriers, constitutes a primordial objective. It is possible that a machine will be "taught" to perform a certain activity together with a man, and each subsequent repetition will result in the extension of the base of knowledge concerning the behavior in given situations. Similarly to human learning, the utmost care has to be maintained, and refraining from an action in case of the uncertainty over the existing conditions is essential. In terms of protective measures, a collaborative robot operating in the assumed system needs to meet such requirements as lifting capacity limit, movement speed limit, collision prevention or immediate stop once the contact with a man is detected.

In the Fig. 1a presumed operating process of the projected system is presented on the example of a manipulation. There are four major stages distinguished, two of which are related to the operation of the system by an operator. The remaining ones concern repeating learnt operations by a robot, the optimization of the operations, and their adaptation to the environmental conditions, and safety in a workspace.

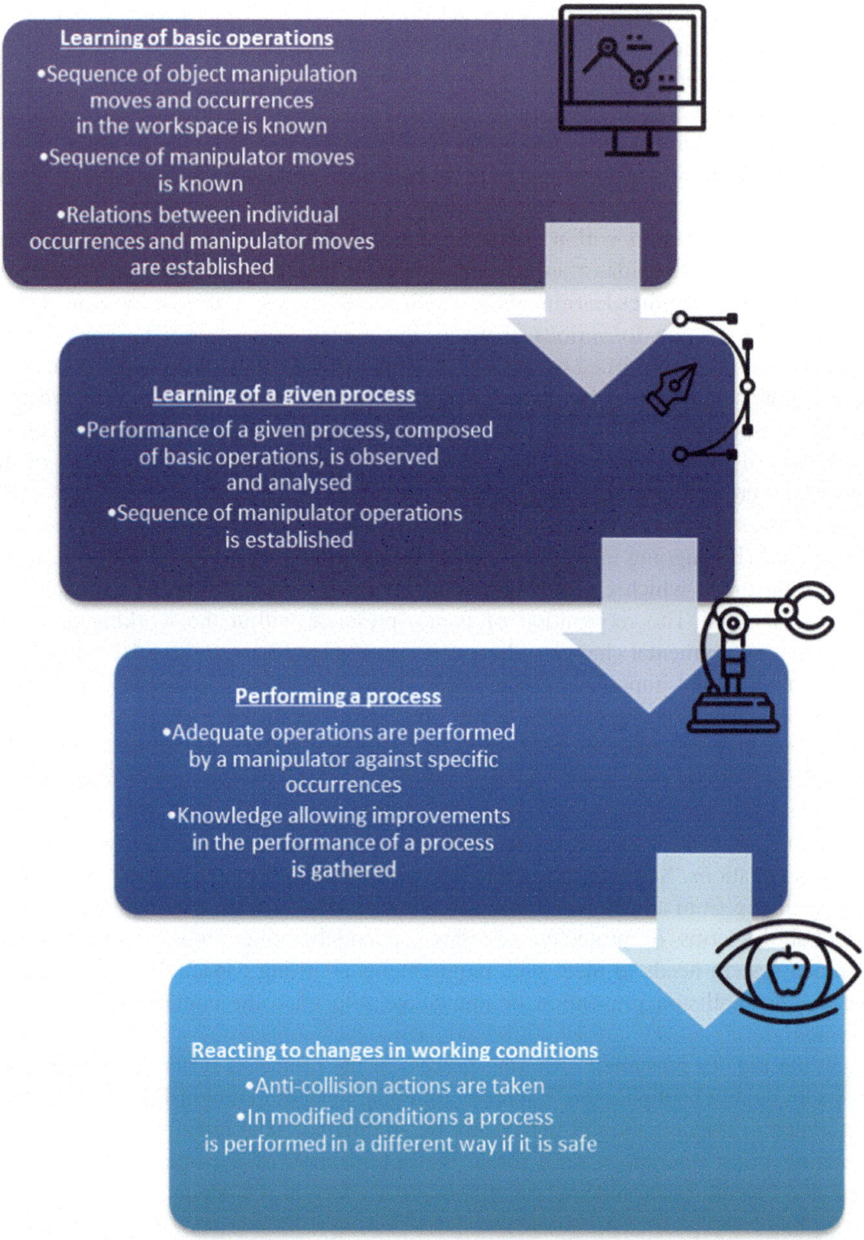

Fig. 1. Diagram of process of learning and performing manipulations in projected system.

4 Technical Realization

The projected system can be divided into a few subsystems, where specified tasks are performed simultaneously. A computer is supposed to constitute a central control station. An adequately developed program should enable the processing and analysis of gathered process data, and executing actions of artificial intelligence; it also needs to act as an interface allowing the operation of the whole system by an operator. What shall constitute a significant part of the research is the choice of an appropriate development environment, permitting the creation of such a computer application. All functions and operation of individual devices need to be integrated, ensuring their stability and swiftness of an action.

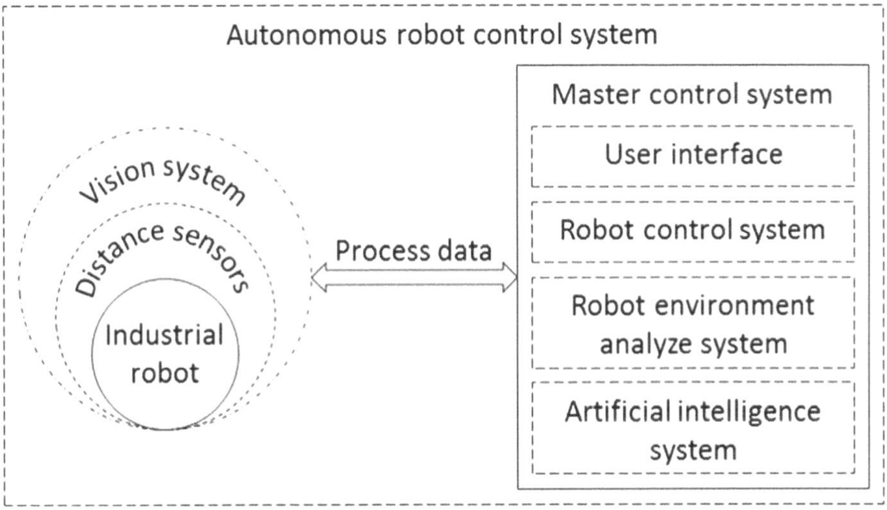

Fig. 2. Diagram of autonomous robot control system.

In the Fig. 2 a diagram presenting individual components of the whole autonomous robot control system is demonstrated.

4.1 Computer Operation of a Robot

The communication between a computer and a robot is possible to be provided with the application of communication protocols. The controller of a robot is equipped with an expansion slot or an expansion card which allow the information exchange with peripheral devices (Fig. 3).

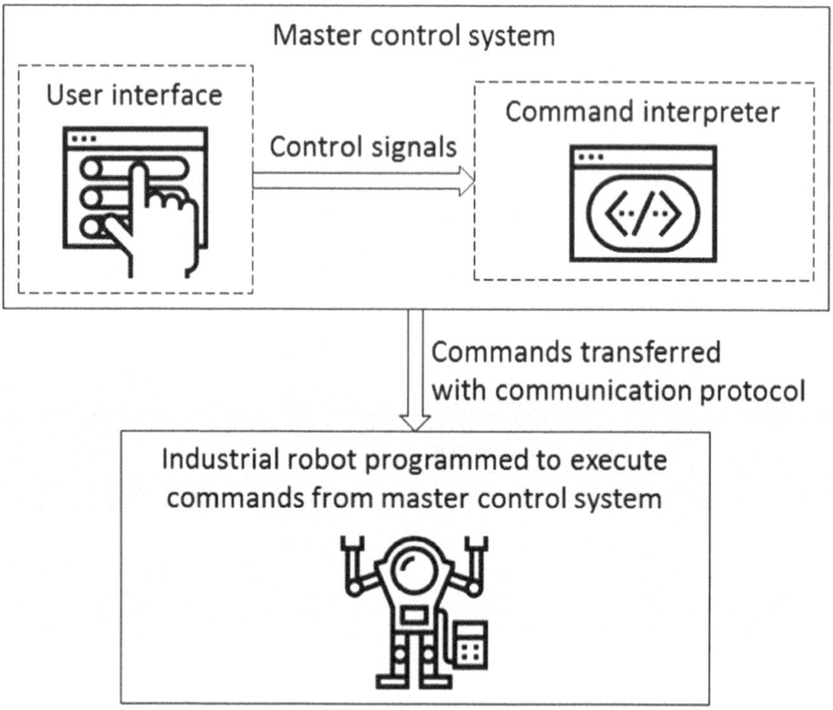

Fig. 3. Diagram of robot control

Depending on the manufacturer of an industrial robot, there are applications or programming libraries available, allowing to manage a robot from the level of a computer. It permits, among others, programming in offline mode or issuing commands concerning movement on an ongoing basis. For real-time control an appropriate program in a robot is required, which is supposed to elicit a certain action of a machine, on the basis of control signals. Moreover, a computer tool allowing the translation of control signals into a robot programming language (a so-called interpreter) is needed.

4.2 Workspace Monitoring

A robot environment monitoring system shall comprise of vision devices, spatial monitoring devices (e.g. laser scanner) and distance sensors (Fig. 4).

Data gathered using such devices are intended to serve to determine the location and trajectory of objects and any other changes in robot workspace. They constitute the foundation in the process of learning the performance of handling operations. Information obtained through environment monitoring permits to adjust robot's movements to various external conditions, prevent collisions and react to unforeseen occurrences caused by, for instance, the presence of men. Setting the rules of the selection, spacing and configuration of individual devices shall play a critical role in the research.

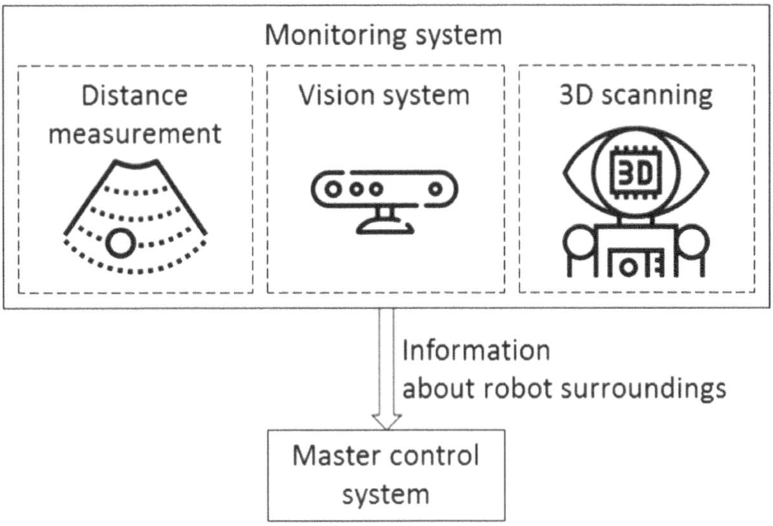

Fig. 4. Diagram of monitoring system.

4.3 Artificial Intelligence Application

It is assumed that the application of artificial neural networks shall be the most reasonable solution in terms of the use of artificial intelligence. They allow to form the mathematical model of reacting to certain input data. If the learning process is divided into learning basic actions and learning processes composed of basic actions, it shall be possible to isolate process data being crucial during the execution of a given operation. Registering and analyzing the course of definite input and output information constitute the foundation for determining the relations between them, and in the present case it concerns the relations between changes in environment and robot's moves. This is how learning performed by artificial neural networks, which may be applied in the projected system, takes place.

Figure 5 presents the diagram of the preparation and flow of information in the machine learning process, to the execution of which in the autonomous robot control system artificial intelligence may be used.

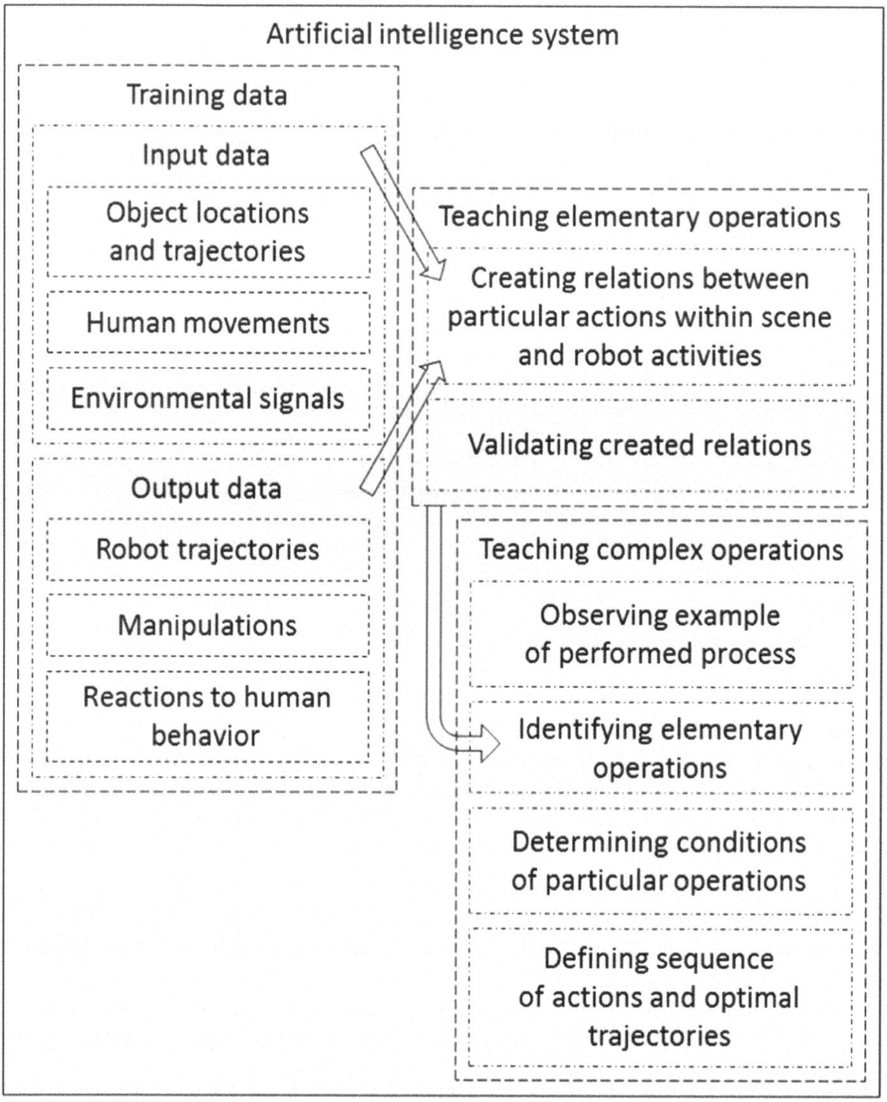

Fig. 5. Diagram of artificial intelligence system.

5 Anticipated Research Result

It is assumed that what shall constitute the result of the research is proving the thesis that cognitive processes typical for human are possible to be translated into the operation of devices, all with the application of the artificial intelligence methods and measurement apparatus. The possibility of implementing the mechanisms of learning the execution of activities through their monitoring and analysis is being considered. The research should show the way of bringing machine intelligence closer to human intelligence in terms of acquiring capabilities of performing operations; besides the studies should demonstrate major problems related to that issue.

Thanks to conducting the research in the suggested way, the field of robotics would develop towards the cooperation of men and machines and the autonomy of the latter. New knowledge in the domain would constitute a foundation for further considerations and conceptions concerning the presence of a robot in man's everyday life. Conducting the research would include considerations of specific types of tasks related to process robotization; they could be in general appertained to the future of the whole discipline due to, for instance, the improvements in terms of safety. Hitherto unknown applications of given technical solutions would be analyzed, what could additionally lead to the deeper exploration of issues in the area of machine learning, machine vision and measurement systems.

References

1. Aydin, Y., Arghavani, N., Basdogan, C.: A new control architecture for physical human-robot interaction based on haptic communication. In: Proceedings of the 2014 ACM/IEEE International Conference on Human-Robot Interaction, HRI 2014, pp. 122–123. ACM, New York (2014)
2. Billingsley, J., Bradbeer, R.: Mechatronics and Machine Vision in Practice. Springer, Heidelberg (2008)
3. Burczyński, T., Cholewa, W., Moczulski, W.: Methods of Artificial Intelligence: [AI METH 2009]. Silesian University of Technology, Gliwice (2009)
4. Commons, M., Grossberg, S., Staddon, J.: Neural Network Models of Conditioning and Action. Routledge, New York (2016)
5. Corke, P.: Robotics, Vision and Control. Fundamental Algorithms in MATLAB® Second, Completely Revised, Extended and Updated Edition. Springer, Heidelberg (2017)
6. Kim, P.: MATLAB Deep Learning: with Machine Learning, Neural Networks and Artificial Intelligence. Apress, New York (2017)
7. Köker, R.: A genetic algorithm approach to a neural-network-based inverse kinematics solution of robotic manipulators based on error minimization. Inf. Sci. **222**, 528–543 (2013)
8. Schwartz, J., Yap, C.: Algorithmic and Geometric Aspects of Robotics. Routledge, New York (2016)
9. Sciavicco, L., Siciliano, B.: Modelling and Control of Robot Manipulators. Springer, Heidelberg (2012)
10. Siciliano, B., Oussama, K.: Springer Handbook of Robotics. Springer, Heidelberg (2016)

Advantages of Using Industrial Sensor Interfaces at the Machine Design Stage

Piotr Michalski(✉)

Institute of Engineering Processes Automation and Integrated Manufacturing Systems, Silesian University of Technology, Konarskiego Street 18 A, 44-100 Gliwice, Poland
piotr.michalski@polsl.pl

Abstract. Capabilities of modern automation systems, especially in the field of industrial networks applications, encourage us to favor them during the process of developing the automation project at machine design stage. Extensive network topologies and simplicity as well as clear structure cause displacement of the classic wiring. This approach simplifies the maximum number of components by means of industrial automation from sensor to PLC. Instead of: multicore cables, intermediate boxes, connectors, terminals or hundreds of cable descriptions, automation systems integrator chooses increasingly clear industrial network bus. A perfect example is the comparison of classical topology and wiring distribution automation equipment in relation to a modern industrial network topology of the AS-interface and IO-Link. Topology simplification is not the only benefit that we get through the selection of industrial networks. Built control systems must often provide the required level of safety. Thus, the designer of the automation system is required to enable the handling of so-called safe I/O modules, safety light curtains, emergency stop buttons, pull-wire switches, etc.

Keywords: Industrial sensors · IO-Link · Industry 4.0 · Machine design

1 Introduction

Industrial networks allow to innovate/increase your workflow, create intelligent systems independent of location, gather and process huge amounts of data from end customers (big data), use advanced analytics, use the cloud, and scale your networks with the creation of virtual systems [3]. Part of IEC 61131 specifies a single-drop digital communication interface (SDCI) technology for small sensors and actuators. SDCI (commonly known as IO-Link), which extends the traditional digital input and digital output interfaces as defined in IEC 61131-2 towards a point to-point communication link. This technology enables the transfer of parameters to devices and the delivery of diagnostic information from the devices to the automation system. This technology is mainly intended for use with simple sensors and actuators in factory automation, which include small and cost-effective microcontrollers specifying the SDCI communication services and protocol (physical layer, data link layer and application layer in accordance with the ISO/OSI reference model) for both SDCI

J. Świder et al. (Eds.): MECHATRONICS 2017, AISC 934, pp. 308–313, 2019.
https://doi.org/10.1007/978-3-030-15857-6_31

masters and devices. This article describes the application possibilities of industrial sensor interface such as IO-Link in the implementation of machines compatible with idea of Industry 4.0. Benefits of new standards such as IO-link at the machine design stage have been described.

2 Classic Approach to the Topic

A common mistake at the machine building stage is the omission of the design phase related to the installation of industrial sensors. This abandonment causes that in the commissioning phase there is a need to provide the machine structure with the appropriate adapter, booms or special holders intended for the installation of automation systems. Quite frequently it is not the machine constructor, but the final integrator company that triggers the commissioning, who decide about placing the control cabinet, the route of electrical wiring and the location of the sensors. Such a situation is unacceptable and deserves to be criticized. Most errors related to the communication of industrial systems originate in the wrong, unplanned position of signal and power cables. The phase of designing the layout of the automation system at the stage of creating the 3D model of the machine seems to be indispensable. According to the Industry 4.0 concept, the virtual twin machine model should be as close as possible to the actual real machine. The stage of modelling the sensor structure of industrial control systems results in well-thought-out solutions, such as special covers or clearances that allow access to sensors in case you need to replace or change settings.

Assuming the correct process of preparation for sensorizing the machine, there are various ways to implement this physical system. Classically, signals from sensors can be fed directly to the control cabinet. The star topology obtained in this way does not assume the need to invest in signal concentrators. This approach works for machines with a low number of sensors. In the case of more sensors (over 10), it involves a lot of wires, rail connectors – a larger control cabinet. Let us consider the fact that a large number of cables requires a large number of cable markers - this process also extends the final commissioning time.

Designers of larger machines use additional intermediate cabinets that aggregate signals from the sensors. The topology thus created causes the groups of sensor wires to meet in the intermediate cabinet, and then, through the rail connectors, pass to one multi-wire cable guided towards the main control cabinet. In this approach, it is easier to take care of laying the wires while the problem associated with a large number of cable markers and the number of wires remain. Greater transparency of the system resulted in a simultaneous increase in the number of components needed to complete the structure. This simple conclusion has become a leaven to promote network solutions, which by increasing the transparency of topologies at the same time reduce the amount of cabling, the number of components and devices and of course number of cable markers. At the same time the end user of the machine gains the ability to diagnose the machine in terms of damage detection in the structure of sensors and actuators. The most well-known interfaces of industrial sensors in Europe include AS-interface and IO-Link [2]. What benefits can be noted at the machine design stage and at the operational stage base on the IO-Link interface?

3 Purpose of the Technology

The single-drop digital communication interface (SDCI) technology for small sensors and actuators (commonly known as IO-Link) defines a migration path from the existing digital input and digital output interfaces for switching 24 V devices as defined in IEC 61131-2 towards a point-to-point communication link. Thus, for example, digital I/O modules in existing fieldbus peripherals can be replaced by SDCI Master modules providing both classic DI/DO interfaces and SDCI. Analog transmission technology can be replaced by SDCI combining its robust ness, parameterization, and diagnostic features with the saving of digital/analog and analog/digital conversion efforts [1].

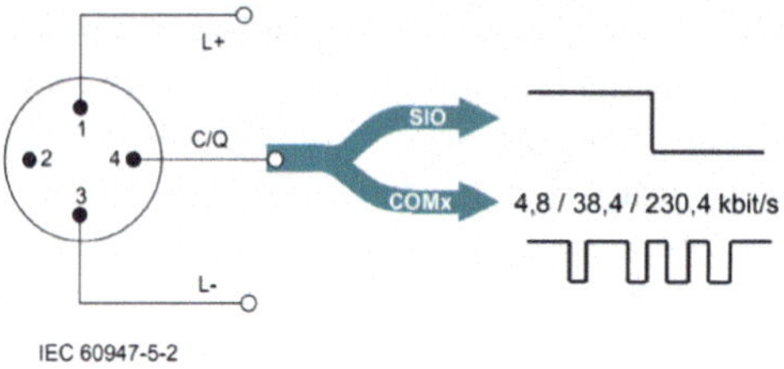

Pin	Signal	Definition	Standard
1	L+	24 V	IEC 61131-2
2	I/Q	Not connected, DI, or DO	IEC 61131-2
3	L-	0 V	IEC 61131-2
4	Q	"Switching signal" DI (SIO)	IEC 61131-2
	C	"Coded switching" (COM1, COM2, COM3)	IEC 61131-9

Fig. 1. SDCI compatibility with IEC 61131-2.

As shown in Fig. 1, the IO-Link concept is based on the behavior of a typical M12 industrial connector. The location of the power pins (pin 1 and 3) has been preserved. Functionality of the basic output pin (pin 4) has been extended – it is possible to use it in two ways: Q – switching signal of digital input (so called SIO – which stands for Standard I/O), and C – codes switching (by mining of digital serial communication base on COM1, COM2 or COM3 which are responsible for different speeds of transmission, respectively 4.8/38.4/230.4 kbit/s).

At the stage of machine design, it is therefore possible to maintain a standard 3-pin industrial sensor cabling system. There is no need for special cabling. In the scope of planning the selection of the appropriate industrial sensor, the purchase of a device equipped with an IO-Link interface should be considered.

Many sensors and actuators nowadays are already equipped with microcontrollers offering a UART interface that can be extended by addition of a few hardware components and protocol software to support SDCI communication. This second operational mode uses "coded switching" of the 24 V I/O signaling line. Once activated, the SDCI mode supports parameterization, cyclic data exchange, diagnosis reporting, identification & maintenance information, and external parameter storage for device backup and fast reload of replacement devices. Sensors and actuators with SDCI capability are referred to as "devices" in this standard. To improve start-up performance these devices usually provide non-volatile storage for parameters [1, 6].

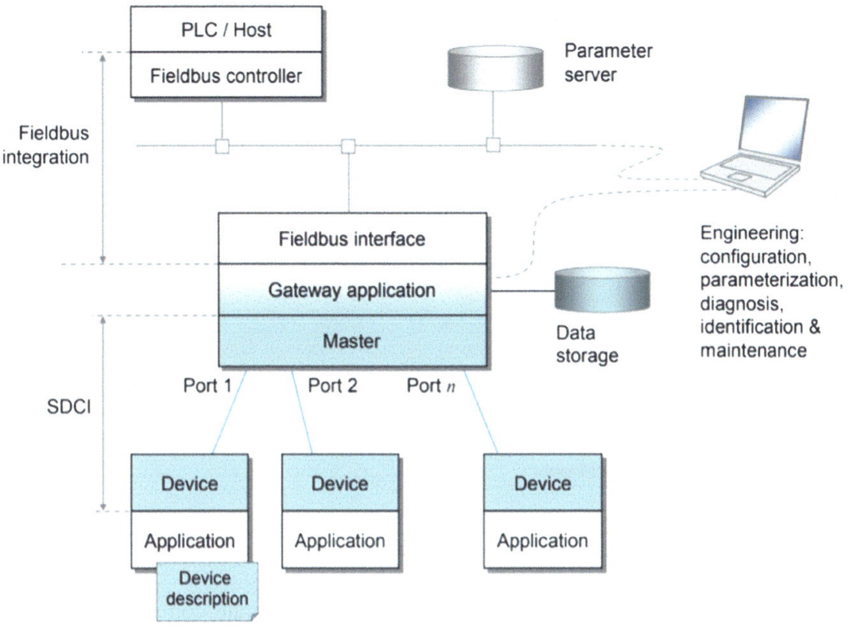

Fig. 2. Positioning within the automation hierarchy.

Figure 2 shows that SDCI devices have to be connected to the concentrator, which is a master of the IO-Link interface. A Master accommodates 1 to *n* ports and their associated data link layers. During start-up it changes the ports to the user-selected port modes, which can be INACTIVE, DI, DO, FIXEDMODE, or SCANMODE. If communication is requested, the Master uses a special wake-up current pulse to initiate communication with the device. The Master then auto- adjusts the transmission rate to COM1, COM2, or COM3 and checks the "personality" of the connected device, i.e. its VendorID, DeviceID, and communication properties. It is typical that master is at the same time the gateway to the fieldbus interface such as: Profibus, Profinet, Ethercat, Powerlink, etc. Gateway application included inside the master electronic allows to support additional connections to data storage system. This simple topology shows that machine designer needs to project the space for master of the IO-Link system, which is already collector of the signals and cable path for power supply and fieldbus interface. From this point of view it is easy to use one collector for different signal, which is different from the equally popular interface such as AS-interface [5]. Also at the stage of project preparation of electrical part it is much more easier to create drawings (the fewer of them the more transparent the structure). At this stage there are also more extra benefits for automation engineers, because IO-Link system supports bidirectional transmissions of cyclic and acyclic data. That fact is very robust because the process parameters of each sensor could be changed from the level of PLC or HDMI device. Figure 3 shows the basic relationship between the nature of data and transmission types. All of the benefits are well visible.

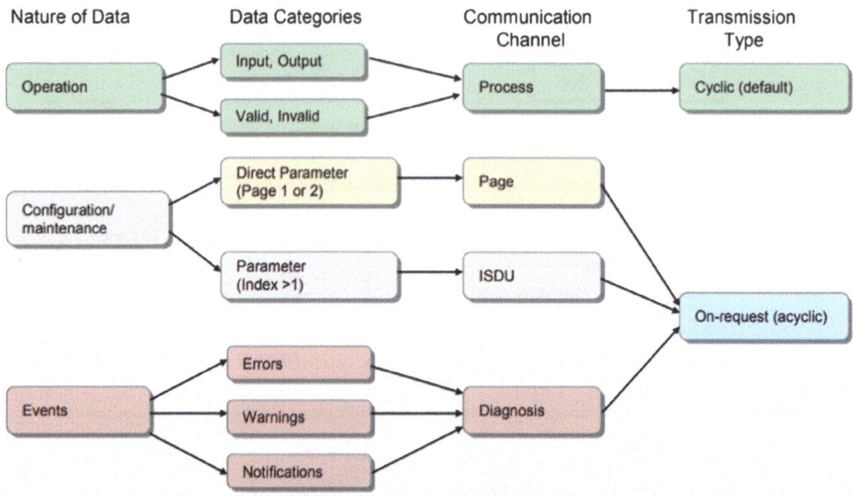

Fig. 3. Relationship between nature of data and transmission types.

Data of various categories are transmitted through separate communication channels within the data link layer, as:

- Operational data such as device inputs and outputs is transmitted through a process channel using cyclic transfer. Operational data may also be associated with qualifiers such as valid/invalid.
- Configuration and maintenance parameters are transmitted using acyclic transfers. A page channel is provided for direct access to parameter pages 1 and 2, and an ISDU channel is used for accessing additional parameters and commands.
- Device events are transmitted using acyclic transfers through a diagnostic channel. Device events are reported using 3 severity levels, error, warning, and notification.

The first octet of a Master message controls the data transfer direction read/write and the type of communication channel. A device may receive process data (out) to control a discrete or continuous automation process or send process data (in) representing its current state or measurement values. The device usually provides parameters enabling the user to configure its functions to satisfy particular needs. To support this case a large parameter space is defined with access via an Index (0 to 65535; with a predefined organization) and a Subindex (0 to 255). The first two index entries 0 and 1 are reserved for the Direct Parameter page 1 and 2 with a maximum of 16 octets each. Parameter page 1 is mainly dedicated to master commands such as device startup and fallback, retrieval of device specific operational and identification information. Parameter page 2 allows for a maximum of 16 octets of Device specific parameters. The other indices (2 to 65535) allow access to one record having a maximum size of 232 octets. Subindex 0 specifies transmission of the complete record addressed by the Index, other subindices specify transfer of selected data items within the record. Within the record, individual data items may start on any bit offset, and their length may range from 1 bit to 232 octets, but the total number of data items in the record cannot exceed

255. The organization of data items within a record is specified in the IO Device Description (IODD). All of those give a very simple and powerful communication system. Using this standard the user of the end machine is supported by process data as well as information about implemented devices and their special parameters or event information which are generated automatically by the device (i.e. errors, warnings or notifications). Additionally the gateway application implemented in the master unit is able to push the process data to the external systems responsible for data processing i.e. for the needs of predictive diagnosis or looking for trends. Also the safety sensors with OSSDe interfaces – equipped with IO-Link communication – can be parameterized via auxiliary tools such as "USB-Masters", then connected to an FS-DI and operated in OSSDe mode. They also can be operated in OSSDe mode on a FS-Master supporting OSSDe. In case these safety sensors are equipped with IO-Link Safety communication in addition, they can be operated in both modes, either OSSDe or IO-Link Safety. This corresponds to the IO-Link SIO paradigm. The concept of IO-Link Safety allows for local safety signal processing (safety functions) if the FS-Master provides a local safety controller.

4 Summary

The spectrum of possibilities that modern technologies bring is very broad, therefore advantages of using industrial sensor interfaces are undeniable. The new solutions are certainly conducive to improving production efficiency due to the possibility of parallel processing of process data by the alarms early warning systems. A well-conducted process of designing a new machine or the process of migration of the control system of an existing machine to a new one, using an approach consistent with the idea of Industry 4.0 may prove very useful for operations carried out in an industrial plant. The success lies in the use of industrial sensor interfaces parallel with the methods based on designing predictive maintenance systems.

References

1. IO-Link Community: IO-Link Interface and System. System Specification V1.1.2 (2013)
2. IO-Link Community: IO-Link Safety System Extension with SMI V1.1 (2017)
3. Jahim, M.: From sensor to the SAP - an MES-dream? Industry 4.0 and IO-Link. In: 14th Branch Meeting of Measurement and Automation Technology - Automation (2013)
4. Kamp, P.: IO-Link integration of IO-Link capable devices into control systems. In: 12th Branch Meeting of the Measurement and Automation Technology – Automation (2011)
5. Michalski, P., Hetmanczyk, M.P.: Implementation of the safety component base on industrial networks. Modern Technologies in Industrial Engineering, Modtech 2015. Book Series: IOP Conference Series – Materials Science and Engineering, vol. 95 (2015)
6. Stripf, W.: IO-Link (single-drop digital communication system) for sensors and actuators. In: Industrial Communication Technology Handbook, 2nd ed. (2015)

Bisection Method for Measuring Integral Nonlinearity of Precision Thermometry Bridges

Aleksandr Mikhal[1] and Zygmunt L. Warsza[2(✉)]

[1] Institute of Electrodynamics, National Academy of Science of Ukraine, Pereohy 56, Kyiv-57 03680, Ukraine

[2] Industrial Research Institute of Automation and Measurement (PIAP), Al. Jerozolimskie 202, 02-486 Warsaw, Poland

zlw@op.pl

Abstract. The highest precision bridges are produced by only a few leading companies in the World. They base on the concept of AC transformer multi-decade bridges developed in the 1960–1970 period. Despite this, their metrological characteristics is not inferior to Quantum-Based Impedance Bridges and allows to do more precise measurements than accuracy of the best standard resistors. Obviously, the usage of such bridges requires certification and regular checking. Therefore, the development of new calibration methods and control methods for main metrological characteristics of the AC transformer thermometric bridges is particularly an important task.

A critical analysis considering existent control methods of the integral nonlinearity of bridges will be provided in the full paper version. The features of their use and their limitations are discussed. A new control method of the integral nonlinearity of precision bridges, named as the bisectional method, was developed and analyzed. This method is based on a selection of points for nonlinearity control, carried out by a dichotomy algorithm. There are following differences of metrological properties between bisection and other methods.

- This method allows measuring an amount of non-linearity at discrete points in a bridge measuring range.
- The control of its nonlinearity is possible in both, the lower and the upper part of a bridge measuring range.
- The method has a very high sensitivity.

A resistance circuit for realization of the bisectional method is proposed. This circuit was tested experimentally for the AC transformer bridge with resolution better than 7 decades. The obtained experimental results and calculated values of the integral nonlinearity of this bridge are almost coincided. Then the ability to control the integral nonlinearity of 0.1 ppm value using this method was proven. In conclusions, the possibility of using a control resistance circuit with the bisectional method for existing AC bridges or for other concept is proposed.

Keywords: Bisectional method · Integral nonlinearity · Measurement error · Precision thermometry bridge

J. Świder et al. (Eds.): MECHATRONICS 2017, AISC 934, pp. 314–325, 2019.
https://doi.org/10.1007/978-3-030-15857-6_32

1 Introduction

In the Roadmap for electrical and magnetic measurements for the years from 2012 [1] as a priority the development of commercial metrological systems based on current tendencies is provided. Improvements in the metrological assurance of temperature measurements are attributed to a new definition of Kelvin degree measurement unit. The former ones are based on physical constants and physical phenomena, which differ from temperature characteristics of specific conductivity for perfectly pure platinum [1]. High expectations are associated with the introduction of quantum-based standards into thermometry. There have been reports on technical characteristics of Quantum-Based Impedance Bridge [2], and currently measurement error constitutes 0.2 ppm. The latter does not exceed the results obtained by conventional methods. Apparently, it will take some time to put modern tendencies into practice. Up to date, precise temperature measurements are based on the International Temperature Scale of 1990 (2000), ITS-90. According to the algorithm, which is embedded into ITS-90, AC transformer bridges with 6–8-decimal decades represent main part of the devices [4–6]. ASL, Measurement International, Hart Scientific and Tinsley are currently amongst major manufacturers of such bridges. Their principles of construction were established in the 60 s and 70 s of the twentieth century. But at present their metrological characteristics is not inferior to the characteristics of the Quantum-Based Impedance Bridge. It is obvious that application of such bridges requires attestation and metrological check. Thus, the development of new calibration methods [3] and, in particular, of control methodology for major metrological characteristics of AC transformer bridges (hereinafter "Bridge") presents to be an important task.

2 Method of Measurement by AC Bridge

In Fig. 1 the equivalent circuit of the precision AC bridge is shown.

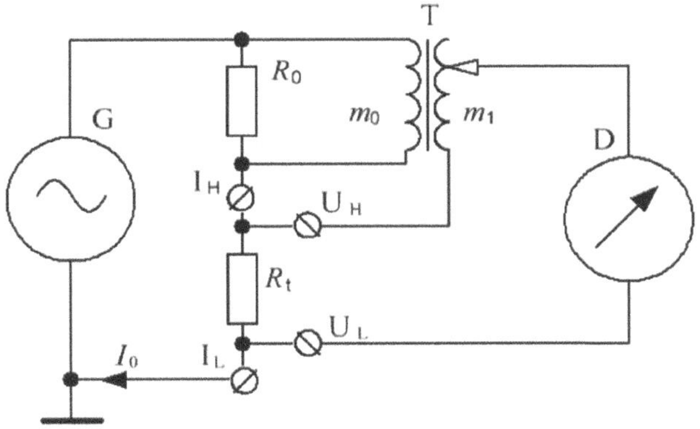

Fig. 1. Simplified equivalent circuit of the AC measurement bridge.

From the condition of the bridge equilibrium (which occurs at detector D indicating zero), the equation of measurement is:

$$R_t = R_0\, m_1/m_0 \tag{1}$$

$$y_{id} = x \tag{2}$$

where: $x = R_t/R_0$ – the ratio of the measured resistance $R_t(T)$ of a platinum thermometer and of the bridge standard resistance $R_0, y_{id} = m_1/m_0$ – the equivalent ratio of the numbers of turns of the induction divider T coils (Fig. 1).

Equation (2) is an ideal relative transfer function of the bridge and is a straight line with the slope equal to 1. Results of the resistance measurement are affected by components of the bridge error: additive (zero) error, multiplicative and nonlinearity error. The actual characteristic $y = f(x)$ is therefore the non-linear function expanded in a series

$$y = f(x) = \sum_{i=0}^{n} a_i x^i = x\,(1+\delta_m)\,(1+\delta_l) + \Delta_a \tag{3}$$

where: a_i – expansion coefficients, $\Delta_a = a_0$ – additive component (zero error), $\delta_m = a_1 - 1$ – multiplicative error and δ_l – linearity error described by series terms as $\delta_l = \frac{1}{a_1}\sum_{i=2}^{n} a_i x^{i-1}$.

The coefficients ai are complex and usually unknown functions of the deterministic (for systematic errors) and random components. They depend on a variety of influencing factors such as: leakage field of signal circuits, transformers and measurement network inside the bridge, inaccurate value of the reference resistance R_0 and bypassing of it by the input impedance of transformer T (Fig. 1), non-ideal characteristics of the magnetic core, the impact of the supply source and of the output and load resistance of decades, etc. Further discussion of the accuracy of precision bridges is applied only to systematic errors, because random errors are much smaller.

If temperature is measured in sub-ranges between their control points as references of the ITS 90 scale by the precision thermometric bridges, then usually the measurements data do not contain the multiplicative error. This is due to the algorithm (4) for calculation of the temperature, which is determined from the value of the W parameter. It is the ratio of bridge readings y/y_V for resistances of the standard sensor Pt immersed in the measured and reference temperature, e.g. the triple point of water. On the basis of Eq. (3) from such two measurements ratio W is obtained

$$W = \frac{y}{y_V} = \frac{x(1+\delta_m)(1+\delta_l) + \Delta_a}{x_V(1+\delta_m)(1+\delta_{lV}) + \Delta_a} \tag{4}$$

Equation (4) shows that when the additive error component Δ_a is equal to zero then the multiplicative component δ_m of the systematic error does not affect the coefficient W. In such a case the greater accuracy can be achieved, and there are two possibilities:

- error Δ_a is less than 0.5 LSB of the bridge indication and can be omitted,
- Δ_a must be known and taken into account when measurement results are calculated.

The additive and linearity errors are the basic metrological parameters of precision of thermometric bridges. Simple methods of their determination will be considered.

3 Reasons for Nonlinearity of AC Bridges

Let us discuss these reasons on the example of typical AC multi-decade thermometric bridge. Figure 2a shows the simplified main part of the bridge circuit with 12-bit inductive voltage divider (IVD) including T_1 and T_2 transformers.

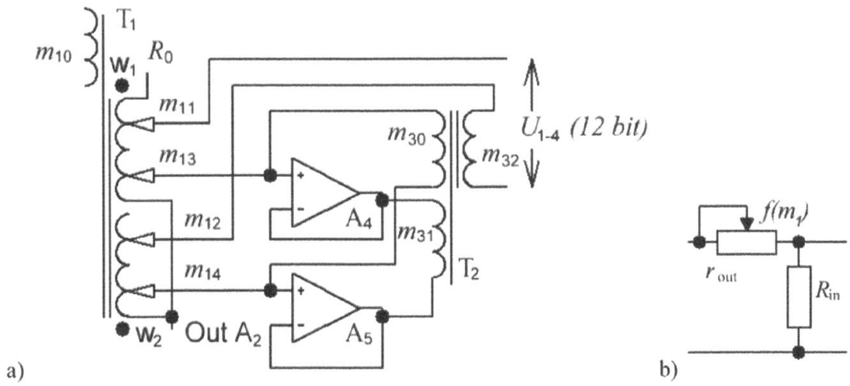

Fig. 2. (a) Simplified circuit of the 12 bit IVD (equivalent to transformer T in Fig. 1), (b) source of the bridge integral nonlinearity.

To reduce the quantization error two-stage transformer voltage dividers with a strong magnetic coupling are used. The secondary winding m_1 of the inductive divider T_1 has several coils. To create the desired ratio of voltages it can be of the decimal or binary format. In binary format, as a rule, the adjustable coils with eight leads (triadic system) or sixteen leads (tetriadic system) are used. Transformers T_1 and T_2 are made on core ring of the amorphous cobalt. Few such transformers are connected in series as cascade. This structure results in adding of voltage settings to all decades. However, there is also the negative effect of a mutual cooperation of the connected decades. Leakage coil inductances are negligible, but resistance of wires – is not. For example, the output resistance rout of m_{13}, m_{14} winding and input resistance R_{in} of transformer T_2 form a voltage divider (Fig. 2b). Output resistance rout of the decade T_1 depends on number of turns m_{13}, m_{14} and is changed when their settings are switched, i.e. $r_{out} = f(m_{13}, m_{14})$. If the transformer cores are made from modern magnetic materials with relative permeability $\mu = (1-2) \times 10^5$, then the input impedance is:

$$R_{in} \approx (j\omega L)^2 / R \quad (5)$$

where: R – primary coil resistance for DC current, L – inductivity of that coil, $\omega = 2\pi f$ – pulsation of AC current of frequency f.

Input resistance R_{in} is (10–20) MΩ, and rout may change in the range 0 to (3–4) Ω. Direct testing of the integral linearity error of the precision AC bridges with the highest resolution of 7–8 digits needs the same level of accuracy set of the standard resistances, including values other than 10 nΩ. Such standards resistors are not yet available. It is the accuracy level of the cryogenic Hall effect resistance standard. This difficulty is avoided by the relative method used for a temperature measurements by precision DC and AC bridge, as described in point 2. These measurements are based on the coefficient W as the ratio of the sensor resistance at two temperatures.

4 Method of Controlling the Bridge Integral Nonlinearity

In general, the nonlinearity of transfer characteristics of the transformer divider may be described similarly as for other analog-to-digital converters, by integral and differential linearity errors. The linearity errors are usually defined as the distance between the real bridge transfer function $y = f(x)$ and the straight line through the boundary points of the bridge range. If transfer coefficients of the decades are changed because of their loadings then the smoothed transfer characteristics of the bridge will be a convex monotonic function, as shown in Fig. 3.

Non-linearity of this function was intentionally exaggerated. For given input and output resistances of the oldest decades of the AC bridge the integral linearity error can reach values (0.1–0.2) ppm.

The concept for controlling integral nonlinearity consists of theoretical, equipment-based and methodological parts.

1. Theoretical part is based on the additive property of linear function. By definition of the linear function we have:

$$f(x_1) + f(x_2) = f(x_1 + x_2) \quad (6)$$

If transfer function $f(x)$ of a bridge is a nonlinear function, then Eq. (6) will be violated. The definition of integral nonlinearity follows from Eq. (6) as an axiom: Integral nonlinearity of a bridge is a difference between its actual (nonlinear) characteristic $y = f(x)$ and idealized straight line which represents full measuring range

$$\Delta_n = f(x_1 + x_2) - [f(x_1) + f(x_2)] \quad (7)$$

A type of function $f(x)$ will be determined by a construction of an inductive divider. As a general rule, nonlinearity of a precision bridge is a small quantity. Thus, a polynomial of the second or third degree can be sufficient for the approximation of function $f(x)$. Nonlinearity of quadratic character is a monotonic convex function.

Electrical model for nonlinearity occurrence is shown in Fig. 2. Output resistance rout of a winding with the regulated number of turns m_1 and input resistance R_{in} of a transformer of the subsequent cascade together form a voltage divider. For modern materials with magnetic permeability $\mu = (1-2) \times 10^5$ the value of input resistance of a two-stage transformer can be (10–20) MΩ. Winding output resistance is a variable quantity which depends on equilibrium value of the number of turns $f(m_1)$. For copper wire with 0.33 mm in diameter resistance can vary from 0 to (1–1.5) Ω. This may result in nonlinearity of more than 0.1 ppm.

Nonlinearity of cubic character is a monotonic function with bending. It has S-like form. Its electric model will be more complex, and it is presented in [11] for DC, and in [9] for AC. Value of such error depends on operating frequency. As a rule, for 8-decade thermometer bridges it does not exceed (0.03–0.04) ppm.

2. Equipment-based part of the concept for controlling integral nonlinearity is based on emulation of thermometer (SPRT) variation. This device is implemented as a set of discrete measures corresponding to the principle of resistance superposition.
3. Methodological part of the concept for controlling integral nonlinearity shall be based on the formation of specific algorithm for measurements and results processing. Such algorithm shall ensure the difference between data pairs on the actual nonlinear characteristic of a bridge and on idealized straight line in the whole measuring range. The problem is that this straight line is virtual, and corresponding points (data) on it shall be obtained in the result of processing of measured data which are received from the controlled bridge.

5 Discussion of Method Used in Practice

Let us firstly discuss the well-known method which is widely used in practice for controlling the integral nonlinearity of the high accuracy thermometric bridges. The scientific basis for this method is presented in [16, 17]. There is also a device (RBC 100) as a commercial proposal [18] for its realization.

The method [16–18] is based on measuring ratios $P_{1,meas} - P_{4,meas}$ of four high accuracy resistors. Their nominal resistances $x_1 - x_4$ are selected as shown in Fig. 3. Hereinafter under x_i we shall understand true value of the ratio $x_i \equiv R_{i,N}/\mathrm{RS}$. The measurement result for such ratio is denoted as $P_{i,meas}$. Based on measurement results for $P_{1,meas} - P_{4,meas}$, 31 data combinations $P_{i,calc}$ are determined. For this purpose, the ideal model of series-parallel connection of base resistors is used [13]. Measurement of $P_{i,meas}$ is carried out for each combination.

Let us assume that conversion function of actual bridge $y = f(x)$ has a quadratic integral nonlinearity, Fig. 3. Upon convex nature of function $f(x)$ nonlinearity measurement results for $P_{i,calc}$ will be above the actual characteristic $f(x)$ in the area between two dashed curves. It is determined because the initial measurement results for $P_{1,meas} - P_{4,meas}$ do not lie on one straight line, and an error has been already embedded in them. It is obvious (by definition) that all measurement results for $P_{i,meas}$ will be on the existing curve $P_{meas} = f(x)$. The essence of the well-known method lies on minimizing the variance of a function. It is represented by Eq. (1), p. 131 [16] or by Eq. (12), p. 1072 [17] and takes the form of:

$$s^2 = \frac{1}{N - \rho} \sum_{i=1}^{N} (P_{i,meas} - P_{i,calc} + dP_i)^2 \qquad (8)$$

where $dP = F(x)$ is a correction equation.

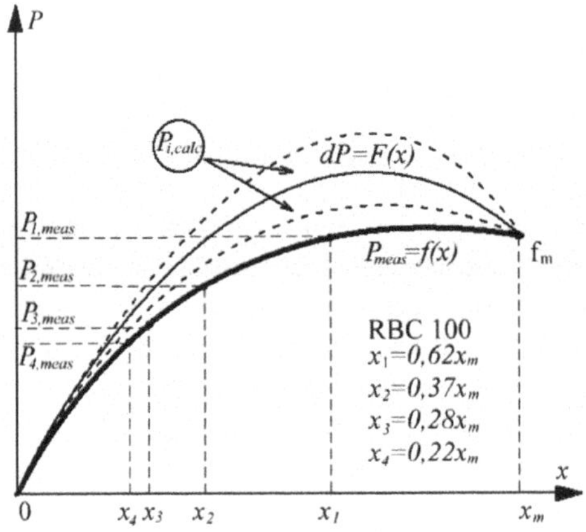

Fig. 3. Illustration of the algorithm for known method.

The graph for this equation will also lie between two dashed curves, Fig. 3. Application of a correction equation does not eliminate nonlinearity due to $f(x) \neq F(x)$. By definition, function $F(x)$ is not a straight line. Hence, the well-known method lacks algorithm according to which it could be possible to determine the difference between the actual characteristic of a bridge and a straight line.

It should be noted that for the ideal (linear) function $P_{meas} = f(x) = x$, the results of initial measurements for $P_{1,meas} - P_{4,meas}$ will be on one straight line. In this case we will have the conditions $dP \to 0$ and $s^2 \to 0$. Thus, the well-known method is only an indicator of nonlinearity. It answers the question whether there is any nonlinearity or not. But it does not answer the question: what is the value of nonlinearity? And it is the major weakness of the well-known method [7, 8, 10, 16–19].

Property of error accumulation is also another drawback of the well-known method. It is caused by bridge noises. Let us assume that we have distribution of random error according to the Gaussian law. Moreover, dispersions of all measurements will have the same uncorrelated quantities. Let us denote them as σ^2. Then a mean square deviation (MSD) of calculation results for each of 35 combinations P_{calc} (according to the number of calculations for i-vi, p. 130 in the paper [16]) will equate

$$\sigma_{calc} = \sqrt{(4\sigma^2 + 12\sigma^2 + 12\sigma^2 + 36\sigma^2 + 16\sigma^2 + 12\sigma^2)} = \sigma\sqrt{92}$$

Let MSD results for each measurement of P_{meas} will also be $\sigma_{meas} = \sigma$. The total number of such measurements is 35. As a result, the total MSD upon determining proceeds in accordance with (7) will equate $\sigma_{total} = \sigma\sqrt{35+92} \approx 11\sigma$. Thus, formation of correction equation $dP = F(x)$ and proceeds evaluation (8) increase both: the noise level and unknown components of systematic errors randomized in GUM as uncertainty u_{B}, by 11 times.

6 Bisection Method

6.1 Methodology for Measuring Nonlinearity

The bisection method is a new method developed by authors [21, 23]. It is developed to measure directly the integral nonlinearity at discrete points within the measuring range, Fig. 4. The points are selected according to the following algorithm.

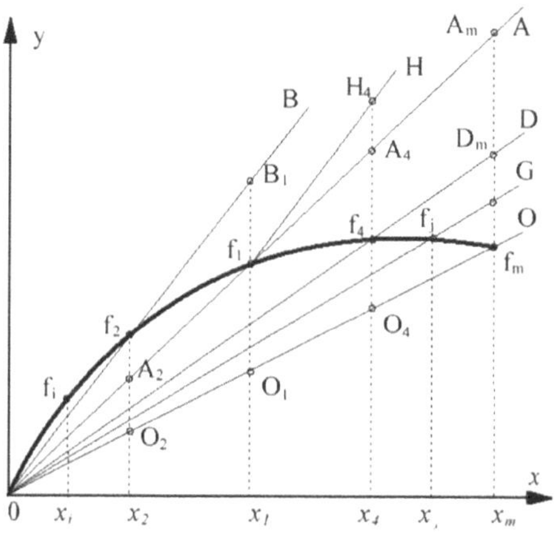

Fig. 4. Illustration to explain the algorithm of a new bisectional method.

At first, measuring range x_m is conventionally bisected (divided in two) and error $\delta_1(x_1)$ is determined at point x_1. In its turn, the lower part of measuring range is also bisected (divided in two) and error $\delta_2(x_2)$ is determined at point x_2. And the same procedure is repeated for each remaining lower part of the measuring range, i.e. each part is bisected and error $\delta_i(x_i)$ is determined at point x_i. The upper part of the measuring range from the point which corresponds to measuring range maximum is also divided in two, and an error $\delta_1(x_4)$ is defined at point x_4. Further, by analogy, each remaining upper part of the measuring range is bisected and an error $\delta_j(x_j)$ is determined at point x_j. In mathematics, such algorithm is named as a dichotomy method [20]. This name seems for us too sophisticated to use in measurement practice and we proposed the simpler name 'bisection method'.

In this new bisection method, the nonlinearity is measured as a difference between two points. This is a pair of ordinates on the actual characteristics of a bridge 0 f_2 f_1 f_4 f_m and on a straight line. A straight line connects the beginning and the end of the actual characteristics 0 O_2 O_1 O_4 f_m, Fig. 4.

For equipment-based implementation of the bisection method an emulator of discretely modified resistance of closely related to Hamon build-up resistor with series-type switching [2, 12–14, 21] is used. If two resistances are closely similar, then their mathematical sum will be almost on one line. These are lines 0 f_1 A_m, 0 f_2 B_1, 0 f_4 D_m, etc. in Fig. 4. Trimming accuracy of resistors in the chain will determine deviation from a straight line. It is an error of nonlinearity calculation. This quantity is obviously of the second order of smallness (an error from an error). In practice it is sufficient to choose resistors with error (0.1–1)%.

Algorithm for calculating nonlinearity begins with the definition of base error $\delta_1(x_1)$ in the middle of the measuring range. The calculation procedures for the lower and the upper parts of the measuring range differ from each other.

For the lower measuring range we propose methodology of nonlinearity calculation in [21]. In general, for a currently i-th point of the first range nonlinearity formula is as follows:

$$\delta_1(x_i) = \sum_{i=1}^{k} \frac{R_C(x_i) - [R_1(x_i) + R_2(x_i)]}{2^{k-i+1} R_{\max}} \tag{9}$$

where: i is the current number of a control point in the first half of the measuring range and k is the total number of control points in this part of the measuring range.

In the upper part of the range, each subsequent segment of the measuring range is also divided in half. But it is combined with previous (basic) one for fixing the current point x_j. For example, we need to add a quarter to a half or three quarters to add an eighth. The abovementioned determines another algorithm for calculating nonlinearity in the upper measuring range.

For a current j-th point of the upper measuring range recurrent formula is given by:

$$\delta_2(x_j) = \delta_2(x_{j-1}) + \delta_1(x_{j+1}) - \frac{R_C(x_j) - [R_1(x_j) + R_2(x_j)]}{R_{\max}} \tag{10}$$

where j is a number of a control point in the upper half of the measuring range.

If we need the last formula to have physical meaning upon (j = 1), we shall use the obvious conclusion. Base errors for the lower δ_1 and the upper δ_2 measuring ranges are same values

$$\delta_2(x_0) = \delta_1(x_1) \tag{11}$$

Thus, calculation shown in Eqs. (10) and (11) allows defining the difference between the actual characteristics of a bridge (curve 0 f_2 f_1 f_4 f_m – Fig. 4) and a straight line 0 O_2 O_1 O_4 f_m (Fig. 4) for any point of the bisectional algorithm. This is integral nonlinearity according to (7).

6.2 Selecting the Number of Control Points

The number of control points can be any. It is true for both upper and lower parts of the measuring range. The number of points is limited only by bridge sensitivity. In practice, maximum value of integral nonlinearity is important. When using bisection method, for its estimation it is enough to have only three control points $x_1 = 0.5x_m$, $x_2 = 0.5x_m - 0.25x_m$ and $x_4 = 0.5x_m + 0.25x_m$ (Fig. 4). Upon nonlinearity of quadratic character at point x_1 we will observe maximum value of nonlinearity. At points x_1, x_2 and x_4 the errors will have the same sign. Upon nonlinearity of cubical character at point x_1 the value of nonlinearity is close to zero, and an error will have maximum value at points x_2 and x_4. It is confirmed by the results obtained in [9, 11]. The criterion for S-like nonlinearity type is a change in the sign of error in x_2 and x_4 points.

6.3 Device for Controlling Nonlinearity

For equipment-based implementation of the method we use an emulator of discretely modified resistance of closely related to Hamon build-up resistor [12]. Major metrological properties for series and parallel connection of four clamping resistance measures are represented in [13]. From [13] it follows that the scheme for the series connection of four terminal resistance measures (Fig. 5a) is simpler than the scheme for parallel connection. Firstly, *it is not necessary* to use complicated procedures to balance resistances of leads [12, 13]. Secondly, *it is not necessary* to solve the problem when "junction of several leads does not form a point, but a small four-terminal resistor" [14]. This problem occurs upon parallel connection and is absent upon series connection of resistors, such as in points are A_2, A_3, etc. in Fig. 5a. For series connection of resistance measures we can define three factors which affect the accuracy of resistances summing. The first factor is the temporary stability of measures on the interval of experimentation (≈1–2 h). For modern resistance standards this parameter can be less than 0.01 ppm.

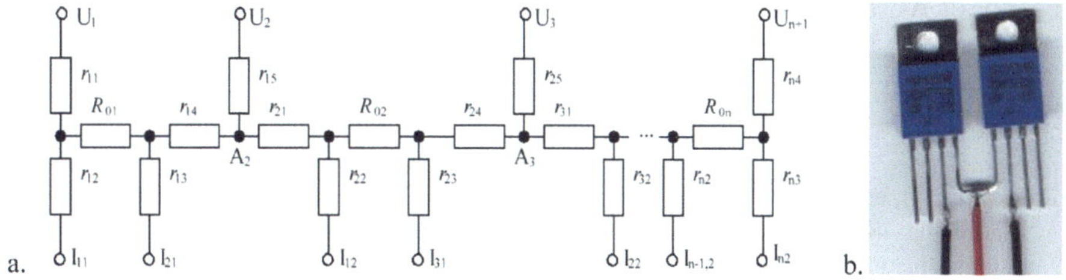

Fig. 5. a. The hardware implementation of the bisectional method, b. Example of realization points for removal of the voltage drop.

The second factor is finite value of temperature coefficient of the leads resistances r_{11}, r_{12}, r_{13}, r_{14}. An error will not exceed 0.04 ppm, if resistances of the leads is small (≈10 mΩ), and if external thermostat (0.01 °C) or on-chip technologies [15] is used.

The third factor is unchangeable (upon switching) form of force lines of electric current in the area of relieving current potential (points A_2, A_3... in Fig. 5a). With proper design this factor can be ignored. The points indicated in Fig. 5a as objects in field theory are three-terminal elements with frozen geometry of electric current force lines. This geometry and, hence, resistance of the local area do not change upon any possible switching option. In Fig. 4b design fragment is shown for summing two four-terminal resistances R_{01} and R_{02}, where red (middle) lead has resistance r_{15}. It is connected to the potential terminal U_2. Black (outer) leads are connected to the current terminals I_{21} and I_{12}, Fig. 5a. Consideration should be given to the fact that some resistors have high resistance of the leads. The resistors USR 4-1220 manufactured by Powertron have the following parameters $r_{11} = r_{14} = 150$ mΩ, $r_{12} = r_{13} = 1.27$ Ω. For Hamon build-up resistor, the best resistors are ones of MP3000-type (ZIP, Krasnodar, Russia) or S102C, VHP 4 from Vishay. They have the same magnitude but smaller resistance of the leads.

6.4 Experimental Results

Experimental verification of bisection method has been carried out upon control of nonlinearity for several modifications of a thermometry bridge [21–23]. Brief specifications of a bridge: measuring range (0–125) Ω; least significant bit – 10 μΩ; noise bandwidth (1–0.02) Hz. Modifications of the bridge have one of three operating frequencies: 975, 325 or 125 Hz. In the model shown in Fig. 1 input resistance of two-stage transformer is defined by expression $R_{in} = (\omega L)^2/r$, where L and r there are inductance and resistance of primary transformer winding correspondingly. Thus, for different bridges input resistance has varied 60 times. Estimated value of integral nonlinearity has been confirmed synchronous to the square of operating frequency. Experimental and estimated values of an error have practically coincided. For operating frequency 125 Hz experimental values of nonlinearity have been within (0.17–0.21) ppm. The mean square deviation of measuring results has not exceeded 0.03 ppm in the noise bandwidth of 0.05 Hz [21]. Thus, bisection method for controlling nonlinearity of AC bridges is suitable for accuracy levels of 0.1 ppm and less.

References

1. Machin, G., et al.: Report of the new Kelvin dissemination workshop: 27th–28th Oct 2010. Metrologia **48**, 68–69 (2011)
2. Palafox, L.: PTB develop quantum-based impedance bridges. Measure **7**(1), 4 (2012)
3. Machin, G., et al.: A European roadmap for thermometry, EURAMET TC-T (2012). http://www.euramet.org/index.php?id=roadmaps. Accessed 30 Aug 2017
4. Kibble, B.P., Rayner, G.N.: Coaxial AC Bridges. NPL Management Ltd., p. 203 (1984)
5. Awan, S., Kibble, B., Schurr, J.: Coaxial Electrical Circuits for Interference-Free Measurements. Institution of Engineering and Technology, London (2011)
6. Uncertainties in the realization of the SPRT Subranges of the ITS-90, CCT-WG3 on Uncertainties in Contact Thermometry, CCT/08-19/rev, 10 July 2009. www.bipm.org/cc/CCT/.../24/D19_rev_WG3_Doc_rev_10July2009.pdf

7. Avramov, S., Oldham, N.: Automatic calibration of inductive voltage dividers for the NASA Zeno experiment. Rev. Sci. Instrum. **64**(9), 2676–2678 (1993)
8. Avramov, S., Oldham, N., Gammon, R.: Inductive voltage divider calibration for a NASA flight experiment. NCSL Workshop Symp. Session **3C**, 225–232 (1993)
9. Hill, J.J., Deacon, T.A.: Theory, design and measurement of inductive voltage divider. Proc. Inst. Elect. Eng. **115**(5), 727–735 (1968)
10. Callegaro, L., Bosco, G.C., D'Elia, V., Serasio, D.: Direct-reading absolute calibration of AC volage ratio standards. IEEE Trans. Instrum. Meas. **52**(2), 380–383 (2003)
11. Zapf, T.L., Chinburg, C.H., Wolf, H.K.: Inductive voltage divider with calculable relative correction. IEEE Trans. Instrum. Meas. **IM-12**(2), 380–383 (1963)
12. Hamon, B.V.: A 1–100 Ω build-up resistor for the calibration of the standard resistors. J. Sci. Instr. **31**(12), 450–453 (1954)
13. Riley, J.C.: The accuracy of series and parallel connections of four-terminal resistors. IEEE Trans. Instrum. Measur. **IM-16**(3), 258–268 (1967). https://doi.org/10.1109/tim.1967.4313632
14. Page, C.H.: Tetrahedral junction error contribution to a series-parallel four-terminal resistor. IEEE Trans. **IM-23**, 5–8 (1974)
15. Joung, W., Gam, K.S., Yang, I., Kim, Y-G.: Uncertainty assessment of resistance thermometry bridges. In: XX IMEKO World Congress, Metrology for Green Growth, Busan (2012)
16. White, D.R.: A method for calibrating resistance thermometry. In: Marcarino, P. (ed.) Levrotto and Bella, Torino, pp. 129–134 (1996)
17. White, D.R., Jones, K., Williams, J.M., Ramsey, I.E.: A simple resistance network for calibrating resistance bridges. IEEE Trans. Instrum. Measur. **46**(5), 1068–1074 (1997)
18. Resistance Bridge Calibrators Models RBC100 m & Rbc400 m, User Maintenance Manual/Handbook. www.isotech.co.uk. Accessed 25 Sept 2017
19. Nicholas, J.V., White, D.R.: Traceable Temperatures. An Introduction to Temperature Measurement and Calibration. Wiley (2001)
20. Dichotomy Method. www.encyclopediaofmath.org. Accessed 25 Sept 2017
21. Mikhal, A.A., Warsza, Z.L.: Unconventional method of determination the nonlinearity of precision thermometric bridges. Pomiary Automatyka Kontrola **59**(1), 19–22 (2013)
22. Mikhal, A.A., Warsza, Z.L.: Simple methods to measure the additive error and integral nonlinearity of precision thermometric bridges. In: TEMPMEKO 2013, Madeira, Portugal, Book of Abstracts, p. 351 (2013) and poster
23. Mikhal, A.A., Warsza, Z.L.: Simple methods to measure the additive error and integral nonlinearity of precision thermometric bridges. In: Progress in Automation, Robotics and Measuring Techniques. Advances in Intelligent Systems and Computing, vol. 352, pp. 157–170. Springer (2015)

Applications of Composite Piezoelectric Transducers in Innovative Mechatronic Systems

Marek Płaczek(✉)

Institute of Engineering Processes Automation and Integrated Manufacturing Systems, Silesian University of Technology, Konarskiego 18A Street, 44-100 Gliwice, Poland
marek.placzek@polsl.pl

Abstract. The work is a report of research works concerning applications of non-classical piezoelectric transducers in mechatronic systems for measuring and analysis of vibrations. Composite piezoelectric transducers called Macro Fiber Composite (MFC) are considered. They are used as sensors in proposed application of a system for traffic monitoring and detection of collisions, based on measurements and analysis of vibrations of the road infrastructure elements. The paper presents assumptions of considered mechatronic systems with MFC transducers, results of carried out measurements and their analysis in case of presented applications.

Keywords: Smart materials · Piezoelectric materials · Vibration · Signal analysis · Traffic monitoring system

1 Introduction

In modern technical devices classical materials are more and more frequently replaced with new ones, which specific properties allow for refinement of the device or even allow the implementation of new, previously unattainable tasks. Among these materials it is essential to include smart materials. Applications growth of piezoelectric transducers is parallel to the process of piezoelectric materials development. Very important step in this process was done in 1996 by NASA when a Macro Fiber Composite (MFC) was invented [1–4]. MFC transducer consists of rectangular piezo ceramic rods sandwiched between layers of adhesive, electrodes and polyimide film. Main benefits of the MFC given by the manufacturer are: increased strain actuator efficiency, damage tolerance, environmentally sealed packages, available as elongators and contractors. The development of transducers allows new applications of piezoelectric materials inter alia, solutions such as the recovery of electric energy from mechanical vibrations (energy harvesting systems) and it improves operation of existing devices. However, designing such modern devices is a challenge for designers, forcing the development of new methods for their design and modeling.

The aim of the presented research work was to analyze the dynamic response of road infrastructure elements such as protective barriers, acoustic screens, lighting

J. Świder et al. (Eds.): MECHATRONICS 2017, AISC 934, pp. 326–336, 2019.
https://doi.org/10.1007/978-3-030-15857-6_33

components, etc. forced by a circular motion. The subject of the work is therefore on the borderline of civil and mechanical engineering and covers a wide range of issues in both areas. The aim was to verify if MFC piezoelectric transducers could work as sensors in proposed mechatronic system for collision detection and control of traffic.

In this work, as modern surveillance of the traffic innovative methods of controlling the road infrastructure are considered, such as lighting, traffic lights or self-control of traffic in the assumed area. Such solutions are being increasingly used and research on their development is widely conducted [5–10]. In times of significant expenditure on environmental protection and the implementation of the Sustainable Development Policy in almost every area of life, the use of this type of intelligent traffic control systems, or the lighting of road infrastructure can have a direct impact on reducing the cost of maintaining the road network, reduce the carbon footprint and to reduce light intensity in areas with significant light pollution. The aim of the research work is therefore to investigate the possibility of using vibration measurement of selected elements of road infrastructure and application of received measurement signals to control this type of systems. As an example above described solution can be proposed to be used to actively control the illumination of the road infrastructure by analyzing the electrical potential caused by moving vehicles.

The hypothesis of the presented research was that there is the ability to infere about the traffic on a given section of road, based on the dynamic response of road infrastructure in this section and the ability to use measurements of the response to control traffic surveillance systems in an innovative way. A key issue was also to obtain the data necessary to indicate the directions of the form of structural optimization of these elements due to the planned use in innovative control traffic surveillance systems and verify the possibility to recover electricity from the vibration of the road infrastructure.

In earlier works of the Gliwice Research Centre, the issues of analysis and synthesis of mechanical vibrating systems were presented [11–13]. Mathematical realization of this problem in terms of graphs and structural numbers was proposed. An effective tool for computer-aided design of this type of systems was proposed. Also the task of analysis of vibrating mechatronic systems, in which piezoelectric transducers were used in the role of dampers or vibration actuators were presented [14–22]. Some of the tasks in the presented work are thus a continuation of the research. Previous author's studies have become the basis of the presented work. These studies may be the basis for further development of mechatronics in the context of applications of piezoelectric elements and mechatronic systems.

1.1 Innovative Traffic Surveillance Systems with Piezoelectric Transducers

It should be mentioned that there is a large interest of the scientific community in the proposed topics. In international application published on 18 November 2010 under the Patent Cooperation Treaty (PTC) with international publication number WO 2010/131126 AI authors of the application present assumptions of the road barrier structure with an integrated system for energy generation and detection and classification of collisions [23]. The idea of the system is that a road-barrier structure is provided with a plurality of piezoelectric modules integrated in a wall and distributed along the latter. The wall is shaped and set in such a way as to be subject to vibrations as a result of the passage of motor vehicles along the road on which the road barrier is installed. The piezoelectric modules are arranged and oriented for exploiting these vibrations in an optimal way in order to produce electrical energy. Furthermore, the piezoelectric modules are arranged in such a way as to be able also to perform, together with the function of generators, also the function of deformation sensors, designed to detect collisions against the road barrier. The modules are connected to an electronic control unit programmed for processing the signals at output from the piezoelectric modules and classifying the detected collisions against the road barrier on the basis of a predetermined scale of importance. The control unit is supplied with the energy generated by the piezoelectric modules. The electronic control unit additionally includes wireless transmission means for transmitting the information regarding the data detected.

Also in European Patent Application number EP1167629A2 a highway crash barrier monitoring system is proposed [24]. In the presented system a highway crash barrier includes a collision sensor that detects when a vehicle collides with the crash barrier. A transmitter is coupled with the collision sensor to transmit a radio frequency signal to a remote location when this collision sensor detects a collision. This radio frequency signal includes a fault message that is forwarded automatically to a user responsible for maintenance of the crash barrier. A camera is controlled in response to the collision sensor to store an image of the crash barrier shortly after the collision.

Another work presents also a system for collisions detection [25]. In this work authors present a system for detection of truck collisions with highway bridges. They mentioned that the detection of collision impact and evaluation of the impact level is a critical issue in the maintenance of a concrete bridge. In the paper, an overheight collision detection and evaluation system is developed for concrete bridge girders using piezoelectric transducers. An electric circuit is designed to detect the impact and to activate a digital camera to take photos of the culpable vehicle. Impact tests and a health monitoring test were conducted on a model concrete bridge girder by using three piezoelectric transducers embedded before casting. From the experimental data of the impact test, it can be seen that there is a linear relation between the output of sensor energy and the impact energy. The health monitoring results show that the proposed damage index indicates the level of damage inside the model concrete bridge girder. The proposed overheight truck–bridge collision detection and evaluation system have the potential to be applied to the safety monitoring of highway bridges [25].

Systems based on piezoelectric transducers applications are also used in railway transport and issues related to the study of the dynamic response of both railway rolling stock and railway infrastructure are the subject undertaken by a number of researchers. Dynamic loads generated during the passage of railway trains are deleterious to the infrastructure, such as bridges or buildings located near tracks, as well as people staying in them. There is however the possibility of their use in systems for condition monitoring of infrastructure components [26–30]. Due to the dynamic loads generated by trains moving with higher and higher speed, in the face of aging and worn out infrastructure, these issues are now becoming extremely important. These problems are discussed among others in the works [27–29, 31], regarding the dynamic impact of high-speed trains on railway bridges and viaducts, as well as acceptance of testing of such facilities under dynamic loading test. Also important are issues of protection of building against vibration generated by the ground and underground rail transportation as well as protection and minimizing the impact of vibrations on both passengers and persons in the buildings surrounding the railway line [30].

In the paper [32] the principles and operation of a system for monitoring loads or state truss railway bridges are presented. Piezoelectric transducers are used as sensors. In this case, the system is based on an analysis of the signals generated by the piezoelectric transducers mounted on railroad tracks before entering the bridge and on the truss bridge elements. The wireless transmission of measured signals to a data center is possible. The system is thus characterized by low cost of installation and lack of significant restrictions on the traffic of railway rolling stock during installation and operation. The amplified and filtered measuring signal can be interpreted in order to identify the load, so weighing the passing train and monitoring the technical condition of the truss bridge. According to the authors, research work aims to create an integrated system for monitoring the technical condition of structures and detect overloaded rail depots. The theoretical assumptions of the authors are supported by the presentation of the results of measurements on a waveform signals recorded by piezoelectric sensors when the train is passing and the reference signal level to the mass of the wagon, determined by a static weight. The values obtained allow clear identification of the type of wagon after previous calibration of the measurement system.

The measurement of the dynamic response of the ground floor of the building located in the vicinity of the Warsaw subway tunnel became the foundation of the system of monitoring, where it is possible to identify trains that generate excessive vibration due to the deformation of worn wheel [30]. The authors present a comparison of horizontal vibration acceleration waveforms on ground floor of the building during the passage of two trains of the same type with different values of radial run out wheels. It has been shown that in the case of rolling stock with deformed wheels it is possible to increase the level of generated vibrations dozen times. It has been proven that there is a possibility to inference about technical condition railway rolling stock based on the analysis of dynamic response of infrastructure forced by its passing. The authors of this study adopted the assumption that such a system can be successfully integrated with the modernized freight wagon and allow monitoring of its technical condition during operation. Attempt to estimate the forces acting on the wheel sets of rail vehicles based on the measured dynamic response is also shown in the paper [33]. Using a developed model of wagon and a finite element method, forces acting on the object of study as a result of deformation of the railway track were determined.

The presented work corresponds to research works that were done by other researchers. The aim was to verify if non-classical piezoelectric composite can be used as sensors in innovative systems designed for traffic monitoring.

2 Measurements of the Dynamic Response of the Road Barrier

The protective barrier type Sp-06 with the places of piezoelectric MFC transducer mounting was considered in this work. Places selected for installation of piezoelectric films are:

- the upper edge of the guide at half the distance between the pillars,
- the flat surface of the guide below stamping - located midway between the pillars,
- the front part of the pillar IPE 140,
- the profile belt, the external centre of the arc, the area located midway between the pillars.

The measurements of the dynamic response of protective barriers were carried out on the highway. The sensors were glued on the elements of the protective barrier that separates the lanes in opposite directions and is located on green belt. In order to perform the measurements a mobile voltage signal acquisition system station was constructed. To build up the system and for the data acquisition equipment and software from National Instruments were used. Four of the MFC piezoelectric films were connected with measuring card NI-9215. Then the measuring card is installed in the module cDAQ-9191. The measuring card with connected piezoelectric transducers is presented in Fig. 1. All elements of a measuring station are:

- a portable computer,
- a battery for power supply,
- a cDAQ-9191 CompactDAQ Chassis,
- the measuring card NI-9215,
- four piezoelectric transducers type MFC M-8514-P1.

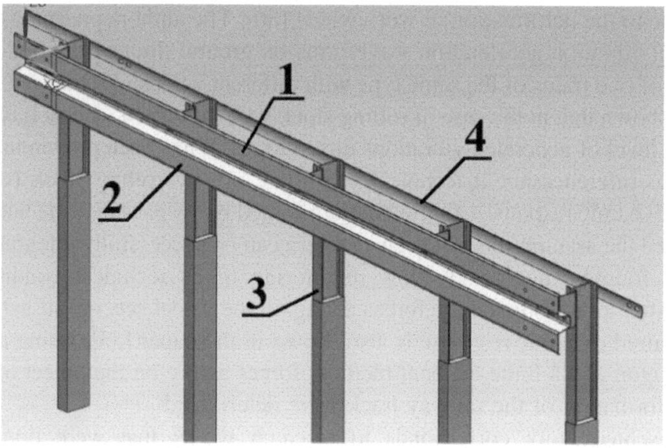

Fig. 1. Distribution of piezoelectric films in the process of dynamic response testing – the barrier type SP-06.

The measurements were carried out simultaneously with the recording of the passing vehicles with the camera. Recorded video was then compared with the voltage waveforms generated by the piezoelectric transducers in order to analyze and interpret the results. The examples of the signals generated by the piezoelectric films are shown in Fig. 2.

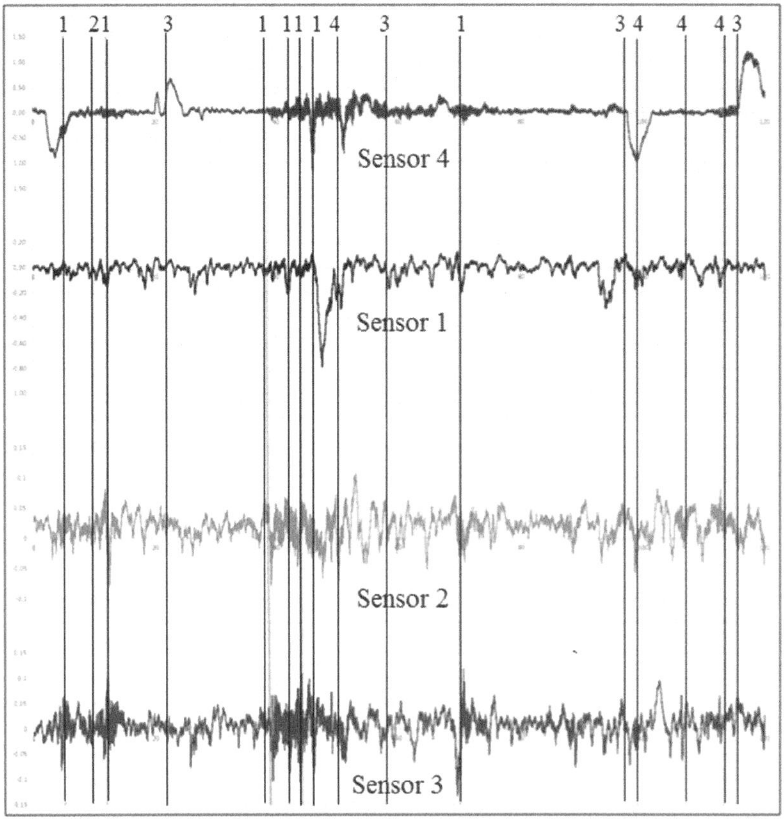

Fig. 2. Distribution of piezoelectric films in the process of dynamic response testing – the barrier type SP-06.

In Fig. 2 the waveforms generated by different piezoelectric transducers are indicated, respectively:

- Sensor 1 – waveform denotes voltage generated by the transducer glued on the upper edge of the guide (number 1 in Fig. 1);
- Sensor 2 – waveform denotes voltage generated by the transducer glued on the flat surface of the guide below stamping (number 2 in Fig. 1);
- Sensor 3 – waveform denotes voltage generated by the transducer glued on the front part of the pillar (number 3 in Fig. 1).
- Sensor 4 – waveform denotes voltage generated by the transducer glued on the profile belt (number 4 in Fig. 1).

Using the vertical lines the moment of passing of different types of vehicles is indicated, respectively:

- the line number 1 denotes passing of a vehicle of weight over 3.5 tons moving along the right lane of the road;

- the line number 2 denotes passing of a vehicle of weight over 3.5 tons moving along the left lane of the road;
- the line number 3 denotes passing of a vehicle of weight up to 3.5 tons moving along the right lane of the road;
- the line number 4 denotes passing of a vehicle of weight up to 3.5 tons moving along the left lane of the road.

During analysis of signals received during each of the tests, it was found that the profile belt is too susceptible element. Numerous spikes of the generated signal voltage were observed reflecting not passing of any vehicle. These signals resulted from weather conditions (wind) and were not results of a vehicle movement. Most precisely the vehicle passing is reflected by the waveform generated by the piezoelectric film mounted on the pillar. Passing of a vehicle of weight over 3.5 tons moving along right and left lanes of the road as well as passing of a vehicle of weight up to 3.5 tons moving along the left lane of the road can be easily detected. Signals generated by all transducers were also affected by vehicles travelling in the opposite direction on the road lanes separated by a green belt.

In the next step the ability to detect impact on the barrier was verified. It was carried out by hitting the barrier guide using a wooden stick with a mass of 0.4 kg and a steel rod of 1 kg at a distance from the place of sensor mounting. In Fig. 3 results obtained during hitting using wooden stick from a distance of 100 m are presented. The impact on the barrier can be easily detected for all piezoelectric sensors.

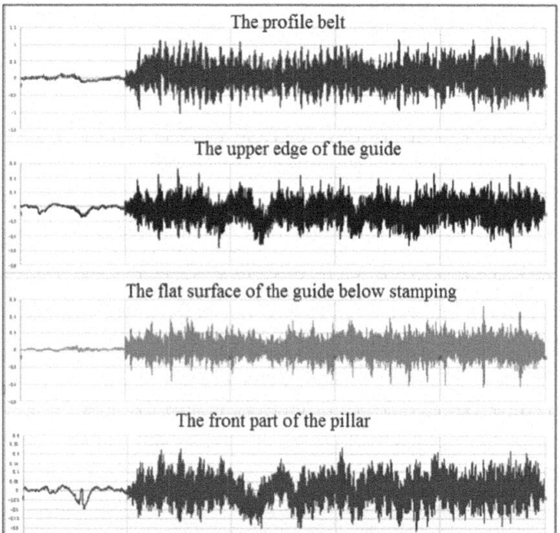

Fig. 3. Detection of impact on the barrier while hitting using wooden stick from a distance of 100 m.

Also the impact on the barrier generated by hitting it using the steel rod can be easily detected even from the distance of 80 m. The obtained results are presented in Fig. 4.

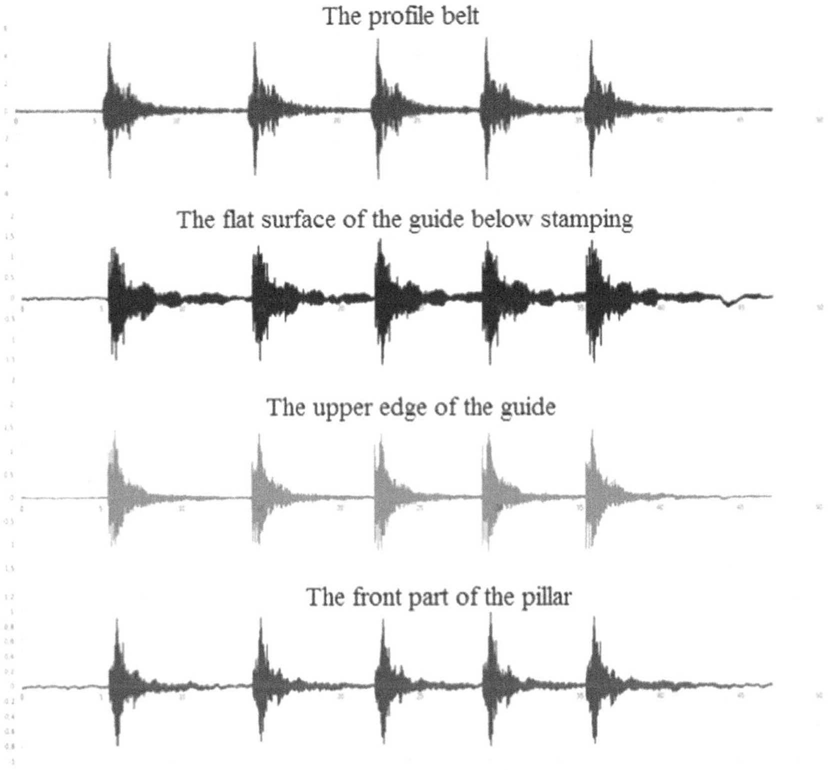

Fig. 4. Waveforms obtained with MFC piezoelectric films during impact a steel rod in the barrier guide – the distance of 80 m.

Taking into account the obtained results it can be assumed that piezoelectric MFC transducers glued on the surface of protective barrier elements or other elements of the road infrastructure can be successfully used as sensors in traffic surveillance systems. It is possible to measure the vibration energy generated by passing vehicles on the infrastructure and measure the dynamic response of these elements in order to analyse the received signals.

3 Conclusions

The work proved that the MFC piezoelectric transducers can be successfully used as sensors in modern traffic surveillance systems. The main task of such systems is to detect collisions and accidents and immediately send the information about their occurrence and exact location to the relevant emergency services. In these situations it is critical to help as soon as possible because health and life of the people involved in accidents is at stake. In the case long road sections away from human settlements quick call for help and precise determination of the place of accident is very difficult and its absence may have dire consequences for participants of the event. This problem is well known to rescue services. In the event of a vehicle collision signals generated by the sensors of the system reach much higher values compared to the normal passing car, so they can be clearly interpreted. Locating the sensors at regular intervals along the entire distance of supervised route allow instant, wireless transmission of information about the situation and its exact location to the management centre and the relevant emergency services. The interpretation of the signals sent by neighbouring sensors also enables quick estimate of the scale of the event, and so determine whether we are dealing with an impact of a single car into the protective barrier or collision of several cars. This information is essential for emergency services in order to plan the action properly.

System configuration should also allow to perform additional functions, such as the lighting illumination of the road immediately in front of an oncoming vehicle and turning it off as soon as no other vehicles are detected. This action allows for huge energy savings and pollution reduction. When an accident happens at night the event area may be lit in a special way that distinguishes it from the rest of the way, allowing for significant simplification of reaching rescue crews at the place of the accident, as well as a warning about it to other road users. Additional features include the ability to monitor traffic on a given part of road, measuring the traffic, the speed of vehicles and the acquisition of data that are extremely important because of the traffic management.

An important advantage of this kind of system is the assembly of its components to existing elements without violating the road surface and the retention of the traffic. The system will also have a very low demand for electricity, which also can be met through the use of renewable energy sources such as solar, wind and even recovery of electric energy from mechanical vibrations using piezoelectric elements making up the system.

The presented results show that there are also a few problems that must be solved. The impact of the vehicles going along the opposite road lane as well as the possibility to detect light vehicles driving on the right lane can be solved by a proper distribution of sensors. In the presented work sensors were glued on the elements of the protective barrier that separate the lanes in opposite directions and is located on green belt. The mentioned problems should be solved by distribution of sensors on all protective barriers (external barriers and barriers on the green belt). Waveforms generated by all sensors in the system can be then filtered and juxtaposed in order to correctly infer the type, location and direction of movement of vehicles. It will be the aim of the author's future work.

References

1. Płaczek, M., Buchacz, A., Wróbel, A.: Use of piezoelectric foils as tools for structural health monitoring of freight cars during exploitation. Maint. Reliab. **17**(3), 443–449 (2015)
2. Płaczek, M.: Modelling and investigation of a piezo composite actuator application. Int. J. Mater. Product Technol. **50**(3/4), 244–258 (2015)

3. Kovalos, A., Barkanov, E., Gluhihs, S.: Active control of structures using macro-fiber composite (MFC). J. Phys: Conf. Ser. **93**(1), 1–7 (2007)
4. Seeley, Ch., Delgado, E., Bellamay, D., Kunzmann, J.: Miniature piezo composite bimorph actuator for elevated temperature operation. In: Proceedings of ASME 2007 International Mechanical Engineering Congress & Exposition, Seattle, Washington, USA, pp. 405–415 (2007)
5. Azpilicueta, L., Falcone, F., Astrain, J., et al.: Analysis of topological impact on wireless channel performance of intelligent street lighting system. Radioengineering **23**(1), 412–420 (2014)
6. Coifman, B., Beymer, D., McLauchlan, P., Malik, J.: A real-time computer vision system for vehicle tracking and traffic surveillance. Transp. Res. Part C: Emerg. Technol. **6**(4), 271–288 (1998)
7. Elejoste, P., Angulo, I., Perallos, A., et al.: An easy to deploy street light control system based on wireless communication and LED technology. Sensors **13**(5), 6492–6523 (2013)
8. Hejun, W., Changyun, M.: Design of intelligent traffic light control system based on traffic flow. In: 2010 International Conference on Computer and Communication Technologies in Agriculture Engineering, vol. 3, pp. 368–371 (2010)
9. Manikandan, G., Srinivasan, S.: Traffic control by bluetooth enabled mobile phone. Int. J. Comput. Commun. Eng. **1**(1), 66–69 (2012)
10. Richu, S.A., Starbell, R.N.: Energy efficient intelligent street lighting system using ZIGBEE and sensors. Int. J. Eng. Adv. Technol. **3**(4), 41–44 (2014)
11. Buchacz, A., Galeziowski, D.: Synthesis as a designing of mechatronic vibrating mixed systems. J. VibroEng. **14**(2), 553–558 (2012)
12. Bialas, K.: Mechanical and electrical elements in reduction of vibrations. J. VibroEng. **14**(1), 123–128 (2012)
13. Dymarek, A., Dzitkowski, T.: Passive reduction of system vibrations to the desired amplitude value. J. VibroEng. **15**(3), 1354–1364 (2013)
14. Buchacz, A., Płaczek, M.: Selection of parameters of external electric circuit for control of dynamic flexibility of a mechatronic system. Solid State Phenom. **164**(1), 323–326 (2010)
15. Buchacz, A., Płaczek, M.: The analysis of a composite beam with piezoelectric actuator based on the approximate method. J. VibroEng. **14**(1), 111–116 (2012)
16. Buchacz, A., Płaczek, M., Wróbel, A.: Control of characteristics of mechatronic systems using piezoelectric materials. J. Theor. Appl. Mech. **51**(1), 225–234 (2013)
17. Buchacz, A., Płaczek, M., Wróbel, A.: Modelling of passive vibration damping using piezoelectric transducers – the mathematical model. Eksploatacja i Niezawodnosc – Maint. Reliab. **16**(2), 301–306 (2014)
18. Buchacz, A., Płaczek, M., Wróbel, A.: Modelling and analysis of systems with cylindrical piezoelectric transducers. Mechanics **20**(1), 87–91 (2014). ISSN 1392-1207
19. Buchacz, A., Płaczek, M., Wróbel, A.: Mathematical algorithm for analysis of piezoelectric stacks with structural damping. Solid State Phenom. **220–221**(1), 385–390 (2015)
20. Płaczek, M.: Dynamic characteristics of a piezoelectric transducer with structural damping. Solid State Phenom. Mechatron. Syst. Mater. IV **198**(1), 633–638 (2013)
21. Wróbel, A.: Kelvin Voigt's model of single piezoelectric plate. J. VibroEng. **14**(2), 534–537 (2012)
22. Zolkiewski, S.: Testing composite materials connected in bolt joints. J. VibroEng. **13**(4), 817–822 (2011)
23. Patent Cooperation Treaty (PTC), publication number WO 2010/131126 AI: Road barrier structure with an integrated system for energy generation and detection and classification of collisions, World Intellectual Property Organization, 18 November 2010

24. European Patent Application EP 1 167 629 A2: Highway crash barrier monitoring system, European Patent Office, 02 February 2002
25. Song, G., Olmi, C., Gu, H.: An overheight vehicle–bridge collision monitoring system using piezoelectric transducers. Smart Mater. Struct. **16**(2), 462–468 (2007)
26. Chłus, K., Krasoń, W.: Simulation of the effect of static and dynamic loads on the strength of the railway platforms. Article Proprietary X Engineering Forum ProCAx, Sosnowiec/Siewierz. 6–9 October 2011. (In Polish)
27. Grebowski, K., Zielińska, M.: Modeling of dynamic interactions of Pendolino train type on structures of historic railway bridges in Poland. Constr. Overview **1**(1), 27–32 (2015). ISSN 0033-2038. (In Polish)
28. Herwig, A., Bruhwiler, E.: In-situ dynamic behaviour of a railway bridge girder under fatigue causing traffic loading. In: Proceedings of the 11th International Conference on Applications of Statistics and Probability in Civil Engineering, ICASP11, Zurich, Switzerland, 1–4 August 2011, vol. 1, no. 1, pp. 389–395 (2011)
29. Oleszak, P., Cieśla, J., Szaniec, W.: Study of the effects of side impacts on railway viaduct lying on the arc. Constr. Architect. **12**(2), 47–54 (2013). ISSN 1800-0665. (In Polish)
30. Stypuła, K.: Selected problems of surface building protection against vibrations generated by underground communication. Min. Geoeng. **3**(1), 351–362 (2009). ISSN 1732-6702. (In Polish)
31. Salamak, M., Łaziński, P., Pradelok, S., Bętkowski, P.: Acceptance testing of railway bridges under dynamic loading test - requirements and practice, design, construction and maintenance of rail transport infrastructure. In: INFRASZYN 2014, Zakopane, 9–11 April 2014, pp. 218–227 (2014). (In Polish)
32. Sekuła, K., Kołakowski, P., Świercz, A.: System for the monitoring of loads and technical state of truss bridges in the railway. In: Condition Monitoring and Evaluation of Technical Construction and Its Vitality, Seminar MONIT, Warsaw, 19 November 2010. (In Polish)
33. Mehrpouya, M., Ahmadian, H.: Estimation of applied forces on railway vehicle wheelsets from measured vehicle responses. Int. J. Veh. Struct. Syst. **1**(4), 104–110 (2009)

Sensor-Less Bilateral Teleoperation System Based on Non Linear Inverse Modelling with Signal Prediction

Mateusz Saków(✉), Arkadiusz Parus, Mirosław Pajor, and Karol Miądlicki

Institute of Mechanical Technology, West Pomeranian University of Technology, al. Piastów 19, 70-310 Szczecin, Poland
mateusz.sakow@zut.edu.pl

Abstract. In the paper a sensor-less control scheme for a bilateral teleoperation system with a force-feedback based on a prediction of an input and an output of a non-linear inverse model by prediction blocks was presented. As a part of the paper a method of a time constant estimation of the prediction block was also presented. The prediction method of an input and an output of an inverse model was designed to minimize the effect of the transport delay and the phase shift of sensors, actuators and mechanical objects. The solution is an alternative to complex non-linear models like artificial neural networks, which requires complex stability analysis and control systems with a high computing power. The effectiveness of the method has been verified on the hydraulic manipulator test stand.

Keywords: Force-feedback · Teleoperation · Non linear inverse modeling · Delay

1 Introduction

Teleoperation systems are playing an important role nowadays because they have been applied in almost all branches of industry. It is hard to find a factory without a device being controlled by a joystick, a keyboard or other type of remote control. However, the remotely controlled device does not need to be limited only to perform a simple set of commands [1]. From the early 60s of the previous century research has been carried out to obtain remote control [2] device operation.

A continuous control system usually becomes unstable when loop delay exceeds one half cycle at any signal frequency for which loop gain exceeds unity [3]. However, the transmission delay is not the only component of an aggregate delay in the teleoperation system. The aggregate delay consists of a processing time of an analog-digital converter, a time of calculation completion by a controller, a processing time of a digital-analog converter and transition states of dynamic properties of actuators, sensors and mechanical objects [3].

J. Świder et al. (Eds.): MECHATRONICS 2017, AISC 934, pp. 337–347, 2019.
https://doi.org/10.1007/978-3-030-15857-6_34

XXI century is a domain of control schemes implementation based on: sliding mode controllers [4], fuzzy logic controllers [5], force-feedback communication channel frequency separation techniques [6], special methods for discretization of a sensor resolution [7], artificial neural networks [8] and adaptive controllers dedicated to a variable and an asymmetric time delay which are complemented with adaptive filtering methods [9]. However, it is important to pay attention that bilateral teleoperation systems feature three types of feedback with the operator: vision-feedback [3], force-feedback [2] and combination of vision-feedback and force-feedback [10, 11]. Remotely controlled devices can be controlled by operator's motion scanners [12] which in a special case are exoskeletons for upper limb [10] and palm [13], by gesture control techniques [14, 15] or by voice control methods [16, 17]. An important classification of bilateral teleoperation systems with force-feedback are systems which use force sensors [18] and devices without force sensors also known as a sensor-less or self-sensing techniques in the telemanipulation field [8, 10–12]. The sensor-less teleoperation systems group usually is based on impedance control methods [19] and the inverse modeling techniques to obtain correct value of force in the force-feedback communication channel [11, 12]. Inverse models are represented frequently by artificial neural networks [8] or nonlinear autoregressive model with exogenous inputs (NARX) [10].

The paper addresses the transport delay problem in a sensor-less control scheme based on a dynamic inverse modeling procedure which allows the system to estimate environmental force affecting the Slave manipulator body. The dynamic inverse model used in the control unit was extended with proposed prediction blocks with its input and output signals. The single prediction block has been a phase shifter with a specific characteristic. The specific characteristics of the prediction block is a strongly linear phase diagram in a useful frequency spectrum which allows the system to predict the manipulator motion with a close to constant time shift. Another important feature is a gain of the block which is close to a unity when system is operating also in a useful frequency spectrum. The proposed nonlinear inverse dynamic model structure which nonlinearity included the modified friction model based on the Stribeck friction interpretation was validated on a 1-DOF hydraulic manipulator test stand. The experiment and simulations confirmed the usefulness and the effectiveness of prediction of an input and an output of the nonlinear inverse model by reducing the time phase shift error between measured and simulated control signals.

The remainder of the paper is arranged as follows: Sect. 2 contains a description of the problem, Sect. 3 contains a description of the linear inverse dynamic model extended with prediction blocks and its simulations, Sect. 4 contains the report from the experiment carried out on the test-stand and key conclusions were summarized in Sect. 5.

2 Problem Statement

The primary issue of using inverse models in a control unit of any device which is always present is transport delay between a cause and an effect [3]. Scientific literature presents solution which can compensate the effect of a transport delay and a phase shift influence caused by actuators, objects and sensors in the communication channel [20]. Likewise, the paper presents an approach for minimizing the effect of a transport delay in the communication channel but based on a prediction of inputs and outputs of an inverse model by a simple phase shifter. The approach minimizes the transport delay effect of a control unit, an actuator, objects and sensors directly in the control unit of a Slave subsystem. Schematic diagram of the Master-Slave system with force-feedback is presented in Fig. 1.

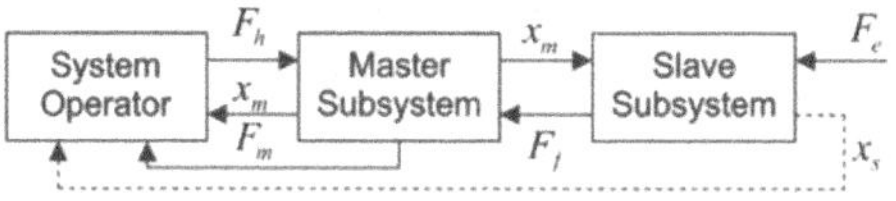

Fig. 1. Master-Slave system analyzed.

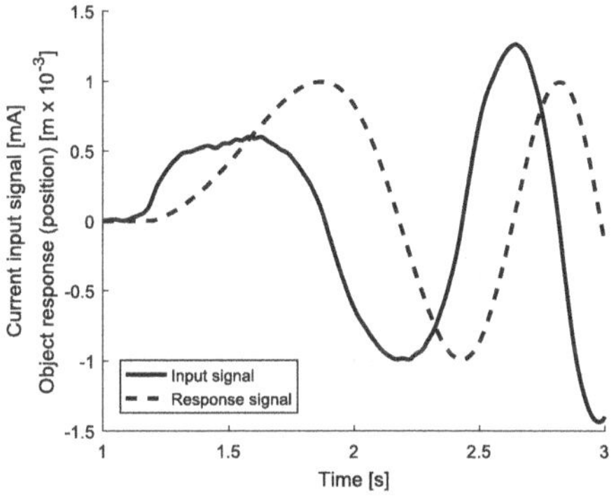

Fig. 2. The transport delay and the phase shift caused by the test stand actuator, object and sensor.

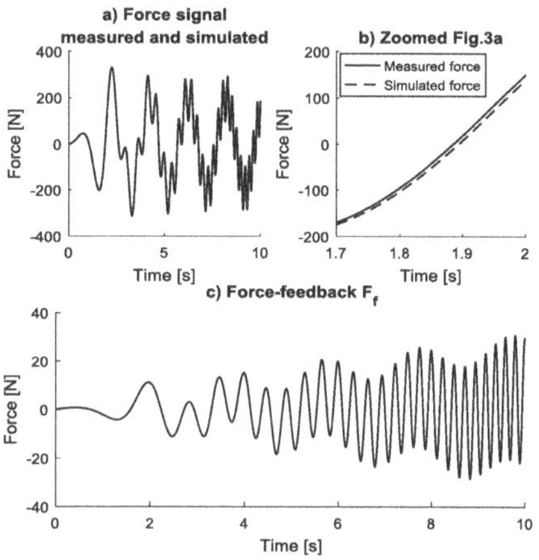

Fig. 3. (a) Measured and simulated force in the force-feedback communication channel, (b) zoomed Fig. 3b, (c) disturbance in the force-feedback communication channel caused by the transport delay.

The system structure contains 3 specific objects: the operator block, master subsystem and slave subsystem. The operator controls the position of the Master subsystem (motion scanner) by affecting it with human force F_h. The Master subsystem transfers its position x_m to the Slave subsystem. The control unit of the Slave subsystem seeks to obtain the position of the Master subsystem by the Slave subsystem $x_s = x_m$. By the force-feedback communication channel F_f environmental force influence F_e is being transferred back to the Master subsystem. The second task of the motion scanner is to deliver the force from the force-feedback communication channel back to the operator $F_f = F_e$. Theoretically, the value of the force $F_m = F_e$ [10], but in practice force F_m delivered by an actuator of the Master subsystem to the operator strongly depends on the inverse model accuracy. In the case of the considered system, 3 types of feedback with the operator could be used: force-feedback F_m (F_f), vision-feedback of Master Subsystem position x_m and vision-feedback of the Slave subsystem position x_s. The Slave vision-feedback can be useful only when both subsystems are in close range or when the image is being transmitted online from the cameras around the Slave subsystem remote placement.

The issue of the aggregate delay is particularly evident when input and output signals of the object are being measured and compared. Data collected during the experiment confirmed the presence of the transport delay which can be seen from 1.1 s to 1.2 s in Fig. 2. The phase shift presents in Fig. 2 is also a natural feature of mechanical objects actuators and sensors because there is no sensors transmittance which is characterized by a gain parameter only. That is why the sensor is shifting in time the measured signal [21]. Design of a control scheme based only on differential equations of motion will cause a delay between measured and simulated signals by the inverse model. The delay between signals will be resulting in a noise appearing in the force-feedback communication channel with an amplitude value that may destabilize the system even during free-motion situation – Fig. 3.

The difference between measured and simulated forces which have been calculated from the control signal - Fig. 3c, is caused by a 10 ms transport delay which is less than identified in the measured data. The seemingly unnoticeable delay has caused a visible deformation of force in the force-feedback channel. The barely visible delay between two signal in Fig. 3a zoomed in Fig. 3b could destabilize the entire bilateral system with force-feedback. The effect of the occurrence is a dangerous noise in the force-feedback communication channel which increases its negative impact on the force sense by the operator especially when system is working with higher frequencies. This feature is a primary concern of the sensor-less control schemes based on inverse modeling techniques but it is also a motivation of research presented in the paper.

3 Inverse Model with Prediction of Input and Output Signals

Theoretical analysis carried out in the previously published paper [10] proved that, the ideal response of a sensor-less control scheme in the force-feedback communication channel is only reachable when system is equipped with an ideal inverse model. In practice, obtaining an ideal inverse model of any subsystem is impossible but there are methods which allow to get closer to the description of a reality. One of these methods is presented in the paper.

In papers [8, 10], the estimated value by the inverse model was a pneumatic air pressure in a chamber of an actuator which has caused a motion of a piston. However, a measurement of the pneumatic air pressure is affected by a significant delay, especially when the pressure sensor is not placed directly in the actuator's chamber [11]. In the paper, it was decided to estimate an environmental force based on the control signal which is well-known, because it is applied directly to the controlled object. The inverse dynamic model in this case estimated control signal which is required for free-motion of a Slave manipulator in a single joint. The next step was a signals subtraction. The estimated control signal was subtracted from the known control signal applied to the object. The estimated control signal in the proposed control scheme was calculated by the inverse model G^{-1} which input was the position x_s in a Slave manipulator joint – Fig. 4.

The control unit scheme in Fig. 4 describes the control unit of the Slave subsystem with a dual channel based communication architecture. In the communication channel two pieces of information are transferred between subsystems, a Master position $x_m(s)$, which is transferred to the Slave control unit and transferred back in an opposite direction the force-feedback value $F_f(s)$. The Slave subsystem consists of a controller $K_c(s)$, an actuator $G_a(s)$, an object $G_o(s)$ and a sensor $G_s(s)$ transfer functions. The actuator transmittance is described by three parameters, a gain K_a, a transport delay T_1^a and a constant parameter of a first order inertia object T_2^a. The object transfer function describes the 1-DOF Slave manipulator body and is characterized by two parameters, a mass M_s and a linear viscous damping h_e used only for simulation and theoretical analysis in Sect. 4. The sensor transfer function is similar to the actuator's transfer function, and is described by three parameters a gain K_s, a transport delay T_1^s and a constant parameter of a first order inertia object T_2^s. The estimation block consists the inverse model G^{-1}, a low-pass filter $G_f(s)$ with a parameter T_f and a gain K_a, same as in the actuator's transfer function. In the case of a known inverse model which describes the Slave subsystem from the Fig. 4 and consisting of an anticipatory object $e^{s(T_1^a + T_1^s)}$. The difference between the object control signal and the estimated control signal in free-motion increased by the gain K_a gain and described by the force $F_f(s)$ in the force-feedback communication channel is equal to the environmental force impact $F_f(s) = F_e(s)$, where s is the Laplace operator. However, the presence of a positive parameter in the exponential functions makes the G^{-1} model which describes the process, impossible for physical interpretation, and without a possibility of an implementation in a control unit.

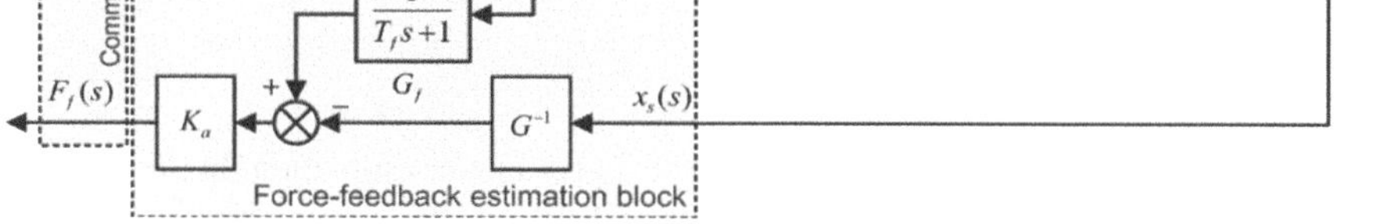

Fig. 4. Control unit scheme of the Slave subsystem with force-feedback estimation block.

To minimize the effect of a transport delay in the force-feedback communication channel, two prediction blocks characterized by an output/input signal ratio $y_p(s)/x_p(s) = (2T_p s + 1)/(T_p s + 1)$ [11], have been implemented into the structure of the inverse model, replacing the exponential function with a positive parameter. Prediction blocks replaced also first order inertia blocks with parameters T_2^a and T_2^s, to minimize the noise of a high order numerical differential calculation. However, to equalize the degree of polynomials in the denominator and the numerator subsequent first-order inertial block with parameter T_d has been added as another low-pass filter. The proposed inverse model is described by an Eq. (1):

$$G^{-1} = \frac{(h_e + M_s s)s}{K_a K_s (T_f s + 1)(T_d s + 1)} \cdot \frac{(2T_p + 1)(2T_p + 1)}{(T_p + 1)(T_p + 1)}. \tag{1}$$

The output signal of the inverse model $K_c^s(s)$ which control unit should apply to the actuator is a signal $K_c(s)$ when the Slave manipulator performs movement in a space without the environmental influence. In practice, the estimated control signal approximates signal controller K_c described by $K_c^s(s) \cong (x_m(s) - x_s(s))K_c(s)$ which is affected by the transport delay. Especially, without using the prediction block or a nonlinear modeling method – Fig. 3. However, to obtain the parameter T_p it is required to find a solution of an equation describing the force-feedback estimation block presented in Fig. 4 under two conditions. First condition concerns situation, when environmental force is not applied to the manipulator's body $F_e(s) = 0$; and second condition when zero value of force in the force-feedback communication is expected $F_f(s) = 0$. Under these two conditions, equation describing subsystem in Fig. 4 takes the following form (2):

$$(x_m(s) - x_s(s))\, K_c(s) G_f(s) + G^{-1}(s, T_p) x_s(s) = 0, \tag{2}$$

where the $G^{-1}(s, T_p)$ is the inverse model (1) as a function of two variables; the Laplace operator and the T_p parameter. Unfortunately, the T_p parameter, even for such a simple example as presented in Fig. 4 is described by a mathematical relationship as a function of the Laplace operator s, $T_p = T_p(s)$. During the research carried out with multiply model structures it was discovered that the optimal value of the T_p has been always obtained for the pulsation as close as possible to zero. Based on this fact it has been proposed to calculate the limit of the T_p as a function of the Laplace operator s when s tends to zero - Eq. (3):

$$T_p = \lim_{s \to 0} T_p(s) \tag{3}$$

Currently, the Eq. 3 is a proposition for the T_p parameter estimation. In the future, it is planned to introduce a proof that will confirm the description which was given in the paper.

Subsystem in Fig. 4 was analyzed during multiple simulations. The comparison of the linear model given by the Eq. (1) and the standard 4th order transfer function with an equal degree of the polynomials in the numerator and the denominator which T_1^a and T_1^s parameters were tuned during an iterative procedure. The simulation research was carried out on the subsystem presented in Fig. 4 and for the following data: $K_c = 100[1]$; $K_a = 1[1]$; $K_s = 1[1]$; $T_1^a = 0.002[\mathrm{s}]$; $T_2^a = 0.002[\mathrm{s}]$; $M_s = 10[\mathrm{kg}]$; $h_e = 1[\mathrm{Ns/m}]$; $T_1^s = 0.002[\mathrm{s}]$; $T_2^s = 0.002[\mathrm{s}]$; $F_e = 0[\mathrm{N}]$; $T_d = 0.0005[\mathrm{s}]$; $T_f = 0.0005[\mathrm{s}]$; and $x_m(t)[\mathrm{m}]$ was a harmonic signal with variable frequency in the range of a 0.1 to 10 Hz. The results are presented in Fig. 5.

Simulation results confirm that the inverse model which structure included prediction blocks is able to predict the control signal with higher accuracy than the same order model obtained by standard identification techniques. The results were obtained even when only inertia and damping parameters have been considered as known. Two prediction blocks during the proposed procedure of identification were able to successfully replace two first order inertia parameters: T_1^a and T_1^s, and also minimize the effect of a transport delay in the system. The results are confirmed by the diagram (a) in Fig. 5, where the inverse model with prediction blocks response is much closer to the original control signal than the standard 4th order inverse model response. This significant difference in identification techniques is especially visible in the force-feedback channel – Fig. 5b. The noise in the force-feedback channel $\mathrm{F_f(t)}$ is barely visible for the proposed model when the 4th order inverse model obtained in the standard identification technique has caused a significant error of the force estimation.

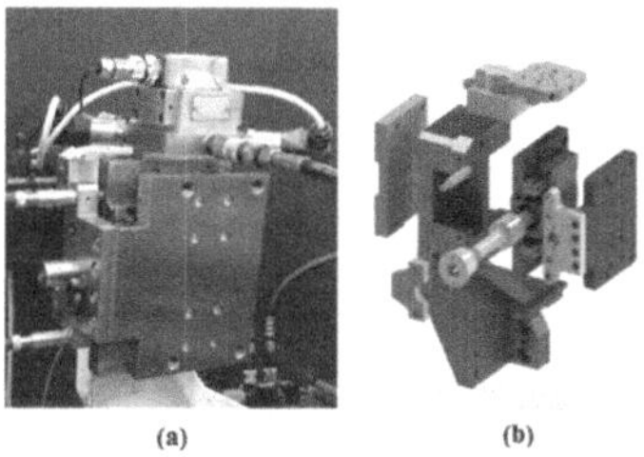

Fig. 5. (a) Experimental test stand (b) exploded view CAD.

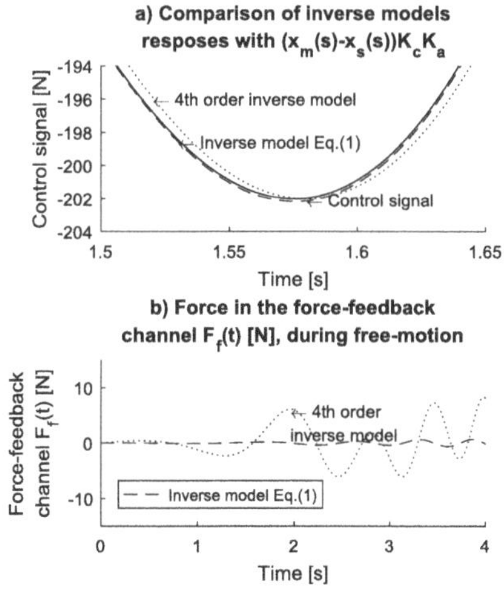

Fig. 6. Subsystem Slave presented in Fig. 4 simulation results (a) force prediction, (b) force-feedback channel.

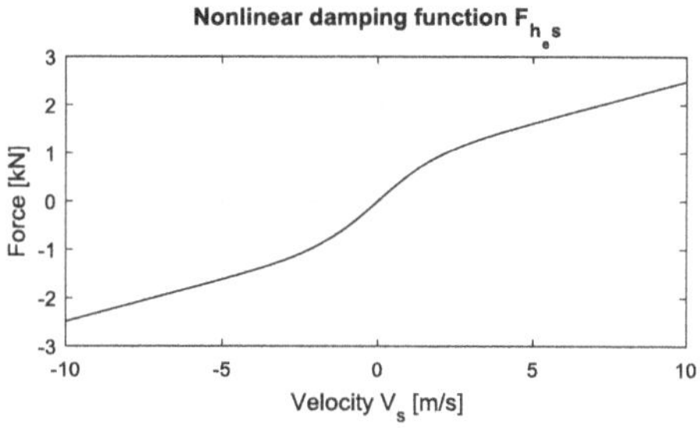

Fig. 7. The nonlinear damping function.

4 Experiment

The method based on a prediction of an input and an output of an inverse model was validated on a hydraulic 1-DoF linear manipulator presented in Fig. 6. The experimental 1-DoF manipulator has been actuated by the 760 s MOOG servo-valve. The mechanical structure was equipped with two independent pressure sensors placed in both piston chambers. The position was measured by two inductive sensors which average signal was used in the control unit. The measured and the estimated control signal by the inverse model was a valve current. The inverse model has a structure presented by the Eq. (1) but included a nonlinear component describing the damping force $F_{h_e s}$, which is given by the Eq. (4):

$$F_{h_e s}(t) = \left((1 + \exp(-v_s(t)))^{-1} - 0.5 + v_s(t)/10 + 0.00045 v_s^2(t) sign(v_s(t))\right) h_e, \quad (4)$$

The proposed damping function described by the Eq. (4) has not been chosen accidently because the noise present in the velocity signal which has caused a significant amplitude of disturbance in the signal when standard Stribeck friction model was used. The nonlinear damping function is presented by diagram in Fig. 7.

The identification of the T_p coefficients was carried out iteratively. For two prediction blocks, the T_p coefficient was identified at 0.0018 during the manipulator movement. Unfortunately, for the remaining transport delay the step change of a friction force described by the Stribeck model is responsible, but it is barely visible amid the noise. The implementation of prediction blocks allowed to minimize the current mean absolute error from 0.55 mA to 0.39 mA when chirp signal response was used as a comparison input signal. This is a significant difference which allowed to reduce the estimation error by a 30%. The control signal estimation ability of the proposed inverse prediction model is presented in Fig. 8.

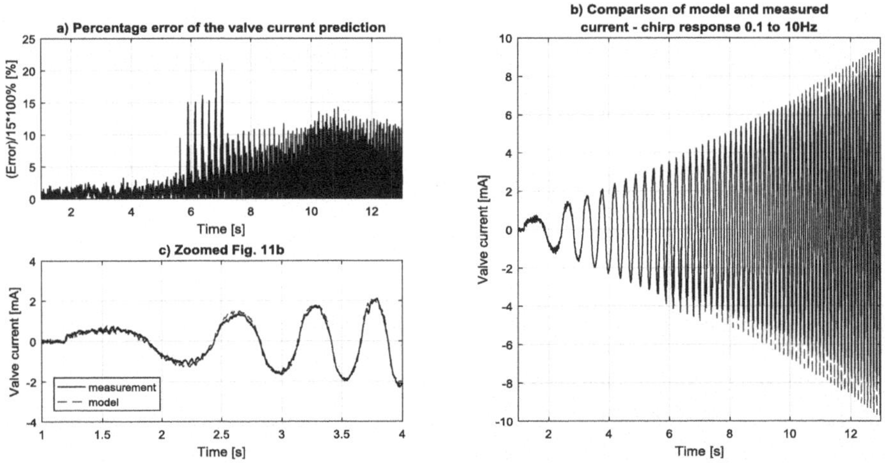

Fig. 8. (a) Percentage absolute error of the valve current relative to a possible maximum current of 15 mA, (b) comparison of model and measured current – chirp response 0.1 to 10 Hz, (c) Zoomed Fig. 8b.

Diagrams presented in Fig. 8 are the best proof of the proper functioning of the proposed method. The advantage of the method, is especially visible for low frequency spectrum, when error is at the level of the noise – below 5% relative to a maximum possible valve current 15 mA. The ability of a prediction allowed the system to maintain the stability even during rigid contact situation task by minimizing the aggregate delay of the system in the force-feedback communication channel. The force in the force-feedback channel during rigid contact task is presented in Fig. 9.

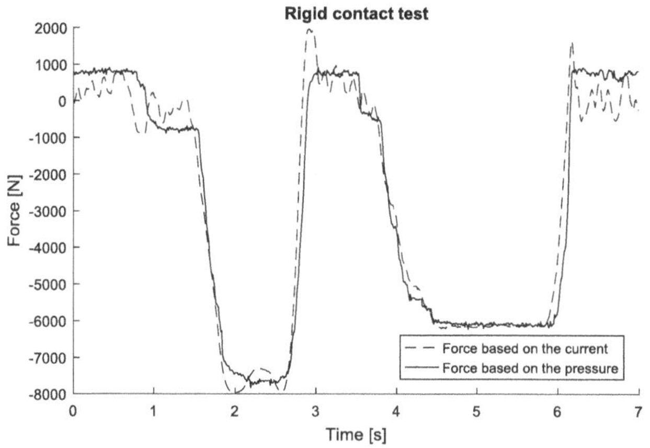

Fig. 9. Force in the force-feedback communication channel during rigid contact test.

During the rigid contact test, the operator was obliged to move the manipulator body until the piston will reach its maxim position and will touch movement boundaries that were made of steel. During the test, the operator forced the remotely controlled manipulator to interact with its movement boundaries with different force interaction twice. The method allowed the system to predict the environmental force up to 200 ms faster than it could be sensed by the pressure sensor during free motion (Fig. 9, time period from 0.5 to 1 s).

System behavior could be considered as a proper even when using a simple 4th order nonlinear inverse model. During the identification procedure of the inverse model out a standard identification procedure of a 4th order inverse model was carried, not based on the signal prediction. Unfortunately, for all of the collected data the model remained unstable and the comparison could not be taken in to account.

5 Conclusions

The paper presents a novel approach to a control design in bilateral and sensor-less teleoperation systems based on the prediction of an input and an output of an inverse model. The technique was based on simple prediction blocks with each prediction block being a phase shifter with specific properties. Simulations and experiments are the proof of an advantage of the technique over standard identification methods. The method allowed significant simplifications of the model while accuracy of the prediction was maintained. The paper presents a proposition of theoretical time coefficient estimation. Unfortunately, for obvious reasons during tests the prediction block parameter were found iteratively. The inverse model was implemented in the control unit of the test stand and series of tests were carried out. Experimental results confirmed that the system equipped with the proposed method is able to predict the environmental force impact with relatively high accuracy.

Acknowledgments. The work was carried out as part of the PBS3/A6/28/2015 project, "The use of augmented reality, interactive voice systems and operator interface to control a crane", financed by NCBiR.

References

1. Miądlicki, K., Pajor, M.: Overview of user interfaces used in load lifting devices. Int. J. Sci. Eng. Res. **6**(9), 1215–1220 (2015)
2. Ferrell, W.R.: Delayed force feedback. Hum. Fact.: J. Hum. Fact. Ergon. Soc. **8**(5), 449–455 (1966)
3. Sheridan, T.B., Ferrell, W.R.: Human control of remote computer-manipulators. In: Proceedings of the 1st International Joint Conference on Artificial Intelligence, pp. 483–494. Morgan Kaufmann Publishers Inc., Washington, DC (1969)
4. Moreau, R., et al.: Sliding-mode bilateral teleoperation control design for master–slave pneumatic servo systems. Control Eng. Pract. **20**(6), 584–597 (2012)
5. Chang, M.-K.: An adaptive self-organizing fuzzy sliding mode controller for a 2-DOF rehabilitation robot actuated by pneumatic muscle actuators. Control Eng. Pract. **18**(1), 13–22 (2010)
6. Polushin, I.G., Takhmar, A., Patel, R.V.: Projection-based force-reflection algorithms with frequency separation for bilateral teleoperation. IEEE/ASME Trans. Mechatron. **20**(1), 143–154 (2015)
7. Hulin, T., Albu-Schäffer, A., Hirzinger, G.: Passivity and stability boundaries for haptic systems with time delay. IEEE Trans. Control Syst. Technol. **22**(4), 1297–1309 (2014)
8. Tadano, K., Kawashima, K.: Development of 4-DOFs forceps with force sensing using pneumatic servo system. In: Proceedings 2006 IEEE International Conference on Robotics and Automation, ICRA 2006, Orlando, FL, USA (2006)
9. Zhai, D.H., Xia, Y.: Adaptive control of semi-autonomous teleoperation system with asymmetric time-varying delays and input uncertainties. IEEE Trans. Cybern. **47**, 3621–3633 (2017)
10. Saków, M., Pajor, M., Parus, A.: Estimation of environmental forces impact on remote control system with force-feedback and upper limb kinematics. Modelowanie Inzynierskie **58**, 113–122 (2016). (in Polish)
11. Saków, M., Parus, A.: Sensorless control scheme for teleoperation with force-feedback, based on a hydraulic servo-mechanism, theory and experiment. Meas. Autom. Monit. **62**(12), 417–425 (2016)
12. Saków, M., Miądlicki, K., Parus, A.: Self-sensing teleoperation system based on 1-DoF pneumatic manipulator. J. Autom. Mob. Robot. Intell. Syst. **11**(1), 64–76 (2017)

13. Zhou, M., Ben-Tzvi, P.: RML glove - an exoskeleton glove mechanism with haptics feedback. IEEE/ASME Trans. Mechatron. **20**(2), 641–652 (2015)
14. Miądlicki, K., Pajor, M.: Real-time gesture control of a CNC machine tool with the use Microsoft Kinect sensor. Int. J. Sci. Eng. Res. **6**(9), 538–543 (2015)
15. Pajor, M., Miądlicki, K., Saków, M.: Kinect sensor implementation in FANUC robot manipulation. Arch. Mech. Technol. Autom. **34**(3), 35–44 (2014)
16. Stuart, K.D., Majewski, M., Trelis, A.B.: Intelligent semantic-based system for corpus analysis through hybrid probabilistic neural networks. In: International Symposium on Neural Networks. Springer, Heidelberg (2011)
17. Stuart, K.D., Majewski, M.: Intelligent opinion mining and sentiment analysis using artificial neural networks. In: International Conference on Neural Information Processing. Springer, Istanbul (2015)
18. Kim, W.S., Hannaford, B., Fejczy, A.K.: Force-reflection and shared compliant control in operating telemanipulators with time delay. IEEE Trans. Robot. Autom. **8**(2), 176–185 (1992)
19. Hyun Chul, C., et al.: Sliding-mode-based impedance controller for bilateral teleoperation under varying time-delay. In: IEEE International Conference on Robotics and Automation, Proceedings 2001 ICRA, Seoul, South Korea (2001)
20. Ge, X., et al.: Analysis of a model-free predictor for delay compensation in networked systems. In: Time Delay Systems, pp. 201–215. Springer, Heidelberg (2017)
21. Kaya, I.: Obtaining controller parameters for a new PI-PD Smith predictor using autotuning. J. Process Control **13**(5), 465–472 (2003)

Time Constant and Model-Free Signal Prediction in Communication Channel of Teleoperation System

Mateusz Saków(✉), Arkadiusz Parus, Mirosław Pajor, and Karol Miądlicki

Institute of Mechanical Technology, West Pomeranian University of Technology, al. Piastów 19, 70-310 Szczecin, Poland
mateusz.sakow@zut.edu.pl

Abstract. In the paper a sensor-less control scheme for a bilateral teleoperation system with a force-feedback based on a model-free prediction in the communication channel by prediction blocks was presented. The prediction block was designed to minimize the effect of the transport delay in the communication channel of bilateral teleoperation system. The single prediction block has been a phase shifter with a specific behavior. The specific behavior of the prediction block is a strongly linear phase diagram in a useful frequency spectrum which allows the system to predict the manipulator motion and the force with a close to constant time shift. Another important feature is a gain of the block which is close to a unity when system operates in the useful frequency spectrum. The solution is an alternative to complex and mostly non-linear methods presented in the literature. The effectiveness of the method has been verified on the hydraulic manipulator test stand under control of many operators.

Keywords: Signal prediction · Remote control · Telemanipulation · Time delay

1 Introduction

Teleoperation systems are playing an important role nowadays because they have been applied in almost all of branches in the industry. It is hard to find a factory without a device being controlled by a joystick, a keyboard or other type of a remote control. However, the remotely controlled device does not need to be limited only to perform a simple set of commands [15]. From early 60s of the previous century research has been carried out to obtain remote control [4, 5] or partially autonomous [27] device operation. The breakdown into remotely controlled and autonomous devices was proposed when scientists discovered the critical value of a delay in the communication channel, which delay had a crucial impact on the stability of the telemanipulation system [27].

A continuous control system usually becomes unstable when loop delay exceeds one half a cycle at any signal frequency for which loop gain exceeds unity [27]. The problem of stability and a counteract of the effect of the delay in the communication channel are addressed by many scientific papers [5, 11, 12, 19]. First methods

J. Świder et al. (Eds.): MECHATRONICS 2017, AISC 934, pp. 348–359, 2019.
https://doi.org/10.1007/978-3-030-15857-6_35

maintaining the stability were the move-and-wait strategy and the deliberate slowdown of operator motion when an environmental object was approached, presented by Ferrell [5]. However, control strategies presented by Ferrell [5] did not give such good results of task time completion like the adaptive control which was introduced in [19]. Later, sensor based control schemes presented in [5], were redesigned and equipped with a shared compliant control method by Kim [12], which included susceptible bodies in a structure of a Master manipulator. Kim also proposed a control scheme of bilateral force control based on a position error between Master and Slave manipulators which has improved transparency of a force in the force-feedback channel [11]. However, none of these control schemes could guarantee the stability of the system when large delays are expected in the communication channel. Only after the modification of the communication channel based on a wave variables it is possible for bilateral teleoperation systems to maintain stability regardless to the delay in the communication channel [26]. During further research wave variables were extended with the passivity formalism [1]. However, a significant improvement of a force projection in the force-feedback communication channel was an introduction of a new architecture of the communication channel; the four-channel architecture [8, 13] which replaced the two-channel architecture [4–6, 11, 12]. The four-channel architecture is characterized by a two-way force and position transfer between Master and Slave subsystems. And finally, the four-channel communication architecture was equipped with the adaptive controller which estimated control parameters of force and position channels simultaneously [33].

XXI century is a domain of control schemes implementation based on: sliding mode controllers [17, 18], fuzzy logic controllers [3], force-feedback communication channel frequency separation techniques [21], special methods for discretization of a sensor resolution [9], artificial neural networks [30] and adaptive controllers dedicated to a variable and an asymmetric time delays which are complemented using adaptive filtering methods [35]. Nevertheless, it is important to pay attention that bilateral teleoperation systems feature of three types of feedback with the operator: vision-feedback [6, 27], force-feedback [5] and combination of vision-feedback and force-feedback [8, 11, 12, 30, 31]. Remote devices can be controlled by operator's motion scanners [4–6, 11, 12, 24, 27] which in a special case are exoskeletons for upper limb [24] and palm [36], by gesture control techniques [16, 20] or by voice control methods [28, 29]. However, the use of a voice control or a gesture control allows to apply only the vision-feedback between the operator and the remotely controlled device. Also an important classification of bilateral teleoperation systems with force-feedback are systems which use force sensor [2, 5, 12, 21] and devices without force sensors also known as a sensor-less or self-sensing techniques in the telemanipulation field [22–25, 30].

Bilateral teleoperation systems with force-feedback undoubtedly are the future development of the robotics next to autonomous systems because many of new scientific papers are addressing the problem of telerobotics and solving its issues. The paper addresses the transport delay problem in the communication channel of a sensor-less control scheme based on a dynamic inverse modeling procedure which allows the system to estimate environmental force affecting the Slave manipulator body [25]. The communication channel of the bilateral teleoperation system was extended with proposed prediction blocks with its input and output signals. The single prediction block has been a phase shifter with a specific behavior. The specific behavior of the prediction block is a strongly linear phase diagram in a useful frequency spectrum which allows the system to predict the manipulator motion and force in the force-feedback channel with close to constant time shift. Another important feature is a gain of the block which is close to the unity when system is operating also in a useful frequency spectrum. The proposed model-free prediction technique in the communication channel by prediction blocks was validated on a 1-DoF hydraulic manipulator test stand. The experiment confirmed the usefulness and the effectiveness of the prediction method by reducing the time shift error between input and output signals in the communication channel.

2 Problem Statement

The primary issue of any teleoperation is always present transport delay between a Master device and a Slave device [27]. Scientific literature presents solution which can compensate the effect of the transport delay and the phase shift influence caused by distance between both devices [7, 34, 35]. Likewise, the paper presents an approach for minimizing the effect of a transport delay in the communication channel but based on a model-free prediction technique which has been based on a simple phase shifter. The approach minimizes the transport delay effect in the communication channel in the control unit of a Master and a Slave subsystem. The schematic diagram of the Master-Slave system with force-feedback is presented in Fig. 1.

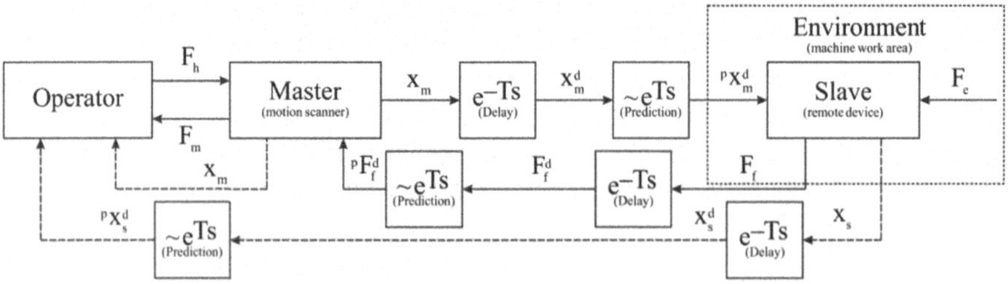

Fig. 1. Bilateral teleoperation system with two possible vision-feedbacks.

The system structure contains 3 specific objects: the operator block which controls the position of the Master subsystem (motion scanner) by affecting it with human force F_h. The Master subsystem transfers its position x_m to the Slave subsystem. In the communication channel appears a transport delay which affects the position x_m^d. The delayed position needs to be predicted ${}^p x_m^d$ by an automation object that is similar to the anticipatory model $\sim e^{Ts}$. The control unit of the Slave subsystem seeks to obtain the position of the Master subsystem by the Slave subsystem $x_s = x_m$. By the force-feedback communication channel F_f environmental force influence F_e is being transfer back to the Master subsystem. In the opposite direction of the communication channel the transport delay is also present. It means that the environmental force also needs to be predicted ${}^{p}F^{d}_{fm}$. The second task of the motion scanner is to deliver the force from the force-feedback communication channel back to the operator $F_f = F_e$. Theoretically the value of the force $F_m = F_e$ [24], but in practice the force F_m delivered by an actuator of the Master subsystem to the operator strongly depends on the inverse model or sensor accuracy. In the case of the considered system, 3 types of feedback with the operator can be used: a force-feedback F_m (F_f), a vision-feedback of the Master Subsystem position x_m and a vision-feedback of the Slave subsystem position x_s. The Slave vision-feedback can be useful only when both subsystems are in close range or when the image is being transmitted online from the cameras around the Slave subsystem placement.

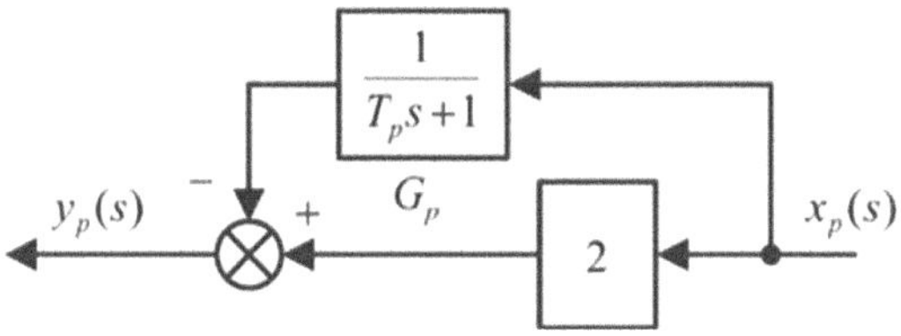

Fig. 2. The prediction block.

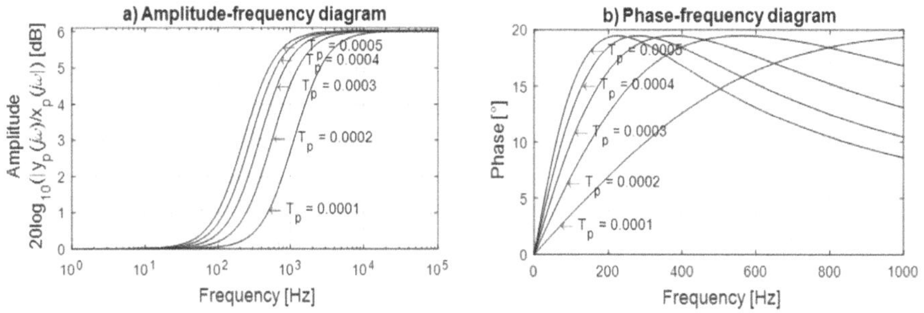

Fig. 3. (a) Amplitude-frequency diagram, (b) phase-frequency diagram, of the prediction block.

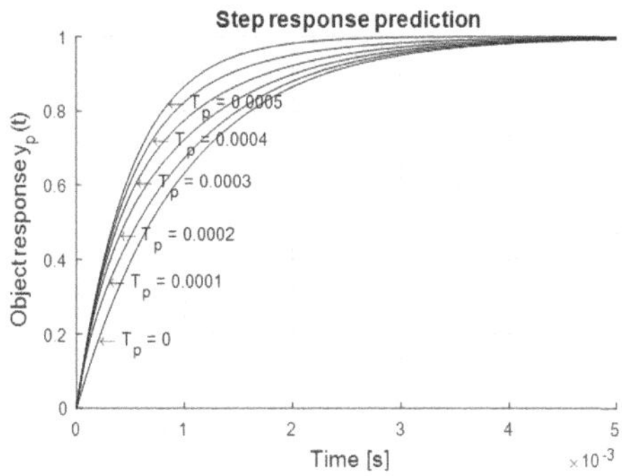

Fig. 4. Step response prediction of a first order transfer function.

The issue of the transport delay is particularly evident when input and output signals of the communication channel are being measured and compared. Data collected during the experiment confirms that even Master and Slave subsystems placed directly in their surroundings are affected by the transport delay. The delay is mostly caused by a discretization process in A/C and C/A converters and a controller's work cycle.

The design of a bilateral teleoperation control scheme which does not consider the present transport delay in the communication channel may result in an unexpected behavior of the system. The delay in a position transfer is a cause of a longer time response of the remotely controlled manipulator, relative to an operator's motion, even during free-motion situation. This feature affects the possibility of precise manipulation in the remote environment. Especially strong impact on the system behavior has a position drift in a contact situation when the Slave manipulator touches the rigid elements of the environment. The transport delay in the position transfer and force/vision feedback my result in a manipulator's end effector damage or destruction of the surrounding elements in the remote environment. The delay which affects the transparency of the value of the force in the force-feedback communication channel may destabilize the system even during free-motion and stiff contact situations [27, 32].

3 The Prediction Block

During the analysis of the previously presented control schemes based on the NARX model [24] and the first introduction of the prediction techniques used in the inverse modeling method [25], in the paper for the communication channel of a bilateral teleoperation system a prediction block with its time constants T_p is proposed. The prediction block and its structure was developed during the analysis of the Smith predictive control schemes [10]. The proposed structure of the prediction block is presented in Fig. 2.

The transmittance characterizing the automation structure in Fig. 2 is described by ratio G_p of output signal $y_p(s)$ and input signal $x_p(s)$ and is given by (1):

$$G_p = y_p(s)/x_p(s) = (2T_p s + 1)/(T_p s + 1) \cong e^{T_p s}, \tag{1}$$

where s is the Laplace operator. Examining the transmittance (1) in the frequency domain attention has to be paid to the amplitude (Fig. 3a) and phase diagram (Fig. 3b) of the prediction block presented in Fig. 3.

The prediction block depending on the T_p constant, is able to linearly shift a phase of an input signal, resulting in a constant time shift in a useful frequency spectrum – Fig. 3b. Unfortunately, the prediction block like any phase shifter is a cause of a gain of the input signal amplitude, but in the useful frequency range the gain is close to a unity Fig. 3a. The useful frequency spectrum is understood to be the achievable for a human motion. Scientific literature gives a limit of 6 Hz [14].

The prediction block in a time domain is a "signal predictor" of the input signal $x_p(t)$ where predicted time depends only on T_p constant. The prediction ability was confirmed also in a simulation of the first order transfer function with T = 0.001 which response is predicted by the prediction block. Simulations results in a time domain are presented in Fig. 4.

Both frequency diagrams in Fig. 3 and the step response in Fig. 4 are the proof of a predictive capabilities of the prediction block. However, the prediction block is sensitive to a noise and changes of a signal derivative sign $x_p(t)$.

4 Experiment

The system based on prediction blocks which were placed in the communication channel of a bilateral teleoperation system was validated on a hydraulic 1-DoF linear manipulator presented in Fig. 5. The experimental 1-DoF manipulator has been actuated by the 760 s MOOG servo-valve. The position was measured by two inductive sensors which average signal was used in the control unit. The environmental force-feedback estimation was based on a sensor-less and self-sensing technique. The identified model was expanded by two prediction blocks and implemented in the control unit of the test stand [25].

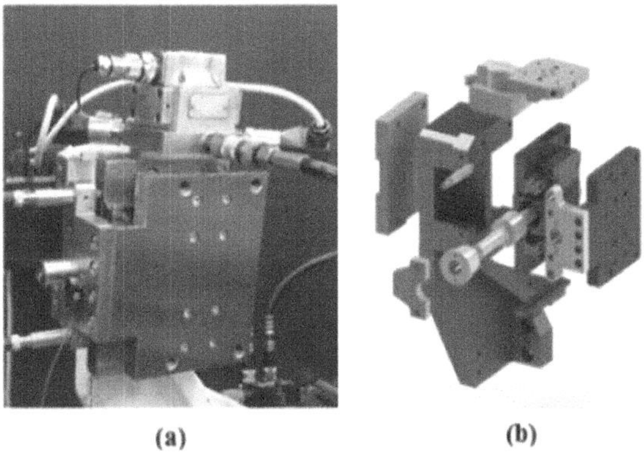

Fig. 5. (a) Experimental test stand (b) exploded view CAD.

During tests system operator was obliged to move the Master manipulator until the Slave manipulator touched the obstruction in a limited time of 10 s. The operator was informed that it is required to touch the obstruction with a force felt in the force-feedback channel, as close as possible to the reference force of 2 N (the hydraulic Slave manipulator force gain was set to 100 according to the force-feedback channel). The vision feedback was provided by a moveable dot on screen which described the position of the manipulator's end effector. Also the vision-feedback of the obstruction was provided. The steel to steel contact, between the Slave manipulator and the obstruction, and use of the two channel communication architecture allowed the experiment to be proceeded only up to 10 ms one-way delay in the communication channel. An increase of a delay caused a significant increase of position drift between both subsystems, which was a primary issue of the operator repulsion by the system during the contact even with the wave variables used in the communication channel.

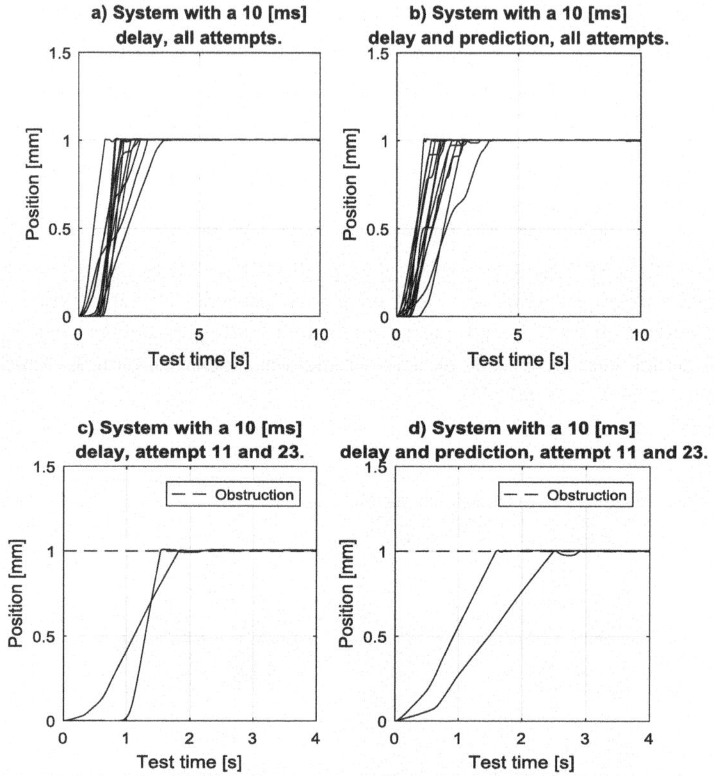

Fig. 6. Comparison of system position tracking during contact.

For the test, 6 independent operators moved the end effector as close as possible to the obstruction and interact with it. Each operator had 5 attempts with 2 types of the control unit. Two control units included a communication channel with a delay and prediction blocks. Unfortunately, each operator was unaware of a type of a control unit, a delay value and had not had any practice before any attempt of the test. The one-way time delay was set to 10 ms. The position results of the Slave manipulator are presented in Fig. 6.

In the case of the position tracking ability by the system, the difference is unnoticeable for most of the operators. However, two of them have noticed that the system with the prediction blocks (Fig. 6b and d) was more robust in terms of slowing down just before the obstruction. In the case of the system with the delay (Fig. 6a, c), there was a strong collision in 90% of attempts. In the case of the system with a prediction serious collisions occurred in 20% of attempts. Notwithstanding, the force tracking ability by the operator was seriously improved - results are presented in Fig. 7.

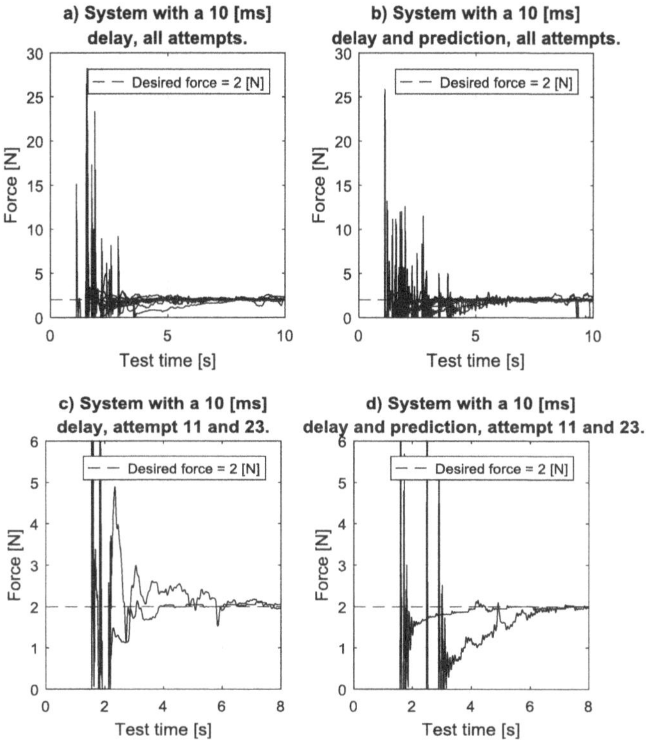

Fig. 7. Comparison of system force tracking during contact.

In the case of the system with a delay, operator has been struggling to maintain reference force in the force-feedback in all attempts. It is confirmed in Fig. 7a and c, when the operator was a cause of low-frequency oscillation after a collision with the obstruction. However, when using the system with prediction blocks, the operator after a contact was able to increase the force of interaction without oscillations to the desired value of 2 N. It is a major improvement because in 80% of attempts, the operator did not overshoot the desired value of force by a significant value of 10%, relative to the desirable value. All the results and conclusions are confirmed by the statistical analysis – Fig. 8.

Histograms in Fig. 8 are a comparison of the force signal samples participation after first 2 s, when the manipulator contact appearance with the obstruction was made. Over 86% of force signal samples in the system with a delay and a prediction (Fig. 8) are below force of 2 N, and their distribution can be considered as a rectangular. This demonstrates the presence of a certain trend in the data, also visible in Fig. 7d. The system with a delay (Fig. 8b) is characterized by a normal distribution around an average of 1.84 N, while the system with a delay prediction was characterized by an average value of force - 1.59 N. This clearly confirms that the system with prediction block was easier to control in terms of force tracking by the operator.

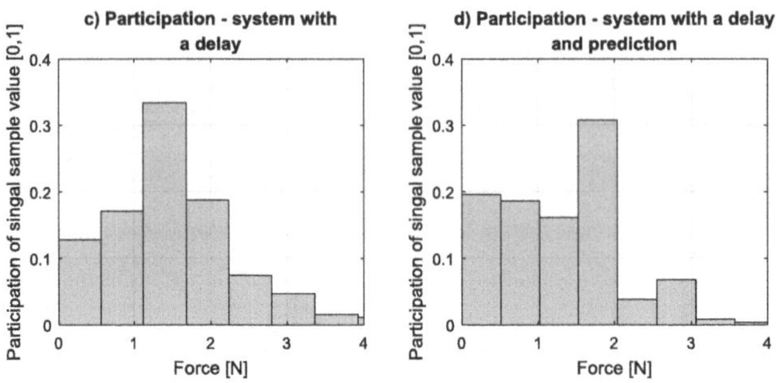

Fig. 8. Dynamic contact statistics - force signal samples participation - 2 s after the first contact.

In the case of a static contact, there was some uncertainty about the behavior of the system. There was a possibility that the noise gain caused by the prediction block would be able to disturb the static contact situation in the system. However, the statistical analysis showed that the operator working with both systems was able to maintain the static force comparatively well. In the case of a mean, there is no significant difference in force signal samples. However, in the case of standard deviation, there is very little significant difference in favor of the system with prediction blocks. Thus, prediction blocks did not disturb the force transparency of the system. Statistical results of the static contact between 6.5 s and 9 s are presented in Fig. 9.

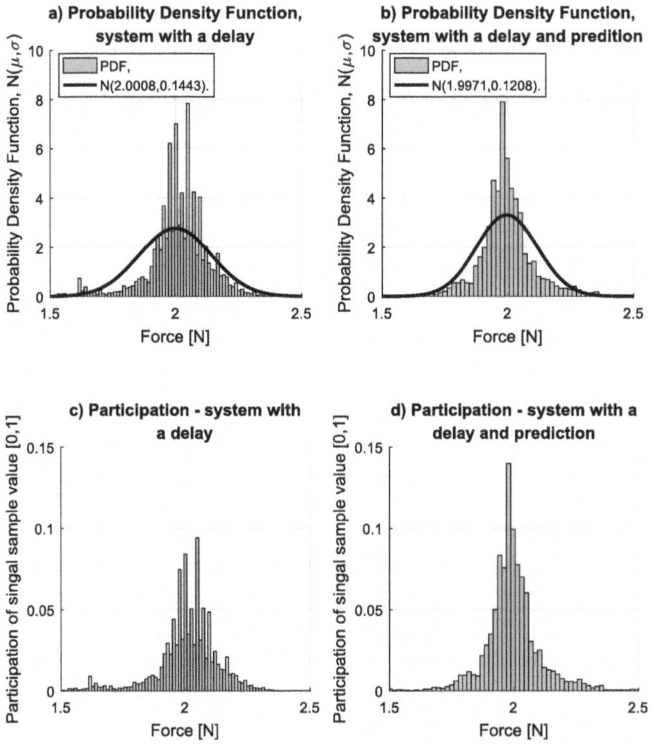

Fig. 9. Static contact statistics - force signal samples participation - [6.5, 9]s contact interaction.

5 Conclusions

The paper presents a novel approach to a control design in bilateral and sensor-less teleoperation systems with a force-feedback. The system was based on the model-free and time constant prediction in the communication channel. The approach was based on prediction blocks which was a phase shifter with specific frequencies characteristics. Multiply experiments are the proof of an advantage of the technique over system without a time constant prediction. The paper also presents a prediction block analysis in a frequency domain. The use of the prediction block significantly improved the force control directly during the dynamic contact part, and did not disturb the system during the static contact situation. All conclusions about the prediction block find its proof in the frequency domain. Moreover, each assumption in the case of the experiment was supported by the statistical analysis with multiply attempts with 6 independent operators of both types of the control unit in the system.

Acknowledgment. The work was carried out as part of the PBS3/A6/28/2015 project, "The use of augmented reality, interactive voice systems and operator interface to control a crane", financed by NCBiR.

References

1. Arcara, P., Melchiorri, C., Stramigioli, S.: Intrinsically passive control in bilateral teleoperation mimo systems. In: 2001 European Control Conference (ECC) (2001)
2. Atashzar, S.F., Polushin, I.G., Patel, R.V.: Projection-based force reflection algorithms for teleoperated rehabilitation therapy. In: 2013 IEEE/RSJ International Conference on Intelligent Robots and Systems (2013)
3. Chang, M.-K.: An adaptive self-organizing fuzzy sliding mode controller for a 2-DOF rehabilitation robot actuated by pneumatic muscle actuators. Control Eng. Pract. **18**(1), 13–22 (2010)
4. Ferrell, W.R.: Remote manipulation with transmission delay. IEEE Trans. Hum. Fact. Electron. **HFE-6**(1), 24–32 (1965)
5. Ferrell, W.R.: Delayed force feedback. Hum. Fact.: J. Hum. Fact. Ergon. Soc. **8**(5), 449–455 (1966)
6. Ferrell, W.R., Sheridan, T.B.: Supervisory control of remote manipulation. IEEE Spectr. **4** (10), 81–88 (1967)
7. Ge, X., et al.: Analysis of a model-free predictor for delay compensation in networked systems. In: Time Delay Systems, pp. 201–215. Springer, Heidelberg (2017)
8. Hastrudi-Zaad, K., Salcudean, S.E.: On the use of local force feedback for transparent teleoperation. In: Proceedings of 1999 IEEE International Conference on Robotics and Automation (1999)
9. Hulin, T., Albu-Schäffer, A., Hirzinger, G.: Passivity and stability boundaries for haptic systems with time delay. IEEE Trans. Control Syst. Technol. **22**(4), 1297–1309 (2014)
10. Kaya, I.: Obtaining controller parameters for a new PI-PD Smith predictor using autotuning. J. Process Control **13**(5), 465–472 (2003)
11. Kim, W.S.: Developments of new force reflecting control schemes and an application to a teleoperation training simulator. In: Proceedings of 1992 IEEE International Conference on Robotics and Automation (1992)

12. Kim, W.S., Hannaford, B., Fejczy, A.K.: Force-reflection and shared compliant control in operating telemanipulators with time delay. IEEE Trans. Robot. Autom. **8**(2), 176–185 (1992)
13. Lawrence, D.A.: Stability and transparency in bilateral teleoperation. IEEE Trans. Robot. Autom. **9**(5), 624–637 (1993)
14. Lichiardopol, S., Van De Wouw, N., Nijmeijer, H.: Control scheme for human-robot co-manipulation of uncertain, time-varying loads. In: 2009 American Control Conference (2009)
15. Miądlicki, K., Pajor, M.: Overview of user interfaces used in load lifting devices. Int. J. Sci. Eng. Res. **6**(9), 1215–1220 (2015)
16. Miądlicki, K., Pajor, M.: Real-time gesture control of a CNC machine tool with the use Microsoft Kinect sensor. Int. J. Sci. Eng. Res. **6**(9), 538–543 (2015)
17. Moreau, R., et al.: Sliding-mode bilateral teleoperation control design for master–slave pneumatic servo systems. Control Eng. Pract. **20**(6), 584–597 (2012)
18. Nguyen, T., et al.: Accurate sliding-mode control of pneumatic systems using low-cost solenoid valves. IEEE/ASME Trans. Mechatron. **12**(2), 216–219 (2007)
19. Niemeyer, G., Slotine, J.J.E.: Stable adaptive teleoperation. IEEE J. Ocean. Eng. **16**(1), 152–162 (1991)
20. Pajor, M., Miądlicki, K., Saków, M.: Kinect sensor implementation in FANUC robot manipulation. Arch. Mech. Technol. Autom. **34**(3), 35–44 (2014)
21. Polushin, I.G., Takhmar, A., Patel, R.V.: Projection-based force-reflection algorithms with frequency separation for bilateral teleoperation. IEEE/ASME Trans. Mechatron. **20**(1), 143–154 (2015)
22. Rakotondrabe, M., et al.: Simultaneous displacement/force self-sensing in piezoelectric actuators and applications to robust control. IEEE/ASME Trans. Mechatronics. **20**(2), 519–531 (2015)
23. Saków, M., Miądlicki, K., Parus, A.: Self-sensing teleoperation system based on 1-DoF pneumatic manipulator. J. Autom. Mob. Robot. Intell. Syst. **11**(1), 64–76 (2017)
24. Saków, M., Pajor, M., Parus, A.: Estymacja siły oddziaływania środowiska na układ zdalnie sterowany ze sprzężeniem siłowym zwrotnym o kinematyce kończyny górnej. Modelowanie Inzynierskie **58**, 113–122 (2016)
25. Saków, M., Parus, A.: Sensorless control scheme for teleoperation with force-feedback, based on a hydraulic servo-mechanism, theory and experiment. Meas. Autom. Monit. **62**(12) (2016)
26. Sheridan, T.B.: Space teleoperation through time delay: review and prognosis. IEEE Trans. Robot. Autom. **9**(5), 592–606 (1993)
27. Sheridan, T.B., Ferrell, W.R.: Human control of remote computer-manipulators. In: Proceedings of the 1st International Joint Conference on Artificial Intelligence, pp. 483–494. Morgan Kaufmann Publishers Inc., Washington, DC 1969
28. Stuart, K.D., Majewski, M.: Intelligent opinion mining and sentiment analysis using artificial neural networks. In: International Conference on Neural Information Processing. Springer, Heidelberg (2015)
29. Stuart, K.D., Majewski, M., Trelis, A.B.: Intelligent semantic-based system for corpus analysis through hybrid probabilistic neural networks. In: International Symposium on Neural Networks. Springer, Heidelberg (2011)
30. Tadano, K., Kawashima, K.: Development of 4-DOFs forceps with force sensing using pneumatic servo system. In: Proceedings 2006 IEEE International Conference on Robotics and Automation, ICRA 2006 (2006)

31. Tavakoli, M., Patel, R.V., Moallem, M.: A force reflective master-slave system for minimally invasive surgery. In: Proceedings of 2003 IEEE/RSJ International Conference on Intelligent Robots and Systems (IROS 2003) (2003)
32. Tomovic, R., Boni, G.: An adaptive artificial hand. IRE Trans. Autom. Control **7**(3), 3–10 (1962)
33. Wen-Hong, Z., Salcudean, S.E.: Stability guaranteed teleoperation: an adaptive motion/force control approach. IEEE Trans. Autom. Control **45**(11), 1951–1969 (2000)
34. Zhai, D.H., Xia, Y.: Adaptive control for teleoperation system with varying time delays and input saturation constraints. IEEE Trans. Industr. Electron. **63**(11), 6921–6929 (2016)
35. Zhai, D.H., Xia, Y.: Adaptive control of semi-autonomous teleoperation system with asymmetric time-varying delays and input uncertainties. IEEE Trans. Cybern. **47**, 3621–3633 (2017)
36. Zhou, M., Ben-Tzvi, P.: RML glove - an exoskeleton glove mechanism with haptics feedback. IEEE/ASME Trans. Mechatron. **20**(2), 641–652 (2015)

Analysis of Selected Factors' Influence on the Specific Range of Modern Jet Transport Aircraft as a Complex Mechatronic System

Robert Sklorz[1], Adrian Zieliński[2], and Jarosław Brodny[3](✉)

[1] Wizz Air Hungary Ltd., Kőér street 2/a, Budapest 1103, Hungary
[2] StorkJet sp. z o. o., Władysława Orkana 8d, 40-553 Katowice, Poland
[3] Faculty of Organization and Management, Silesian University of Technology, Roosevelta 26-28, 41-800 Zabrze, Poland
jaroslaw.brodny@polsl.pl

Abstract. Modern jet transport aircraft shall be considered as complex mechatronic system. Looking at both mechanics and electronics it is a complicated technical device. In order to meet safety and efficiency goals jet transport aircraft are equipped with recording systems including flight data recorders such as FDR (Flight Data Recorder) and QAR/DAR (Quick/Data Access Recorder). DAR serves for purpose of operator's customized data recording which allows detailed studies and analytics such as flight path optimization. This article describes method which utilizes DAR data to determine aircraft performance during cruise. Research was conducted based on real Airbus A320 DAR data. The main objective was to determine relations between recorded flight parameters and aircraft specific range, which is directly related to fuel consumption and key parameter for cruise efficiency. Analysis included assessment of data recording quality influence on the data source and obtained results. Method described allows also to verify manufacturer's performance model, guarantees, monitor tail-specific aircraft cruise performance, as well as determine optimum altitudes, speeds and estimate relation of weight and aircraft fuel consumption. From mechatronics point of view it may be assumed that connection of mechanical and electronic elements together with flight path governing parameters allows control of key flight parameters, which in result allows their optimization and enhances efficiency.

Keywords: Mechatronic system · Aircraft · Specific Range · Recording system · Flight data

1 Introduction

Cruise performance of jet transport aircraft defines its range, altitude capabilities and fuel consumption. Along with take-off and landing performance, range will limit operator's network span. Fuel consumption will influence economical profitability through direct operating cost and payload capabilities [1]. Cruise performance is therefore one of the key elements in an operator's business.

J. Świder et al. (Eds.): MECHATRONICS 2017, AISC 934, pp. 360–376, 2019.
https://doi.org/10.1007/978-3-030-15857-6_36

As a consequence of relations indicated above, precise determination of aircraft cruise performance based on actual flight data is important from several operational and economical aspects:

- verification of the manufacturer's guarantees of aircraft performance,
- verification of the manufacturer's performance model defined as "book level",
- monitoring of cruise performance and establishing performance deterioration level,
- determination and checking of optimum operating speeds and altitudes,
- estimating relation of weight and fuel consumption.

Historically there have been multiple successful attempts to determine aircraft cruise performance using flight data [2, 3]. The proposed method in conjunction with nowadays higher flight data recording quality gives the opportunity to improve quality of determined cruise performance data. This is essential in order to precisely answer the problems related to the aspects listed above (Fig. 1).

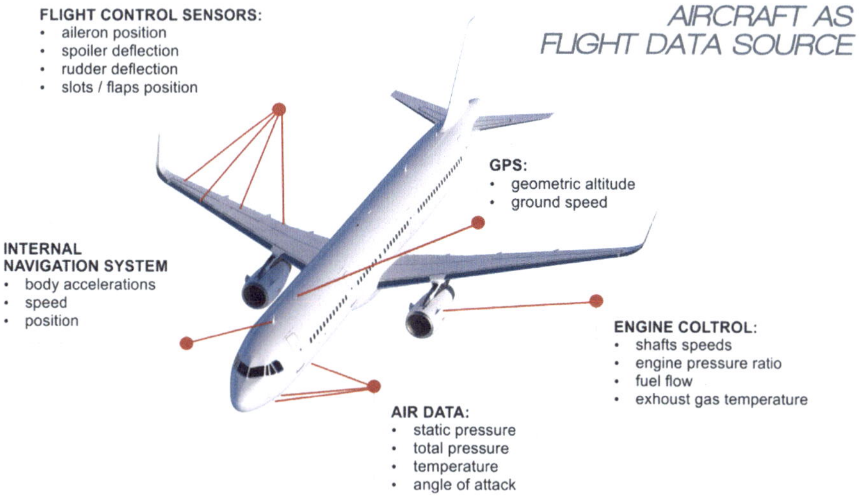

Fig. 1. Primary flight data parameters recorded in-flight (own source).

2 Data Sources

Nowadays typical jet transport aircraft utilizes thousands of parameters originating from various sensors. The most important safety related parameters are recorded on Flight Data Recorder or Digital FDR. Among them there are parameters related to flight path, environment, engine operation, pilot inputs and systems which are crucial to flight safety. Parameters recorded on FDR are also recorded on Quick Access Recorder which enables much quicker read-out, required mainly for Flight Data Monitoring or maintenance related purposes. Another newly introduced option for recording and storage is a wireless QAR, which is a fully automated system that transmits data automatically via mobile telephone network. Currently almost all jet transport aircrafts are equipped with Quick Access Recorders. In addition to QAR on the most of jet transport aircrafts Data Access Recorder is available. DAR enables configuration change of recorded parameters regardless of FDR/QAR (which are defined by manufacturer and certification requirements) [4].

FDR/QAR/DAR parameters are recorded with defined frequency and precision. These together with number of recorded parameters will be limited by the storage capabilities. Normally QAR/DARs are configured in such way to enable recording of 128 or 256 12-bit "Words per second" Wps, but also option for 1024 Wps exists. Obviously each of the sensors or data acquisition units will have its own frequency and precision. It is important to balance the configuration of parameters in a way that required precision and frequency is sustained while the storage capabilities or read-out process are not penalized.

In relation to the subject these factors will have direct impact on quality of the results. In order to achieve the highest quality, utilized dataset was recorded by DAR with 1024 Wps and high frequency – at least 1 Hz or higher depending on parameter. Precision of the recorded parameters was close to the one at originating data bus. In some cases reduction of precision was beneficial as it had no impact on the final result, but reduced data storage requirements.

It is worth underlining that many of parameters recorded by QAR originate from ARINC 429 data bus, for which all binary coded numeric values must be checked to determine parity error, no Digital Information Transfer System update, Sign Status Matrix SSM Failure Warning, SSM Test and SSM No Computed Data. On QAR there is no SSM and therefore the above errors are recorded instead of values, most of times as alternating codes. In order to obtain the highest quality (Table 1).

Table 1. Parameters utilized for the study.

Flight path	Environment	Engine	Other
Normal acceleration	Total air temperature	EPR	Gross mass
Longitudinal acceleration	Barometric altitude	EGT	Centre of gravity
Bank angle	GPS altitude	Fuel flow	Air conditioning pack flow
Ground speed	Mach number		Electric generator load
Vertical speed	Indicated airspeed		
Latitude	Wind speed		
Longitude	Wind direction		
Heading			

3 Flight Data Characteristics

In this article cruise performance is considered during unaccelerated, straight and level flight as defined in performance related publications [5]. Due to nature of environment and need of flight management it often happens that these criteria are not met. During typical flight aircraft encounters disturbances causing changes in flight parameters originating mainly from:

- crossing of pressure areas,
- flying between different air masses,
- flying along or trough jet streams,
- up and downdraughts,
- turbulence.

Additionally pilot or Auto Flight System changes flight level or speed for purpose of flight management. All of the listed phenomena or inputs cause the situation that it is rare to observe perfectly unaccelerated, straight and level flight during actual operations. Especially longitudinal acceleration or gross vertical velocity are rarely equal to 0 during real flight. It is important to underline that the magnitude of disturbances vary largely.

In order to represent the cruise performance it is important to select these segments of cruise flight which are as close as possible to unaccelerated, straight and level flight. Otherwise achieved results will be affected by high scatter and disable a possibility of determining cruise performance with high quality.

Figures 2, 3, 4, 5 and 6 present behavior of flight parameters during 2 kinds of cruise segments – one considered as stable (solid line) and one randomly selected (dashed line). Timeframe of both segments is equal to 100 s. High correlation is observed between Fuel Flow and Engine Pressure Ratio.

Fig. 2. Time series of one engine fuel flow.

Fig. 3. Time series of one Engine Pressure Ratio.

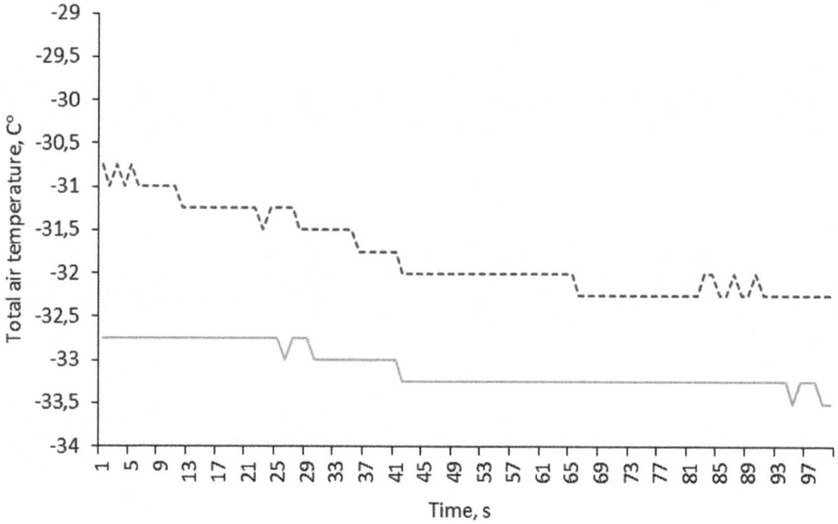

Fig. 4. Time series of total air temperature.

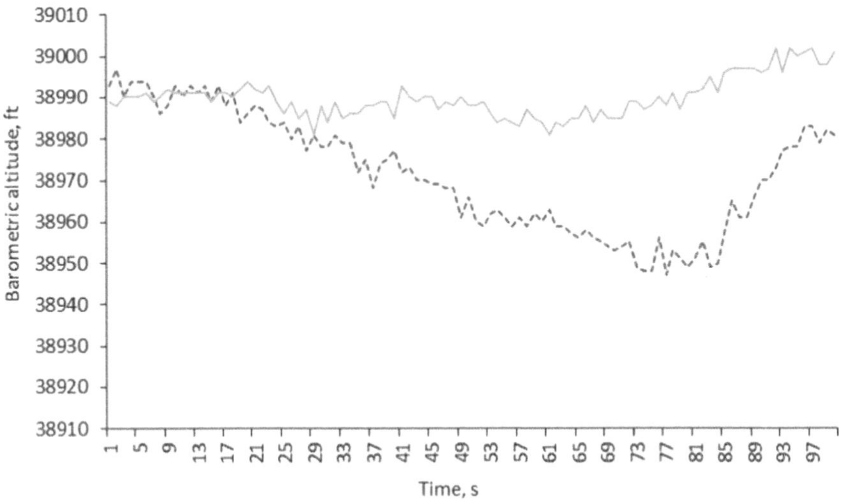

Fig. 5. Time series of barometric altitude.

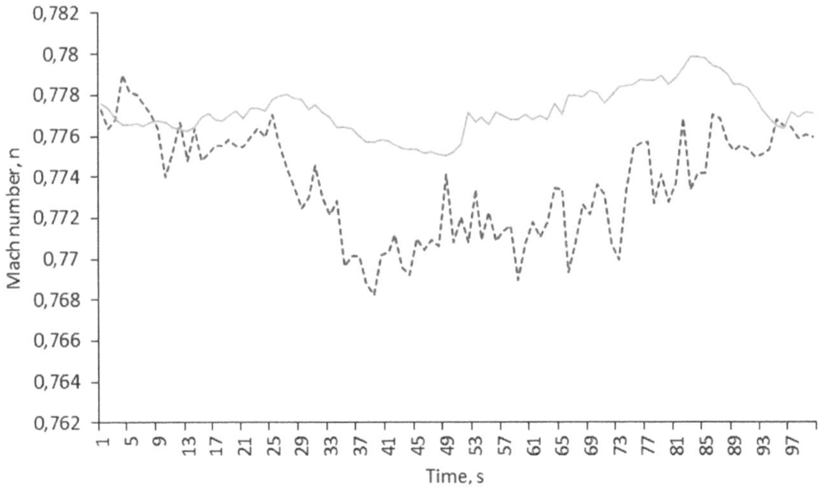

Fig. 6. Time series of Mach number.

It is worth paying attention to the engine thrust control parameter - Engine Pressure Ratio "EPR" behavior, which shape has sinusoidal character. The reason behind it originates from Auto Thrust System behavior on considered aircraft type – Airbus A320, which controls the target EPR. For comparison data from Boeing 737 aircraft was analyzed. Parameter controlling the thrust – N1 was stable (no sinusoidal character) during stable cruising condition.

4 Stable Segment Quality

Raw data recorded by DAR contains hardware faults and environmental perturbations which negatively affects quality of determined cruise performance – increase scatter. It is therefore important to apply measures allowing filtering out incorrect or low quality data. In the presented study this has been achieved by implementing algorithm which calculates quality of every segment and selecting segments which have the highest quality.

The stable segment quality is defined as the sum of weighted flight parameters variances.

The selection of flight parameters should be balanced between two criteria:

- Flight parameters relation to stability.
- By definition stable cruise segment is level and unaccelerated flight. Parameters such as Mach number, vertical acceleration, vertical speed, roll, and temperature will directly influence stability. Attention should be also paid to parameters related to the engine operation & control such as Low Pressure Shaft speed, Fan Inlet pressure, Low Pressure Outlet pressure, Active Control Clearance or Compressor Stator angle.
- Multicollinearity.
- Having in mind that final result of quality is sum of parameter variances and data point values are average of observed segment effects related to multicollinearity should be considered. Selecting multiple parameters which indicate the same value or are closely related should be avoided as it may result in selection of segments where unusual behavior of variable affects multiple parameters and leads to over-optimistic stability or on contrary causes that quality of good segment is poor.

During the study effects of various parameters compositions on stable segment quality were evaluated. It has been found that parameters composition suggested in similar methodology (Table 2) includes too many engine parameters which are closely correlated. Taking into account behavior of the Auto Thrust System the stable segment quality value was the highest where engine parameters behaved unusually, flattening the regular sinusoidal behavior. Their unusual behavior is most of times caused by influence of environmental parameters. Figure 7 presents histogram of influence on stable segment quality for set of flight data parameters typically used for the purpose of stable segment selection.

It is worth noticing that approx. 50% share in the total quality is contributed to Engine Pressure Ratio "epr" and Fuel Flow "ff" which are closely correlated (see Fig. 5). Analogically adding more engine control related parameters such as Fan Inlet pressure P2 would cause that the final stable segment quality is ruled more often by factors affecting engine stability than entire flight stability.

Table 2. Stable segment parameter composition by Speyer [6].

Parameters	Unit	DP limit Initial values	Readjusted DP limit
IVV	ft/min	100	300
FPAC	g	0.012	0.03
ROLL	Deg	0.8	4
TAT	Deg C	1.1	2
N2	%	0.9	0.9
EGT	Deg C	18	15
VACC	g	0.03	0.1
MACH	–	0.008	0.01
N1	%	1.6	1.6
FF	KG/H	100	100
PT2	PSIA	0.05	0.1

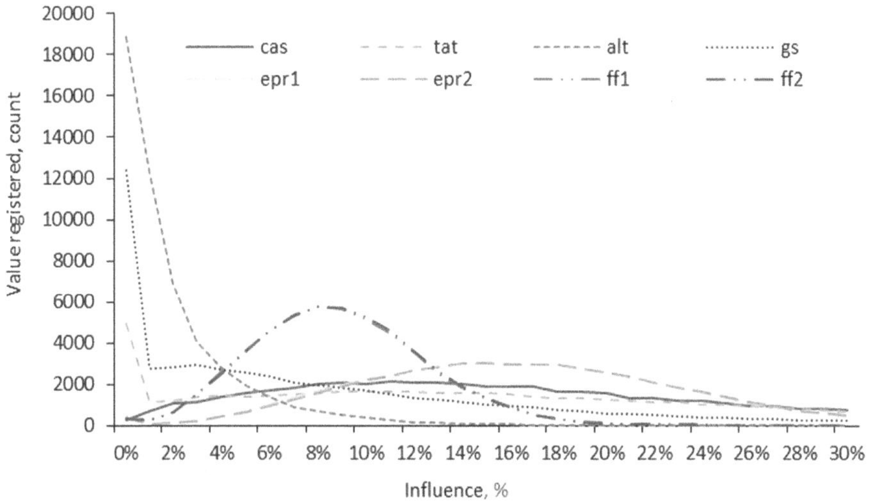

Fig. 7. Histogram of particular parameters on stable segment quality.

5 Tail Number Specific Cruise Stability

Each aircraft has different average cruise performance results. The sheer difference between aircraft with the best and the worst stable segment quality is as high as two times the value of best performing one. Comparison was made in order to identify from where this disproportion comes from.

Table 3. Comparison of statistical indicators for worst and best performing aircraft.

Aircraft	Count	Min	Max	Average	Median	Std. dev.	First quartile	Third quartile
x_worst	302	0.007	24.326	0.587	0.094	2.442	0.051	0.187
x_best	641	0.011	6.475	0.225	0.085	0.607	0.049	0.164

Standard deviation and highest value recorded for *x_worst* with quartiles and median are not so far from *x_best* as seen in Table 3 which implies that there is a problem with abnormal observations. Finding outliers with IQR method, presented in Table 4, shows that there is not much difference in outliers count between *x_best* and *x_worst*.

Table 4. Percentage number of outliers calculated with IQR method.

	k = 1.5		k = 3	
	Count	Percentage	Count	Percentage
Aircraft 1	43	14.238	24	7.947
Aircraft 2	92	14.352	49	7.644

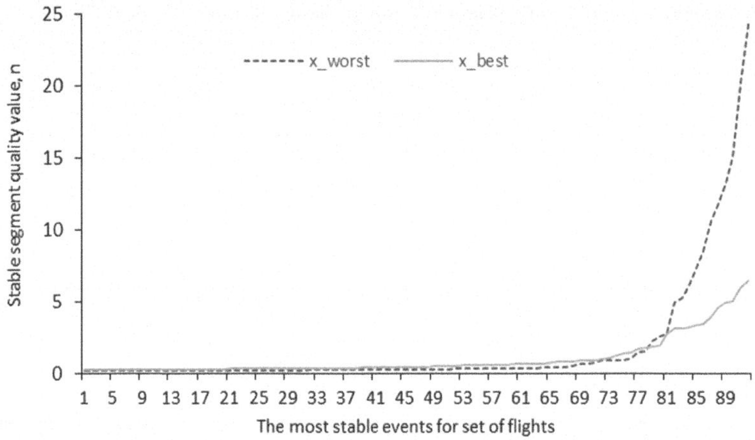

Fig. 8. Lowest to highest results of two aircraft (own source).

As shown in Fig. 8, problem is not with the number of outliers but values of 10 individual flights that had the worst result highly influencing whole results. It is therefore important to filter out outliers before analysis of the results. There was no co-relation found to any technical malfunction in presented example.

6 Tail Number Specific Cruise Stability

Precision of recorded parameters plays an important role in variance calculation. As indicated in (2) precision of parameters recorded on FDR/QAR is reduced in order to meet the data storage requirements and at the same time enable accident investigation. Analysis (see Fig. 9) indicates that the recorded parameter should meet more stringent requirements in order to be used for the purpose of the subject. Mach number parameter was used as an example.

Difference in the precision may affect variation calculation and also significance of the final result. Taking into account precision of maintaining constant speed by auto flight system the desired precision of the Mach number is 0.001. FDR/QAR will normally store that parameter on 8 bits. For subsonic aircraft this will result in precision of 0.004. Precision of Mach number parameter on the originating data bus is according to the reviewed specification equal to 0.0000625. In order to compare the data sources, DAR was configured to store Mach number with maximum precision.

Figure 9 presents 1 min, equal to 60 samples of Mach number recording, originating from QAR, DAR (configured for maximum precision) and value series rounded to 0.001.

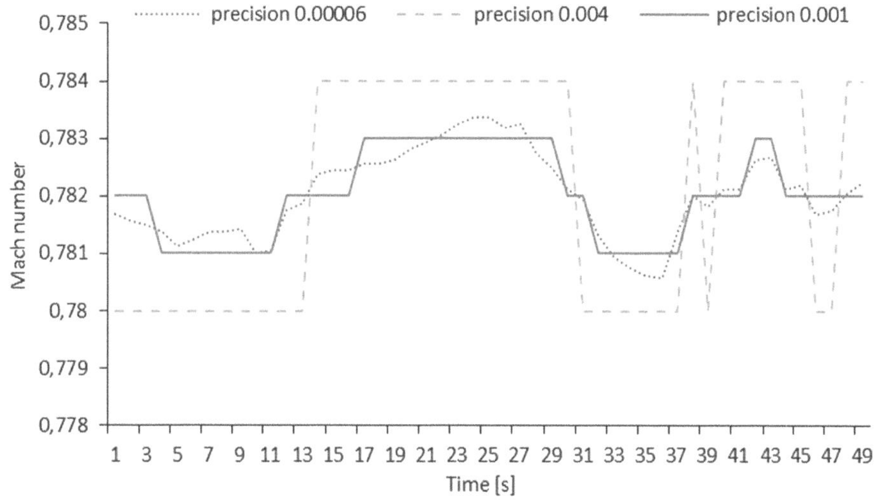

Fig. 9. Impact of precision on data recording.

Table 5. Difference in percentage, between variance of given datasets.

	Precision 0.0006	Precision 0.001	Precision 0.004
Variance	0.5596xE−6	0.5914xE−6	4.0751xE−6
Difference in percentage		5.68%	628.18%

There is high difference in variance between original precision and 0.004 where it got sextupled (see Table 5). Difference for precision 0.001 was not far off from maximum precision possible for DAR.

7 Results

Results present average cruise performance of Airbus A320 aircraft based on 550,000 stable segments generated from 53,000 flights. Results are grouped into aircraft with standard wing tip fence and with “Sharklets” installed. Data from 8000 flights did not produce stable segments due to lack or short cruise phase.

7.1 Variants and Gross Mass Effect on Cruise Performance

Airbus A320 operates in 2 variants. With and without “Sharklets”, which are similar to winglets. Figure 8 presents average Specific Range for given set of aircraft’s gross mases, both Sharklet and standard “Wing Tip Fence” (Fig. 10).

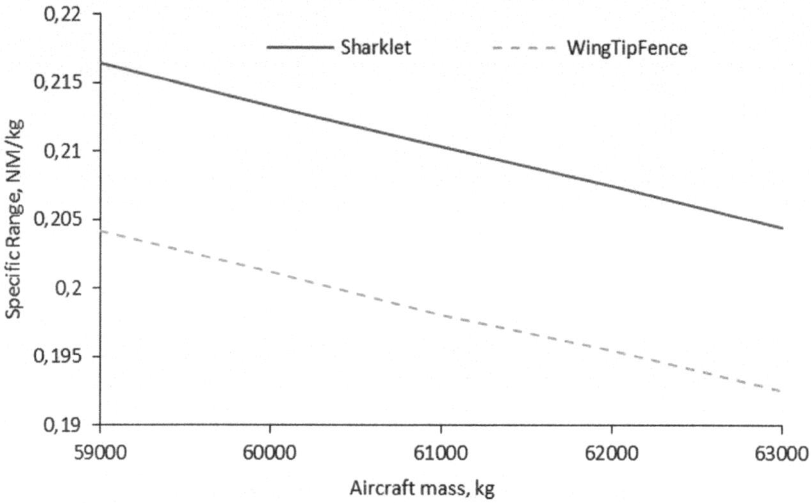

Fig. 10. Aircraft mass and Specific Range relation.

Similarity in function angle between 2 variants and close to linear representation of each indicates good data quality. Difference in Specific Range between 2 variants is 6%. It should be noted that data points are based on measurements from 2 variants of fleets with different age. In order to precisely indicate effect of “Sharklets” the difference should be divided into 3 elements:

- installation of sharklet itself,
- engine degradations,
- airframe deterioration

where especially engine degradations has close relation to engine age. Based on observed data it is not possible to clearly distinguish the reasons of differences. Gross mass difference of 1 ton during cruise affects the SR by 0.003 Nm/kg, approx. 1.5%. Taking into account average True Air Speed, each extra ton of mass will increase fuel consumption by 30 kgs on both fleets.

7.2 Air Temperature

Air temperature impacts Specific Range in two ways (Fig. 11): first it is related to engine thermodynamic efficiency and secondly it affects True Air Speed for given Mach number.

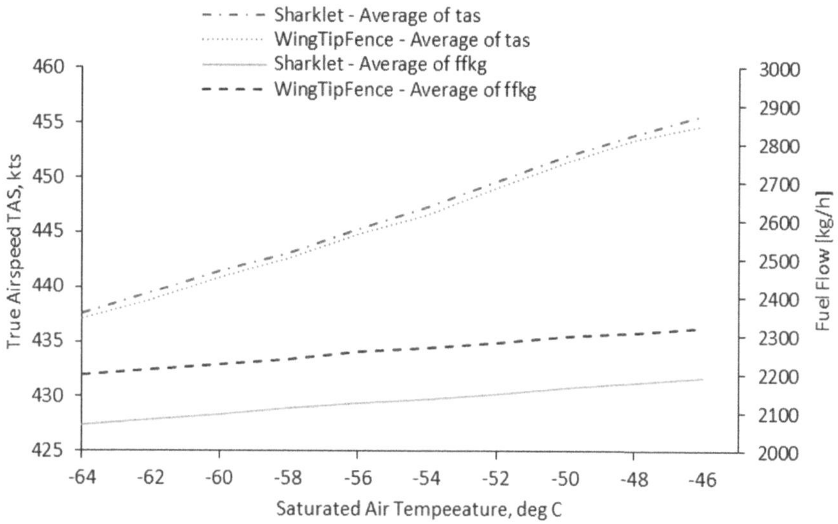

Fig. 11. Effect of air temperature on true airspeed & fuel flow.

The difference in fuel consumption between Sharklet and standard "Wing Tip Fence" aircraft is as described in the point above. The reason why data from Sharklet fleet indicates higher TAS most likely originates from the fact that optimum altitude of Sharklet aircraft is higher and therefore they are more often operated at higher altitude, achieving higher TAS. If data points are filtered only for given altitude, TAS for Sharklet and Wing Tip Fence aircraft are the same. The fuel consumption difference is approx. 65 kg per hour for 10 °C temperature change.

The combined effect of TAS & engine efficiency change in relation to temperature is presented in Fig. 12.

The most optimal area of operation is the one with lower air temperature. Magnitude of engine thermodynamic efficiency is higher than magnitude of TAS change. Total impact of temperature change on SR is approx. 0.9% per 10 °C, which is approx. 18 kg of fuel for constant average distance covered within an hour of flight. Data beyond −64 and −46 was removed due to low population.

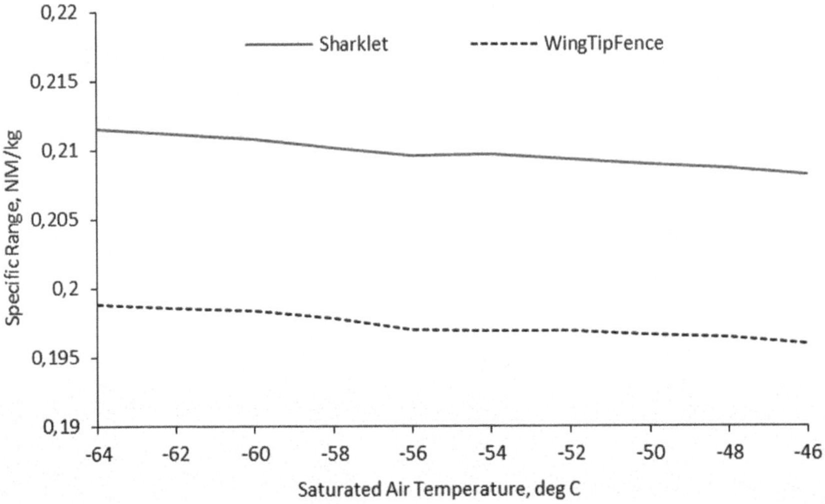

Fig. 12. Relation of SAT & Specific Range.

7.3 Optimal Altitude

Optimal altitude is defined as altitude at which Specific Range is the highest. Optimal altitude considered at given point is a function of Mach, air temperature, gross mass and wind. For the purpose of this study wind is considered as equal to 0. Figure 13 presents average Specific Range for gross mass of 61 tons.

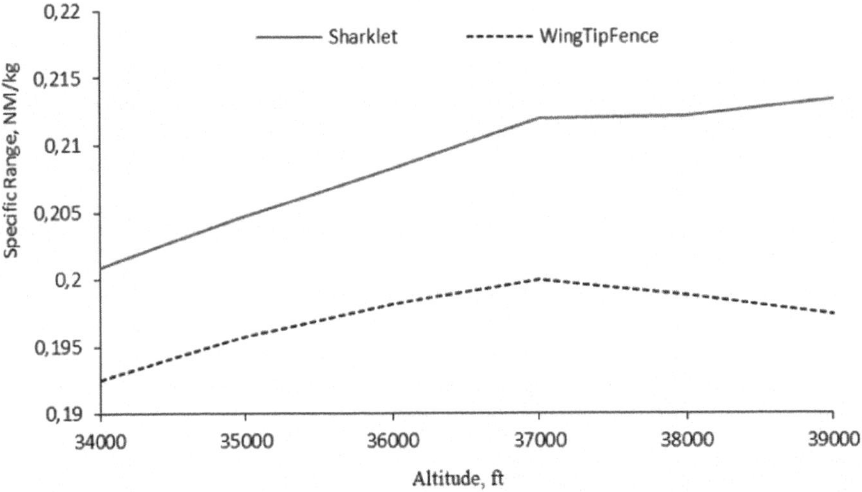

Fig. 13. Relation of altitude and SR, GM 61t, SAT between −50; −60, Mach 0.76; 0.78.

For Sharklet aircraft the optimal altitude is at 39,000 ft and for Wing Tip Fence variant it is 37,000 ft. This benefit can be contributed to Sharklet only. Figure 14 presents the same function for Gross Mass of 63 tons.

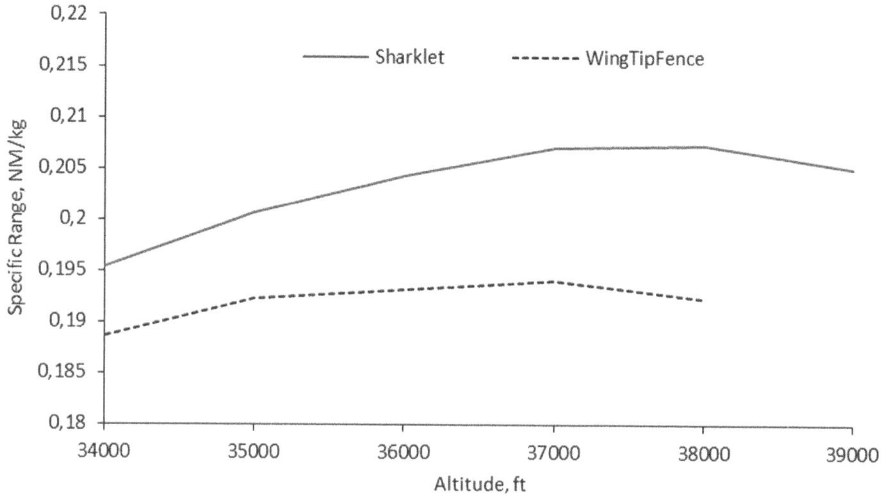

Fig. 14. Relation of altitude, GM 63t, SAT between −50; −60, Mach 0.76–0.78.

It is visible that with higher Gross Mass, the SR values for Sharklet aircraft flattens at 38,000 ft and eventually decreases at 39,000 ft. There were no flights recorded by Wing Tip Fence variant aircraft for this Gross Mass of 39,000 ft.

7.4 Mach Number

Specific Range is highly influenced by Mach number. Figure 15 presents specific range as function of Mach number, for Airbus A320 V2527-A5 with standard Wing Tip Fence at 60, 61, 62 and 63 tons. The temperature range was limited to SAT between −50 °C and −62 °C. Altitude is limited to 36,000 ft (FL360). The total number of samples was almost 10,000, however for Mach number 0.76 there were only 550 samples in total.

SR is optimum at Mach 0.78 and at M 0.77 the SR has a negative fall. This is confirmed for the entire range of considered Gross Masses. Lack of parallelism between the SR lines most probably originates from too small number of data samples. For comparison Fig. 16 is the analogical graph based on airline's performance engineering program.

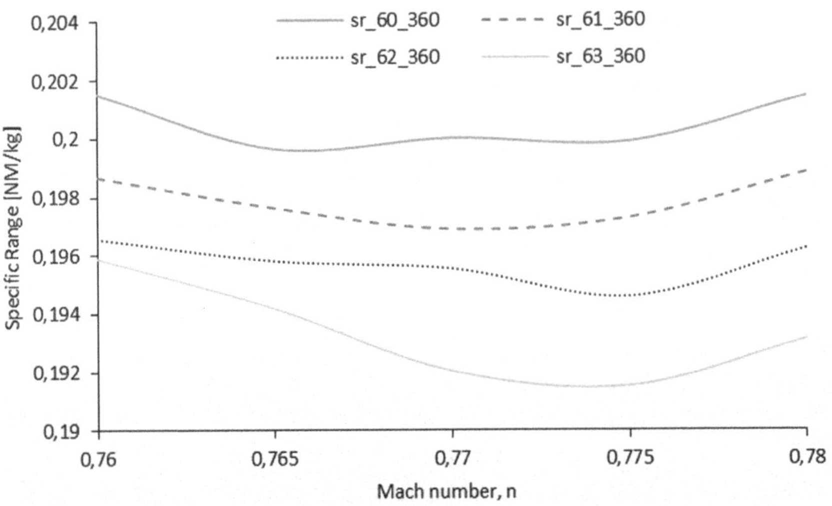

Fig. 15. SR chart FL360, A320 V2527-A5 Wing Tip Fence.

The SR functions are flat and continually decreases towards Mach 0.78. Such difference between the SR values for actual and theoretical data, especially the fact that based on actual data SR increases at Mach 0.78 instead of decreasing, has high consequences on actual optimal Mach number and may contribute to unnecessary fuel consumption penalties.

Figure 17 presents analogical SR chart for Sharklet variant.

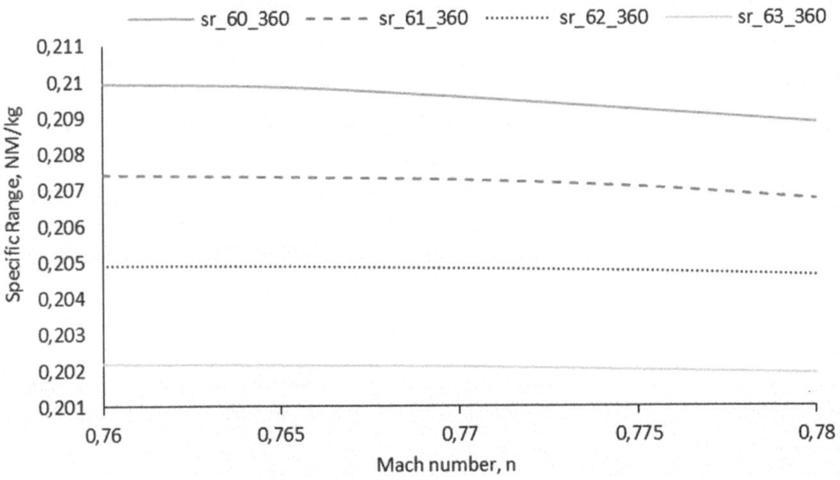

Fig. 16. SR chart FL360, A320 V2527-A5 Wing Tip Fence.

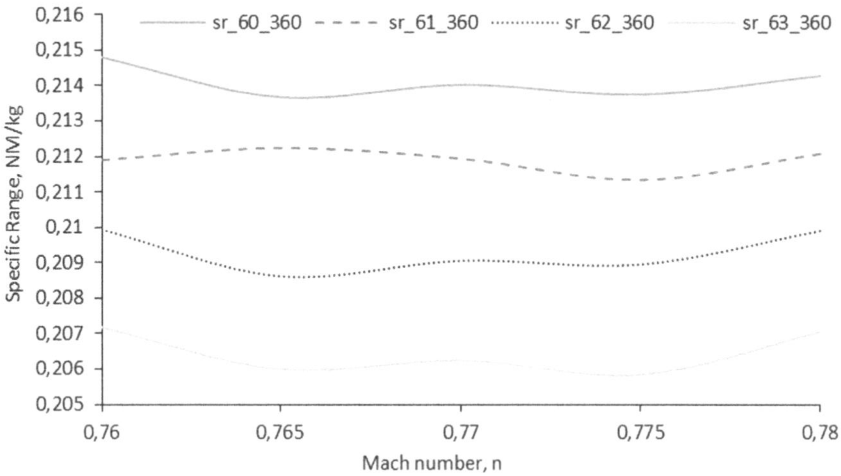

Fig. 17. SR chart FL370, A320 V2527-A5 Sharklet.

Magnitude of differences between SR values for Sharklet aircraft at given Gross Mass is lower than for Wing Tip Fence aircraft. Except Gross Mass of 61 tons it is visible that Mach range between 0.765 and 0.775 has results in lower SR values than Mach 0.76 and Mach 0.78.

8 Conclusion

Cruise performance of aircraft equipped with QAR can be determined. The precision of results will largely depend on the population of recorded flights. Especially with regards to SR as function of Mach number it has been found that points consisting of less than 100 samples resulted in unexpected values. Population of 50,000 flights and more for given aircraft variant is expected to resolve data precision issues.

Proper design of stable segment quality algorithm is a key issue in achieving representative results. Too many parameters may cause that stable segments quality is the highest because of co-relation with other root parameters rather than due to the true high stability. It seems that solution proposed in [6] and which has formulated ground for future developments in the area of Aircraft Performance Monitoring is overly fitted by engine related parameters.

Apart from higher population of flights further improvement in aircraft cruise performance determination should be possible by incorporation of data mining methods [7]. This would allow to select cruise data more freely with less regard to specific weight, altitude, temperature and other parameters, but with higher values of stable segment quality.

References

1. Obert, E.: Aerodynamic Design of Transport Aircraft, pp. 11–12. IOS Press BV, Amsterdam (2009)
2. Senzig, D.A., Fleming, G.G., Iovinelli, R.J.: Modeling of terminal-area airplane fuel consumption. J. Aircraft **46**(4), 1089–1093 (2009)
3. Trani, A., Wing-Ho, F., Schilling, G., Baik, H., Seshadri, A.: A neural network model to estimate aircraft fuel consumption. In: 4th AIAA Aviation Technology, Integration and Operations (ATIO) Forum, pp. 2004–6401. AIAA, Chicago, 20–22 September 2004
4. Campbell, N.: The evolution of flight data analysis. In: ASASI Paper (2007)
5. Hull, G.: Fundamentals of Airplane Flight Mechanics. Springer, Heidelberg (2007)
6. Speyer, J.: Auditing aircraft cruise performance in airline revenue service. In: 7th Performance and Operations Conference, Cancun (1992)
7. Baklacioglu, T.: Modeling the fuel flow-rate of transport aircraft during flight phases using genetic algorithm-optimized neural networks. Aerosp. Sci. Technol. **49**(2016), 52–62 (2016)

Investigation of Newly Developed Microwave Heated Moisture Analyzer Measurements of Ketchup and Milk Samples in Climatic Chamber

Jakub Szałatkiewicz(✉), Mateusz Kalinowski, and Roman Szewczyk

Industrial Research Institute for Automation and Robotics,
Al. Jerozolimskie 202, 02-486 Warsaw, Poland
jszalatkiewicz@piap.pl

Abstract. The article presents evaluation of newly developed microwave moisture analyzer for moisture content measurements, for drying and heating applications. This microwave device operates at 2.45 GHz and consists of waveguide line and resonant chamber optimized with the use of microwave simulation software. The work presents results of tests of developed system, carried out in climatic chamber, allowing air humidity and temperature adjustment in broad range. Experiments were carried out in 20 °C to 40 °C and humidity from 40% to 80%. The study showed that, using developed microwave moisture analyzer, it is possible to dry and measure moisture content in a given sample significantly faster than in traditional methods. This relates also to extreme environmental conditions, in which the system was validated and proved its operating parameters. The results were in line with reference method of determining moisture content in samples.

Keywords: Moisture analyzer · Microwave dryer · FEM · Microwaves · Waveguide · Resonant cavity · 2.45 GHz

1 Introduction

Nowadays development of microwave systems involves modelling energy distribution in order to determine geometry which is best suited for the application. Such modelling is mainly conducted with the use of advanced software. The research presented in this article on microwave moisture analyzer, Elmer [1] software with Vectorial Helmholtz module was used in modelling resonant cavities and waveguides. Its usability was previously confirmed in studies [2].

The aim of the research project was to develop new microwave based moisture analyzer allowing fast and reliable moisture analysis in problematic samples. In samples such as ketchup it is hard to measure moisture with typical moisture analyzers based on heating or IR radiation. Using microwave moisture analyzer, difficult to measure products and samples can be analyzed in shorter time than usually. That was the reason why development of microwave heating and drying application was carried out.

J. Świder et al. (Eds.): MECHATRONICS 2017, AISC 934, pp. 377–383, 2019.
https://doi.org/10.1007/978-3-030-15857-6_37

In such market targeted research, the usability for end user of the new device needs to be taken into account – this factor had impact on choosing diameter of cavity and its construction. Operating frequency was chosen to be 2.45 GHz, as microwave equipment e.g. magnetrons, waveguides etc. for this frequency are widely available on market. Also 5.8 GHz was considered as an alternative, but it was abandoned due to lack of availability and higher costs. Furthermore in 5.8 GHz frequency, modelling revealed that there are issues relating to impractical multimode setup at diameters longer than wavelength in resonant cavities.

Presented 3D geometry on Fig. 1 visualizes internal volume of waveguides and resonant cavity without surrounding walls.

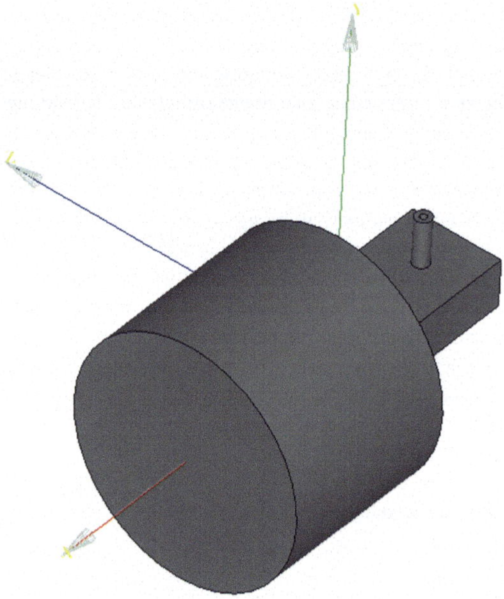

Fig. 1. View of exemplary 3D model of cavity used for modeling.

Different cavity parameters – diameter and height - were investigated. Input power in every case was the same. Basing on data-analysis of simulation results, the magnitude of electric field in given space was calculated. Highest value of electric field for every cavity was studied and, basing on those results, cavities were chosen for further investigation. Visualization of magnitude of electric fields was carried out with Paraview [3]. Exemplary visualized simulation results are presented in Figs. 2 and 3.

Figure 2 presents electric field distribution in cavity with preferred dimensions, but not optimal electric field distribution. Highest magnitude is located near walls or in waveguide entrance, and these maxima are concentrated on small areas.

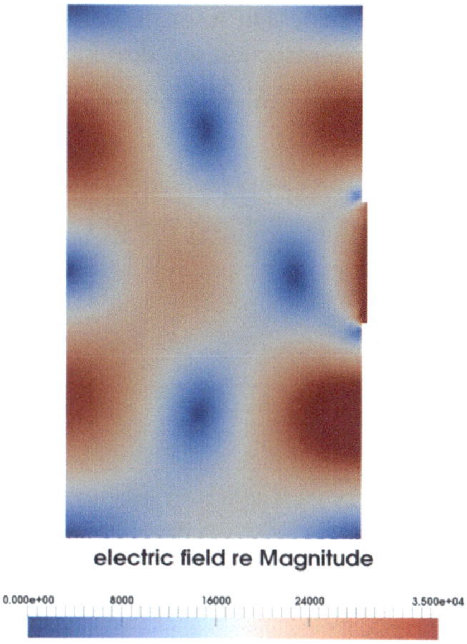

Fig. 2. Example of cavity with appropriate dimensions and high electric field magnitude (V/m), but not suitable electric field distribution. Section view on symmetry plane.

Figure 3 presents cavity with similar dimensions and good electric field distribution. It is stretched and placed in correct distance from the walls. However, electric field magnitude is not as high as in cavities.

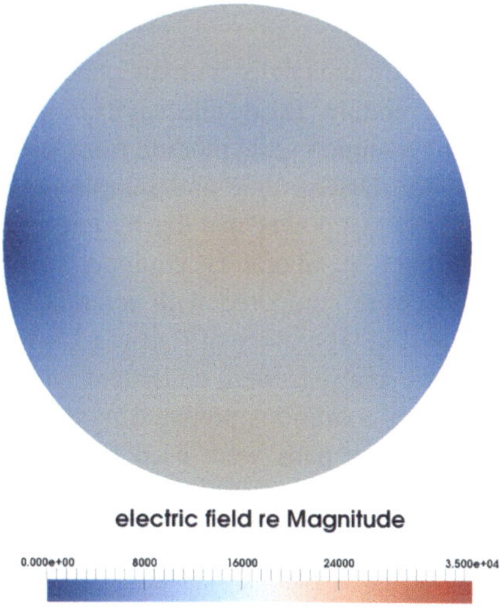

Fig. 3. Example of cavity with appropriate dimensions and suitable electric field distribution (V/m), but low electric field magnitude.

Basing on modeling of waveguide and resonant chamber, appropriate cavity and waveguide were chosen, then the construction was developed and machined for testing. Microwave humidity analyzer which was developed and used for tests is shown in Fig. 4.

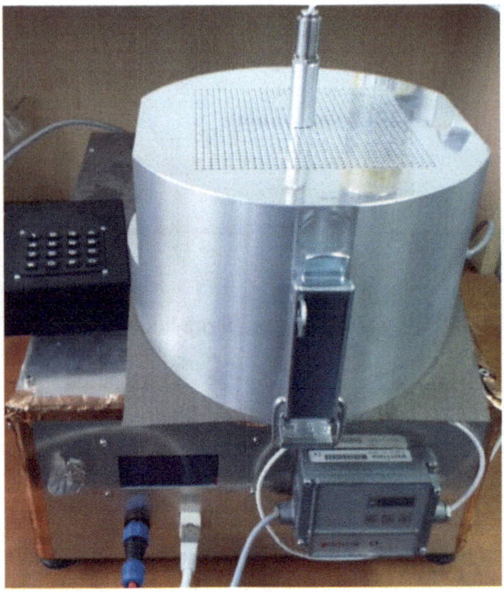

Fig. 4. Photography of tested design of moisture analyzer.

2 Materials and Methods

The test stand consisted of microwave cavity with attached waveguide and magnetron, power supply unit and scale module. Inside the cavity a weighing pan was placed, which was connected to high resolution scale module from Radwag. The test stand was placed inside climatic chamber. Drying time and sample mass were measured using installed scale interface. Microwave power was set by remote keypad.

Testing took place in PIAP, in laboratory equipped with climatic chamber, on chosen products. The results were compared with reference method of determining moisture content, which is used in laboratory of Radwag Company. The results for the developed microwave moisture analyzer were consistent with reference method, even in non-typical difficult operating conditions prepared in climatic chamber.

For best results two glass fiber pads were used for each trial. Every time scale system was tarred with new pads to avoid errors. Each sample was homogenized and placed between these pads. Then pads with evenly distributed material were placed on weighing pan inside microwave cavity. Climatic chamber was closed and environmental conditions were set. Then measurement began: at the same time microwave power was switched on and moisture analysis program was started. The measurement ended when mass change was less than 1 mg for 10 s.

Reference method used in Radwag Company:

1. Stir sample.
2. Weigh the necessary amount.
3. Spread sample evenly.
4. Set temperature to 110 °C.
5. Dry for 2.5 h.
6. Measure mass with accuracy of 0.1 mg.

Mass of each sample was about 1 g. Adequate power for this setup was determined to be 600 W. The results presented in this article were obtained using samples of food from the same package.

3 Results

For the experiment, ketchup and milk were chosen [4]. Every product was tested in 6 different environmental conditions. Air temperature was set to 20 °C and 40 °C, and humidity to ca. 40%, 60% and 80%. Measurement results for ketchup are shown in Table 1 and milk results are in Table 2.

Table 1. Moisture content of ketchup dried with developed moisture analyzer in climate chamber.

Start mas	Dry mass	Moisture content	Air temperature	Air humidity
g	g	%	°C	%
0.958	0.232	**75.78**	20	72
0.954	0.248	**74.00**	20	61
1.0008	0.2585	**74.17**	20	40
1.132	0.293	**74.12**	40	78
1.008	0.261	**74.11**	40	58
0.8721	0.2231	**74.42**	40	36

Table 2. Moisture content of milk dried with developed moisture analyzer in climate chamber.

Start mas	Dry mass	Moisture content	Air temperature	Air humidity
g	g	%	°C	%
1.058	0.128	**87.90**	20	72
0.97	0.123	**87.32**	20	58
1.2895	0.1705	**86.78**	20	44
1.043	0.133	**87.25**	40	72
1.195	0.123	**89.71**	40	57
1.085	0.1172	**89.20**	40	42

Both product samples used in tests were also checked in Radwag [5] laboratory accordingly to their reference method. The results from reference method are presented in Table 3.

Table 3. Moisture content of ketchup and milk obtained with reference method.

Product	Moisture content %
Ketchup	**73.69**
Milk	**88.02**

Drying time for the presented samples were between 1:30 and 3:00 min. Overall ketchup was drying faster than milk. This can be attributed to lower moisture content in ketchup.

For every sample moisture analysis there are two phases of drying process. During the first one, which takes usually less than 60 s, water quickly evaporates. In the next phase evaporation of water significantly slows down, and final drying in some cases can take up to 10 min. This phenomenon can be linked to two factors. Water content in sample drops which slows receiving energy. Also air inside cavity becomes saturated with steam, which slows down evaporation of remaining water. Figure 5 shows exemplary plot with weight loss by evaporation during one of moisture content analysis. Mass is on vertical axis whereas time is on horizontal axis.

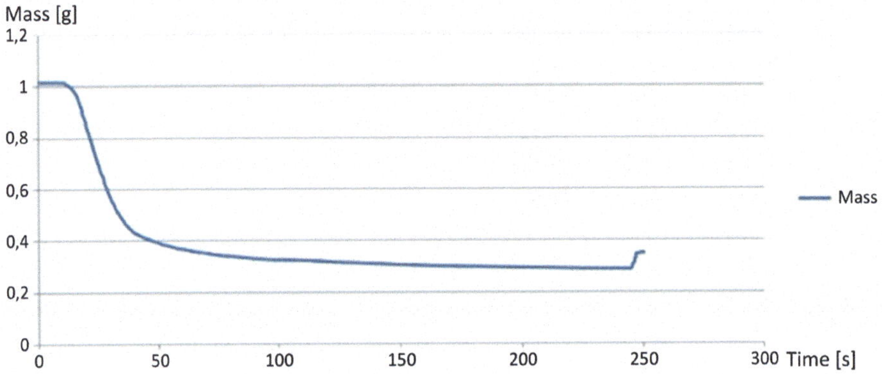

Fig. 5. Plot of sample mass during drying process.

4 Discussion

Results obtained in the presented research are consistent across different experiment conditions set by climate chamber, although slightly different from reference results. Generally tests indicated lower moisture content in milk comparing to the reference method. For ketchup it was opposite - the moisture content result was slightly higher than for the reference analysis. However, the presented results from microwave method need to be analyzed also for possible uneven and local burns of sample during experiments. Also some difficulties occurred during microwave drying tests in climate chamber where the whole test stand was placed and operated. For example few ketchup samples were overheated and then slightly burnt up. This was caused because power was too high and there was no method for temperature and sample condition monitoring during tests. To solve this issue, the use of pyrometer to measure temperature of the dried sample and modification of control system is required.

To further shorten drying times, better ventilation should be considered but the drying time of 3 min is satisfactory and several times shorter than for the other methods.

The research presented in this article, in scope of operation of microwave moisture analyzer in climate chamber, in humidity conditions from 40% to 80%, and air temperature from 20 °C to 40 °C proves the advantage of this approach over the other types of moisture analyzers. This is achieved because the energy is being transferred by microwaves directly into the water contained in dried sample, not as it is in other types of dryers, to the surface of sample.

That is why the developed microwave moisture analyzer system allows fast and accurate moisture content measurement based on water evaporation by microwave heating even in such extreme conditions as 40 °C and 80% humidity.

References

1. ELMER. https://www.csc.fi/web/elmer. Accessed 07 Mar 2017
2. Szałatkiewicz, J., Szewczyk, R., Budny, E., Kalinowski, M., Kataja, J., Råback, P., Ruokolainen, J.: Advances in FEM based modeling of waveguide and waveguide systems for microwave applications, using newly developed open source software. In: Recent Advances in Systems, Control and Information Technology. Advances in Intelligent Systems and Computing, vol. 543, pp. 172–186. Springer, Heidelberg (2016). https://doi.org/10.1007/978-3-319-48923-0_22
3. ParaView. http://www.paraview.org/. Accessed 07 Mar 2017
4. Krokida, M., Maroulls, Z.: Effect of microwave drying on some quality properties of dehydrated products. Drying Technol. **17**, 449–466 (1999)
5. RADWAG Company. www.radwag.com. Accessed 07 Mar 2017

Computer Aided Planning of Adept Six-300 Robot Trajectories

Tadeusz Szkodny(✉)

Institute of Automatic Control, Silesian University of Technology, Akademicka 16 Street, 44-100 Gliwice, Poland
Tadeusz.Szkodny@polsl.pl

Abstract. This work presents the robot motion planning Adept Six-300 using the program *Tobject* calculates coordinates of position and orientation of the manipulation object, observed by the two cameras Edimax IC-7100P. The method of determining the geometric parameters occurring in the equations of the kinematics of the gripper and manipulation object was presented. In order to geometric calibrate these cameras original algorithm that does not require calculation of the projection matrix or fundamental matrix was used. Thanks to this used algorithm is faster and more accurate than methods with these arrays. The input of these algorithms are characteristic points coordinates of the manipulation object, obtained by these two cameras. On the basis of these coordinates the program *Tobject* calculates the trajectory of: (a) approaching, (b) capturing, (c) moving the observed object to a predefined point, and (d) returning the gripper to the starting position. In discretization point of the trajectory this program calculates coordinates the joint variables of the manipulator. In the calculations multiple solutions of inverse kinematics are taken into account, including solutions in singular configurations. Among these multiple solutions which minimizes the square indicator of quality is chosen. The program *Tobject* is installed in external computer constituting a server that cooperates with cameras and robot controller, by means of the Ethernet.

Keywords: Intelligent robots · Robot vision systems · Robots

1 Introduction

Calculations of the position and orientation of the manipulation object based on the video information from the cameras enable automatic approach to this object and grasp it. Computer programs realizing such calculations are one of the basic components of computer intelligence of robots [1].

In this paper of such program called *Tobject* results are presented. This program was tested on a robot Adept Six-300 laboratory station. The robot is equipped with a gripper WSG-50 with controlled grip force. The external computer *SerwerView Suite*, two cameras *Edimax IC-7100P*, the controller *CX SN 2000-00212*, the manipulator *Adept Six-300* and worktable constitute this laboratory station. The external computer constitutes a server executing the master control and is connected with cameras and

J. Świder et al. (Eds.): MECHATRONICS 2017, AISC 934, pp. 384–394, 2019.
https://doi.org/10.1007/978-3-030-15857-6_38

robot controller via Ethernet (see Fig. 1). In the server among others programs including *Tobject* and *Adept Desk Top* are installed.

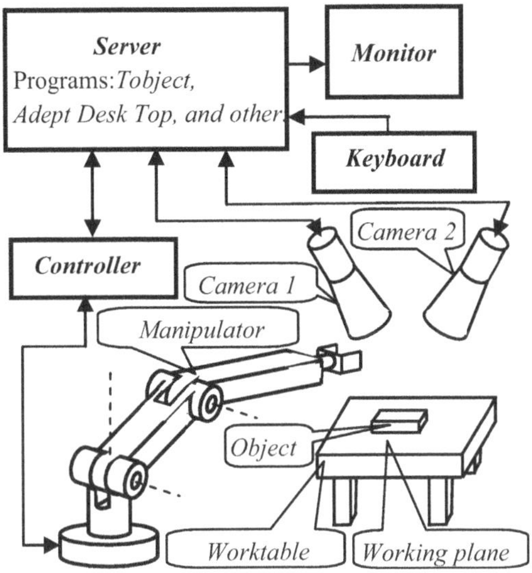

Fig. 1. The laboratory station.

The program *Tobject* geometric calibrates the *Camera1* and *Camera2*, and calculates the desired trajectory of the gripper to the object observed by the cameras. The geometric calibration method [2] used here is faster and more accurate than in [3–5]. In these works, to geometric calibration of cameras are used geometric projection matrix **P** or fundamental matrix **F**. For the calculation of the matrices **P** and **F** the iterative minimally-square method is applied, which requires a large number of calculations and it is not always stable numerically. In the program *Tobject* were used the algorithms for geometric calibration cameras [2], in which were used an iterative calculation of the linearized system 9 equations with 9 unknowns. In these calculations the starting point is determined from roughly measured position coordinates of the cameras. Thanks to that the calculations are fast and accurate.

The program *Adept Desk Top* allows sending the calculated required trajectory coordinates (computed in the program *Tobject*) to the controller realizing the slave control of robot. Furthermore, the program *Adept Desk Top* allows observation of controller records on a monitor. This makes it possible to read the coordinate values of joint variables and Cartesian coordinates of selected manipulator points.

The second section presents the characteristics of the program *Tobject*. Determination of geometrical parameters appearing in the kinematic equations of the station and gripper are given in the third section. The fourth section describes the calculation of the sample desired trajectory of the gripper and its realization. The summary of this work is the fifth section.

2 The Characteristics of the Program *Tobject*

The program *Tobject* consists of many sub-programs. The main window of the program shows Fig. 2. From the window, one can run different subprograms.

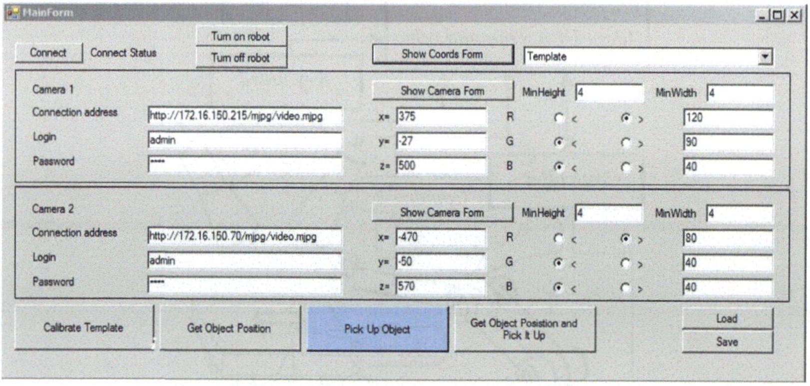

Fig. 2. The main window of the program *Tobject.*

After turning on the *Server* you must run the program *Adept Desk Top* by clicking the mouse on the icon of the program. The window connecting process this program with the robot controller opens. The window setting the speed servo manipulator opens. Next we turn on the program *Tobject* and the main window as in Fig. 2 appears on the monitor.

The buttons *Show Camera Form* for *Camera1* and *Camera2* allow observation of template image from these cameras and the appropriate setting of the cameras positions. In addition, these buttons allow set the setting template image and color filters thresholds RGB appropriately.

Pressing the button *Show Coords Form* allows among other things to display matrix $\mathbf{T}_{c1}$ and $\mathbf{T}_{c2}$.

When you press the button *Get Object Position* on the monitor window appears asking you to put the manipulation object in the working plane. When you place an object on the working surface in this window appears images of colored markers of the object. Then the images are processed and in the end matrix $\mathbf{T}_{ob}$ is calculated, describing the location and orientation of the manipulation object in the station frame. After these calculations appears on the monitor calculated matrix $\mathbf{T}_{ob}$.

When earlier the power of robot was turned off, we have to turn them. Having turned the power by pressing the appropriate button in window of the program *Adept Desk Top* we calibrate the robot.

After calibration the robot, we press the button *Connect*. This connects the program *Tobject* with controller of the robot and the description to the right to of the button *Connect* from *Connect Status* on the *Connected* changes.

Pressing the button *Pick Up Object* causes approach of the gripper to the observed manipulation object, grasping and raising on the height defined in the program *Tobject*.

When you press the button *Get Object Position and Pick It Up* appears window, where the coordinates of the target point of the raised manipulation object you can define. Then the robot moves the manipulation object to a defined target point and returns to the starting position.

The robot can also be turned on or off using the corresponding buttons *Turn on robot* or *Turn off robot*.

The button *Load* is used to load thresholds RGB filters from a previously saved file. The saving the current thresholds color filters into file allows the button *Save*.

3 The Determination of the Geometric Parameters

The program *Tobject* described in Sect. 2 calculates the position and orientation of the frame $x_{ob}y_{ob}z_{ob}$, associated with the object manipulation (observed by 2 cameras), in the station frame xyz. The result of this calculations is the matrix $\mathbf{T}_{ob}$. To calculate the coordinates of the desired joint variables $\Theta_{1d}(s) \div \Theta_{6d}(s)$ of robot a matrix $\mathbf{T}_{6d}(s)$ is needed. The parameter s is the length of the path of the gripper, calculated from the start point S. We assume that the gripper on the closer to the object, before grasping, is located in point F. At the start point S the parameter $s = 0$, at the point F $s = s_F$. With the gripper is associated a frame $x_7y_7z_7$. The frame $x_7y_7z_7$ relative to the frame $x_{ob}y_{ob}z_{ob}$, associated with the object observed by the cameras matrix ${}^{ob}\mathbf{T}_{7d}(s)$ describes. The Eq. (1) (Eq. (6) in [6]) the matrix $\mathbf{T}_{6d}(s)$ describes.

$$\mathbf{T}_{6d}(s) = \mathbf{T}_s\mathbf{T}_{ob}{}^{ob}\mathbf{T}_{7d}(s)\mathbf{E}^{-1}. \tag{1}$$

The matrix $\mathbf{T}_s$ describes the station frame xyz in the robot base frame $x_0y_0z_0$. The matrix $\mathbf{E}$ describes the frame $x_7y_7z_7$ in the frame $x_6y_6z_6$, associated with the sixth link of manipulator (working link).

At this point F the gripper should get closer to the manipulation object in order to grasp it well. At the point the position and orientation of the object frame $x_{ob}y_{ob}z_{ob}$ relative to gripper frame $x_7y_7z_7$ describes a matrix $\mathbf{E}_\mu$ [6, 7]. Thus we can write the Eq. (2)

$${}^{ob}\mathbf{T}_{7dF} = {}^{ob}\mathbf{T}_{7d}(s_F) = \mathbf{E}_\mu^{-1}. \tag{2}$$

From the Eqs. (1) and (2) it results that to the description of the matrix $\mathbf{T}_{6d}(s)$ we need to know the matrices: $\mathbf{T}_s$, ${}^{ob}\mathbf{T}_{7d}(s)$, $\mathbf{E}$ and $\mathbf{E}_\mu$. The matrix $\mathbf{T}_{ob}$ we get from the program *Tobject*, the matrices ${}^{ob}\mathbf{T}_{7d}(s)$ and $\mathbf{E}_\mu$ we define. While matrices $\mathbf{T}_s$, and $\mathbf{E}$ should be determined for a given laboratory station.

3.1 Parameters of the Station Frame

On a laboratory station the base frame $x_0y_0z_0$ and station frame xyz are associated with the working plane. On working plane are planes x_0y_0 and xy, and axes z_0 and z are directed perpendicularly to the top, as in Fig. 3.

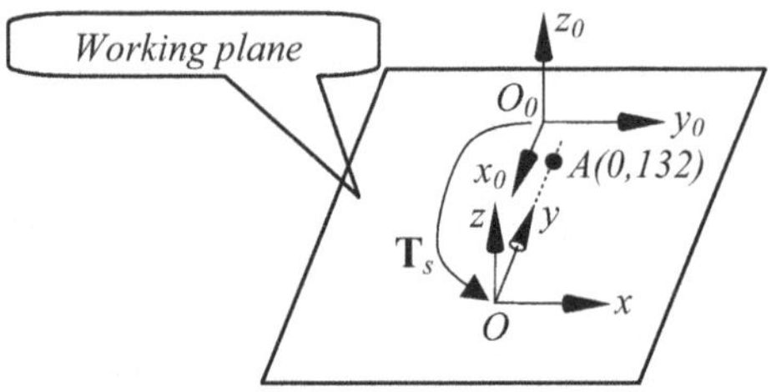

Fig. 3. The base frame $x_0y_0z_0$ and station frame xyz on the working plane. A, O – measurement points.

The gripper frame $x_7y_7z_7$ is illustrated in Fig. 4a. The gripper fingers are moving along the axis x_7. The origin O_7 is at the center of closed fingers.

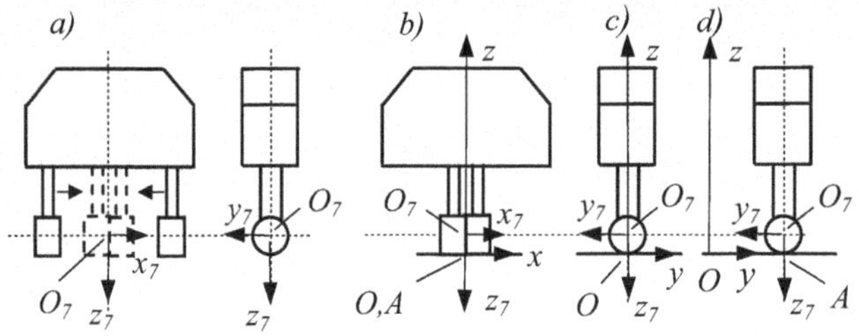

Fig. 4. The gripper and station frames; (a) the gripper frame, (b) the gripper frame on measurement points O and A, (c) on the point O, (d) on the point A.

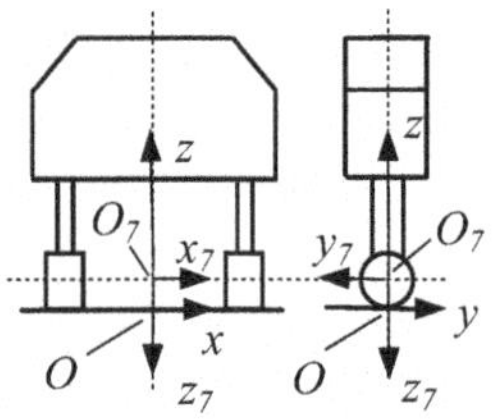

Fig. 5. The gripper on the axis x.

Before determining the matrix $\mathbf{T}_s$ we set the template on the station frame xyz in such a way that x-axis is approximately parallel to the axis y_0 and y axis is opposite to the axis x_0. Then, using the robot controller (program *Adept Desk Top*) we set the gripper on the measuring point O so that the axis z_7 is directed vertically down, and the axis x_7 parallel to the x axis. We set the gripper with open fingers so that the plane y_7z_7 passed through point O and the plane x_7z_7 is directed through the axis x (see Fig. 5). Next we passes the gripper up slightly, with the unchanged orientation. We close the fingers and passes the gripper to the point O as in Fig. 4b, c. Thus we read from the robot controller position coordinates of the beginning O_5 of the fifth link frame in the base frame. By $\vec{r}_5(O)$ is denoted vector describing this point. Then we move the gripper to point A, with constant orientation (see Fig. 4b, d). We read from the robot controller coordinates of the vector $\vec{r}_5(A)$ describing the position of the point O_5, also in a base frame. If the coordinates x_0 of vector $\vec{r}_5(A)$ and $\vec{r}_5(O)$ are different, we correct template setting so that this difference disappears. Thanks to this correction axes x and y_0 are parallel, and axis y is opposite to axis x_0. After the setting of the template we must correct the orientation of the gripper by setting it on the x-axis, as in Fig. 5. Next we raise the gripper slightly above the surface xy, close fingers and we set the gripper on the point O, as shown in Fig. 4b, c.

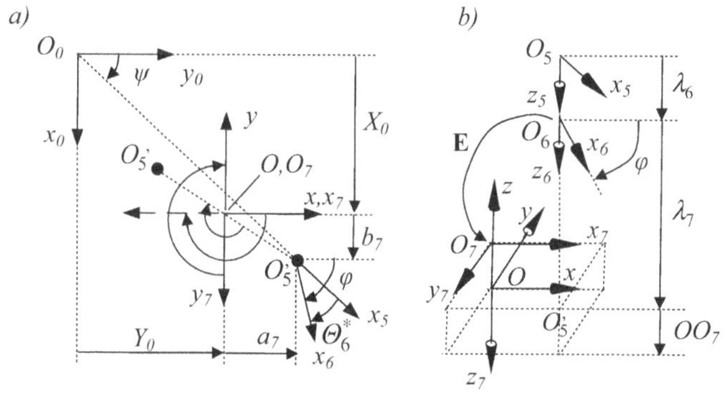

Fig. 6. The gripper set on point O as in Fig. 4 (a) top view, (b) side view.

The gripper is attached to the manipulator link 6-*th* so that the axes of z_6 and z_7 are parallel. The axes z_5 and z_6 are also parallel [8], thus axes z_5, z_6 and z_7 are parallel. Therefore, after setting the gripper at the measuring point O as shown in Fig. 4b, c axes z_5, z_6 and z_7 are directed the down, as shown in Fig. 6b. From Fig. 6 result the following dependences (3) on the coordinates of the point O_5 above the point O_5', in the base frame $x_0y_0z_0$:

$$x_{0O5'} = X_0 + b_7, y_{0O5'} = Y_0 + a_7, z_{0O5'} = \lambda_6 + \lambda_7 + OO_7. \tag{3}$$

After raising, rotate the gripper by 180° around the axis z_7 and set the gripper on the measuring point O, a vertical projections of axis z5 and z6 will be at the point O_5'' (see Fig. 6a). Then the coordinates of the point O_5 above the point O_5'' are as follows:

$$x_{0O5''} = X_0 - b_7, y_{0O5''} = Y_0 - a_7, z_{0O5''} \approx z_{0O5'}. \tag{4}$$

For the gripper WSG-50 OO_7 = 10 mm, and the parameter λ_6 = 90 mm [8]. From the Eqs. (3) and (4) we obtain the coordinates X_0, Y_0 and Z_0.

$$X_0 = 0.5(x_{OO5'} + x_{OO5''}), Y_0 = 0.5(y_{OO5'} + y_{OO5''}), Z_{OO5'} = 0.5(z_{OO5'} + z_{OO5''}). \quad (5)$$

Coordinates $x_{OO5'}$, $y_{OO5'}$ and $z_{OO5'}$ describe the position of the point O_5 above the point O_5' and are read from the robot controller (program Adept Desk Top). Similarly we read coordinates $x_{OO5''}$, $y_{OO5''}$ and $z_{OO5''}$ of the point O_5 above the point O_5''. From Fig. 6 and the Eq. (5) we obtain the matrix Ts, describing station frame xyz in a base frame x0y0z0.

$$\mathbf{T}_s = \begin{bmatrix} 0 & -1 & 0 & X_0 \\ 1 & 0 & 0 & Y_0 \\ 0 & 0 & 1 & 0 \\ 0 & 0 & 0 & 1 \end{bmatrix}. \quad (6)$$

3.2 Parameters of the Gripper

The matrix **E** describes the gripper frame $x_7y_7z_7$ in the working link frame $x_6y_6z_6$. From Fig. 6 results the transformation (7) of the frame $x_6y_6z_6$ to the frame $x_7y_7z_7$.

$$\mathbf{E} = Rot(z, -\varphi)Trans(-a_7, -b_7, \lambda_7), \varphi = \psi + \Theta_6^*. \quad (7)$$

Parameters a_7, b_7 and λ_7 can be determined from Eqs. (3) and (4).

$$a_7 = 0.5(y_{OO5'} - y_{OO5''}), b_7 = 0.5(x_{OO5'} - x_{OO5''}), \lambda_7 = Z_0 - \lambda_6 - OO_7). \quad (8)$$

The angle Θ_6^* can be read from the controller, and the angle ψ results from Fig. 6a.

$$\psi = atan2(x_{OO5'}, y_{OO5'}). \quad (9)$$

4 The Calculations of an Exemplary Desired Trajectory and Its Realization

After turning on the server and pressing the button *Calibrate Templates* (Fig. 2) the program *Tobject* has done geometric calibration of two cameras, based on images of 24 template points [6]. The result of this calibration are matrices $\mathbf{T}_{c1}$ and $\mathbf{T}_{c2}$, describing the location and orientation of the *Camera1* frame and *Camera2* frame respectively, in the station frame xyz. After pressing the button *Show Coords Form* (Fig. 2) and selecting the appropriate section calculated matrices appeared on the monitor.

4.1 The Position and Orientation of the Manipulation Object

After calibration of the cameras the manipulation object was located on the working plane (plane xy). The object is a plate with a handle as in Fig. 7. Four colored markers in the form of a circles with a diameter of 15 mm are situated on the surface the plate.

The circles yellow, green, pink and blue are markers of the characteristic point *P1*, *P2*, *P3* and *P4* respectively. Characteristic points are the centers of the corresponding circles respectively. Figure 7 shows the frame $x_{ob}y_{ob}z_{ob}$ associated with the plate.

After pressing the button *Get Object Position* (Fig. 2) the program *Tobject* retrieved images of the object from the two cameras and the calculated matrix $\mathbf{T}_{ob}$, based on the coordinates of the characteristic points $P1 \div P4$. After these calculations the matrix $\mathbf{T}_{ob}$ appeared on the monitor, as shown below.

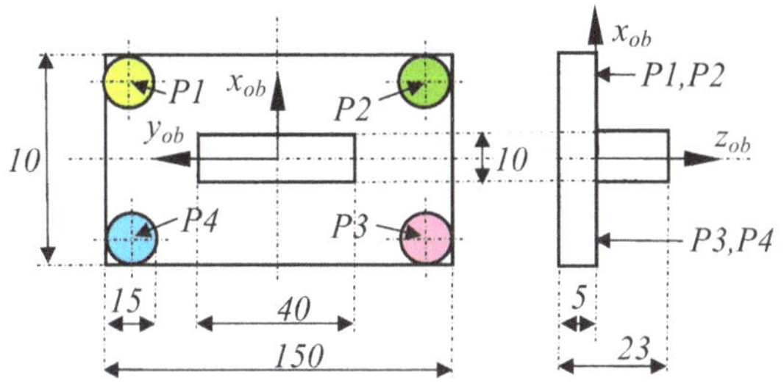

Fig. 7. The manipulation object.

$$\mathbf{T}_{ob} = \begin{bmatrix} 0.7012 & 0.7077 & -0.0000 & -1.9279 \\ -0.7130 & 0.7065 & 0.0000 & -15.0623 \\ 0.0001 & 0.0001 & 1.0000 & 4.8557 \\ 0 & 0 & 0 & 1 \end{bmatrix}. \tag{10}$$

4.2 The Exemplary Desired Trajectory of the Manipulator

We assumed that the desired trajectory of the gripper to an object is a straight line which coincides with the axis z_{ob} [6]. The gripper starts to move from the point *S* and ends at point *F* above manipulation object, with the orientation of the frame $x_7y_7z_7$ as in Fig. 8.

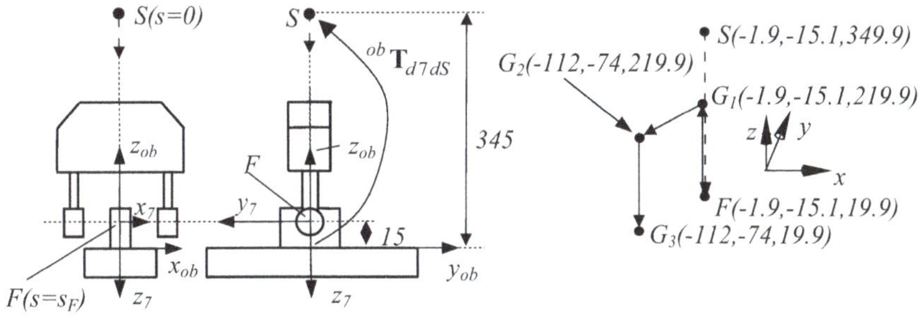

Fig. 8. The desired trajectory of the gripper. Points are described by the coordinates of the frame *xyz,* in mm

It is assumed that the point S is at a height of 345 mm above the plane $x_{ob}y_{ob}$. The straight lines FG_1, G_1G_2, and G_2G_3 are desired trajectory after grasping the manipulation object. Point G_1 lies on the straight line FS. We assume the length of section FG_1 equals 20 cm. The coordinates x and y of point G_3 are defined by user. The other coordinates of points $G_1 \div G_3$ are calculated by program *Tobject*.

The matrix $\mathbf{E}_\mu$ describes the object frame $x_{ob}y_{ob}z_{ob}$ in the station frame $x_7y_7z_7$ at the point F, for which $s = s_F = 330$ mm. Thus, from Fig. 8 results $\mathbf{E}_\mu$ as in the Eq. (11). The desired gripper trajectory SF corresponds to the desired working link trajectory, described by the matrix $\mathbf{T}_{6d}(s)$, calculated from the Eq. (1). For the calculation of the matrix $\mathbf{T}_{6d}(s)$ we need to determine earlier matrices $\mathbf{T}_s$ and $\mathbf{E}$. After setting the gripper on the measuring point as in Fig. 6, coordinates appearing in the Eq. (5) are read from the robot controller. $x_{OO5'} = 400.002$ mm, $y_{OO5'} = 6.996$ mm, $z_{OO5'} = 246.987$ mm; $x_{OO5''} = 487.982$ mm, $y_{OO5''} = 7.002$ mm, $z_{OO5''} = 246.992$ mm. From the robot controller the angle $\Theta_6^* = -42.001°$ was read. For these coordinates we get from Eqs. (6)–(9) matrices as below.

$$\mathbf{E}_\mu = \begin{bmatrix} 1 & 0 & 0 & 0 \\ 0 & -1 & 0 & 0 \\ 0 & 0 & -1 & 15 \\ 0 & 0 & 0 & 1 \end{bmatrix}, \mathbf{T}_s = \begin{bmatrix} 0 & -1 & 0 & 443.992 \\ 1 & 0 & 0 & 6.999 \\ 0 & 0 & 1 & 0 \\ 0 & 0 & 0 & 1 \end{bmatrix}, \mathbf{E} = \begin{bmatrix} 0.6820 & 0.7313 & 0 & 32.173 \\ -0.7313 & 0.6820 & 0 & 30.001 \\ 0 & 0 & 1 & 146.989 \\ 0 & 0 & 0 & 1 \end{bmatrix}. \tag{11}$$

4.3 The Realization of the Desired Trajectory of the Manipulation Object

After pressing the button *Connect* (Fig. 2), and next *Pick Up Object* the section SF was divided into sections of 1 mm length, next the matrix $\mathbf{T}_{6d}(s)$ and desired joint variables have been calculated at these points of division. Such computed joint variables were sent to the robot controller. After that controller has led the gripper to point S and oriented it as in Fig. 8 ($x_7 \| x_{ob}$, $-y_7 \| y_{ob}$, $-z_7 \| z_{ob}$). Then the gripper was led straight to the point F. After reaching the point F gripper caught the plate observed by both cameras and lifted it up by 20 cm, declared in the program *Tobject* (the point G_1 in Fig. 8).

After pressing the button *Get Object Position and Pick Up It Up* a window appeared, where you can define the coordinates x and y of the point, where the plate was to be located (the point G_3 in the Fig. 8). After entering these coordinates (see Fig. 8) the coordinates of the point G_2 were calculated by this program. Next the matrix $\mathbf{T}_{6d}(s)$ was calculated on the straight sections FG_1, G_1G_2 and G_2G_3. Orientation gripper on those sections of the trajectory is changeless, that is, yeah same as in point F.

These sections were divided in the same way as section SF, and desired joint variables have been calculated at these points of division. For these calculations have been applied algorithm of solving the inverse kinematics presented in [8]. In this algorithm a geometric interpretation of all formulas in closed form are used. These interpretations ensure getting all the possible solutions, including the infinite multiplicity of solutions in singular configurations. Among these various solutions the chosen solution minimizing the square indicator of quality. The algorithms [8] are simple because do not require very complex calculations resulting from equations described by the matrices of size 12×12, requiring the calculation of eigenvalues of the matrix, as in the work [9].

After these calculations, the robot moved the plate from the point G_1 to G_3 through G_2, opened gripper fingers, and returned to the point S.

5 Summary

The experimental studies presented in this paper were conducted using a stepper mode of the program *Tobject*. In this mode, the calibration of the cameras, determination of coordinates of the object, coming to this object, picking up and putting this object in defined place are initiated by pressing corresponding buttons. With this mode it's possible to analyze taking the successive steps.

The experimental studies have confirmed the correctness of algorithms presented in this work. The research shows that the calculation error coordinates of the object is not greater than 0.8 mm.

This program can work in continuous mode. In this mode, after starts of the calibration camera the program automatically tracks the presence of the object. Upon seeing the object itself determines its coordinates, is coming to the object, picks up and puts it in place, defined it before the calibration of cameras. Then the gripper back to the point S and waiting for appearance of the next object. The continuous mode allows external computer (running as a server) to control the movement of the gripper (with the robot controller) to the observed object without human intervention. Therefore program *Tobject* is an essential component of computer intelligence of robot Adept Six-300.

References

1. Siciliano, B., Khatib, O.: Springer Handbook of Robotics, pp. 212–214. Springer, Heidelberg (2008). Part C, Chap. 9.2
2. Szkodny, T., Łęgowski, A.: The decreasing of the 3D position errors in the system of two cameras. In: Lecture Notes in Computer Science, LNAI, vol. 9246, Part III, pp. 285–298. Springer, Cham (2015)
3. Luchowski, L.: Mobile stereovision and second – degree surfaces, Chap. 2. Institute of Theoretical and Applied Informatics Polish Academy of Science, Gliwice, Poland, pp. 11–21 (2012)
4. Tomaka, A.A.: The multimodal analysis of image for non-invasive orthodontic, Chap. 2. Institute of Theoretical and Applied Informatics Polish Academy of Science, Gliwice, Poland, pp. 23–33 (2013). (in Polish)
5. Cyganek, B.: Computer processing three-dimensional images, Chap. 3. Academic Publishing House EXIT, Warsaw, Poland, pp. 34–79 (2002). (in Polish)
6. Szkodny, T.: Application of vision information to planning trajectories of Adept Six-300 robot. In: 2016 International Conference on Advanced Robotics and Mechatronics (ICARM), pp. 88–94. IEEE Explore, Macau (2016)

7. Szkodny, T.: Basics of Robotics, Chap. 3, 2nd ed. Silesian University of Technology, Gliwce, Poland, pp. 56-67 (2012). (in Polish). ISBN 978-83-7335-945-1
8. Szkodny, T., Łęgowski, A.: Planning and realization of singular trajectories of Adept Six-300 robot. In: Lecture Notes in Computer Science, LNAI, vol. 9245, Part II, pp. 309–320. Springer, Cham (2015)
9. Manocha, D., Canny, J.: Efficient inverse kinematics for general 6R manipulators. IEEE Trans. Robot. Autom. (Inst. Electr. Electron. Eng.) **10**(5), 648–657 (1994)

Control of Selected Operational Parameters of the Scraper Conveyor to Improve Its Working Conditions

Jerzy Świder[1(✉)], Krzysztof Herbuś[1], and Kamil Szewerda[2]

[1] Institute of Engineering Processes Automation and Integrated Manufacturing Systems, Faculty of Mechanical Engineering, Silesian University of Technology, Gliwice, Poland
jerzy.swider@polsl.pl
[2] Laboratory of Modelling Methods and Ergonomics, Institute of Mining Technology KOMAG, Gliwice, Poland
kszewerda@komag.eu

Abstract. The concept of controlling operational parameters of the conveyor, utilising the ability to adjust extension of telescopic trough of the conveyor and the ability to adjust the value of supply voltage frequency for drive motors is described. The control algorithm was designed to prevent unfavourable dynamical phenomena and dangerous loosening of the chain during the conveyor's operation. On the basis of the rules describing developed algorithm, value of supply voltage frequency for the drive motor, located both at the return end and at the discharge end is adjusted. Another adjusted parameter is the extension of telescopic trough on the return drive. By adjusting parameters listed here, unfavourable loosening or excessive tension of the scraper chain are eliminated. Moreover, adjusting the value of the supply voltage frequency for drive motors may provide proper cooperation between them, preventing situations when one of the motors, the through scraper chain, loads the other one. In addition, changing frequency of voltage enables adjusting the motor's rotary speed, and at the same time controlling the speed of the chain with the scrapers. Applying the control algorithm will allow to improve conveyor's operation. It will also decrease its failure frequency and its wear and tear, by reducing frictional wear.

Keywords: Mining industry · Chain conveyor · Armoured face conveyors (AFCs) · Numerical simulations

1 Introduction

Armoured face conveyors (AFCs) are the main machines used in longwall complexes in hard coalmines. AFCs are installed along the longwall and their length may reach up to 500 meters. Conveyor's trough is used as the track for the longwall shearers. AFCs cooperate also with powered roof support, securing the roof of longwall face. However, transportation of run-of-mine from the longwall face to beam stage loaders (BSL) and further haulage is the main task of AFC. Due to uneven load on the conveyor, related to

J. Świder et al. (Eds.): MECHATRONICS 2017, AISC 934, pp. 395–405, 2019.
https://doi.org/10.1007/978-3-030-15857-6_39

diversified stream of run-of-mine coming from the longwall shearers, unfavourable dynamic phenomena are observed during its operation. These phenomena intensify with increasing the length of the conveyor when elastic elongation of the scraper chain occurs. Loosening of the scraper chain is unfavourable. Such situation increases the risk of a loosened chain link or scraper getting stuck, which may cause serious failure of the conveyor. Avoiding the loosening of the chain requires adjusting its tension, however too much tension will result in increasing frictional wear. To improve operating conditions of the conveyor, constant control of the chain tension and its drive motors is essential in maintaining specified operating parameters. AFCs currently operating in Polish mines are either not equipped with devices to control operational parameters, or the devices' ability to control is very limited [1, 4, 7]. The concept of control operational parameters of the conveyor has been developed in the recent years. State-of-the-art conveyors are equipped with electric motor frequency converters or with drive motors integrated with frequency converters. It is also possible to use telescopic return end to adjust length of the conveyor. Implementing these solutions allows to adjust the voltage supply frequency for the driving motors and to adjust extension of the telescopic trough on the return end. By adjusting the parameters it is possible to control changes in tension of the scraper chain to minimise the risk of loosening or excessive tensioning. Change of these parameters enables changing tension of the scraper chain to minimise the risk of the chain loosening or its excessive tension, change in proportion of load to driving motors to improve their cooperation as well as change in conveyor speed to adapt it to the current need [1, 4, 5]. The concept of the algorithm to control the supply voltage frequency of the motor on the return end and to control the extension of the return drive for keeping the correct tension of the chain and for and to ensure better cooperation of driving motors is presented. Additionally, a possibility of setting one of three assumed speeds of the chain depending on expected load to driving motors was included. The described method for controlling the operational parameters of the conveyor is based on the signals from the transducers recording chain tension at the point of the return drive and the discharge drive as well as on measurements of the current in driving motors.

2 Computational Model of AFC

Computational model of RYBNIK 850 scraper conveyor was developed to prepare and to test the algorithm for controlling the operational parameters of the conveyor. The model consists of 3 modules: a physical model of the conveyor, a model of driving motors with frequency converters as well as with a model of control system.

The physical model of the conveyor was developed in the environment of the software used for calculation of dynamics and kinematics of multibody systems. The model consists of rigid bodies recreating the conveyor trough, driving drums scrapers and links of the scraper chain. The rigid bodies were connected by geometric constrains, elastic-dumping elements as well as by defining the contact parameters between the selected bodies. Besides, vectors of forces and torques acting during conveyor operation were defined in the physical model. Also models of chain tension transducers at the places, where the chain leaves the driving drum at the return end and the discharge end were included in the conveyor model. These tension transducers enable detecting chain loosening or its excessive tension on both drives.

The model of electric motors was developed in MATLAB/Simulink environment and it consisted of two separate models of asynchronous electric motors of power 315 kW each. Each of the motors was equipped with a frequency converter which enabled control of motor's operation.

Rules describing the developed algorithm for controlling the operational parameters of the conveyor were implemented in the model of the control system. This model was also developed in MATLAB/Simulink environment.

During numerical simulations, the co-simulation technology integrating the software environments for analyses of kinematics and dynamics of multibody systems (MSC Adams) with MATLAB/Simulink programme was used. For the mentioned integration, the signals used in communication between modules of the computational models were defined [2, 3, 5, 6]. The following parameters: the rotational speed of the motor shaft, signals from transducers of the scraper chain tension and the extension of the telescopic trough are the output signals from the physical model of the conveyor. The torque generated by the driving motors, change in extension of the return drive and the force loading upper line of the scraper chain were the input signals to the module of the conveyor physical model. The rotational speed of the motor shafts and the frequency of the supply voltage of each motor were the input signals to the module of the driving motors model. The torque generated by motors and current in each motor were the output signals from the driving motors model. Current recorded in driving motors, the signal from transducers of the scraper chain tension, the frequency of motors supply voltage and the extension of the return drive were the input signals to the model of the control system. The signal for increasing or reducing the extension of the return drive and the signal increasing or reducing the frequency of driving motors supply voltage are the output signals from the model of the control system. The structure of the computational model of AFC with the I/O signals of each module is presented in Fig. 1.

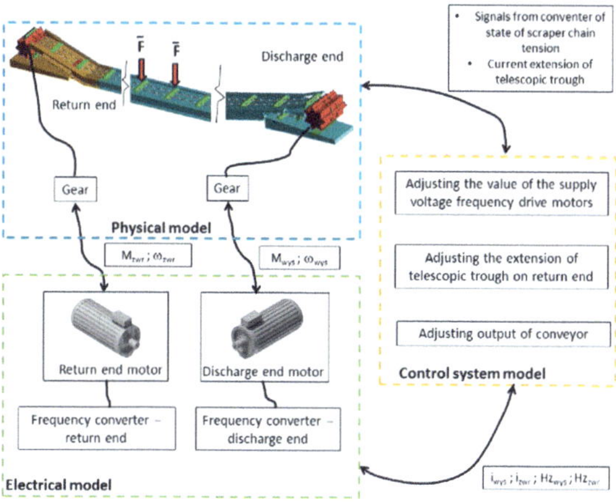

Fig. 1. The structure of the computational model of AFC together with the signals enabling the integration of the modules of the computational model [6]

The developed computational model enabled the numerical simulation, results of which were used in development of the algorithm for control of AFC operational parameters. The model is also used in the numerical simulations aiming at the verification of the proper functioning of the developed algorithm.

3 Algorithm for Controlling the AFC Operation

The idea of the algorithm for controlling the AFC operational parameters is to maintain proper tension of the scraper chain, and especially elimination of the situation of the chain loosening at the place, where it leaves the driving drums at return or discharge end. Ensuring the proper cooperation of electric driving motors i.e. avoiding the situation, when one of the motors loads the other one (brakes it) through the scraper chain is the next aim of the control system. An adjustment of AFC output to the current feeding rate is another effect of controlling the operation parameters. The possibility of changing the frequency of driving motors supply voltage and changing the extension of the telescopic trough at the return drive was used in the control algorithm. The control algorithm of the operational parameters can be divided into two parts – first one (blue dotted line in Fig. 2) is used to control the tension of the scraper chain and to control the distribution of the power between the motors, second one (orange dotted line in Fig. 2) is used to control the speed of the scraper chain to adjust it to the current load need. The block diagram presenting the concept of the control algorithm is presented in Fig. 2.

The scraper chain tension is recorded by transducers installed at the places of leaving the drums of both return and discharge drives. Signals from the chain loosening detection transducer on the return drive (eLZ) and the discharge drive (eLW) as well as from the transducer detecting excessive tension on the discharge end (eNW) are 0–1 signals. The signal from the given transducer means that the scraper was detected within its set range, that is the information about the chain loosening or its excessive tension depending from which the transducer the information is sent. Fourth transducer is installed on the return drive and it measures the scraper pressing force on the conveyor's trough (eFz). The greater pressing the force is, the greater the scraper tension is. In the developed algorithm, the following four levels of force pressing the scarper to the conveyor's trough were assumed: no pressure, small pressure (accepted during operation), high pressure and very high pressure. Signals from all four chain pressure transducers are sent to the control system for further processing. Besides, currents in conveyor driving motors (i_{wys}; i_{zwr}) are recorded and are sent to the control system and further processed. The chain tension on the return drive (SZ) and the discharge drive (SW) is interpreted on the basis of the information from traducers of detecting the chain tension. The ratio of loads of the motors (SP) informing about the correct cooperation between both motors. The proportion is identified on the basis of quotient of currents in both motors. Conveyor operating conditions are identified on the basis of SZ; SW; SP signals according to the rules of the control algorithm by assigning the current condition of the scraper chain on the return drive (SZ) and the discharge drive (SW) as well as the proportion of the load to both motors (SP) to one of previously defined 37 cases of the conveyor operating conditions. Depending on the state of the conveyor operation, basing on the algorithm rules, the signal about decreasing or increasing frequency of the discharge motor's (ΔHz wys 1) and return motor's (ΔHz zwr 1) supply voltage and the extension of the telescopic trough (Δextension) is generated. According to the rules describing the control algorithm, in some specified states of the conveyor's operation, the proportion between frequencies of voltage supplying the motor of the discharge drive and the motor of the return drive can be changed to improve the cooperation of both motors.

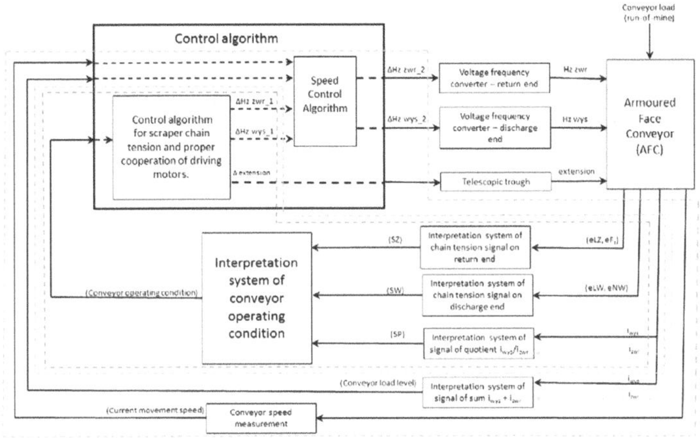

Fig. 2. Block diagram of the algorithm for control of AFC's operational parameters [6]

The current conveyor load level is assessed on the basis of the sum of currents in both driving motors (i_{wys}; i_{zwr}) sent to the control system. An increase of the sum means that the increase of the load to the conveyor by run-of-mine and a decrease of the sum means the decrease of the load. The signal with the information about the current speed of the scraper chain is also delivered to the control system. In the developed algorithm three levels of the conveyor operational speed were assumed: the low speed of the chain, the medium speed of the chain and the high speed. In the case, when there is low level of the conveyor load or when the feeding stream is stopped, the speed of the chain is reduced to the low level. When the load to the conveyor is medium, the speed of the scraper is speeded up to the medium level. When the conveyor is highly loaded the conveyor operation is changed to the high speed of the chain – maximum capacity of the conveyor. The speed of the chain is controlled by change in the rotational speed of driving drums, which is changed by change in the frequency of the driving motors' supply voltage. In the developed algorithm, when the conveyor operates with the low speed of the chain, the frequency of the driving motors supply voltage is 25 Hz, when the speed of the chain is medium, the frequency is 35 Hz and during operation with the high speed of the chain, the frequency is 50 Hz. However, in the developed control system it is possible to establish any number of possible levels of the scraper chain speed. Supply voltage frequencies resulting from the assumed operational speeds are changed if necessary in such a way to keep the same disproportion between frequencies of voltage in the motor of the return drive (ΔHz zwr 2) and the discharge drive (ΔHz wys 2) as well as the distribution of power between the motors if such disproportion results from the first part of the algorithm controlling the chain tension. The detailed description of the part of the algorithm, which is responsible for controlling the AFC output is described in [5]. Both interpretations of the chain tension, of the load to the motors and its proportion as well as operations associated with change operation parameters are made in a loop with a certain time interval. This time interval can differ for different types of conveyor as well as for different types of driving units. That is why the control algorithm should be adjusted individually for each conveyor.

4 Results of Analysis of the Algorithm Use

The developed AFC control algorithm was verified by numerical simulations. During simulations, the conveyor operated in a horizontal position, without the inclination, and it was subjected to a variable load of the upper chain. The character of the load is shown in Fig. 3.

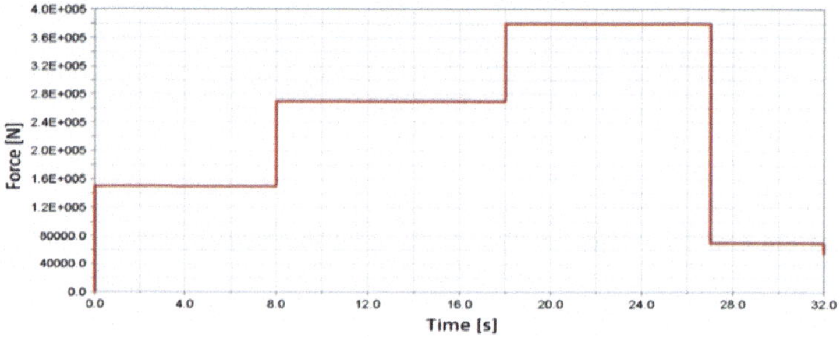

Fig. 3. The character of the load to the upper chain [6]

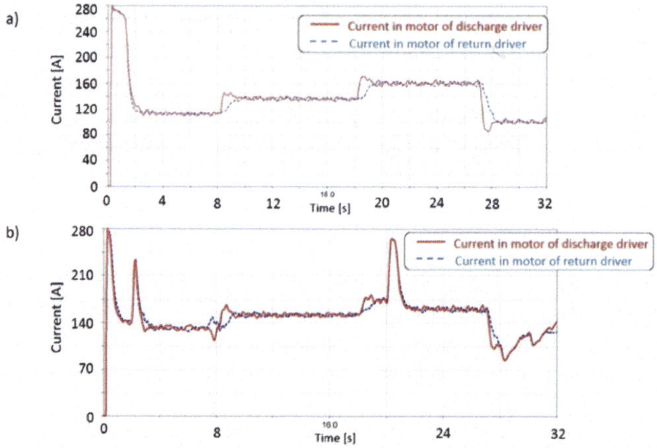

Fig. 4. Currents in motors recorded during simulations (a) without the control system, (b) with the control system [6]

During numerical simulations currents of the driving motors were recorded. The recorded values are shown in Fig. 4 for the simulation without the control system (a) and with the control system (b).

The frequency of supply voltage of the motor and the extension of the telescopic trough on the return end were regulated during the numerical simulations according to the control algorithm. The way of changing these values is shown in Fig. 5.

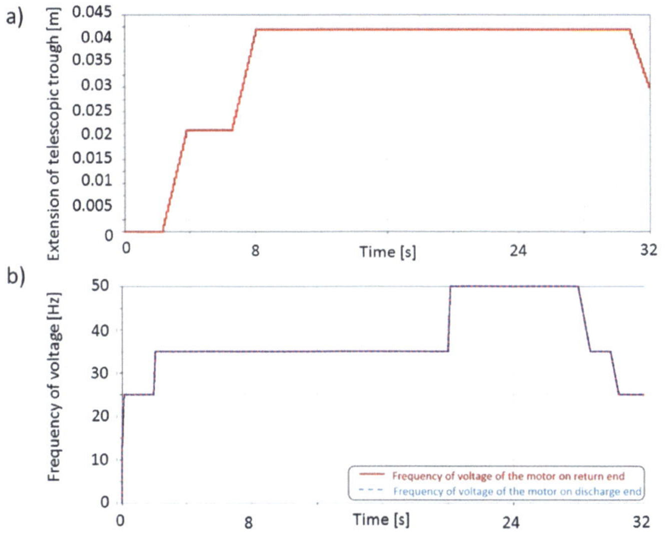

Fig. 5. Results of regulation (a) the extension of the return end, (b) the frequency of the motors supply voltage [6]

The forces recorded by the chain tension transducer on the return end during numerical simulations with and without the control system are shown in Fig. 6.

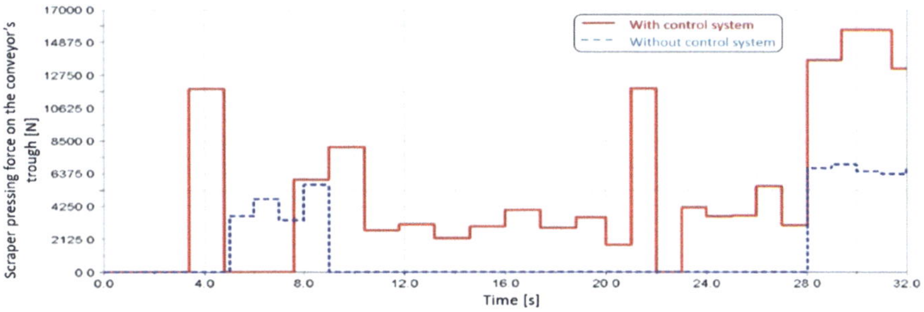

Fig. 6. The force recorded by the transducer of the chain tension signal on the return end [6]

The recorded signals of the chain tension transducer and currents in motors were used to define the AFC working conditions. The comparison of these conditions with the simulations carried out with and without the control system is shown in Fig. 7. The work state numbered by 1 and 10 indicate the case in which the chain tension was considered as correct. The chain loosening at the return end occurred in the cases numbered 28, 31 and 29. In the case 31, the high chain tension on the discharge drive was recorded. In the case 29 excessive load to the return drive motor was observed in the comparison to the discharge drive motor. In the case 19, too much chain tension on the return end of the conveyor was reported.

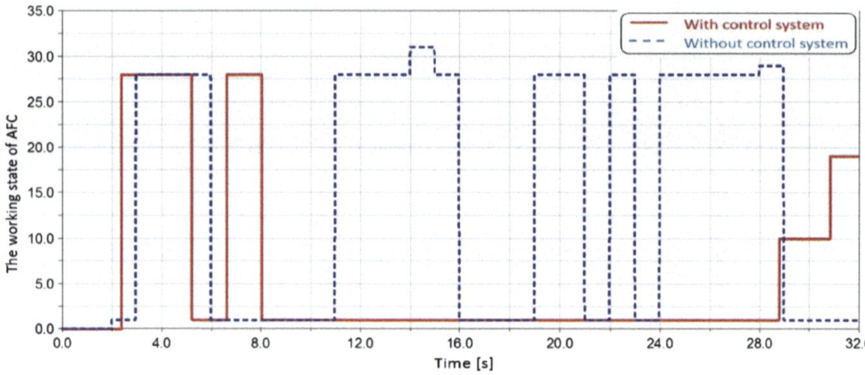

Fig. 7. The working state of AFC [6]

Total time at which the chain loosening or the excessive tension of the chain were observed was calculated, taking into account both motors, with the use of defined working conditions. The calculated time is shown in Table 1.

Table 1. The calculated time of the chain tension type during 32 s of simulation [6]

Working conditions	Without the control system	With the control system
Loosening the scraper chain on the return end	16 s	4.2 s
Excessive tension of the scraper chain on the return end	0 s	1.2 s
Excessive tension of the scraper chain on the discharge end	1 s	0 s
Status of correct work	14 s	24.2 s

The second part of the algorithm adjusted the speed of the scraper chain movement to the load of the conveyor. Figure 8 shows the graph of the rotational speed of the return drive drum for the simulation with and without the speed control system.

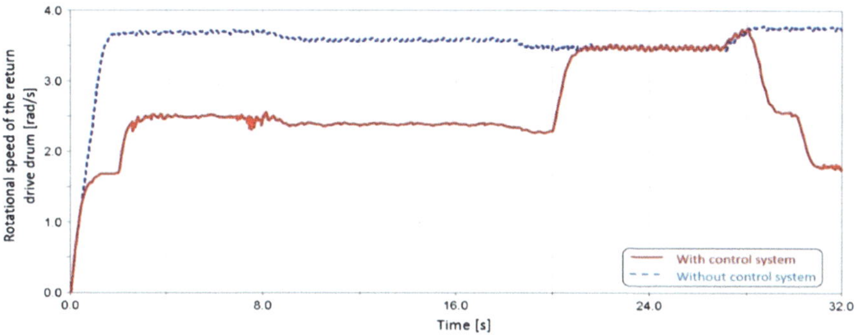

Fig. 8. The rotational speed of the return drive drum, recorded during simulations [6]

Analysis of the simulation results, where the selected parameters of the conveyor operation were controlled, allowed to conclude as follows:

- Changes in operational parameters enabled reducing the uneven load to the conveyor motors resulted from changes in the load to the conveyor's upper chain.
- Peaks of the current in motors, observed at 2nd and 20th seconds of the simulation, are associated with change of the set value of the scraper chain speed and its increase.
- On the return end, the improvement of working conditions of the scraper chain was observed. This improvement is related to increase in the chain tension. This change has reduced the risk of a dangerous chain loosening on leaving the return drum. Total time of the chain loosening on the return end was shortened. This time was reduced from 16 s during the simulation without the control to 4.2 s during the simulation with the control system.
- Total time of the correct operation of the scraper chain on both conveyor ends increased. During simulations without the control system, the correct chain operation was equal to 14 s, while during the simulation with the control system, total time of the correct chain operation was 24.2 s.
- Possibility of changing the operational parameters of the conveyor allowed eliminating the excessive chain tension at the discharge end (during the simulation without the control system 1 s of such chain tension was recorded). This change resulted in the excessive chain tension on the return end, which lasted about 1.2 s.
- The control of the conveyor operational parameters allowed adjusting the scraper chain speed to the current load of the upper chain. This adjustment is possible by changing the set speed and in the result changing the rotational speed of the drive motors. An increase in the load of the conveyor to 380 kN (20th second of the simulation) resulted in an increase of the frequency of the supply voltage of the driving motors, which was related to the set high speed option of the scraper chain. After reducing the load of the conveyor to 70 kN (approx. 31.5th second of the simulation), the frequency of the motors supply voltage was reduced twice, resulting in a value of 25 Hz for the motors supply voltage and corresponding to a slow speed option of the chain.

5 Conclusions

The computational model of AFC, which was used in numerical simulations for the purpose of development of the algorithm for controlling the selected operational parameters was presented. The developed algorithm consists of two parts. First part is used to control the extension of the telescopic trough at the return end and to control the frequency of the driving motors supply voltage to ensure the proper tension of the scraper chain and the proper cooperation of both motors. The main function of this part of the algorithm is to avoid the situation when the scraper chain is loosened at the place of leaving the driving drums on both return and discharge ends. Elimination of the cases, when one of the motor is loaded, through the scraper chain, by another one at the opposite end of the conveyor is another function of this part of the algorithm. Second part of the algorithm was developed to enable adjusting the conveyor's output to the current need. The control of the conveyor's output is realised by the regulation of the frequency of the driving motors supply voltage which results in change of the drum rotational speed. The change in drums rotational speed changes the linear speed of the conveyor.

Both parts of the algorithm are integrated and their functionalities complete each other. Control signals from the first part of the algorithm are taken into account in the second part of the algorithm.

Use of the developed control algorithm can significantly improve the operation of AFC and in consequence extend life of the conveyor's components and subassemblies. The control of the conveyor rotational speed enables maintaining the proper tension of the scraper chain which reduces components frictional wear as well as emission of heat and noise resulting from movement of the scraper chain. Besides, the use of the control system minimizes time of the scraper chain loosening and minimises probability of the chain failure as well as the dynamic overload to the chain causing excessive wear of the conveyor components. Finally, numerical tests with the use of the computational model presented in the paper are planned to confirm benefits resulting from using the control algorithm during the conveyor operation.

Acknowledgments. The article is a result of the project realised within the doctoral grant financed by the KOMAG Institute of Mining Technology.

Calculations were made on the computers of Centrum Informatyczne Trójmiejskiej Akademickiej Sieci Komputerowej (Academic Computer Centre in Gdansk - TASK).

References

1. Development of a virtual model of a discrete model of the conveyor with the control test carried out numerical simulation model. The report of the research project ICON: system load balancing drives, high-performance scraper conveyors. Path software In-Tech, Gliwice (2012, unpublished)
2. Herbuś, K., Ociepka, P.: Analysis of the Hexapod Work Space using integration of a CAD/CAE system and the LabVIEW software. In: IOP Conference Series: Materials Science and Engineering, vol. 95 (2015). https://doi.org/10.1088/1757-899x/95/1/012096

3. Herbuś, K. Ociepka, P.: Integration of the virtual 3D model of a control system with the virtual controller. In: IOP Conference Series: Materials Science and Engineering, vol. 95, p. 012084 (2015)
4. Suchoń, J.: Mining scraper conveyors. The construction and use. Institute of Mining Technology KOMAG, Gliwice (2012)
5. Szewerda, K., Świder, J., Herbuś, K.: Concept of the algorithm controlling the output of AFC. Maszyny Górnicze (Min. Mach.) **4**, 93–103 (2016)
6. Szewerda, K.: The method of parameterisation and selection of algorithms for AFC's control. The Ph.D. thesis, KOMAG Institute of Mining Technology (unpublished)
7. Ultra-Long Face Conveying: As longwalls get longer AFC challenges increase. International Mining, pp. 36–37, September 2007

Power Quality in the "Shore to Ship" System – The Improvement of the Unbalanced Voltage Factor

Dariusz Tarnapowicz(✉) and Sergey German-Galkin

Maritime University of Szczecin, Waly Chrobrego 1-2, 70-500 Szczecin, Poland
{d.tarnapowicz, s.german-galkin}@am.szczecin.pl

Abstract. The emission of air pollution in ports originating from land sources (factories, cars, trains) has been drastically reduced in the last twenty years by implementing strict emission standards. The pollution generated by ships at berth in ports is currently unacceptable. One of the effective ways to reduce the negative impact of ships at berth in ports on the environment is to use "Ship to Shore" (STS) which stands for a supply of electricity to the ships from the local power stations. It makes it possible to disable and be independent from autonomic ships' diesel generator sets in port.

The problem of improving the quality of power supply requires an interdisciplinary preparation within the field of electrical engineering and continuous updating of knowledge. The quality of electricity supplied in the STS system is related to the safety of use of marine systems and economic functioning of the ship. The requirements related to the power quality are determined by the appropriate rules of ship classification societies. In order to determine the quality of electricity, the quality parameters are used. One of the parameters that is important for power quality assessment is the voltage unbalance. This paper presents a method aimed to improve the voltage unbalance factor within the STS system with the use of power electronic active filters.

Keywords: Air pollution from ships in port · Shore to Ship system · Voltage unbalance factor · Transformation of the coordinate system · Shunt Active Filter

1 Introduction

Marine transportation has a great impact on the world economy. Due to sophisticated systems of automatics being used in shipbuilding industry, the demand for energy supply on ships is constantly growing. It contributes to the increase of air pollution, which is a relevant issue for ships at berth in ports. Gas emission analyses carried out in ports reach conclusion that marine ships are the key source of air pollution here [1]. Limiting air pollution allows for decreasing the harm to environment and, consequently, harm to the health of port cities' inhabitants. In order to minimize emission of air pollution to the atmosphere it is possible to switch off autonomous ship power generating systems (Diesel Generators) which are the main source of air pollution. It can be realized through connecting a ship to electrical on land grid (using STS – Shore-to-Ship system).

J. Świder et al. (Eds.): MECHATRONICS 2017, AISC 934, pp. 406–416, 2019.
https://doi.org/10.1007/978-3-030-15857-6_40

In fact, creating STS system limiting the influence of the ships at berth in ports on the natural environment is not easy in technical terms. Energy infrastructure of port installation necessary to supply ships must be designed so that it can serve different types to ships. It is complicated since the ships all over the world are equipped with different nominal parameters of electrical networks. The lack of normalization of frequency in Europe, Asia (50 Hz) and America (60 Hz), as well as different low nominal voltages (<1 kV) and middle nominal voltages (3.3 kV, 6.6 kV, 11 kV) used on ships result in the fact that port energetic infrastructure must be complex.

STS system infrastructure should be designed to enable adjustment of the parameters of energy in local electrical grids to the network of ships berthing in ports. STS system should be fully controllable, flexible and reliable. High quality of electricity supplied from the onland grid is particularly important. The specificity of ship's network implies specific requirements [2–4].

The main problem while building STS system was its standardization to make it fully universal. The idea to supply the ships at berth with the electricity from the on land grids was written in the EU law in 2006 [5]. In 2012 standardization of STS system was established. International Electrotechnical Commission formulated and published instructions for the constructors of STS systems. IEC/ISO/IEEE 80005-1 [6] document published in July, 2012 stipulated requirements, designers' support, practicality and functionality of the connection ensuring health of the crew and port employees during shore-to-ship connection.

Figure 1 presents a schematic diagram of STS system including EU recommendations and IEC standard.

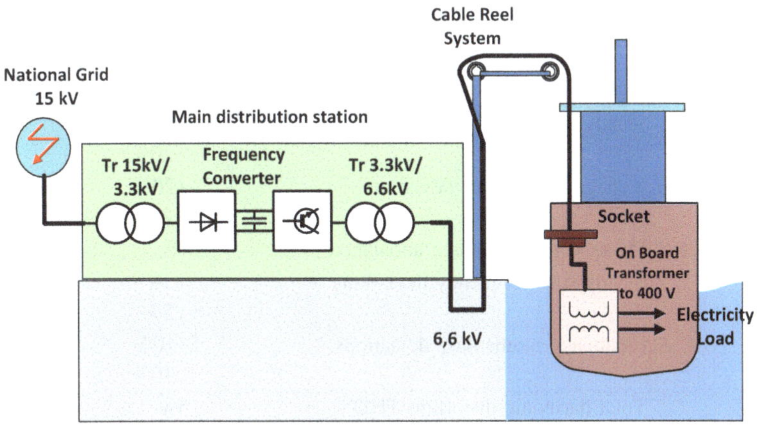

Fig. 1. The idea of the STS system in accordance with the recommendations of the EU and IEC Standard.

It is characteristic for this topology to install all the devices transforming electricity in one place, called main distribution station. Power electronic converter supplying a few ships at once is the key element of the system.

Apart from adjusting electroenergetic onland grid to the ship's network, which is the basic task STS system has to perform, it also has to supply electricity of high quality.

2 The Quality of Electricity in STS System; Voltage Unbalance

The shore-to-ship supply system must be flexible. The ship electric system must give a possibility of selecting voltage frequency. The voltage level can be adjusted using additional transformers on board of a vessel or directly on land. Standardizing the connection is significant in order to serve as many ships as possible. Shore-side electricity must be effective, fast and safe.

Ships supplied from the onland grid at berth in the port should have continuity of supply provided to them. A temporary lack of supply or any voltage disruptions could result in irreversible consequences to ship systems and eventually to the ship personnel.

Electricity undergoes degradation under the influence of electromagnetic disturbances i.e. actual values differ from nominal ones. The quality of electricity is defined as a group of numeric values also named as quality parameters of electricity. These quality parameters in electrical onland grids as well as in the ship networks are restricted by certain provisions. The provisions for vessels are established by the classification societies and IEC global standards.

Limit values of quality indexes for vessels are defined by the IEC provisions [7]. Table 1 presents supply voltage quality parameters according to IEC standards.

Table 1. Supply voltage requirements according IEC standards.

Parameter	Value
Steady-state voltage deviations	+6% −10%
Voltage momentary deviation	+20% −20%
Phase to phase voltage unbalance	3%
Steady-state frequency deviations	+5% −5%
Frequency momentary deviations	+10% −10%
Total harmonic distortion THD	5%
A single distortion (any harmonic greater than)	3%
Voltage transients recovery time	1.5 s
Frequency transients recovery time	5 s

Voltage unbalance is one of many parameters describing the quality of supply voltage. This article presents a method of improvement of supply voltage unbalance index in STS system. Voltage unbalance is a state, in which effective values of phase voltages or phase shifts between the consecutive phases are not equal. Supply voltage unbalance is measured with the voltage unbalance factor which, for three-phase voltage systems can be presented in the form of the following equation [8]:

$$U_{UBAL}\% = \sqrt{\frac{1 - \sqrt{3 - 6\beta}}{1 + \sqrt{3 - 6\beta}}} \cdot 100\% \tag{1}$$

where:

$\beta = \frac{U_{AB}^4 + U_{BC}^4 + U_{CA}^4}{(U_{AB}^2 + U_{BC}^2 + U_{CA}^2)^2}$,

U_{AB}, U_{BC}, U_{CA} – effective values of first harmonics.

Voltage unbalance is caused mainly by asymmetrical one-phase outputs as well as by three-phase receivers working as one-phase units. Electrical heaters of fuel containers can serve as an example of such receivers in ship systems. The results of such unbalance relate to safety and economy of the whole system operation on the ship. Among the most frequent problems caused by the voltage unbalance there are:

- induction engines – engine torque change caused by the change of the shape of rotating field (elliptical shape instead of circular shape),
- power losses in transmission lines,
- synchronous power generators – excessive heating,
- transformers – additional losses which lead to oversizing of transformers.

3 Improving the Rate of Voltage Unbalance in the STS System Using Shunt Active Filter

The power factor for AC electrical network is defined as the ratio of active power to total power:

$$PF = \frac{P}{S} \tag{2}$$

where: PF – power factor, P (W) – active power, S (VA) – total power.

The power factor can take values from 0 to 1. Active power (P) is a useful work done in the circuit for certain time. Total power (S) is the product of RMS (Root Mean Square) voltage and current in the circuit. Total power can exceed active power for the following reasons:

- load is active-reactive,
- load is nonlinear causing the appearance of harmonic components in the current,
- load is asymmetric in the three-phase network.

The difference between total and active power can be caused by one or several of the listed reasons.

Total power S (VA), active power P (W), reactive power Q (Var) and power of distortion or asymmetry of T (Vars) are related by the known relation.

$$S^2 = P^2 + Q^2 + T^2 \tag{3}$$

where: Q – reactive power, T – distortion or asymmetry power.

One of the key directions in ensuring energy efficiency and security at present is building intelligent self-contained power systems. These are flexible, reconfigurable systems, which not only meet our own needs but may also be connected to the system of distributed electricity generation. In the networks of centralized power supply problems of maintaining maximum energy performance are being solved by the development and introduction of active filters. These filters generate the whole range of non-active components of power into load and thereby reduce, by 20–25%, the load of feeders and generating systems.

The functional diagram of the Shunt Active Filter (SAF), control symmetry voltage, connected to the supply mains, is shown in Fig. 2.

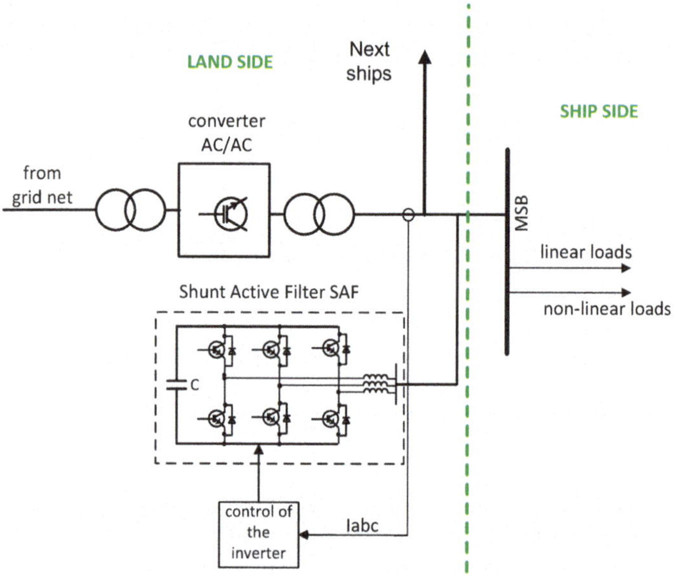

Fig. 2. The STS system diagram with SAF to improve the unbalanced voltage factor

The power part of SAF is a three-phase semiconductor converter switch connected to the mains by AC terminals via inductors. A capacitor C is placed in the DC link of SAF. The power part control is executed by a relay-operated controller ensuring a sliding mode for current. The calculated required active load current and the current in the network are compared at the input of a relay-operated controller. As a result, SAF generates all inactive components of the current (power) load into the network.

The main problem in constructing SAF is to calculate active components of load current. This is achieved through an adequate transformation of coordinates which allows us not only to simplify mathematical description but also to synthesize efficient algorithms for SAF control. Coordinate transformation is widely used in modern theories of instantaneous power [9, 10]. The transformation from the coordinate system (*d*, *q*, 0) to a system coordinates (*d*, *k*, *r*) [11] leads to an error in calculation of the active current (active power) in case of load dissymmetry.

Figure 3 shows the idea of converting the coordinate system *d*, *q*, *0* into coordinate system *d*, *k*, *r*.

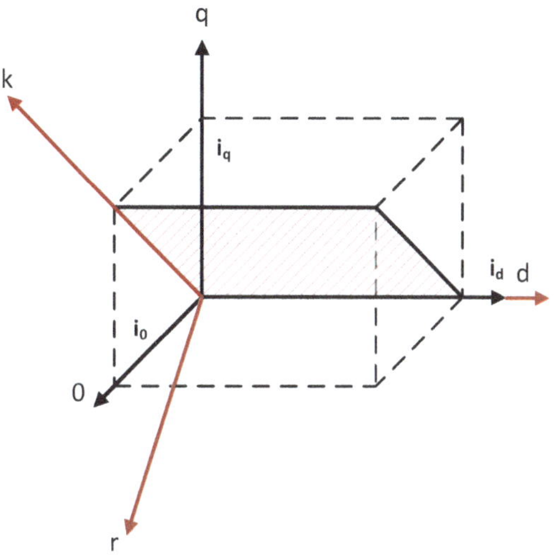

Fig. 3. Converting a coordinate system d, q, 0 in the co-ordinates d, k, r.

The system of coordinate *d*, *k*, *r* is turned with an angle (with) respect (to) the system of coordinate *d*, *q*, *0*. The projection of the current vector on the axis r must be equal to zero.

This allows the electromagnetic processes in the SAF to be described in the plane and use methods of analysis of the resultant vector and "smooth component".

Conversion of the currents coordinate system *d*, *q*, *0* in coordinate system *d*, *k*, *r* is carried out in accordance with the Eq. (4).

$$\begin{bmatrix} i_d \\ i_k \\ i_r \end{bmatrix} = \begin{bmatrix} 1 & 0 & 0 \\ 0 & \frac{|i_q|}{i_{k0}} & \frac{i_0}{i_{k0}} \\ 0 & -\frac{i_0}{i_{k0}} & \frac{i_q}{i_{k0}} \end{bmatrix} \cdot \begin{bmatrix} i_d \\ i_q \\ i_0 \end{bmatrix} \tag{4}$$

where: $i_{k0} = \sqrt{i_q^2 + i_0^2}$.

It should be emphasized that the transformation of the currents from coordinate system d, q, 0 into the coordinate system d, k, r must be invariant under power, which in turn for symmetric source requires equality of currents in the original and the transformed coordinate system. Mathematical notation for this condition reduces to Eq. (5).

$$\sqrt{I_d^2 + I_q^2 + I_0^2} = \sqrt{I_k^2 + I_d^2} \quad \text{or} \quad I_k = \sqrt{I_q^2 + I_0^2} \tag{5}$$

In the coordinate system d, k, r electromagnetic processes in the SAF can be described in the complex plane, taking the real axis of d, and for the imaginary – axis k.

The basis of the mathematical description of electromagnetic processes in the SAF is a method of the space vector. When the description of electromagnetic processes is carried out with the use of space vector method, then functional diagram of SAF is converted to the form shown in Fig. 4.

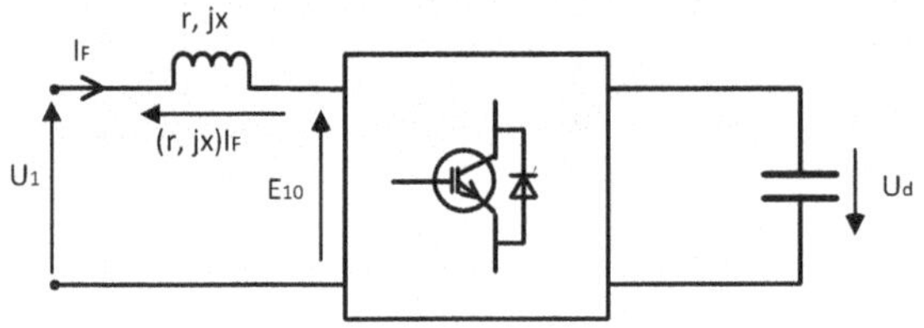

Fig. 4. The functional diagram of SAF.

In this function diagram the converter has one side connected to a three-phase AC voltage ($\bar{U}_1$) through the throttle, and the other – to the capacitor in the network direct current. On the network side the mathematical description of the SAF can be represented as:

$$\bar{U}_1 = \bar{E}_{10} + r\bar{I}_F + jx\bar{I}_F \approx \bar{E}_{10} + jx\bar{I}_F \tag{6}$$

where: $\bar{U}_1$ – the score vector voltage, $\bar{E}_{10}$ – the resulting vector of back emf SAF for first harmonic, r – resistance, taking into account the resistance of the source, throttle and two conductive semiconductor devices SAF, $x = \omega_1 L$ – reactance's of choke on the mains frequency.

The current of the network has an active and reactive (inductive) components. In order for the current of network to be purely active, current SAF should be compensated and capacitive reactive component of the load current. It implies a fundamental conclusion that for capacitive current in the SAF back emf $\bar{E}_{10}$ exceeds voltage $\bar{U}_1$.

In the rotating coordinate system *d*, *k*, *r* all electromagnetic state variables in Eq. (6) are constant, so that the analysis of the SAF can carry similar analysis of a dc-to-dc converters. When there are the sliding control currents in the steady mode, the current SAF pulses with high carrier frequency with respect the current setpoint $I_F = I_r$.

Closed expressions for the modulation factor and the voltage on the capacitor as a function of current SAF can be expressed as:

$$m = \frac{1}{1 + I_F(pu)}, \quad U_d(pu) = 2(1 + I_F(pu))^2 \tag{7}$$

where: $U_d(pu) = \frac{U_d}{U_1}$ – relative average voltage across the capacitor,
$I_F(pu) = I_r(pu) = \frac{xI_F}{U_1}$ – relative current of reactive load.

The reactive load current, which can be compensated SAF is $I_F^*(pu) = 0.3 - 0.7$.

Thus at given reactive load current according to Eq. (7) is defined by one of the main variables – the voltage on the capacitor filter.

After defining capacitor voltage on the filter it is easy to calculate the maximum voltage across the semiconductor elements of SAF.

4 The Simulation Research on the Shunt Active Filter in STS System

Three-phase unbalanced network with parameters containing SAF and determination of the control signals in the coordinate system *d*, *q*, *r*, is implemented in the program environment MATLAB/Simulink. Topology simulation model which is consistent with the configuration shown in Fig. 2 is presented in Fig. 5.

Three-phase unbalanced network has the following parameters:

$$U_1 = 6600\,\text{V}, \omega_1 = 314\,\text{rad/s}, I_{F.\max} = 425\,\text{A}$$

Parameters of SAF, calculated in accordance with the proposed method, are the following: $f = 5$ kHz, $I_F(pu) = 0.47$, $m = 0.68$, $U_d(pu) = 4.4$ ($U_d = 29$ kV), $L = 0.023$ H ($X_L = 7.8\ \Omega$).

Figures 6 and 7 show currents and voltages network in the power supply STS system without the use of SAF (Fig. 6) and with the use of SAF (Fig. 7). The results show full compensation inactive power and asymmetry in the supply network.

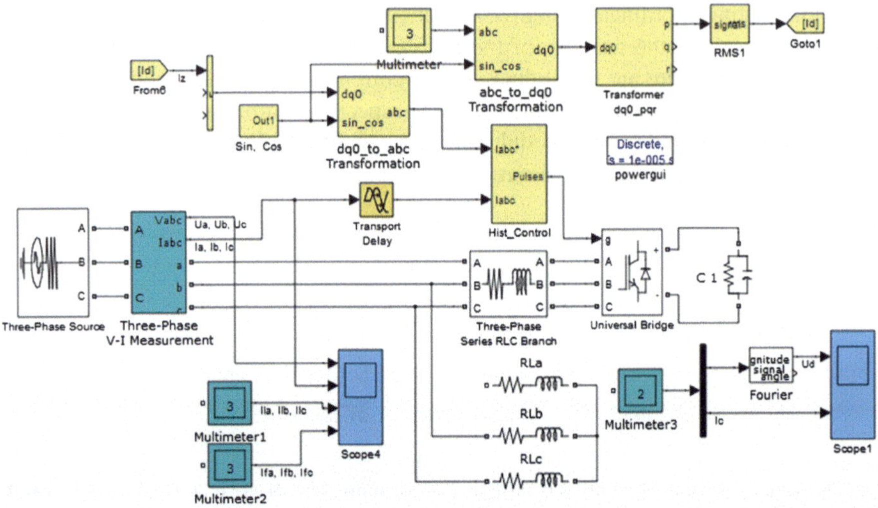

Fig. 5. The model of the SAF.

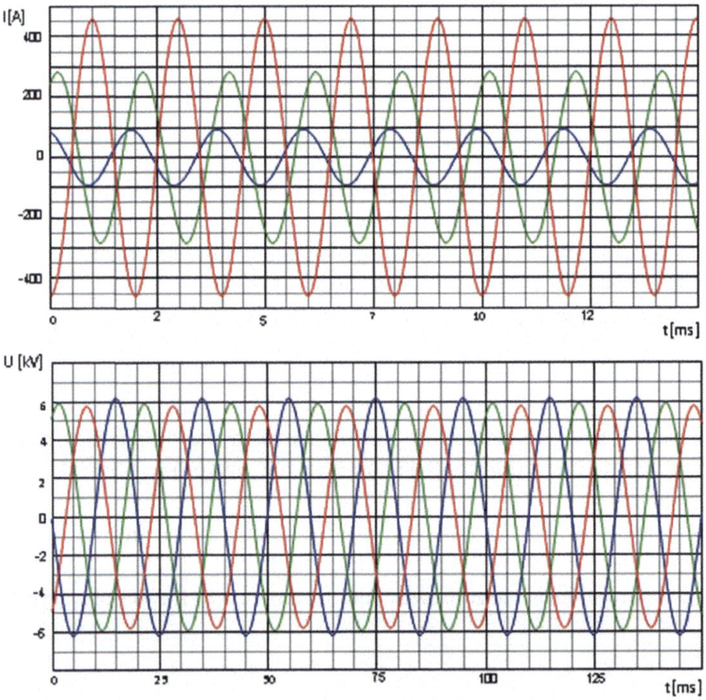

Fig. 6. Currents and voltages network in the power supply STS system without use SAF.

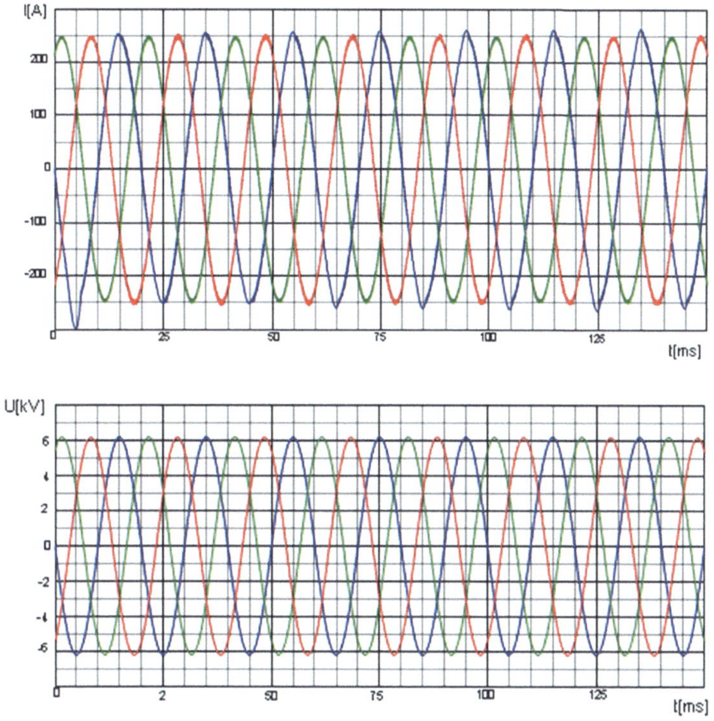

Fig. 7. Currents and voltages network in the power supply STS system with use SAF.

5 Conclusions

Power quality is of great importance for the reliability of marine systems and the economic operation of a ship. The safety of a marine system is directly related to the quality of electricity and additional operating costs will increase with the deterioration in the quality of energy. The issue of power quality becomes more important from the point of view of the correct operation of all of the ship systems.

In the last years, many ports in Europe have been expanding their port infrastructure of the "Shore to Ship" system. The use of "Shore to Ship" for ships at berth in ports, allowing for reduction of pollution emission, requires solutions to many technical problems. High quality of electricity supplied from shore to ship is an important issue. The results of the research presented in this paper confirm the effectiveness of the proposed method to improve one of the most important power quality parameters which is voltage unbalance.

References

1. Tarnapowicz, D., Borkowski, T.: Shore to Ship – Alternative Power Supply of Ships in Port. Wydawnictwo Naukowe Akademi Morskiej (2014). ISBN-978-83-64434-03-7
2. Matuszak, Z., Nicewicz, G.: Analysis of marine electric power plants loads. In: Modern Methods of Construction Design. Lecture Notes in Mechanical Engineering, pp. 273–280 (2014)
3. Nicewicz, G., Tarnapowicz, D.: Autonomous photovoltaic electricity generating system for ships mooring at ports as an environmentally friendly solution. In: GeoConference on Energy and Clean Technologies. International Multidisciplinary Scientific GeoConference-SGEM, SGEM 2014, vol. 1, pp. 41–48 (2014)
4. Tarnapowicz, D.: Synchronization of national grid network with the electricity ships network in the "shore to ship" system. Manag. Syst. Prod. Eng. **3**(11), 9–13 (2013). ISSN 2299-0461
5. EU (2006/339/EC) Commission Recommendation of 8 May 2006 On the promotion of shore-side electricity for use by ships at berth in Community ports (Text with EEA relevance)
6. International Electrotechnical Commission, IEC/ISO/IEEE 80005-1 pt.: International Standard Utility connections in port – part 1: High Voltage Shore Connection (HVSC) Systems – General requirements, Edition 1.0, July 2012
7. IEC 61000-4-30: Testing and measurement techniques – power quality measurement methods. CEI/IEC (2003)
8. Urbański, K.: Pomiary i rejestracja asymetrii napięcia w obwodach trójfazowych. Pomiary Automatyka Kontrola **6**, 100–102 (2006)
9. Akagi, H.: New trends in active filters. In: Conference on Proceedings of EPE 1995, Sevilla, pp. 0.017–0.026 (1995)
10. Park, R.H.: Two reaction theory of synchronous machines generalized method of analysis-part I. AIEE Trans. **48**, 716–727 (1929)
11. Kim, H., Blaabjerg, F., Back-Jensen, B.: Instantaneous power compensation in three – phase systems using p-q-r theory. IEEE Trans. Power Electron. **17**(5), 701–710 (2002)

Mathematical Modelling and Selecting the Parameters of Magnetic Circuit of Disk Torque Converter

Arkadiusz Tomas[1], Tomasz Trawiński[2(✉)], and Arkadiusz Mężyk[3]

[1] KOMAG Institute of Mining Technology, Pszczyńska 37, 44-100 Gliwice, Poland
[2] Department of Mechatronics, Silesian University of Technology, Akademicka 10A, 44-100 Gliwice, Poland
tomasz.trawinski@polsl.pl
[3] Institute of Theoretical and Applied Mechanics, Silesian University of Technology, Konarskiego 18A, 44-100 Gliwice, Poland

Abstract. The concept and mathematical model of torque converter using the magnetic field of permanent magnets and electromagnetic induction for torque transmission is presented. Different examples of known types of magnetic gears are discussed and patent survey is presented. The torque converter has changeable dynamic and kinetic gear ratio. Magnetic flux density in air gap of new torque converter for five different arrangement of permanent magnets is analyzed. Mathematical field-circuit model of a torque converter is presented. Resultant flux density distribution is regarded as input quantity for field-circuit model of this converter. Results of simulation in a form of the converter operational characteristics are given.

Keywords: Energy conversion · Gears · Circuit analysis · Finite element methods

1 Introduction

Magnetic gears are the mechanisms, which transmit torque through magnetic coupling. The first concepts of magnetic gears were developed at the beginning of 20th century. It was a power transmitting device [1] and a front magnetic gear, in which teeth of toothed gear were replaced by ferrite magnets [2]. The use of, the ferrite magnets in magnetic gears at the beginning resulted in transmission of low torques (in comparison to classical toothed gears). The use of stronger neodymium magnets [3] in magnetic gears increases the amount of transmitted torques. The volumetric torque density i.e. quotient of transmitted torque to a total volume of magnetic gear, is one of the parameters enabling to assess performance of magnetic gear. For the worm magnetic gear the torque density g_m is expected to be between 0.74 to 2 Nm/dm^3 [4, 7], but for angular magnetic gear the torque density g_m is 3 Nm/dm^3 [4]. In radial (external) gear the torque density g_m is up to 1.6 Nm/dm^3 [4]. One-step or two-step cycloid magnetic gear is another solution of magnetic gears. One-step cycloid magnetic gears are characterized by low gear ratio and in such a gear g_m reaches the value of 290 Nm/dm^3

J. Świder et al. (Eds.): MECHATRONICS 2017, AISC 934, pp. 417–426, 2019.
https://doi.org/10.1007/978-3-030-15857-6_41

[5, 6]. For cycloid magnetic gear, in two-step manufacture of high gear ratio torque density, g_m is equal to 75 Nm/dm^3 [7]. In this case practical values of g_m are within the range 45–90 Nm/dm^3 [4, 7].

Magnetic one-step internal radial gear [4] (also named a concentric gear [7]), of low gear ratio can reach the volumetric torque density g_m in the range between 70 to 100 Nm/dm^3 [7]. All the above-mentioned magnetic gears have a constant gear ratio. The presented gears, excluding planetary gears with rotating magnetic field, use direct interaction of permanent magnets to transmit torque. In the planetary gear with rotating filed, an intermediate element in a form of ferromagnetic rods [4, 8] or outer toothed ring for magnetic field modulation is used [17]. Change in number of poles generated by the magnets of input rotor is the task of this intermediate element. The field acts on outer rotor, setting it into motion [4, 8]. Disk variant of magnetic planetary gear with rotating field was also developed. Research work on magnetic gears with variable gear ratio that require external power supply was also carried out [9–12]. In a typical solution, such gears are designed as the variants of magnetic planetary gear with rotating field, in which electromagnetic coils with external power supply are added to control the rotational speed of outer rotor. This outer rotor, depending on required rotational speed is braked by magnetic fields generated by coils. Complex control systems, examining transmitted torque and current slippage occurring in the working gears are also offered [18]. The solutions described in patents CN104767357 [19] and CN105006951 [20] use the effect of induction and are combinations of systems with permanent magnets and coils, where the coils have ferromagnetic cores and are connected with use of brushes and slip rings with external control system. In the solution described in patent CN104767357 [19] the interaction between disk with permanent magnets and disk with coils is realized through the control system. In the solution described in patent CN105006951 [20], the current generated in the system of coils is processed out of gear and transferred to the second system of coils, which cooperates with a disk/cylinder receiving the drive. The solution, in which the torque is transmitted between rotating component with coils with cores and the component with inductive cage through the intermediate component, where the coil winding is supplied with DC, is presented in patents US2654849 [21] and US2790917 [22]. Variant system with inductive cage induced by two magnetic poles and with the next inductive cages of different resistance, where each cage is mechanically coupled with one stage of the gear, is described in patent US2864016 [23]. Depending on the load the next converter stages are actuated.

One of the authors is the co-author of the patent PL402445 [24] entitled “Induction torque converter”, where changeable magnetic field generated by rotating magnets is the source of induced electromotive force and torque in the receiving component in a form of winding, preferably inductive cage. The presented solution refers to the solution described in the above-mentioned patent, but the cage is replaced by the system of coils and the second disk with permanent magnets is added. Thus, the conducting component, in which voltage is induced, is between the magnets. The described solution, unlike the known solutions, uses a lot of poles, what enables to determine the number and volume of harmonics of B_z component of induction. The use of windings (coils) without ferromagnetic core is important. The yokes of disks play a role of the assembly component. The solution is a special case of coreless transducer, which has no movable core, where eddy current could be induced. Unlike many known solutions, the windings are not power-supplied. It is important to use mechanically coupled magnets on both sides of winding, which enables better use of energy of magnetic field.

The suggested concept of electromagnetic torque converter (of disk design) assumes using the active intermediate element in a form of coils as the element transmitting torque between disks with permanent magnets.

2 Design of Electromagnetic Torque Converter

Electromagnetic torque converter is a device, which plays a role of gear of controlled (variable or settable) kinematic and dynamic ratio. The main components of the converter are as follows:

- excitation system – its role is to generate alternating magnetic field in the result of rotational movement caused by an external torque,
- intermediate element – its role is to act on magnetic field generated by the excitation system in such way to enable changing the number of pole pairs,
- receiving system – its role is to convert driving torque by conversion of magnetic field energy.

Block diagram of electromagnetic torque converter is presented in Fig. 1. Magnetic field excitation system (in Fig. 1 marked by 1) consists of one or two disks with permanent magnets deployed on their inner parts.

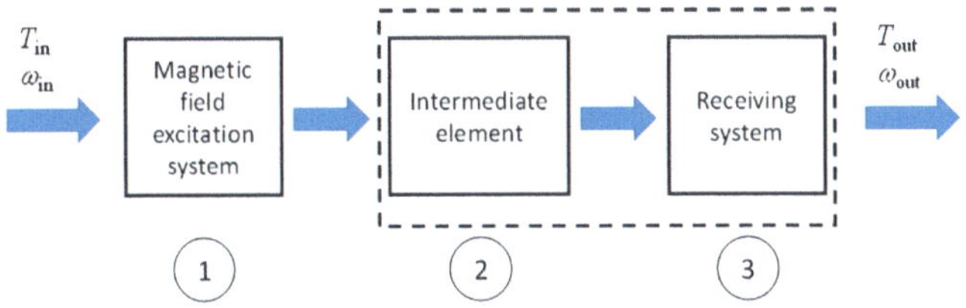

Fig. 1. Block diagram of magnetic/electromagnetic torque inverter.

The disks can move freely against each other or can be mechanically coupled. External driving torque acts on the excitation system. In the case, when disks are mechanically coupled, the receiving system (in Fig. 1 marked by 3) consists of a dozen or so coils spatially arranged in such a way to form a disk. In this case, the receiving system plays also a role of intermediate element (in Fig. 1 marked by 2). In the case, when the external torque drives only one disk of magnetic field excitation system, the ferromagnetic elements and winding of intermediate element change number of pole pairs of the generated magnetic field, and the second disk with permanent magnets takes a role of the receiving element converting energy of magnetic field into mechanical energy. The version of electromagnetic converter of the following arrangement: excitation system of magnetic field will consist of two disks with mechanically coupled permanent magnets and receiving system will consist of coil winding (playing the role of intermediate component) will be further mathematically analyzed.

2.1 Mathematical Model of Electromagnetic Torque Converter

Magnetic field generated by rotating permanent magnets in air gap of electromagnetic converter has a significant impact on its operational characteristics. Technical solutions presented so far are focusing on such spatial arrangement of permanent magnets to obtain magnetic field close to the sinusoidal one. Problem of such spatial deployment of permanent magnets to obtain magnetic field significantly different from a sinusoidal one, and they will be discussed later. In literature, we can find articles indicating a possibility of operation of electromagnetic converters (with strongly deformed magnetic field in air gap) with few different rotational speeds [14–16]. Bearing this fact in mind, possibility of operation of a torque converter with few different rotational speeds i.e. few kinematic and dynamic gear ratios, will be the expected result. That is why mathematical analysis was made in few stages. At the first stage, field models enabling analysis of distribution of magnetic field in air gap of torque converter, were formulated. The analysis was made to check:

- Content of harmonics of magnetic flux density (in volume of side of winding of the receiving system).
- Selection of optimal point of operation of permanent magnets, after taking decision about their spatial arrangement on front surfaces of disks.

At the second stage, the circuit mathematical model of the receiving system composed of a disk with coil winding was formulated. Calculation results of averaged magnetic flux density from the first stage were used at this stage.

2.2 Calculation of Spatial Distribution of Magnetic Field in the Air Gap of Converter

Testing the impact of decreasing the number of poles as well as dimensions of magnets and design of a single pole on harmonics of averaged flux density in the area occupied by the winding of torque converter was the analysis objective. Excitation system of magnetic field of torque converter consists of two disks placed coaxially against each other with segmental permanent magnets deployed on their internal face surfaces. The following variants of spatial distribution of permanent magnets were tested in-situ:

- variant 1–30 magnets (dimensions 5 × 15 × 40 mm) symmetrically arranged on internal face surface of each disk,
- variant 2–24 magnets (dimensions 5 × 15 × 40 mm) symmetrically arranged on inner face surface of each disk,
- variant 3–18 magnets (dimensions 5 × 15 × 40 mm) symmetrically arranged on inner face surface of each disk,
- variant 4–60 magnets (dimensions 5 × 5 × 40 mm) symmetrically arranged on inner face surface of each disk, where each of three magnets forms one segment pole,

- variant 5–48 magnets (dimensions 5 × 5 × 40 mm) symmetrically arranged on internal face surface of each disk, where each of three magnets forms one segment pole.

Comsol Multiphysic computer program was used in calculations of spatial distribution of flux density. Neodymium permanent magnet of magnetic remanence B_r equal 1.2 T (N37UZ Shin-Etsu Co. Ltd. rare earth magnet, H_C = 891 kA/m, BH_{max} = 271 kJ/m^3) was used in analysis, while the analysis itself, due to symmetry, was limited to the sections of excitation circuit as it is presented in Fig. 2. The magnetic circuits of yokes (which are part of converter disk) are marked with 1, and permanent magnets are marked with 2 Standard Soft Iron material (without losses) was selected as the material of yokes from the Comsol library. Analyzed sections were discretized using 400,000 spatial components – tetrahedrons, which resulted in a total number of degrees of freedom of models up to 70,000. Boundary condition of periodicity of vector magnetic potential was assumed on both sides of excitation circuit (on the left and on the right side), formed after cutting the section off. A cuboid representing the side of winding was entered to the analyzed geometry (for all analyzed variants) to determine the averaged flux density in the area occupied by a single side of coil winding of a disk of receiving component of torque converter. The length of this cuboid is the same as the length of permanent magnets of disks of excitation circuit (i.e. 40 mm), its width is equal to 5 mm and its height is equal to 8 mm. So the coil may consist of 33 wires of 1 mm diameter. For such dimensions of cuboid (one of sides of coil winding) the two-sided air gap of 1 mm width is formed between permanent magnets and face surface of winding, on each side of winding. Position of the cuboid between disks of excitation circuit, for variant 1 of torque converter, is presented in Fig. 2.

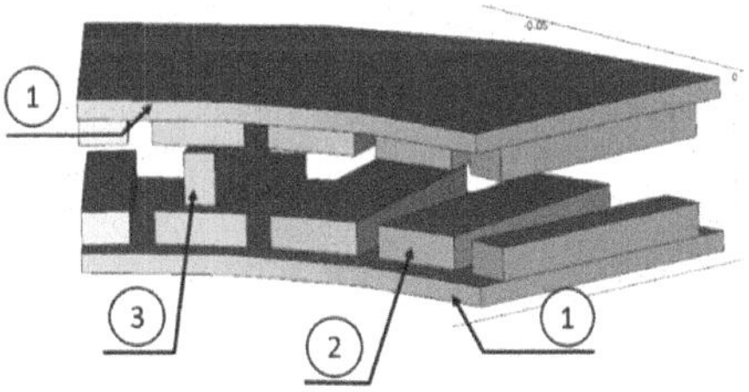

Fig. 2. Magnetic circuit of yokes – (1), one of segment permanent magnets – (2), cuboid representing one of winding sides – (3).

The cuboid was angularly moved by 0.5 or 1 degree between permanent magnets. After each change of position, the calculations of spatial distribution of flux density were made and average flux density in the cuboid area (one side of winding coil [13]) was calculated according to the following formula:

$$B_{zav} = \frac{1}{v_{cu}} \iiint_v B_z dv_{cu} \tag{1}$$

where: B_z – component of z axis of vector of magnetic flux density **B**, v_{cu} – volume of a single side of cuboid, v – volume.

The obtained results of averaged flux density in the volume of a single side (of cuboid) of coil winding are presented in Fig. 3.

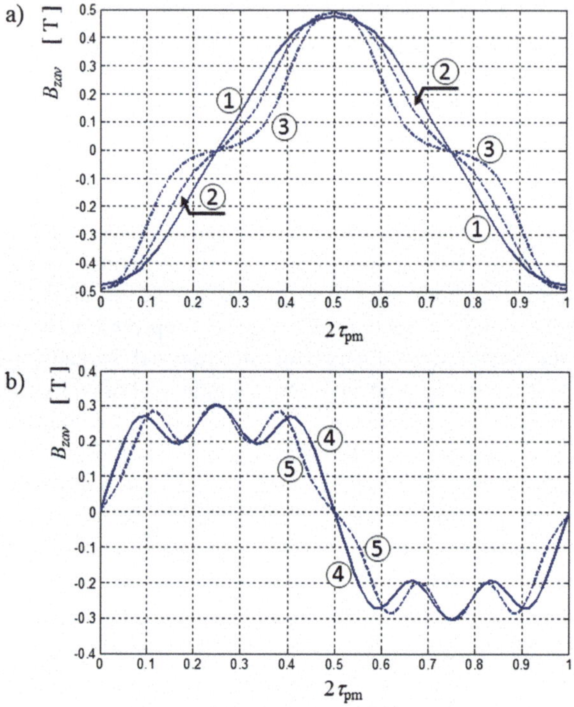

Fig. 3. Distribution of averaged flux density in the volume of a single side in the range of two poles pitches: (a) Variant 1 – (1) (24°), Variant 2 – (2) (30°), Variant 3 – (3) (40°), (b) Variant 4 – (4) (36°), Variant 5 – (5) (45)

Percentage share of harmonics of flux density averaged by volume in one side of coil winding, referred to the basic harmonic, is given in Table 1.

Table 1. Percentage share of higher harmonics of flux density.

Variant	3rd	5th	7th	9th	11th	13th
1	0.74	1.91	0.02	0.03	0.01	–
2	13.40	3.83	0.75	0.17	0.04	–
3	34.00	0.86	3.27	0.97	0.09	0.13
4	25.30	19.60	6.42	0.89	0.28	–
5	16.00	1.84	16.80	2.47	1.55	0.18

By entering the coefficients of induced voltage and the coefficient of torque to the mathematical model of converter it will have the following final form [25, 26]:

$$\frac{d}{dt}\begin{bmatrix} i_1 \\ i_2 \\ i_3 \end{bmatrix} = \begin{bmatrix} L_{11} & L_{12} & L_{13} \\ & L_{22} & L_{23} \\ sym & & L_{33} \end{bmatrix}^{-1} \left(-\begin{bmatrix} R_1 & 0 & 0 \\ & R_2 & 0 \\ sym & & R_3 \end{bmatrix}\begin{bmatrix} i_1 \\ i_2 \\ i_3 \end{bmatrix} - (\omega_m - \omega_o)\begin{bmatrix} k_{e1l} + k_{e1r} \\ k_{e2l} + k_{e2r} \\ k_{e3l} + k_{e3r} \end{bmatrix} \right) \quad (2)$$
$$\frac{d}{dt}\omega_o = \frac{1}{J_o}\left(-\sum_k i_k \sum_i (k_{ti}(\vartheta) + k_{ti}(\vartheta + \vartheta_0)) - T_L\right)$$

where: i_k – current of k-th phase, R_k – resistance of k-th phase, k_{eil}, k_{eir} – induced voltages coefficients in left and right sides of coils, L_{ii} – self-inductance of a given phase, L_{ij} – mutual inductances between i-th and j-th phase.

3 Implementation of Mathematical Model and Simulations of Torque Converter

Mathematical model of the components of electromagnetic torque converter was implemented in MATLAB/Simulink computer program. Block diagram of the model is presented in Fig. 4. System, which generates the wanted spatial distribution of (averaged) induction (in the volume of a single side of winding coil) is included in "ke synthesis" block. Coefficients of induced voltage calculated in this block are transferred to "Electric system of windings" block, which represents mathematical model of windings of the receiving component [25, 26]. Additionally, values of angular speed of disks of excitation circuit and angular speed of receiving component (and loading component) are transferred to this block.

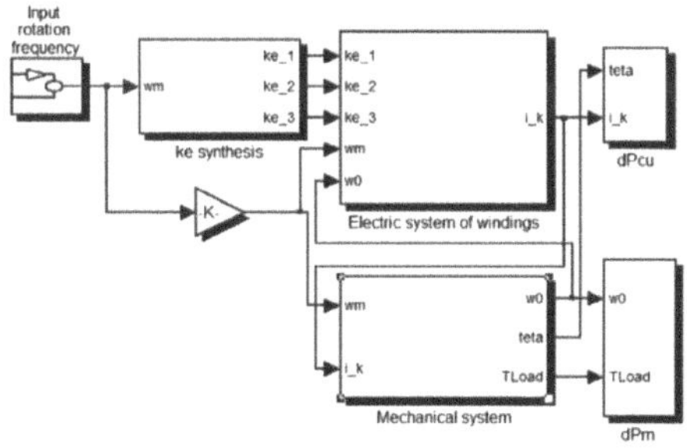

Fig. 4. Implementation of the mathematical model of torque converter in a form of block diagram.

Calculated values of current in windings are transferred to the following blocks: "Mechanical system" block (representing the model of mechanical system of receiving component and a loading component), in which electromagnetic torque and equation of motion are calculated, and "dPcu" block, in which power losses in windings are calculated. Mechanical power of load is calculated in "dPm" block. Trajectory of increase of rotation frequency of disks of excitation system is set in "Input rotation frequency" block. Checking the operational characteristics, i.e. load torque vs. rotational speed of converter, at constant angular speed of disks of excitation system, was the objective of numerical calculations. During simulations, the rotation frequency of disk was assumed to be from 20 Hz to 50 Hz and it increased to maximum value in 1 s at constant acceleration. Obtained characteristics for electromagnetic torque converter (variant 1) is given in Figs. 5, 6 and 7, i.e.: external characteristics of torque converter (representing changes of torque transferred by the receiving component in a function of rotational speed – in Fig. 5), characteristics of transferred power (in a function of slip of the receiving component – in Fig. 6) and characteristics of losses in windings of the receiving component – in Fig. 7.

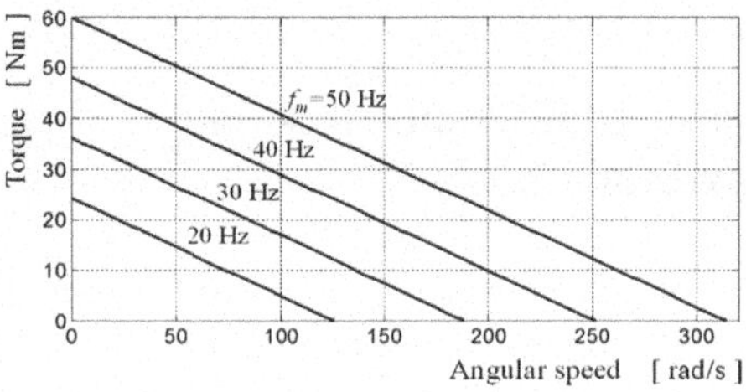

Fig. 5. Transferred torque in a function of angular speed of torque converter (variant 1), for different rotation frequencies of excitation system.

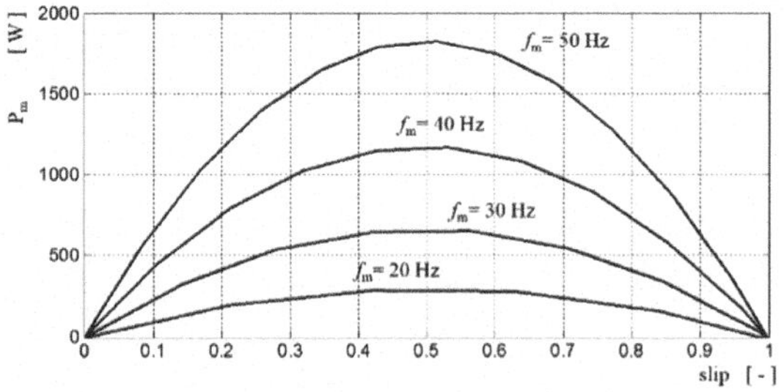

Fig. 6. Transferred power in a function of slip of torque converter (variant 1), for different rotation frequency f_m of excitation system.

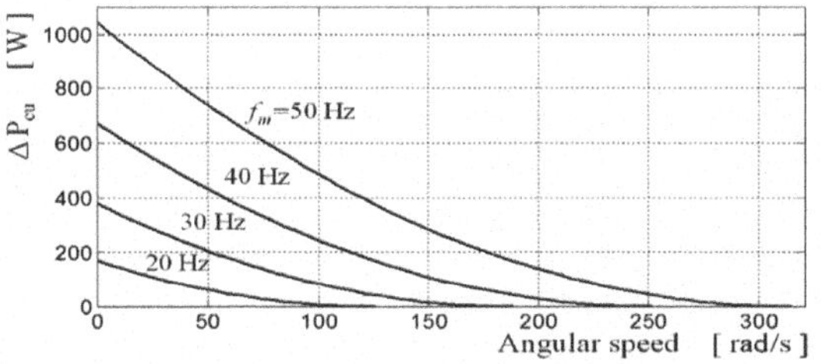

Fig. 7. Power losses in windings of torque converter (variant 1), for different rotational frequencies of excitation system.

Presented curve refers to all analyzed variants of torque converter. Maximum efficiency is in the case of torque converter loaded with a torque (i.e. converter transferring the torque) equal to half of the starting torque. Due to inclination (and shape of external characteristics of torque converter), the working point at which maximum power is transferred, is for half of angular speed of the receiving component. At this point, the efficiency is $\eta = 0.947$ (electromagnetic efficiency). After including the mechanical losses (friction in bearings, ventilation losses, etc.) total efficiency will be lower. It is expected to be $\eta = 0.85$.

4 Summary

The following conclusions can be drawn from the resulted characteristics. Transferred torque depends both on input and output rotational speed of the converter. Increase of input speed causes proportional increase of transferred torque. The torque decreases linearly with an increase of output speed. Transferred power depends on input rotational speed (excitation speed) and output rotational speed. Transferred power increases (linearly) with increase of input speed, while its dependence on output speed has a form of parabola with maximum value for half of output of the maximum possible speed. Maximum losses are for zero speed of the receiving component and they decrease with increase of speed of this component. This relationship is hyperbolic.

Moving the magnetic poles in excitation system apart causes increase in the number of induction harmonics. However, at the same time, the basic harmonic of magnetic field decreases and, in the result, power transferred by electromagnetic torque converter also decreases. Phenomena [14–16] showing the possibility of operation of torque converter at several different rotational speeds were not observed during numerical experiments for induction harmonics presented in Table 1.

References

1. Armstrong, C.G.: Power transmitting device. US Patent No. 687 292 (1901)
2. Faus, H.T.: Magnet gearing. U.S. Patent 2 243 555 (1941)
3. Trout, S.R.: Permanent magnets based on the lanthanides Raw Materials, Processing and Properties. www.spontaneousmaterials.com/Papers/Koreapaper.pdf. Accessed 2016
4. Li, X., Chau, K.T., Cheng, M., Hua, W.: Comparison of magnetic-geared permanent-magnet machines. Prog. Electromagn. Res. **133**, 177–198 (2013)
5. Jørgensen, F.T., Andersen, T.O., Rasmussen, P.O.: The cycloid permanent magnetic gear. IEEE Trans. Ind. Appl. **44**(6), 1659–1665 (2008)
6. Li, K., Bird, J., Kadel, J., Williams, W.: A flux focusing cycloidal magnetic gearbox. In: Magnetics Conference (INTERMAG), IEEE Transactions on Magnetics (2015)
7. Mężyk, A., Tomas, A.: Przekładnie magnetyczne – nowa jakość w transmisji momentu obrotowego, Przegląd Mechaniczny, vol. 10, pp. 40–44 (2013). (in polish)
8. Atallah, K., Howe, D.: A novel high-performance magnetic gear. IEEE Trans. Magn. **37**(4), 2844–2846 (2001)

9. Atallah, K., Calverley, S.D., Howe, D.: Design, analysis and realization of a high-performance magnetic gear. In: IEE Proceedings-Electric Power Application, vol. 151, no. 2 (2004)
10. Jian, L., Chau, K.T.: Design and analysis of a magnetic-geared electronic-continuously variable transmission system using fine element method. Prog. Electromagn. Res. **107**, 47–61 (2010)
11. Chen, M., Chau, K., Liu, C.: Design and analysis of an advanced magnetic variable gear for hybrid electric vehicles. In: Magnetics Conference (INTERMAG) (2015)
12. Shah, L., Cruden, A., Williams, B.W.: A variable speed magnetic gear box using contra - rotating input shafts. IEEE Trans. Magn. **47**(2), 431–438 (2011)
13. Trawiński, T., Kluszczyński, K., Kołton, W.: Lumped parameter model of double armature VCM motor for head positioning system of mass storage devices. Przegląd Elektrotechniczny **87**(12b/2011), 184–187 (2011). (in polish)
14. Trawiński, T., Pilch, Z., Burlikowski, W., Kluszczyński, K.: Operation fundamentals of the electromagnetic torque generator part I - properties of the pulsating torque. In: Technical and Economic Aspect of Modern Technology Transfer in Context of Integration with European Union, vol. 2, pp. 255–260 (2004)
15. Pilch, Z., Burlikowski, W., Trawiński, T., Kluszczyński, K.: Operation fundamentals of the electromagnetic torque generator part II - design process. In: Technical and Economic Aspect of Modern Technology Transfer in Context of Integration with European Union, vol. 2, pp. 261–265 (2004)
16. Trawiński, T., Kluszczyński, K.: Symbolic calculations - tool for fast analyzing poliharmonic models of squirrel-cage motors. Pr. Inst. Elektrot. **216**, 117–129 (2003)
17. Zhang, X., Liu, X., Chen, Z.: A novel coaxial magnetic gear and its integration with permanent-magnet brushless motor. Trans. Magn. (2016). https://doi.org/10.1109/TMAG.2016.2520363
18. Montague, R., Bingham, C., Atallah, K.: Servo control of magnetic gears. IEEE/ASME Trans. Mechatron. **17**(2), 269–276 (2012)
19. Fuxing, Q., Hai, H.: Winding type permanent magnet coupling transmission device. Patent CN104767357 (2015)
20. Fuxing, Q., Hongtao, L.: Magnetic coupling transmission device running at constant power and adjustable speed. Patent CN105006951 (2015)
21. Trofimov, L.A.: Electric-torque converter. U.S. Patent no. 2654849 (1953)
22. Trofimov, L.A.: Electric-torque converter. U.S. Patent no. 2790917 (1957)
23. Waltsheff, D.D.: Electromagnetic transmission torque converter. U.S. Patent no. 2864016 (1958)
24. Mężyk, A., Tomas, A., Fice, M.: Induction torque converter. Patent PL402445 (2013)
25. Tomas, A., Trawiński, T.: Modelowanie charakterystyk pracy indukcyjnego przemiennika momentu. Masz. Gór. R. **34**(2), 64–82 (2016). (in polish)
26. Tomas, A., Trawiński, T.: Możliwości zastosowania indukcyjnego przemiennika momentu w napędach maszyn przeróbczych. Innowacyjne i przyjazne dla środowiska techniki i technologie przeróbki surowców mineralnych. In: Bezpieczeństwo - jakość - efektywność. Komeko 2016. Monografia. Praca zbiorowa. Red. Adam Klich, Antoni Kozieł. Instytut Techniki Górniczej KOMAG, Gliwice: Instytut Techniki Górniczej KOMAG, 2016, dysk optyczny (CD-ROM), pp. 206–226 (2016). (in polish)

Modelling the Anthropomorphic Mechanical Hand

Jakub Turek, Marek Daniszewski, Przemysław Wolnicki, Tomasz Machoczek, and Paweł Jureczko(✉)

Institute of Theoretical and Applied Mechanics, Faculty of Mechanical Engineering, Silesian University of Technology, 18A Konarskiego Street, 44-100 Gliwice, Poland
Pawel.Jureczko@polsl.pl

Abstract. The article presents results of the anthropometric hand design. It describes its kinematic chain and mechanical structure. Furthermore the ranges of motion for specific joints are being shown. Moreover, examples of other mechanical hand solutions are being presented.

Keywords: Mechanical design · Anthropomorphic robotics · Mechanical modeling

1 Introduction

First mechanical extremity counterparts were constructed centuries ago in order to let disabled people without limbs to keep some of the able-bodied human dexterity. Fast development of the electronics in 20th century allowed mankind to introduce more sophisticated prostheses light weight enough not to become a burden for user. The wide variety of the anthropomorphic robotics applications makes this branch attractive, but due to its extensive area of interest, the article focused on the mechanical design of the human-like hand.

2 Process of Human-Like Mechanical Hand Modeling

2.1 Main Development Goals for Our Anthropomorphic Hand Prototype

During development of the project, it was essential to keep as much similarity to a human movement as it was possible. The target was to develop the mechanical platform, which together with control system, would allow an operator to operate human like robotic arm remotely. This was motivated by the idea of providing universal tool for people to do their dangerous work in hazardous environment without taking risk (Fig. 1).

J. Świder et al. (Eds.): MECHATRONICS 2017, AISC 934, pp. 427–434, 2019.
https://doi.org/10.1007/978-3-030-15857-6_42

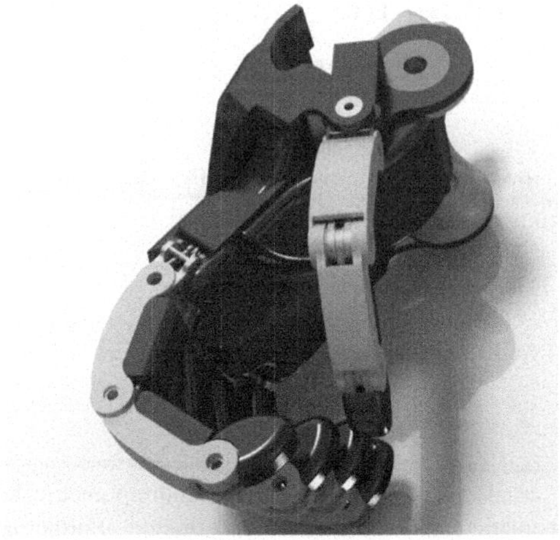

Fig. 1. Rendered image of the hand prototype.

2.2 Concept of the Hand Construction

Modelling such a complex structure is a demanding task. One of the most vital aspect of hand design is to create kinematic chain that would allow to maintain human like mobility. This task requires to transmit power to specific part of the hand (for example finger joints). The first step was to define desired movements in kinematic chain of the hand (Fig. 2).

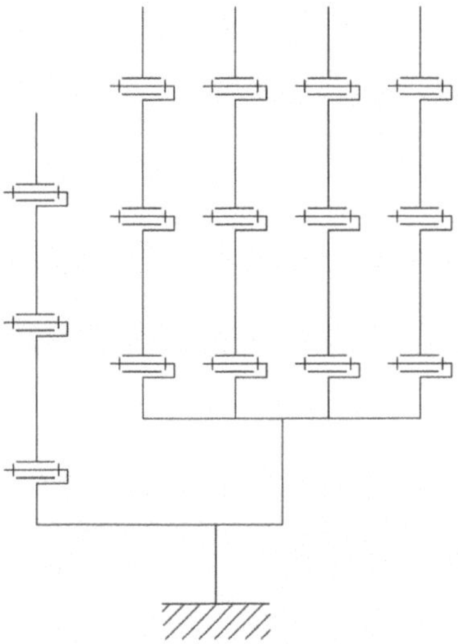

Fig. 2. The kinematic chain of the hand.

2.3 The Finger Construction

The mobility of the used mechanism should allow to mimic human movement, and the most important moving parts of the hand are fingers, thus it was essential to choose suitable solution. The chosen finger construction is complex, which is a disadvantage, but it allows to control phalanges precisely, and has compact design.

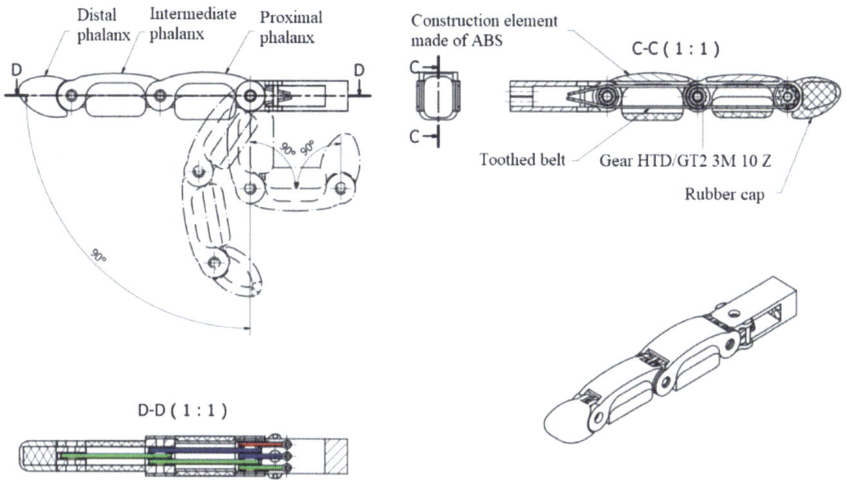

Fig. 3. Construction of the finger.

The color of the lines in Fig. 3 corresponds to the power transmission to specific phalanx: red moves proximal phalanx, blue intermediate phalanx, and green moves distal phalanx. The toothed belt has been used is HTD 3M. Toothed belts are transmitting torque well and do not have slippage when the direction of rotation changes. Thanks to the use of synchronous belts, it is easy to control inclination of the phalanxes.

The way in which the power is transferred to individual phalanxes, forces the control system to add the angle of rotation of the previous joint, so that the angle of inclination of the specific phalanx corresponds to the given angle.

Transferring the torque to the last gear wheel in spite of several gear ratios, where movement resistance occurs and thus decreasing the torque value, is an easy task. Approximately, the final torque is 85–95% compared to the initial one. The synchronous wheels are mounted on a common shaft and are separated by appropriate spacers, which prevent the belt from changing position.

In each joint there is a wheel connected to a specific phalanx and from zero to two double gear wheels responsible for transferring the torque to further parts of the finger.

The tension of the toothed belt is very important for its proper work, so there are openings in each element designed to adjust the tension. The width of the cord due to limited space is 2 mm. The sections thus obtained were applied to each of the fingers.

Such a width of the toothed belts is not available on the market, so the belt was cut from 9 mm to 2 mm. The belts thus obtained were applied to each of the fingers.

Belts at the knuckle, the first of the finger joints, are terminated with a string running in the armour. Through the armour, the string reaches the engine that drives the particular segment (Fig. 4).

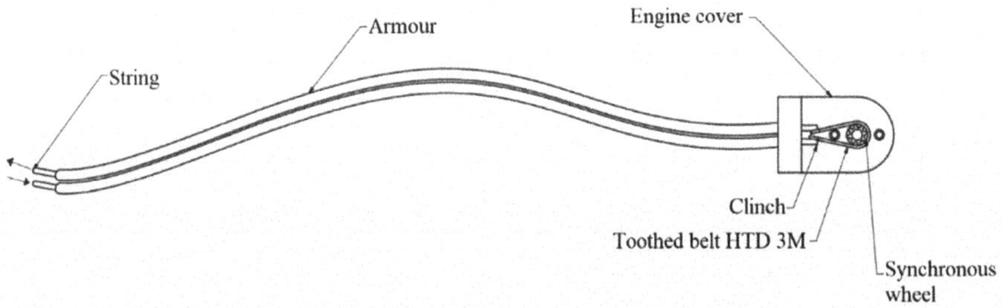

Fig. 4. The way the string is led to the engine.

2.4 The Knuckle Design

The finger is able to rotate around knuckle in horizontal plane (Fig. 5).

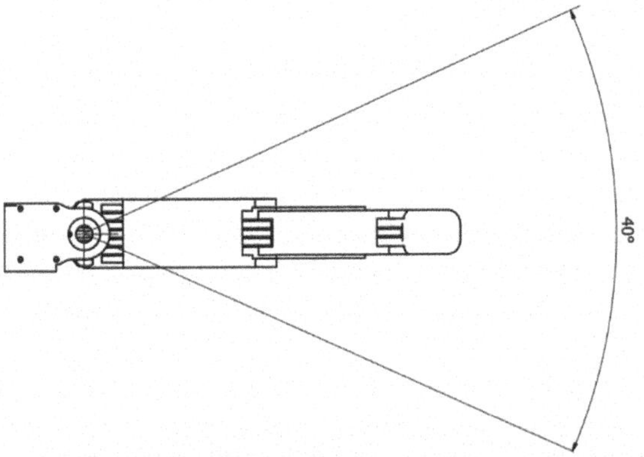

Fig. 5. The finger range of motion in horizontal plane.

To achieve this type of motion the brace was introduced to the structure of the knuckle. The change of the angle between finger and knuckle changes the distance between pulley and the insertion of the armour, so without this brace it would be impossible to rotate in this plane (Fig. 6).

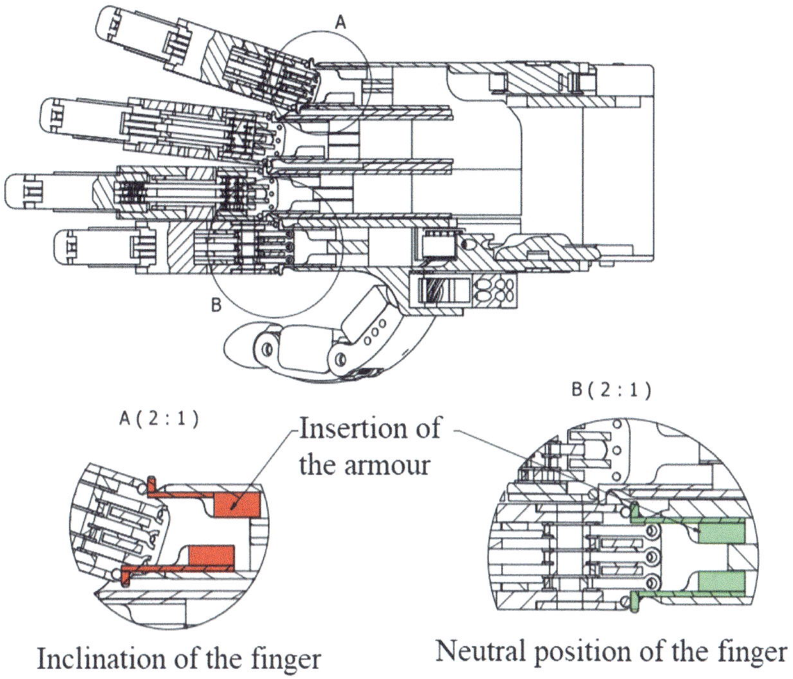

Fig. 6. Cross section of the hand showing the brace mechanism.

The armour is a tube made of the PTFE with 4 mm of outer radius and 2 mm of inner radius. Thanks to the good sliding properties of this material, the additional movement resistance is reduced. 40 armours run from the fingers to the part of the forearm, where the engines driving motion of the fingers are placed. The space inside the hand allows the armours to move freely relatively to each other during the wrist rotation. The inner side of the hand not only protects the armours, but – thanks to the gum it is made of – it imitates human skin allowing to grip items.

2.5 Wrist Construction

The wrist is designed in such a way as to provide enough room for the armour, while allowing adequate anthropometric movement. The motors that drive the rotational movement of the wrist are shown in Fig. 7.

To ensure that the joint would remain slim, two types of torque transition were used. One of them is based on the toothed belts, which was described before. Engine 1 in Fig. 7 provides horizontal rotation of the hand. The gear ratio is 1:3 and generates 3.4 Nm of torque and based on experiments, that allows to lift 3 kg of load at the 70 mm arm. The vertical rotation is provided by the same kind of engine that was described in the first case, although the gear is not alike. The first gear is toothed belt with 1:1.3 gear ratio, and then the torque is transmitted through shaft to the conical gear wheel with 1:2.5 gear ratio. Each joint is supported by two bearings. These are respectively two F608 ZZ and two 61803 ZZ bearings.

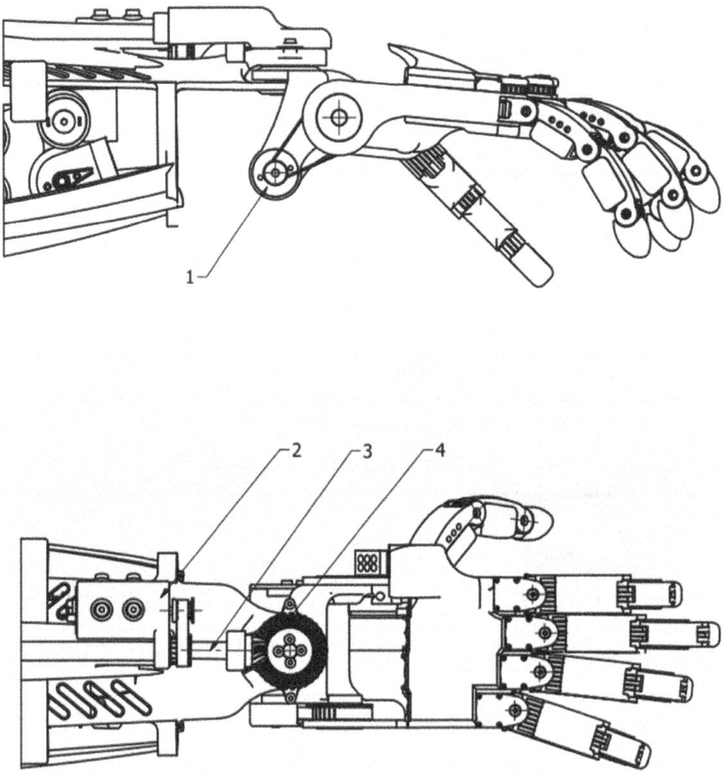

Fig. 7. Positions of the drives at the wrist.

3 Existing Projects for Comparison

Most of the other human-like hands constructions are limited to extremely technologically advanced and expensive laboratory prototypes. Others are very simple and do not meet requirements in terms of grasp possibility. These devices are different in the way of control and construction, and this is one of the most vital part of the project, because of the differences between power transmissions to the parts of fingers. Simple mechanical hands are based on the bands or pushrods. More complex devices use gears in joints to provide motion [1]. Below are some examples of structural solutions of the artificial upper limb.

3.1 FESTO ExoHand

One of the examples of the anthropomorphic mechanical hand is FESTO's ExoHand. This pneumatic-based hand is controlled by wearable exoskeleton glove. The robot is mimicking operator's movement, and thus user can remotely grab or manipulate objects. Based on the force feedback, person can feel what the robot is gripping. Thanks to that feature, it can be used for example as the interface between the real and virtual worlds [2] (Fig. 8).

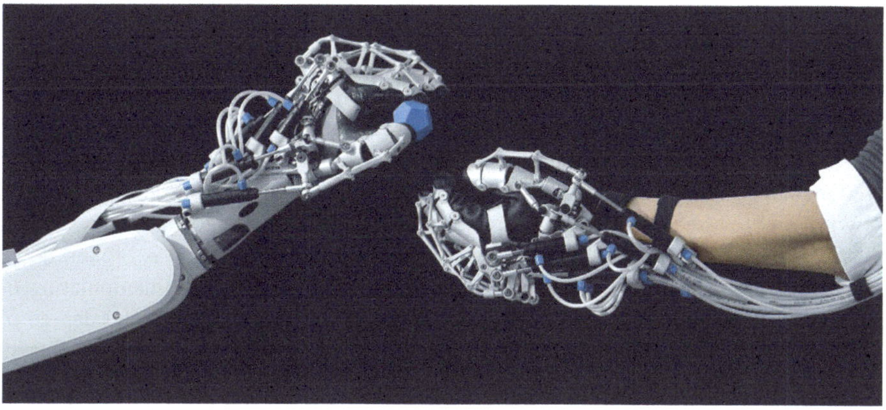

Fig. 8. FESTO ExoHand.

3.2 Shadow Dexterous Hand

This example of anthropomorphic hand developed by Shadow Robot Company is very advanced. Absolute position, force sensor, and 20 actuated degrees of freedom provide very precise movement and can successfully mimic real man movement. Based on 24 joints, every movement range including thumb is very similar to the natural human hand [3] (Fig. 9).

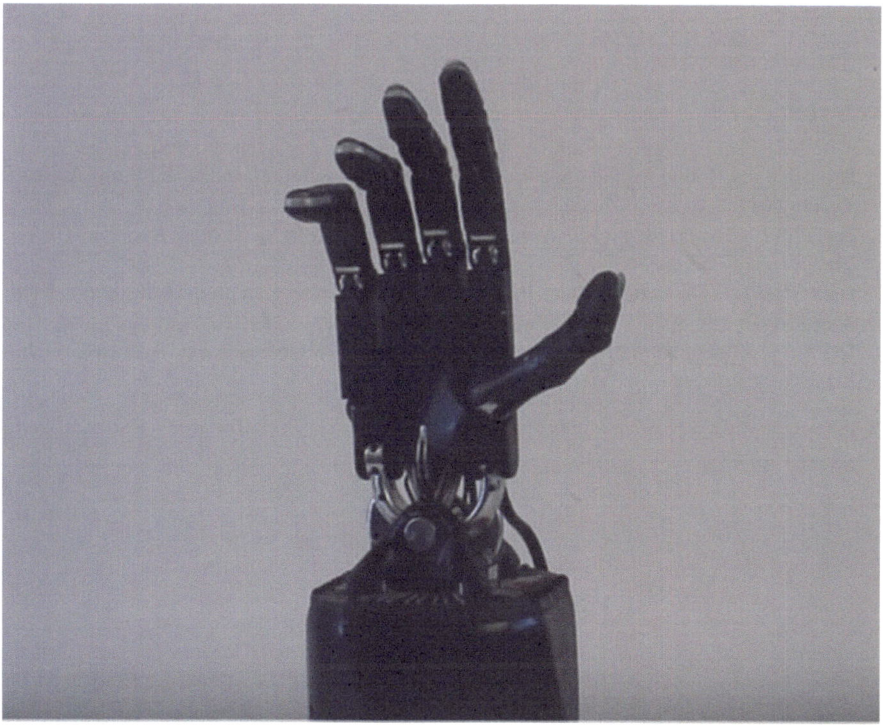

Fig. 9. Shadow Dexterous Hand.

3.3 NASA Robonaut Hand

Many devices are not designed to be manipulated or grasped by regular robots. NASA developed the Robonaut Hand, due to the need of universal, compact tools in space. It was designed to fit all required places and operate extravehicular activity tools. Also, the Robonaut Hand was meant to exceed the human range. It has fourteen degrees of freedom and consists of a forearm in which the motors and drive electronics are placed, a two degree of freedom wrist, and twelve degrees of freedom for a hand with five fingers. The hand allows to use tools based on the idea that the hand is broken down into two sections, one of them responsible for gripping and second manipulating or actuating given object. The grasping part of the hand consists of two, one degree of freedom fingers while dexterous set consist of two three degrees of freedom fingers (pointer and index), and a three degrees of freedom thumb [4].

4 Conclusions

This project has potential for further development. There are still some possibilities to increase anthropometric aspect of its construction like adding additional degrees of freedom to the thumb, or making every phalanxes' length more accurate. The already existing arm that was not covered in this paper is going to have some thorough changes. One of the most vital things is to create advanced control system, that would be safe for people near the robot, and that would successfully imitate human natural movement.

References

1. The Midtrans Robotic Hands Section. http://mindtrans.narod.ru/hands/hands.htm. Accessed 07 July 2017
2. The FESTO ExoHand. https://www.festo.com/group/en/cms/10233.htm. Accessed 07 July 2017
3. The Shadow Dexterous Hand. http://www.shadowrobot.com/products/dexterous-hand/. Accessed 07 July 2017
4. The NASA Robonaut 1. https://robonaut.jsc.nasa.gov/R1/sub/hands.asp. Accessed 07 July 2017

Uncertainty Analysis of the Two-Output RTD Circuits on the Example of Difference and Average Temperature Measurements

Zygmunt L. Warsza[1(✉)] and Adam Idźkowski[2]

[1] Industrial Research Institute of Automation and Measurement (PIAP), Al. Jerozolimskie 202, 02-486 Warsaw, Poland
zlw@op.pl

[2] Faculty of Electrical Engineering, Bialystok University of Technology, Wiejska 45D Street, 15-351 Bialystok, Poland

Abstract. This paper focuses on the expression and the evaluation of results in the case of indirect multi-parameter measurements. A theoretical basis for determining the estimates of values, the uncertainties and the correlation coefficients of indirectly obtained multi-measurand, which are processed from data of the simultaneously measured variables, is shortly presented. It is considered on the examples of a difference and an average of two-temperature indirect measurement by using the RTD circuits, which are used in the laboratory and industrial temperature control systems. The accuracy of measurements depends on the errors of two RTDs as well as two temperature values and the structure of signal conditioning circuits. If a measured temperature difference is small, the high relative accuracy cannot be obtained when two independent measuring circuits are used. The metrological parameters can be better when RTD sensors are into a common two-output measuring system. Therefore, some examples of measuring circuit structures with two RTDs are considered. The numerical example is presented.

Keywords: Measurement circuit · Temperature difference · Sensor · Uncertainty · Multi-measurand

1 Introduction

The simultaneous measurements of several statistically related, i.e. correlated, quantities are performed in metrology and measurement systems, mechatronics and other disciplines. Basing on the digitally processed data of m variables, which are directly measured on the input, the n other variables (observables) can be determined indirectly on the output, if their mutual relation is known. Also, it is important to know about correlation coefficients of output quantities if these variables would be jointly processed in further steps.

J. Świder et al. (Eds.): MECHATRONICS 2017, AISC 934, pp. 435–446, 2019.
https://doi.org/10.1007/978-3-030-15857-6_43

1.1 Theoretical Background About the Uncertainty of Multivariable Measurements

The accuracy of evaluated output multi-measurand depends on the statistical uncertainties of given input multi-measurand, as well as on the accuracy of their processing. Final rounding of indirectly obtained data of output multi-measurand should depend on the uncertainty of the input data [1].

There are two types of the uncertainty components in the indirect multi-dimensional measurements:

- u_A (Type A) calculated by a statistical method from a data sample,
- u_B (Type B) estimated by heuristic methods from knowledge about a measurement process - see "Guide to the expression of uncertainty in measurement (GUM)" [2].

If the uncertainty $u_A \ll u_B$ then it is enough to provide the necessary instrumental resolution and accuracy for measurement of input values and to determine cross-links to the output values. If u_A are of the same level or higher than u_B, even when all environmental effects interacting on input measurements are carefully eliminated, then one should achieve maximum attention to use the information obtained in the experiment. It should not be partially lost in the processing of random input data and in the rounding of the obtained results. The number of observations should be as large as possible to minimize the statistical type uncertainty u_A [3].

1.2 A Few Words About RTD Sensor Temperature Measurements

Resistance temperature detectors (RTDs) are commonly used in temperature measurement instruments and systems. They are connected to the input converters, e.g. the imbalanced bridge [4], by 2-, 3-, or 4-wires. This circuit transforms the temperature dependent changes of RTD resistance into an analogue or digital signal, useful for further processing. Then after conditioning, this signal is sent to the controller and final control device to bring the controlled variable temperature to the proper value [5].

In the highest accuracy temperature measurements taken in metrological laboratories, also within the cryogenic temperature range, the special resistance temperature detectors (SRTD) are used and the measurements are mainly based on the DC or AC automatic balanced bridges [6].

In a number of practical problems and research there is also a need of a simultaneous measurement of two interrelated temperature parameters, e.g. small temperature differences at different points of an object and one temperature or their function. Such needs occur in heating to measure the temperature difference at inlet or outlet of a heating medium and the average temperature of a heated subject, in the study of temperature distribution within heating or cryogenic chamber and in medicine for continuously monitoring of temperature and temperature differences in the various points of human body. If a measured temperature difference is small, when using two independent measuring circuits with temperature sensors the high accuracy of measurements cannot be obtained.

1.3 Related Works

In this work a few basic structures of the two-variable measurement systems with RTD platinum sensors are presented, i.e. the simplified active circuit, the classic bridge resistance circuit and the unconventional one-mesh resistance network, supplied from two current sources. The paper focuses on the temperature measurements when both temperature sensors are connected into a common two-output measuring system. An average and a difference of two measured temperatures are considered as a multi-measurand.

The accuracy of these systems were carried out and compared in previous works [7, 8, 11, 12]. However, it was done in a different way. Analysis was performed for the two-component models of errors and uncertainties of sensors (i.e. their initial value and changes with the sensor resistance). The relationships between systematic errors and the uncertainties of two outputs of circuit were considered in [4] and [7]. The limiting errors and uncertainties of the average and difference temperature measurements with the use of such type of circuits were presented in [8, 12]. The results were obtained for the measuring systems within initial calibration and without recalibration, e.g. in the case of continuous operation of the system, when a rapid replace of the sensor "in situ" is needed without removing the whole circuit from the measured object.

2 Basic Formulas of a Single RTD Element

The processing function of resistance thermometers as temperature sensors is represented by a power series

$$R(T) = R_0 \cdot \{1 + \varepsilon(T)\} = R_0\{1 + \alpha \cdot (T - T_0) + \beta \cdot (T - T_0)^2 + \ldots\}[\Omega] \quad (1)$$

where R_0 – the resistance of thermometer at initial temperature T_0, e.g. 0 °C; $T - T_0$ – temperature increase which is relative to T_0; ε – the relative increase in resistance at temperature T; $\alpha(T_0)$, $\beta(T_0)$ – the temperature coefficients of resistance R.

The temperature coefficients α and β are determined by a material and its cleanliness, and the R_0 value – by its size, shape and resistivity at an initial temperature T_0.

If the resistance R of a RTD changes in a moderate range under the influence of temperature, a linear function can be used and the derogations from a line can be considered as the systematic errors of initial resistance R_0 and its relative increase ε, i.e.

$$R(T) = R_0(1 + \varepsilon) = R_0\{1 + \alpha_T(T - T_0)\}[\Omega] \quad (2)$$

If for some practical reasons it is more convenient to use other temperature T_0' than T_0 given in sensor technical data, e.g. in the begin or middle of the measured temperature range, then

$$R(T) = R_0' \cdot (1 + \varepsilon') = R_0' \cdot \{1 + \frac{R_0}{R_0'}\alpha_T(T - T_0')\}[\Omega] \quad (3)$$

As it is usually stated for many sensors and measuring instruments, the sensor accuracy can be specified by the two-component method [7]. For any variable resistance R_i, given in (4), the absolute error is Δ_i. Hence the relative error δ_{Ri} can be expressed by two relative errors: initial error $\delta_{i0} = \Delta_{i0}/R_{i0}$ and error of increment $\delta_{i\varepsilon} = \Delta_{i\varepsilon}/(R_{i0}\varepsilon_i)$

$$\delta_{Ri} = \frac{\Delta_i}{R_i} = \delta_{i0} + \frac{\varepsilon_i}{1+\varepsilon_i}\delta_{i\varepsilon}[-] \quad (4)$$

The RTD's produce heat when a current is passed through them. Thin film RTD's are more susceptible to self-heating, therefore 1 mA current should not be exceeded. Thus, the voltage is a function of temperature, current and other parameters of sensor:

$$v(t) = i \cdot R_0 + i \cdot R_0 \cdot \alpha \cdot T \text{ [V]} \quad (5)$$

where i – electric current.

3 Temperature Difference Measurement in Two-Channel Circuit

For measuring the temperature difference at least two thermometers are used. Taking into account the characteristics of them the equations are obtained

$$v_1(t) = i_1 \cdot R_{01} + i_1 \cdot R_{01} \cdot \alpha_1 \cdot t_1 \quad (6)$$

$$v_2(t) = i_2 \cdot R_{02} + i_2 \cdot R_{02} \cdot \alpha_2 \cdot t_2 \quad (7)$$

where: $t_1 = T_1 - T_0$, $t_2 = T_2 - T_0$ – temperatures measured by the sensors defined as the increases from initial temperature.

If the currents of two sensors are stable over time and voltages (6), (7) are linearly processed by the input circuits then the outputs $V_1(t)$, $V_2(t)$ of these systems in the general case are as follows

$$V_1(t) = B_1 + K_1 \cdot t_1 \quad (8)$$

$$V_2(t) = B_2 + K_2 \cdot t_2 \quad (9)$$

where: B_1, B_2 – the initial values of output signals for temperature T_0; K_1, K_2 – the processing coefficients for two voltages; T_1, T_2 – temperature measured by both sensors.

Then voltages depending on temperature for both sensors (6), (7) and signals (8) and (9) at the outputs of their linear input transducers are described by two linear functions. Their parameters are processing (sensitivity) coefficients K_1 and K_2 and the parameters B_1 and B_2.

When two thermometers are used to measure the temperature difference $T_D = t_1 - t_2$, which is positive for $t_1 > t_2$, we obtain an expression for the difference of sensor signals in two forms:

$$\begin{aligned} V_1 - V_2 &= K_1 \cdot t_1 - K_2 \cdot t_2 + B_1 - B_2 = K_1(t_2 + t_D) - K_2 t_2 + B_1 - B_2 \\ &= (K_1 - K_2)\, t_2 + K_1 t_D + (B_1 - B_2) = (K_1 - K_2)\, t_1 + K_2 t_D + (B_1 - B_2) \end{aligned} \tag{10}$$

The Eq. (10) is then a linear function not only limited to the same temperature difference t_D with sensitivity coefficient K_1 or K_2, it is a function of t_1 or t_2 as well. When you setup zero for both sensors at temperature t_1 or t_2, i.e.

$$(K_1 - K_2)t_2 + (B_1 - B_2) = 0 \text{ or } (K_1 - K_2)t_1 + (B_1 - B_2) = 0. \tag{11}$$

Then this formula (14) is simplified to the form:

$$V(t_D) = K_1 t_D \text{ or } V(t_D) = K_2 t_D. \tag{12}$$

If sensitivity factors $K_1 = K_2$ of the sensors are identical then the Eq. (12) will be independent from temperature T and equal to zero if $B_1 = B_2$. In case of non-identical sensors, the fulfillment of both conditions $K_1 = K_2$, $B_1 = B_2$ may occur while processing of sensor output voltage and it requires at least two regulations in a measuring system. It is, e.g. a zero adjustment for the initial temperature range T_0 or different predetermined temperature and a gain adjustment in one of the channels at a different temperature.

This can be realized in a measuring system of temperature difference with two separate measuring channels before subtraction of signals depending on the voltage of both sensors [8, 12]. It is recommended that zero and sensitivity adjustments were independent and situated either in one of these channels or each in the different channel (Fig. 1).

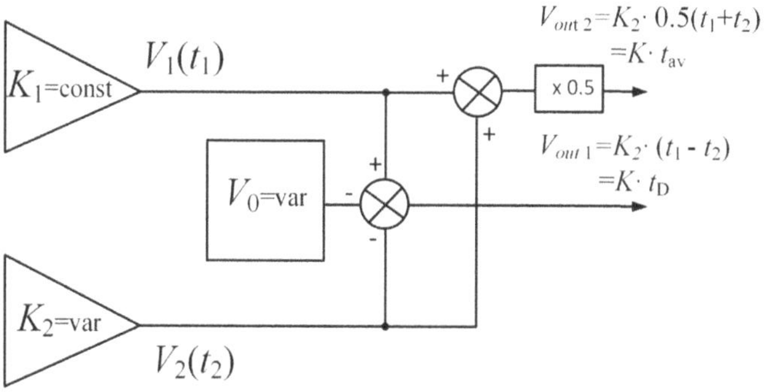

Fig. 1. Sensitivity and offset nulling in a two-channel circuit for measuring a difference t_D and an average t_{av} of two temperatures.

The average $K \equiv 1/2(K_1 + K_2)$ can be assumed as the value of sensitivity. Then for $K_1 > K_2$ one receives $K_1 = K + \Delta K$ and $K_2 = K - \Delta K$. Taking into account that $t_1 = t_2 + t_D$ the formula (10) is expressed as

$$\begin{aligned} V_1 - V_2 &= (K + \Delta K) \cdot t_1 - (K - \Delta K) \cdot t_2 + B_1 - B_2 \\ &= K\, t_D + \Delta K (t_1 + t_2) + (B_1 - B_2) \end{aligned} \tag{13}$$

In temperature difference measurements as a fundamental equation one can assume the simplified formula

$$V(t_D) = K_D \cdot t_D + B_D. \tag{14}$$

Then a second part of Eq. (13): $\Delta K(t_1 + t_2) = 2\,t_2 \Delta K + \Delta K\, t_D = 2\,t_1 \Delta K - K\, t_D$ has two components. They, in different ways, have influence on the intercept B_D and the sensitivity K_D, depending on whether the same meter reading t_D is set at both sensors placed in a temperature t_2 or t_1, i.e.

$$\text{for } t_2 \quad K_D = K + \Delta K \text{ and } B_D = B_1 - B_2 + 2\Delta K\ t_2, \tag{15}$$

$$\text{for } t_1 \quad K_D = K - \Delta K \text{ and } B_D = B_1 - B_2 - 2\Delta K\ t_2. \tag{16}$$

Such a calibration can be made automatically in case of a laboratory measurement of temperature difference t_D when at least one of the sensors is permanently mounted. Because the factors K_D and B_D are proportional to the current flowing through the sensor, the setting current can be used in the calibration process. However, we should take into account that the adjustment, by changing a sensor current, makes the change in proportion of both components in the formula (8) or (9).

If both thermometers are calibrated at the same temperature and the current flowing through one of them varies for the voltage setting $V(t_D) = 0$ then $B_D = 0$ is obtained.

Then the dependence of a voltage difference on a temperature difference is a linear function of one factor K_D, i.e. $V(t_D) = K_D \cdot t_D$. A specific temperature difference between the sensors should be set and the occurring difference of both voltages should be measured to determine the numerical value of this factor. For example, one of the RTDs is placed in an environment with temperature t_2, and the other remains at t_1 wherein the first sensor is calibrated.

4 Example Structures of the 2D Temperature Measuring Circuits

4.1 The Measurement of a Difference and an Average of Temperatures in the Loop with a Common Supply Current

The simplified active circuit with RTDs for measuring a difference and an average of two temperatures is presented in Fig. 2. It is based on the well-known circuit (Anderson's loop) with sensors connected in series and is supplied by the common stable current source J. The settings of a zero signal $V_0(t = 0) = 0$ for the initial temperature T_0 and the settings of sensitivity K for at least one of temperature channels should be located in the second pair of amplifiers. In most of temperature measurements with middle accuracy in laboratory, industrial and other applications this type of circuit brings a satisfactory solution.

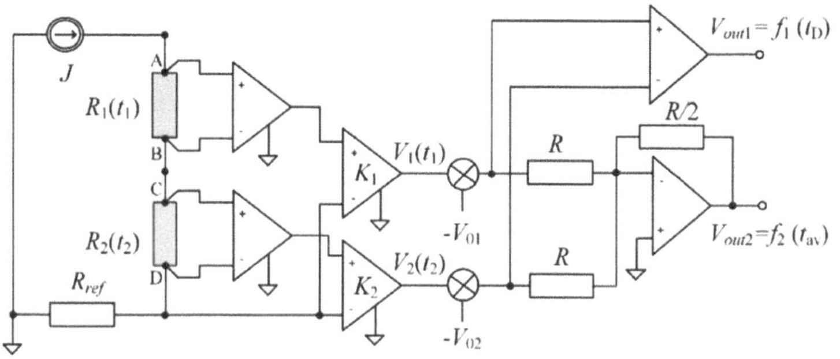

Fig. 2. The simplified active circuit with two 4-wire RTDs for measuring a difference t_D and an average t_{av} of temperatures t_1, t_2.

4.2 A Differential Amplifier with Classic Bridge-Circuit

The idea of a two-sensor circuit used for measuring of temperature difference t_D with a differential voltage comparator is presented in Fig. 3. The lower input of a comparator (at point D) is connected to a voltage of the sensor R_2 in temperature t_2 and another input (at point C) is connected to the sensor R_1, which is temperature t_1. The amplifier circuit is used to detect a temperature difference t_D as the output voltage becomes a function of the changes in active arms R_1 and R_2 of the resistive bridge. Resistors R_3 and R_4 enable to set the zero and the range of measured temperature difference.

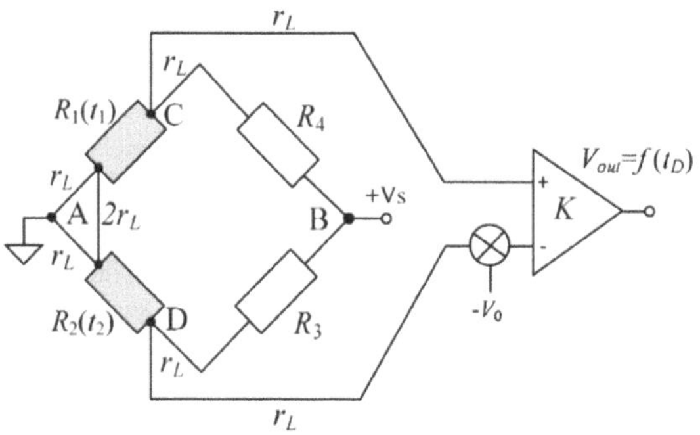

Fig. 3. The classic bridge-circuit with two 4-wire RTDs and a differential amplifier for temperature difference measurement.

This circuit can directly measure the difference of two temperatures $t_D = t_1 - t_2$ but its reading and accuracy measures depend on T_0 as well. Then, the additional circuit to measure one of these temperatures should be added. For example, a variable resistor can be used instead of the sensor R_2. Resistors R_3 and R_4 usually are a matched divider pair or very low temperature coefficient devices.

4.3 The Unconventional Double-Current Circuit

The 4R single-loop network unconventionally supplied, as it is presented in Fig. 4, can be also used to measure a difference and a sum of temperature increments.

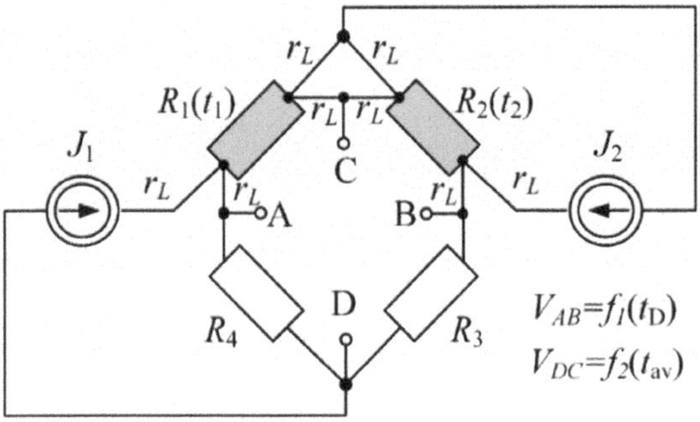

Fig. 4. An idea of two-output resistance to voltage converter (the double-current circuit 2J).

The original concept is presented in [9] and verified experimentally [10]. The circuit is powered by two equal stable current sources $J_1 = J_2 = J$ connected in parallel to opposite arms of the 4R loop or by one source J switched between these arms. In this type of circuits, the acronym 2J was proposed. This circuit has two output voltages V_{AB} and V_{CD} on diagonals (Fig. 4). The balance conditions for each diagonal are different, i.e. $V_{CD} = 0$ if $R_{10}\ R_{20} = R_3\ R_4$ and $V_{AB} = 0$ if $R_{10}\ R_4 = R_{20}\ R_3$. Then both balance conditions occur when the products of pair of neighboring resistances, connected to each output, are equal. This 2J circuit is simultaneously balanced in two diagonals if $R_{20} = mR_{10}\ R_{40} = mR_{10}$ and $R_{30} = R_{10}$. Only if all initial resistances $R_{10} = R_{20} = R_{30} = R_{40}$ are equal ($m = 1$) then the 4R loop circuit is balanced. Also, it occurs for $R_{10}\ R_{30} = R_{20}\ R_{40}$, as it is in the case of a bridge classically supplied. Accuracy analysis of the 2J circuit was published in [11].

Two transfer resistances (r_{AB}, r_{CD}) of the 2J circuit depend on the resistance increments in different way. For all variable resistances, $|\varepsilon_i| \ll 1$; $\varepsilon_i \approx \alpha_{Ti}(T_i - T_0) = \alpha_{Ti}\, t_i$, $\alpha_{Ti} = \alpha_T$, the output voltages are

$$V_{AB} = Jr_{AB} = Jr_{AB0}f_3\{\varepsilon_i(t_i)\} \cong J \cdot r_{AB0} \cdot \alpha_T(t_1 - t_2 - t_3 + t_4), \tag{17}$$

$$V_{CD} = Jr_{CD} = Jr_{CD0}f_4\{\varepsilon_i(t_i)\} \cong J \cdot r_{CD0} \cdot \alpha_T(t_1 + t_2 - t_3 - t_4), \tag{18}$$

where: $r_{CD0} = \frac{R_{10}R_{20}}{\Sigma R'_{i0}}, \quad r_{AB0} = \frac{R_{10}R_{40}}{\Sigma R''_{i0}}$.

When the circuit in Fig. 4 has only two variable resistances R_1, R_2. If the nominal parameters of RTD sensors are the same, i.e. $R_{20} = R_{10}$, $\alpha_{T1} = \alpha_{T2} \equiv \alpha_T$, the resistances are $R_4 = R_{10}$, $R_3 = R_{10}$, then $r_{\mathrm{CD0}}=r_{\mathrm{AB0}} = \frac{1}{4}R_{10}$, $V_{\mathrm{AB0}} = 0$, $V_{\mathrm{CD0}} = 0$, $K_{\mathrm{AB}} = K_{\mathrm{CD}} = K$ and

$$V_{AB} = 0.25J \cdot R_{10}\alpha_T(t_1 - t_2) = K \cdot t_{\mathrm{D}} \tag{19}$$

$$V_{CD} = 0.25JR_{10}\alpha_T(t_1 + t_2) = 2K \cdot t_{\mathrm{av}}, \tag{20}$$

where: t_{av} – average temperature.

The output voltages of the 2J circuit are proportional to a difference and a sum of temperature increments t_1, t_2, starting with the initial temperature T_0. The sensitivities K of both outputs are the same for $R_{i0} = R_{10}$ only. The above mentioned information about the necessity to use RTD sensors with equal temperature coefficients of resistance $\alpha_{T1} = \alpha_{T2}$ is also actual in this case. It is possible to adjust zero and sensitivity K in this circuit by means of varying R_3 and R_4.

5 The Uncertainty of Temperature Difference and Average Measured Simultaneously (Two Sensors in One Circuit)

5.1 Theoretical Background of the Multivariable Measurements

The input measurand is formulated by a random vector $\boldsymbol{X} = [X_1, X_2, \ldots, X_m]$ and the output measurand by a vector $\boldsymbol{Y} = [Y_1, Y_2, \ldots, Y_n]$. These vectors of dimensions m and n, respectively, express the multi-dimensional distributions of $\boldsymbol{X}$ and $\boldsymbol{Y}$. The relation between them is presented by the generalized formula

$$\boldsymbol{X} = F(\boldsymbol{Y}). \tag{21}$$

If F is a linear operator, then $n \leq m$.

The basic structure of the numerical estimation of the multi-measurand should contain averaged components of the random vector and a description of the multidimensional scatter region. If this region can be defined by a model of joint n-dimensional probability distribution then, for a given probability density $\boldsymbol{p}(X_1, \ldots, X_n)$ of n-dimensional normal distribution, such region takes the form of a n-dimensional hyper-ellipsoid. Its center is located at the end of the average vector [3]. The relation between covariance matrices of hyper-ellipsoids of the output and input measurand (in a linear approximation of the observables in the vicinity of the end of mean vector $\bar{\boldsymbol{X}}$) is expressed as

$$\boldsymbol{c}_{\mathrm{Y}} = \boldsymbol{S}\,\boldsymbol{c}_{\mathrm{X}}\,\boldsymbol{S}^{\mathrm{T}}, \tag{22}$$

where $\boldsymbol{S} \equiv \partial \boldsymbol{Y}/\partial \boldsymbol{X}$ is the matrix of sensitivity coefficients for nonlinear functions [2]. The matrix $\boldsymbol{r}$ of the correlation coefficients (correlator) is defined by the relation

$$\boldsymbol{r} = \boldsymbol{\sigma c \sigma}^{\mathrm{T}}, \tag{23}$$

where $\boldsymbol{\sigma}$ is the diagonal matrix of standard deviations.

A multivariable distribution is normal if matrices $\boldsymbol{c}$ and $\boldsymbol{r}$ are positive definite, i.e. their eigenvalues λ_i, which are roots of characteristic equations det[$\boldsymbol{c} - \lambda \mathbf{1}$] = 0, and det[$\boldsymbol{r} - \lambda \mathbf{1}$] = 0, respectively should be positive [3].

According to Law of Propagation of Uncertainty (LPU) [2] the uncertainty of sum and average of two measured temperatures by means of two-channel measuring system can be presented by using the formulas (21)–(23). Both output quantities t_D, t_{av} depend on the same (two) input quantities

$$t_D = \frac{V_{out1}}{K_1} = t_1 - t_2, \tag{24}$$

$$t_{av} = \frac{V_{out2}}{K_2} = \frac{t_1 + t_2}{2}. \tag{25}$$

A covariance matrix $\boldsymbol{c_Y}$ is

$$\mathrm{c_Y}(t_D, t_{av}) \cong \left[\frac{\partial \mathrm{Y}}{\partial \mathrm{X}}\right] \cdot \mathrm{c_X}(t_1, t_2) \cdot \left[\frac{\partial \mathrm{Y}}{\partial \mathrm{X}}\right]^T = \begin{bmatrix} c_{11} & c_{12} \\ c_{21} & c_{22} \end{bmatrix} = \begin{bmatrix} \sigma_{t_D}^2 & \mathrm{cov}(t_D, t_{av}) \\ \mathrm{cov}(t_{av}, t_D) & \sigma_{t_{av}}^2 \end{bmatrix}, \tag{26}$$

$$\mathrm{c_Y}(t_D, t_{av}) = \begin{bmatrix} \sigma_{t1}^2 + \sigma_{t2}^2 & 0.5 \cdot (\sigma_{t1}^2 - \sigma_{t2}^2) \\ 0.5 \cdot (\sigma_{t1}^2 - \sigma_{t2}^2) & 0.25 \cdot (\sigma_{t1}^2 + \sigma_{t2}^2) \end{bmatrix}. \tag{27}$$

Jacobian matrix contains the partial derivatives of the scalar components of $\boldsymbol{Y}$ with respect to the scalar elements of $\boldsymbol{X}$. Then correlation coefficient is:

$$\rho_{xy} = \frac{\mathrm{cov}(t_D, t_{av})}{\sigma_{r_1} \cdot \sigma_{r_2}} = \frac{c_{12}}{\sqrt{c_{11}} \cdot \sqrt{c_{22}}} = \frac{\sigma_{t1}^2 - \sigma_{t2}^2}{\sigma_{t1}^2 + \sigma_{t2}^2}, \tag{28}$$

The result of t_D (or t_{av}) measurement can be presented as the mean and the surrounded area (ellipse) with 99% or 95% level of confidence.

5.2 Numerical Example

Let us consider a set of measurements obtained from a measuring system (2J circuit). In Table 1 there are the values of input quantity $\boldsymbol{c_X}$, which is assigned a bivariate Gaussian PDF.

Table 1. The results of two temperatures obtained from a measuring system (2J circuit).

Sample no.	t_1 (°C)	t_2 (°C)
1	21.141	22.099
.	21.139	22.105
.	21.139	22.103
.	21.140	22.103
.	21.142	22.105
.	21.146	22.104
.	...	...
100	21.137	22.101
	$\bar{t}_1$	$\bar{t}_2$
Mean	21.140	22.102
	σ_{t1}	σ_{t2}
Std. dev.	0.0026	0.0020

After using the formulas (24)–(27) the matrix c_Y is obtained:

$$c_Y(t_D, t_{av}) = 10^{-6} \cdot \begin{bmatrix} 10.45 & 1.28 \\ 1.28 & 2.61 \end{bmatrix}. \tag{29}$$

After using the formula (28) the correlation coefficient is 0.245.

For a condition $c_{11} > c_{22}$ and correlation coefficient ρ_{xy} near zero, the axes of the ellipse (contours of the joint PDF for the input quantities t_1, t_2) are parallel to the coordinate axes with the major axis parallel to the horizontal axis. In the case of matrix (29) the correlation ρ_{xy} is greater than 0. Therefore, the ellipse has a slope [1–3].

The results obtained by using Gauss's formula (GUM) and Monte Carlo Method (MCM) – sample size 100000, were similar, i.e.

- $t_D = -0.962$ °C, $\sigma(t_D) = 0.0033$ °C, MCM coverage interval $t_D = (-0.970, -0.954)$ °C for $k = 2.4$ and $p = 99\%$ or $t_D = (-0.9684, -0.9556)$ °C for $k = 2$ and $p = 95\%$;
- $t_{av} = 21.621$ °C, $\sigma(t_{av}) = 0.0016$ °C, MCM coverage interval $t_{av} = (21.617, 21.625)$ °C for $k = 2.4$ and $p = 99\%$ or $t_{av} = (21.618, 21.624)$ °C for $k = 1.8$ and $p = 95\%$.

6 Conclusions

The general idea of temperature difference accuracy analysis is presented. The way of estimating t_D (and t_{av}) accuracy is explained and it can be expanded for the measurements conducted with the use of various circuits. There is presented the theoretical background of the 2D measurements. After all, there are illustrated three following circuits measuring simultaneously a difference and an average of two temperatures with the use of RTDs such as:

– the well-known bridge-circuit to detect temperature difference t_D,
– the loop of two sensors supplied by a common current source with processing sensor voltages by operational amplifiers (modification of Anderson's loop),
– the unconventional circuit supplied by two equal current sources (shortly named as the 2J circuit).

The active two-channel circuit (Fig. 2) enables to adjust individually each channel to a particular temperature, however it provides less accuracy when measuring a temperature difference than other circuits. The 2J circuit (Fig. 4) can be used directly to measure a temperature difference and an average temperature. It seems it is more accurate than the active two-channel circuit.

It is important to note that the uncertainties of t_D and t_{av} measurements do not only depend on both measured temperature values and their standard deviations (type A uncertainty), as it is presented in the numerical example. They depend also on a type of structure and the error (type B uncertainty) of signal conditioning circuit.

References

1. Warsza, Z., Ezhela, V.: About evaluation and publication of multivariate measurement data – current status and needs. Pomiary Automatyka Robotyka **10**, 68–76 (2011)
2. Evaluation of measurement data – Supplement 2 to the Guide to the expression of uncertainty in measurement – Extension to any number of output quantities. http://www.bipm.org/utils/common/documents/jcgm/JCGM_102_2011_E.pdf. Accessed 31 Aug 2017
3. Warsza, Z.: Evaluation and numerical presentation of the results of indirect multivariate measurements. In: Advanced Mathematical and Computational Tools in Metrology and Testing IX, pp. 418–425. World Scientific Publishing Company, Singapore (2012)
4. Warsza, Z., Idzkowski, A.: Errors and uncertainties of imbalanced bridge-circuits as primary converters for RTD sensors. Advances in Intelligent Systems and Computing, vol. 393, pp. 397–410, Springer, Heidelberg (2016)
5. Mikhal, A.A., Warsza, Z.L.: Simple methods to measure the additive error and integral nonlinearity of precision thermometric bridges. Advances in Intelligent Systems and Computing, vol. 352, pp. 157–170, Springer, Heidelberg (2015)
6. Weier, S., Shafi, M.A., McMahon, R.: Precision calorimetry for the accurate measurement of losses in power electronic devices. IEEE Trans. Ind. Appl. **46**(1), 278–284 (2010)
7. Warsza, Z., Idzkowski, A.: Relations of limited errors and uncertainties type B of the bridge as conditioner circuit for sensors of broadly variable resistances. Przeglad Elektrotechniczny – Electr. Rev. **90**(12), 83–86 (2014)
8. Warsza, Z., Idzkowski, A.: About accuracy of the simultaneous measurement of temperature difference and average using transducers with Pt sensors. Przem. Chem. **96**(2), 380–385 (2017)
9. Warsza, Z.: Bridges supplied by two current sources – new tool for impedance measurements and signal conditioning. In: Proceedings of IMEKO-TC 7 Symposium, Cracow, pp. 231–236 (2002)
10. Walendziuk, W., Golebiowski, J., Idzkowski, A.: Comparative evaluation of the two current source supplied strain gauge bridge. Elektronika ir Elektrotechnika **22**(6), 33–38 (2016)
11. Warsza, Z., Idzkowski, A.: Accuracy analysis of the 2D resistance-to-voltage converters for resistive sensors. Meas. Autom. Monit. (MAM) **61**(10), 460–464 (2015)
12. Warsza, Z.L., Puchalski, J.: Matrix estimation of uncertainty of indirect multi-parameter measurements with examples. Meas. Autom. Robot. Part 2 31–40 (2018). (in Polish). 1014311/PAR_228/31-40

Interactive Controller Aiding the Process of Upper Limb Rehabilitation

Piotr Wodarski(✉), Marek Gzik, Miłosz Chrzan, Andrzej Bieniek, Barbara Łopacka, and Robert Michnik

Department of Biomechatronics, Faculty of Biomedical Engineering, Silesian University of Technology, Akademicka 2A, 44-100 Gliwice, Poland
{piotr.wodarski,marek.gzik,milosz.chrzan, andrzej.bieniek,robert.michnik}@polsl.pl, barbara.lopacka@gmail.com

Abstract. Interactive devices supporting the rehabilitation of upper limbs are becoming more and more popular among physicians and physiotherapists. It has been observed that the motivation increases in patients while doing exercises aided by computer systems. A controller developed in this research supports the process of diagnosis and therapy. It has been made in the form of a glove worn on the upper limb. The device enables the patient to control an application in the virtual world of computer graphics by detecting the movements of the limb and flexion of two selected fingers. This work encompasses the concept of the device, documentation of the prototype and verification of its functioning. The verification was conducted in two stages: the first one focused on the verification of the operation of tensometric flex sensors, and the second on the verification of indications of an orientation sensor. The verification tests determined the hysteresis of tensometric sensors and a mean-square error as well as a correlation coefficient between different system orientation values measured by means of various measuring systems. The verification enabled the determination of uncertainty of parameters measured as well as functionality of the device.

Keywords: Virtual Reality Technology · Supporting rehabilitation processes · Interactive controller

1 Introduction

Rehabilitation aims at the improvement of the level of human physical ability and thus the improvement of life quality. A traditional therapy is conducted in many stages which are of a different nature: healing, pedagogical, vocational, psychological or social. The fact whether somebody needs rehabilitation, and to what extent, is decided by a board of specialists on the basis of complex examination results [1]. While selecting a specific rehabilitation method it is of the utmost importance to consult a physician or a therapist in order to determine the type of disease and its cause. As far as the arm rehabilitation is concerned, in most cases an efficient therapy is provided by movement methods, which include a whole range of exercises. Such exercises can be done by the patients themselves or with the assistance of therapists in order to improve or restore movement capacity in the area of upper limbs.

J. Świder et al. (Eds.): MECHATRONICS 2017, AISC 934, pp. 447–454, 2019.
https://doi.org/10.1007/978-3-030-15857-6_44

In the process of rehabilitation patients may also use special devices which can either replace the entire missing limb or partially cover a given organ to support the recovery of the lost sensation. Application of such devices enables, among other things, the performance of a mirror therapy, which is very popular in the USA and Western Europe [3, 7]. This therapy is particularly used in patients with paresis or after the upper limb amputation. In front of a mirror, patients cover the sick limb and use the healthy limb to make movements while watching its reflection. This creates an effect of illusion thanks to which the patients may change the way of reception of stimuli by their brain. The reflection of the healthy limb is perceived as a real image of the missing limb or the disabled limb. This therapy is an effective method that enables the restoration of lost functions of the limbs or makes it possible to deal with other issues resulting from such conditions. There are special devices dedicated to the mirror therapy of the upper limb available on the market, such as Internet video cameras, Kinect, 3D optic systems, inertial systems, etc. [2, 9]. Application of these devices transports the users into the virtual world so that they could see their mirror reflection, for example in the form of an avatar. Different exercises are then done, including the motion of pronation and supination of the forearm, clenching fingers, lifting up arms, etc. The results of the tests which examined the interactive devices therapy show that this kind of therapy is more efficient than classic exercises [4, 6, 10].

Interactive controllers supporting the rehabilitation process may also be used in the case of exercises done with the PNF (Proprioceptive Neuromuscular Facilitation) method. This method encompasses a wide range of movements made by the upper limb, which can be implemented into a scenario of a therapeutic game. Such a game makes the user execute certain movements which activate particular parts of muscles [10, 11]. There are also various types of prostheses available on the market. Not only do they replace missing limbs [5] but also enable the performance of exercises involving the movement of both limbs. It is possible due to the application of bio-electrical sensors [8].

There are no universal devices on the market which would serve the purpose of assessment of motor dysfunction of the upper limb with the possibility to use them for the rehabilitation of upper limbs. However, there are many prostheses that replace the whole missing limb. Such devices are often uncomfortable, heavy and onerous for the users. There is still a great need for new devices enabling the rehabilitation of upper limbs, particularly hands and fingers.

2 Research Objective

The objective of the research was to develop a concept and a prototype of a mechatronic device dedicated to the support of diagnostics and therapy in the scope of upper limbs. The research included testing of the device functionality and verification of the correctness of the measured parameters.

3 Device Structure

In order to meet the demand for the development of new devices aiding the rehabilitation of upper limbs, a model of an interactive device in the form of a glove was created. The purpose of this device is to restore a certain scope of movement in the area of the upper limb. The solution developed may also be used as a controller in the form of a computer mouse to control computer applications dedicated to rehabilitation support.

The structure of the device consists of the following components: two tensometric flex sensors 1.12 × 6.3 mm – SparkFun, an accelerometer and a gyroscope in the form of the MPU6050 sensor, a Bluetooth set with the HC-06 module, a controller in the form of microcontroller AVR Atmega 32u4 (on Arduino ProMicro microcontroller board), which was programmed in 'Arduino IDE' environment. Additionally, the device is equipped with the source of supply in the form of PowerBank as well as an elastic retractable microUSB wire connecting the device with the power source. All the elements are placed in casings and implemented into an elastic glove, as shown in Fig. 1. In the glove design a soft and stretching fabric (neopren) was used, which provides comfort while wearing the device on the forearm.

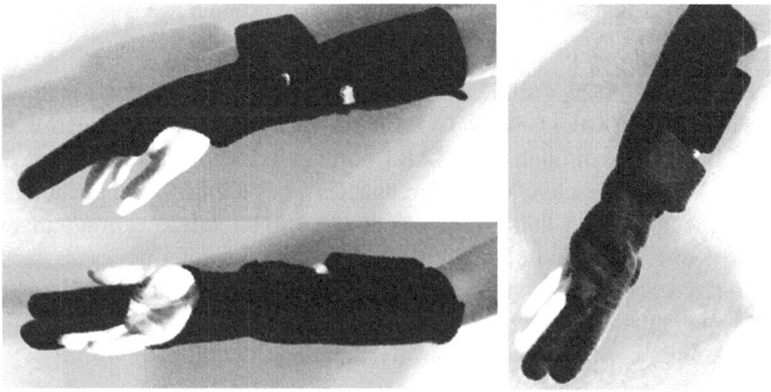

Fig. 1. View of the glove with a visible pocket for PowerBank.

Individual components were connected according to the diagram presented in Fig. 2.

A special software program was developed for the device and after compilation it was uploaded into the microcontroller memory. Its algorithm includes the following steps:

1. initialization of the microcontroller,
2. initialization of internal components,
3. initialization of communication with embedded peripheral modules,
4. configuration of the MPU6050 sensor,
5. configuration of the HC06 module,

6. configuration of interrupts, IRQ (interrupt request line) information on orientation and acceleration measured by the MPU6050 sensor,
7. cyclic readout of voltage from tensometric dividers,
8. cyclic transmission of data through the bluetooth module.

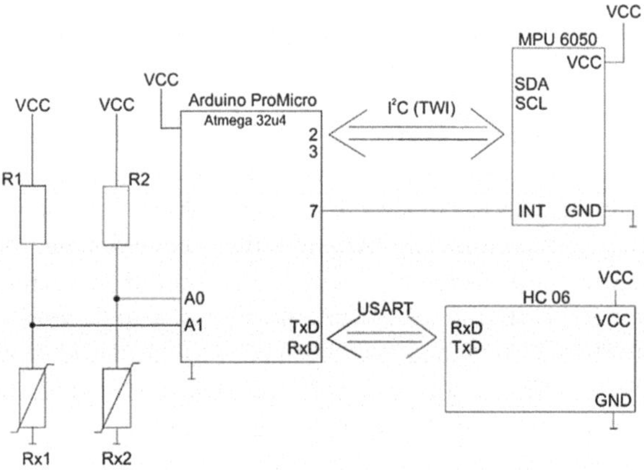

Fig. 2. Connection diagram of the electronic part.

The HC06 module is visible in computer as controller of the glove and it is compatibility with HID standard and COM port. The exchange of information from the microcontroller is done through the USART port at the speed of 115200 bauds/second.

The developed interactive controller enables controlling the applications which support the diagnosis and therapy of the disabled. The software program makes it possible to connect the controller with a graphic engine offering the possibility of visualization of a given application in Oculus 3D system. This can be done to serve the purpose of rehabilitation as well as further research on the systems aiding diagnostics and rehabilitation of patients with motor dysfunctions.

4 Methodology of Verification Tests

In order to test the correctness of the controller's operation two series of tests were performed. The first stage focused on the examination of the operation of flex sensors. According to the information provided by the manufacturer, these sensors are of a linear characteristic. The change of the working point could take place due to the elements used in the system and as a result of their tolerance. It was thus necessary to test linearity of the whole system along with the sensors. In order to do that the measurements of the voltage values at a given angle were performed. Ten measurements of flexion and extension were conducted for each of two fingers. The fingers were flexed between the intermediate phalanx and proximal phalanx. The angle was measured with a goniometer, at subsequent 0, 30, 45, 60, 90°. The values recorded during the tests were used to generate the diagrams of the courses of voltage for flexion and extension. The next step included the determination of occurrence of hysteresis during flexion and extension of flex sensors.

The second series of tests encompassed the verification of the measurements of orientation indications provided by the MPU 6050 sensor implemented in the device. In order to do that, one of the sensors of the MVN Biomech systems of Xsens company was mounted on the glove. The sensor was placed above the other one at the wrist of the examined person. In order to verify the system operation, a computer application was developed in the environment of virtual reality. This application was written on the basis of Unity 3D graphic engine and consists in catching blocks which fly towards the participant. Moving their hand in the right direction, the participant directs the application controlling element at the appearing blocks. The process of catching the blocks takes place through clenching the hand where the sensors are placed.

To determine the accuracy of the measured values, three two-minute measurements of the acceleration of the upper limb were performed using the computer application. During the execution of exercises, the orientation of the sensors in space was recorded in the application. Then, the results were synchronized with the measurements obtained from the MVN Biomech sensor. The next step was to calculate the mean-square error and Pearson correlation coefficient between the courses recorded from two systems.

5 Results

The first series of tests included the recording of the changes of voltage values during the flexion of fingers using a goniometer as well as the calculation of mean values. On their basis the diagrams of the courses of voltage for flexion and extension were generated (Fig. 3).

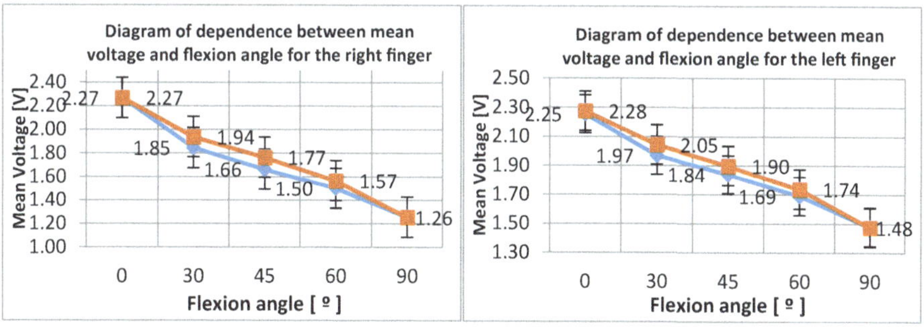

Fig. 3. Diagrams of dependence between mean voltage and flexion angle for the right and left fingers.

A maximum difference between the values obtained for flexion and extension of the left finger amounts to 0.07 V, whereas in the case of the right finger it amounts to 0.1 V. As far as percentage values are concerned, in relation to the whole measuring scope for the left finger (which amounts to 0.8 V for the left finger and 1.01 V for the right finger) this value equals 8.7% of the scope, while it amounts to 9.8% for the right finger.

Some examples of the results of the second series of tests for the first trial were presented in Fig. 4 in the form of diagrams of orientation angles on the X, Y, Z (sagittal, transverse and vertical) axes.

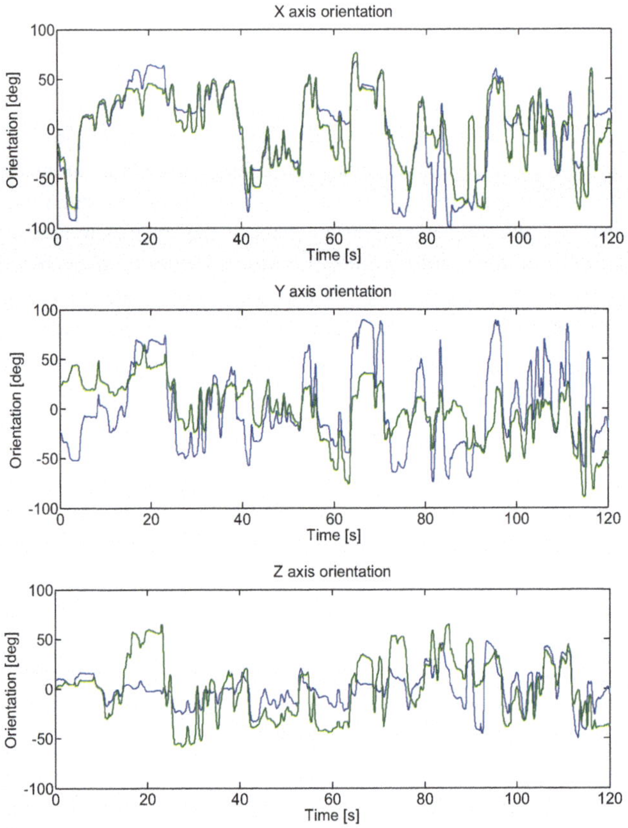

Fig. 4. Diagram of orientation of sensors during testing in relation to X, Y and Z axes (values in green for the developed system, values in blue for the MVN Biomech system)

In the next phase the mean-square error and Pearson correlation coefficient were calculated between the courses of two systems. The Pearson correlation coefficient and the RMS for the compared systems have been presented in Table 1.

Table 1. Values of correlation coefficient and mean-square error for subsequent orientation measurements.

	X axis			Y axis			Z axis		
	Study 1	Study 2	Study 3	Study 1	Study 2	Study 3	Study 1	Study 2	Study 3
Pearson correlation coefficient	0.84	0.54	0.65	0.55	0.89	0.79	0.61	0.73	0.77
RMS	43.27	47.65	49.5	35.55	33.91	31.96	29.09	25.04	25.79
RMS mean	46.81			33.81			26.64		
RMS SD	2.61			1.47			1.76		

6 Discussion

Verification of the developed device aiding the diagnostics and therapy of the forearm and fingers shows its utility on condition of the compliance with the determined measuring accuracy and functionality of individual components. In the first series of tests, the highest percentage difference was found between the dispersion of voltage values measured on the tensometric bridge. The dispersion of values resulted from the existing hysteresis of resistance in relation to the functions of flexion for the tensometric flex sensors. The dispersion in the designed system does not exceed 10%, which translates into a 10% error of the sensor flexion. This value may be acceptable in the case of solutions connected with detection of the movement of fingers and the determination of their approximate location, for instance in the solutions connected with therapeutic games using the Virtual Reality Technology. Thanks to the device it is possible to detect the movement of the patient's fingers in the computer application, which entails a therapeutic outcome.

The second stage of tests aimed to compare the operation of the authors' controller with the MVN Biomech system as well as to verify the correctness of parameters, such as orientation and location. In a specially designed application, the scenario which motivated the patient to execute certain movements, it was necessary to move the hand in such a way as to control its avatar in order to catch elements appearing on the screen. The orientation values recorded during the tests for both sensors have a very similar shape in their graphic comparison as well as high correlation calculated by means of Pearson correlation coefficient. High correlation was obtained for all the conducted measurements. The diagrams also show the shift of the orientation values in relation to each other. The differences in the diagrams may result from the fact that sensors were not placed perfectly on the hand of the tested person. In order to test if this value is constant, the RMS coefficient was determined. This coefficient is frequently used in biomechanics to analyze the differences, for instance between the courses of anatomic angles in joints. The value of standard deviation for the calculated RMS coefficients is rather small (the maximum measured is 2.6° for the mean value 46.81°), which indicates constancy of the measured error between the courses. Such an error may result from the difference in the initial orientation of sensors before the test, which is subject to no change during all tests.

7 Conclusions

The designed and developed model of the interactive controller may be dedicated to people who as a result of an accident, illness or impairment suffer from the paresis of the upper limb. Verification of the research in the scope of functionality and repeatability of the measurements allowed the researchers to determine the controller's utility as well as its ranges and uncertainty of measurement. The developed device is very universal. It may be used as a measuring device, a controller for games or as a system for rehabilitation making it possible to determine the existing motor dysfunctions.

References

1. Guberek, R., Maxime, T.R., Heidi, S., Levin, F.M.: Motor learning in children with hemiplegic cerebral palsy and the role of sensation in short-term motor training of goal-directed reaching. Dev. Med. Child Neurol. **55**(12), 1121–1128 (2013)
2. Gzik, M., Wodarski, P., Jurkojć, J., Michnik, R., Bieniek, A.: Interactive system of engineering support of upper limb diagnosis. In: Innovations in Biomedical Engineering. Advances in Intelligent Systems and Computing, vol. 526, pp. 115–123. Springer (2016). ISBN 978-3-319-47153-2, ISSN 2194-5357
3. Jaspers, E., Desloovere, K., Bruyninckx, H., Klingels, K., Molenaers, G., Aertbelien, E., Gestel, L., Feys, H.: Three-dimensional upper limb movement characteristics in children with hemiplegic cerebral palsy and typically developing children. Res. Dev. Disabil. **32**, 2283–2294 (2011)
4. Koziol, L.F., Budding, D.E., Chidekel, D.: Sensory integration, sensory processing, and sensory modulation disorders: putative functional neuroanatomic underpinnings. Cerebellum **10**, 770–792 (2011)
5. Medynski, C., Rattray, B.: Bebionic prosthetic design. In: Myoelectric Symposium (2011)
6. Michnik, R., Jurkojć, J., Wodarski, P., Gzik, M., Bieniek, A.: The influence of the scenery and the amplitude of visual disturbances in the virtual reality on the maintaining the balance. Arch. Budo **10**(1), 133–140 (2014). ISSN 1643-8698. aaICID: 1105314
7. Opara, J., Rycerski, W., Szczygieł, J., Mazurek, J., Wardejn, R., Kucińska, A.: Innowacyjne zastosowanie wirtualnej rzeczywistości w unowocześnianiu terapii lustrzanej w niedowładzie kończyny górnej. Analiza przypadku (2016)
8. Pasquina, P.F., Evangelista, M., Carvalho, A.J., Lockhart, J., Griffin, S., Nanos, G., McKay, P., Hansen, M., Ipsen, D., Vandersea, J., Butkus, J., Miller, M., Murphy, I., Hankin, D.: First-in-man demonstration of a fully implanted myoelectric sensors system to control an advanced electromechanical prosthetic hand. J. Neurosci. Methods **244**, 85–93 (2015)
9. Riener, R., Harders, M.: Virtual Reality in Medicine. Springer, London (2012). ISBN 978-1-4471-4010-8
10. Schilbach, L., Wohlschlaeger, A.M., Kraemer, N.C., Newen, A., Shah, N.J., Fink, G.R., Vogeley, K.: Being with virtual others: neural correlates of social interaction. Neuropsychologia **44**, 718–730 (2006)
11. Yalon-Chamovitz, S., Weiss, P.L.: Virtual reality as a leisure activity for young adults with physical and intellectual disabilities. Res. Dev. Disabil. **29**, 273–287 (2008)
12. Zuniga, J., Katsavelis, D., Peck, J., Stollberg, J., Petrykowski, M., Carson, A., Fernandez, C.: Cyborg beast: a low-cost 3D-printed prothetic hand for children with upper-limp differences. BMC Res. Notes (2015)

Double Physical Pendulum with Magnetic Interaction

Mateusz Wojna(✉), Jan Awrejcewicz, and Grzegorz Wasilewski

Department of Automation, Biomechanics and Mechatronics, Lodz University of Technology, Łódź, Poland
mateuszwojna94@gmail.com

Abstract. The paper is devoted to experimental investigation of a system consisting of a double physical pendulum with magnetic interaction caused by a pair of permanent magnets repelling each other – one mounted on the end of a second link of a pendulum and one mounted in the body of the setup. To the experimental rig of double physical pendulum system with the first body periodically forced a constructed magnetic interaction forces measurement system is added. The work consists of investigation on how magnetic interaction forces depend on pendulum's links relative angle. Constructed measurement system consists of two links shafts encoders and one force sensor – extensometer beam with an output signal amplifier. Three analog signals are read by acquisition device, sent to connected PC and processed in LabVIEW measurement software programs. The dynamics of the system is also observed. Regular and chaotic behaviors of the system are recorded depending on the frequency of the motor. The work is the first approach to electromechanical system of the double physical pendulum with magnets. Such system is a unique construction. The magnetic forces measurement system is constructed and tested. The regular and chaotic behaviors of the system are shown and discussed.

Keywords: Magnetic and mechanical system · Double pendulum · Magnets · Experiment and construction

1 Introduction

1.1 Object of Investigations

This paper is a first approach to dynamics of a unique electromechanical system. An object of presented investigations is a double physical pendulum with magnetic interaction caused by a pair of permanent magnets: one mounted at the end of second link of the pendulum and one mounted in the body of an experimental rig. The magnets are oriented so that repelling force exists between them. The considered system is unique for a few reasons. Firstly, there is no paper which describes a double physical pendulum with magnets or electromagnets. Additionally pendulum is constructed so that each link is able to do multiple revolutions around its axis of revolution. Furthermore, a motion of the first link of pendulum is forced by a specially constructed motor.

J. Świder et al. (Eds.): MECHATRONICS 2017, AISC 934, pp. 455–464, 2019.
https://doi.org/10.1007/978-3-030-15857-6_45

As mentioned above the considered system has unique features. In the literature may be found dynamic analysis of simpler systems which consist of single pendulum with magnetic interaction. Simple example of EMS is described in [1]. In this paper a system consisting of a pendulum, a permanent magnet and an AC electromagnet is considered. A magnet is mounted at the end of pendulum and additionally on its body. There is presence and relative relations of chaos and parametric resonance investigated. Analytical results are confronted with numerical simulations and experimental studies.

In [2] a system with a single physical pendulum placed in the magnetostatic field (two repelling magnets) and forced sinusoidally is considered. A system dynamics is experimentally studied. Computer simulations results are shown. A measurement system consists of two encoders. Both the experimental studies and the numerical simulations indicate the presence of chaotic and regular behavior of the system (depending on parameters of the system). The following parameters are defined: forcing frequency, relative distance of magnets and relative magnets orientation in horizontal plane. Additionally there peaks of magnitude, hysteresis and bistable states are observed.

Paper [3] considers a system with physical pendulum. An the end of the pendulum a permanent magnet is mounted. A magnet interacts with another one placed in the body. The magnets are close when a pendulum is placed downwards. In this work physical and mathematical models are presented. The correctness of models has been proved by simulations of bifurcations generated by the estimated coefficients for several frequencies. Simulations show also that estimated coefficients of one frequency can be used to predict system behavior in case of other forcing frequencies.

The authors of [4–6] describe experimental and numerical analysis of a system consisting of triple physical pendulum. Such construction with a possibility of recording multiple revolutions of every link of triple pendulum is a unique construction. The system is strongly nonlinear and there can be observed phenomena among others: regular and irregular behaviors, bifurcations, coexisting periodic, quasiperiodic, chaotic and hiperchaotic solutions. In considered system the motion of first link is forced by a specially constructed motor. The motor consists of two immovable stators and two rotors. Both parts of motor are symmetric and electrically coupled, but optical commutator is placed only on the one stator's plate. Such drive eliminates possible misalignments of construction. More detailed description of forcing motor and an entire measuring system can be found in [4]. Parameters of the system have been identified by the sum of squares of deviations minimalization of signals from experimental and numerical data. A high compatibility of experimental and simulation results has been obtained.

2 Experimental Rig

Existing experimental rig of double physical pendulum with magnets is augmented with moment of force of magnetic interaction measurement system. The system enables to measure how the force depends on links relative angle. Entire setup is shown in Fig. 1.

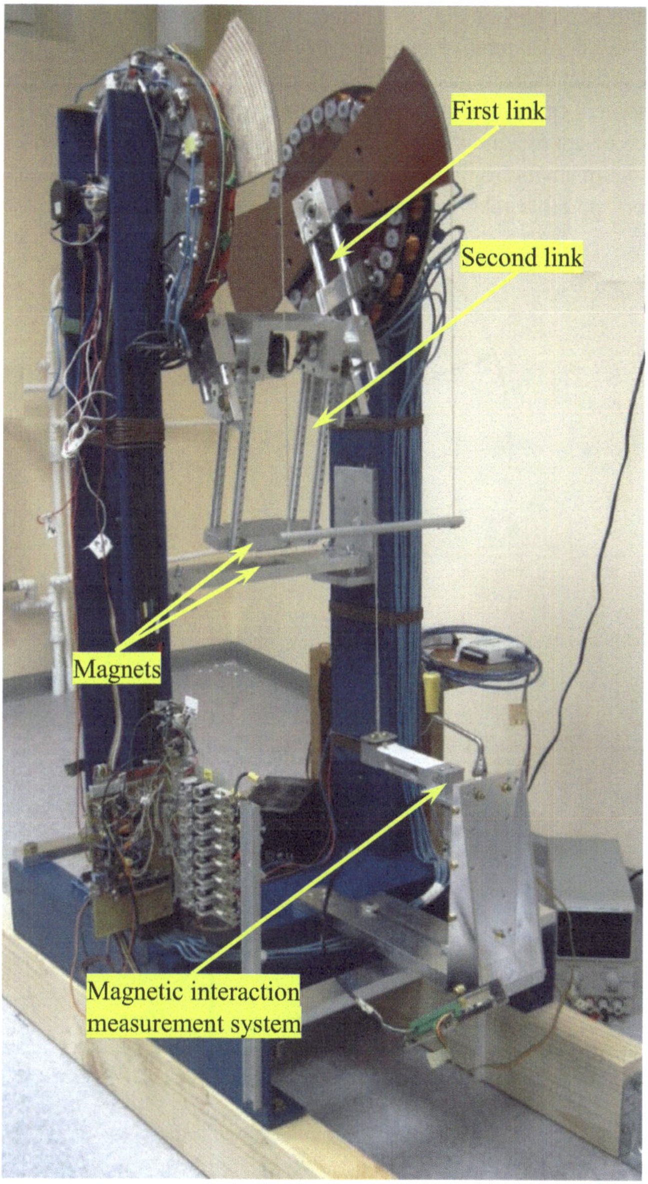

Fig. 1. Experimental setup with magnetic interaction measurement system.

A physical model and a scheme of the setup is presented in Fig. 2. The first link of pendulum is mechanically coupled with an extensometer beam by a ties system. To measure a force acting on the extensometer beam and calculate the moment of force the second link has been stiffly coupled relatively to the first one in known angular position. Constructed measurement system enables us to couple the pendulum's links together, which is shown in Fig. 3. Thanks to such construction there are 19 possibilities for links angular relative coupling in the range of ±40°. During the measurement the extensometer beam has been moved down pulling the tie at the same time. In this way a moment of force acts on the first link of pendulum trying to overcome the repelling force of approaching magnets. Such measurements have been carried out by approaching the magnets from the right and the left side of the experimental setup, and including every possible relative angular coupling of links.

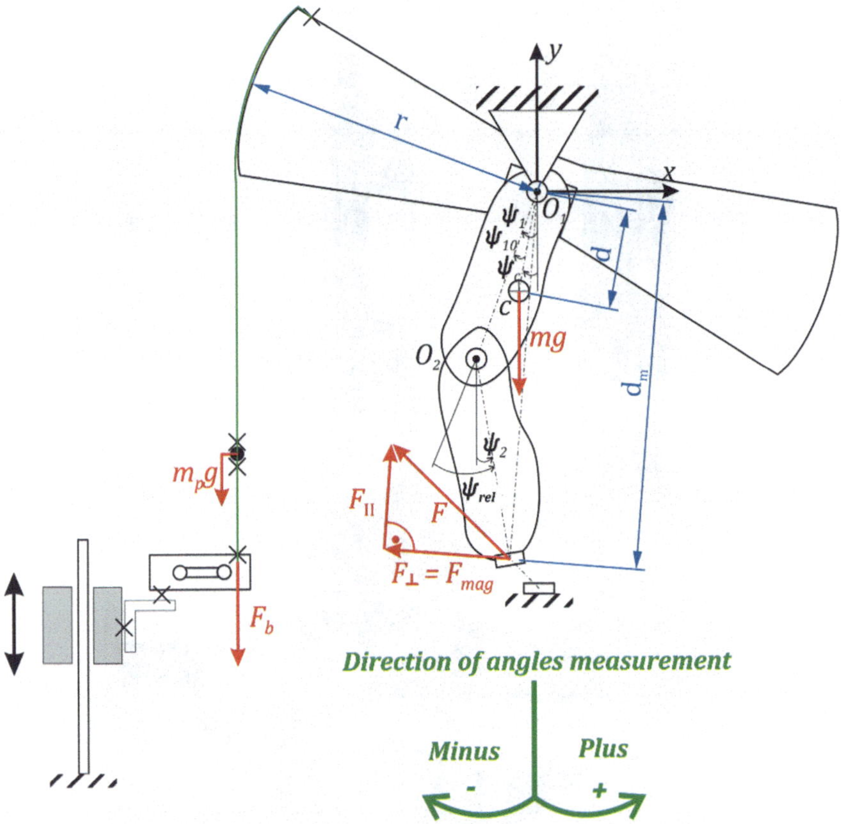

Fig. 2. Physical model of the setup.

The force acting on the extensometer beam has a constant arm r. Thanks to it the moment of force acting on the first link of pendulum due to magnets interaction has been easily obtained. Read values of moment of force have been modified and supplemented by known values of weight of the tie system and the gravity force of pendulum. Thus, only moment of force due to magnets interaction has been considered.

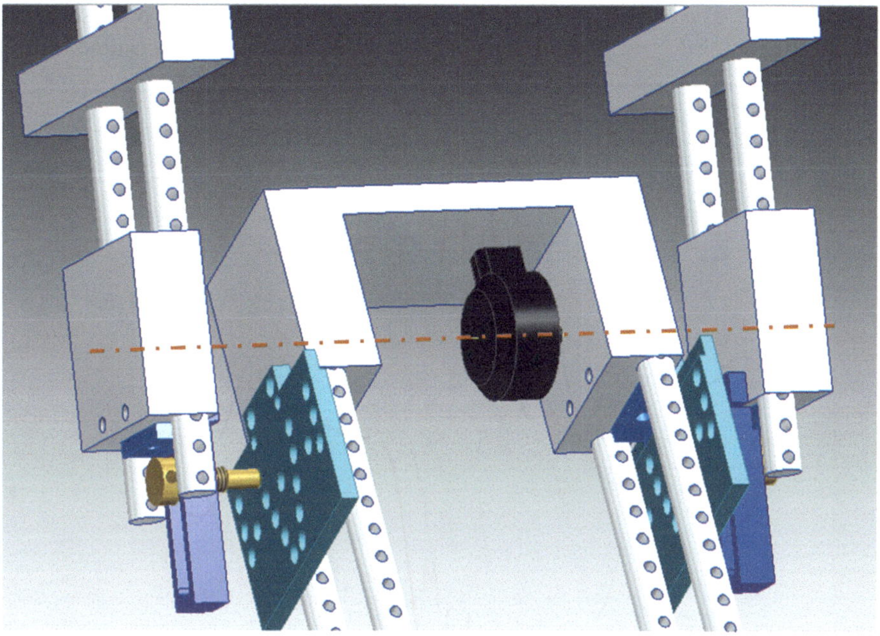

Fig. 3. Relative angular pendulum's links coupling system.

2.1 Sensory System

During the studies three signals from three sensors have been used and processed. Following sensors have been implemented: an extensometer beam (a force measurement) and two rotary-pulsed transducers (an angular positions measurement). The sensors have been connected to NI data acquisition device and read by PC computer. Recording, readout and data processing have been led in LabVIEW environment.

3 Experiment

Conducted experiments include two main parts. Firstly, approximations of moment of force caused by magnets as a function of first link absolute angle for possible relative coupling have been obtained. Furthermore, time series of pendulum's links have been recorded for a few different frequency of forcing.

3.1 Moment of Force of Magnets Characteristics

Figure 4 illustrates polynomial approximations curves of moment of force caused by magnets obtained during the experimental study. To have a full view of moment's course for one specific relative coupling the characteristics of measurements form right and left setup side should be combined. It may be noticed that the highest values of moment exist for the smallest relative coupling angles absolute values, i.e. for

$\psi_{rel} = 0°, \pm 3°$. The highest values are about 13.5 Nm. For comparison purposes it can be noticed that for $\psi_{rel} = \pm 35°$ moment M_{mag} reaches the value about 5 times lower.

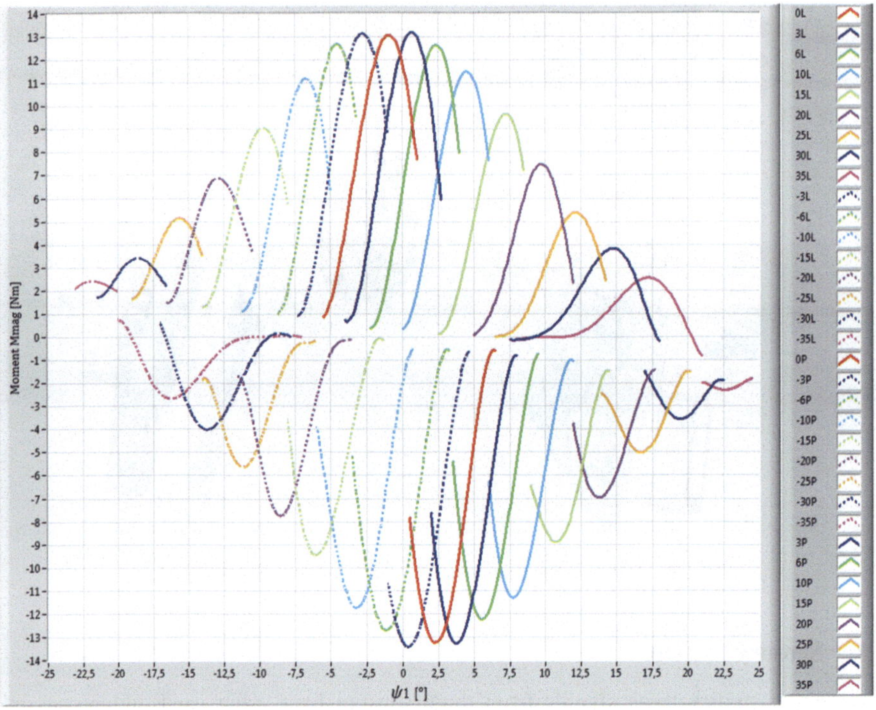

Fig. 4. Moment of force due to magnets characteristics in function of an angle of the first pendulum's link (e.g. 15*L* label in legend indicates the measurement conducted from Left side of setup with $\psi_{rel} = +15°$).

Not approximated values measured form right and left side of the setup with $\psi_{rel} = 0°$ are presented in Fig. 5. It can be easily noticed that the moment values are symmetric and change from positive to negative values nearby $\psi_1 = 0°$.

The obtained characteristics will be necessary in the further stage of studies. In dynamic analysis a mathematical model of the system will be presented. The crucial component of the equations of motion is reserved for the moment of magnetic interaction. The considered characteristics will enable us to obtain the coefficients of function describing magnetic interaction.

3.2 Pendulum's Time Series Recording

In this chapter exemplary time series of double pendulum motion with magnets are presented. External pendulum forcing is realized by rectangular function of torque (see Fig. 6) set by generator connected to the pendulum motor control circuit.

Below there are time series of regular and irregular behavior provided. The preset parameter was frequency f.

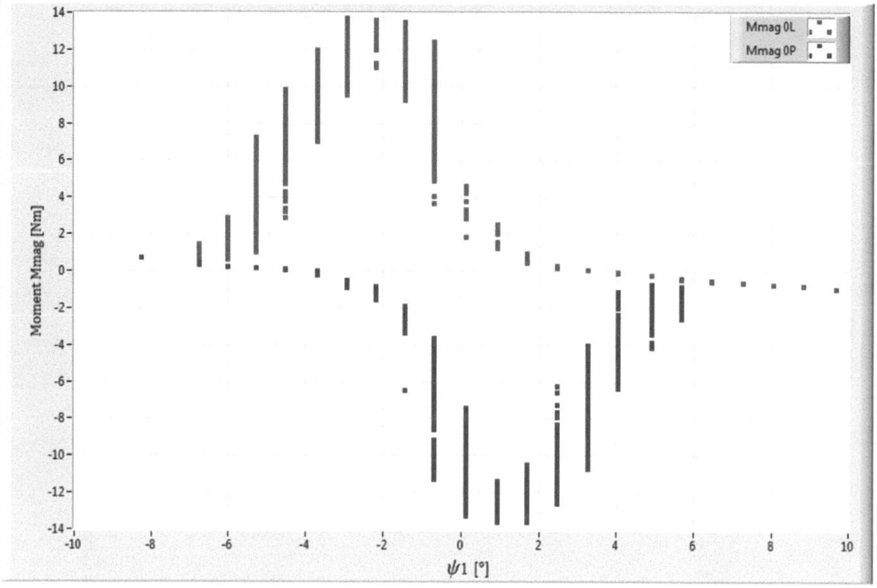

Fig. 5. Not approximated moment of force due to magnets in function of an angle of the first pendulum's link for $\psi_{rel} = 0°$

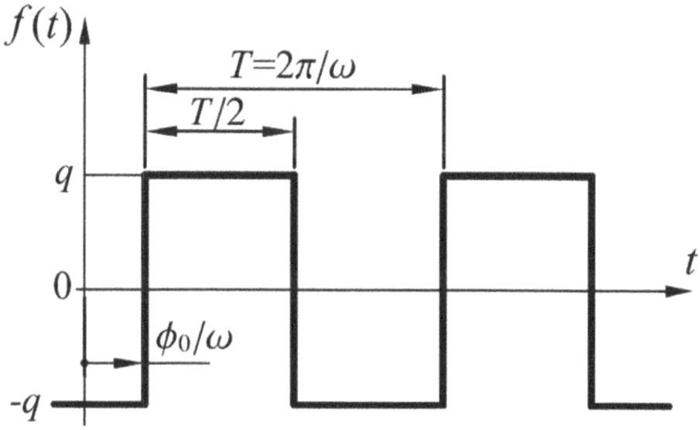

Fig. 6. External forcing motor torque function

Figure 7a presents regular motion of both links with the frequency $f = 0.1$ Hz. Enlarged part of these time series is shown in Fig. 7b. There a kind of regularity may be easily noticed. With the value of $f = 0.5$ Hz chaotic motion can be observed (see Fig. 7c). For such frequency multiple revolutions do not occur. The behavior, where multiple revolutions of each link can be observed is presented in Fig. 7d, e with the forcing frequency $f = 1$ Hz. The second link of pendulum does about 20 revolutions while the first one does only one revolution. The motion is chaotic. With frequency $f = 1.8$ Hz the motion with quite long temporary irregular stage is detected (Fig. 7f and g). After firstly 90 s of temporary stage, a regular behavior of a system exists. The amplitudes are small and links are in the opposite phases.

a)

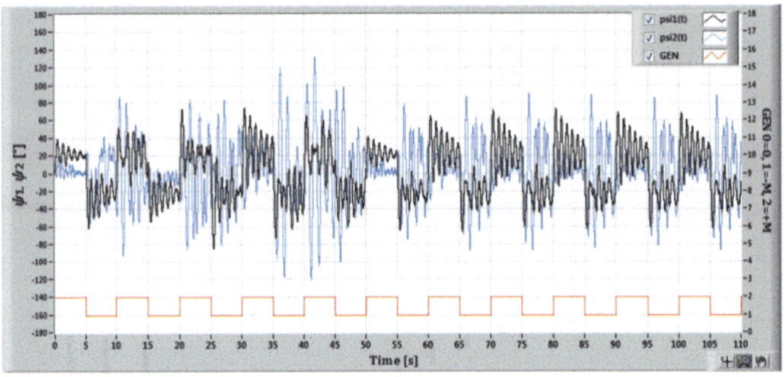

b)

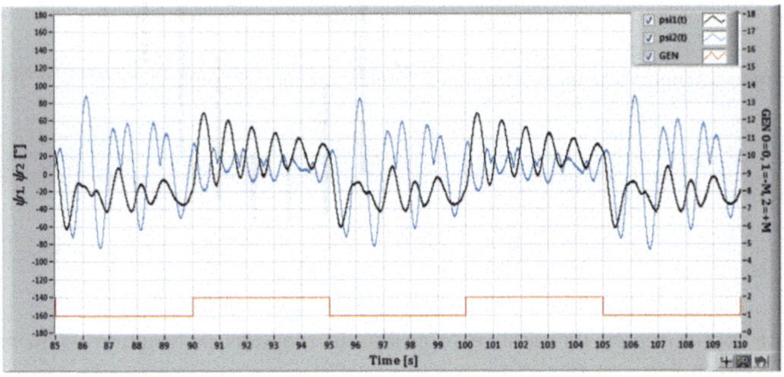

c)

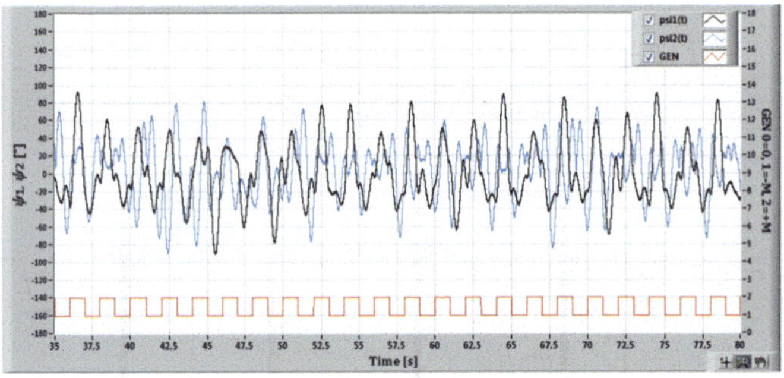

d)

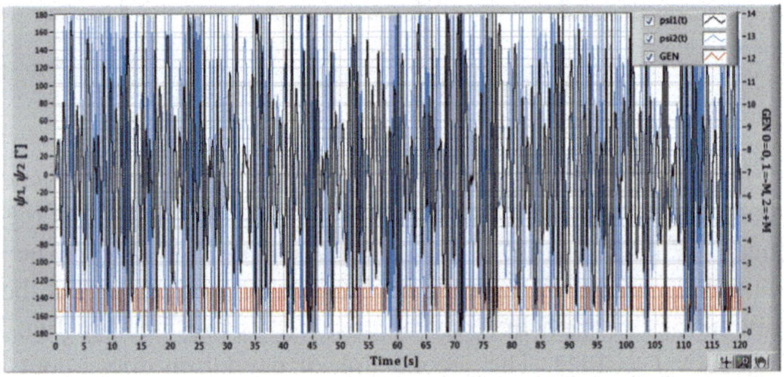

Fig. 7. Exemplary time series of motion of pendulum, (a) and (b) forcing frequency $f = 0.1$ Hz; (c) $f = 0.5$ Hz; (d) and (e) $f = 1$ Hz; (f) and (g) $f = 1.8$ Hz.

e)

f)

g)

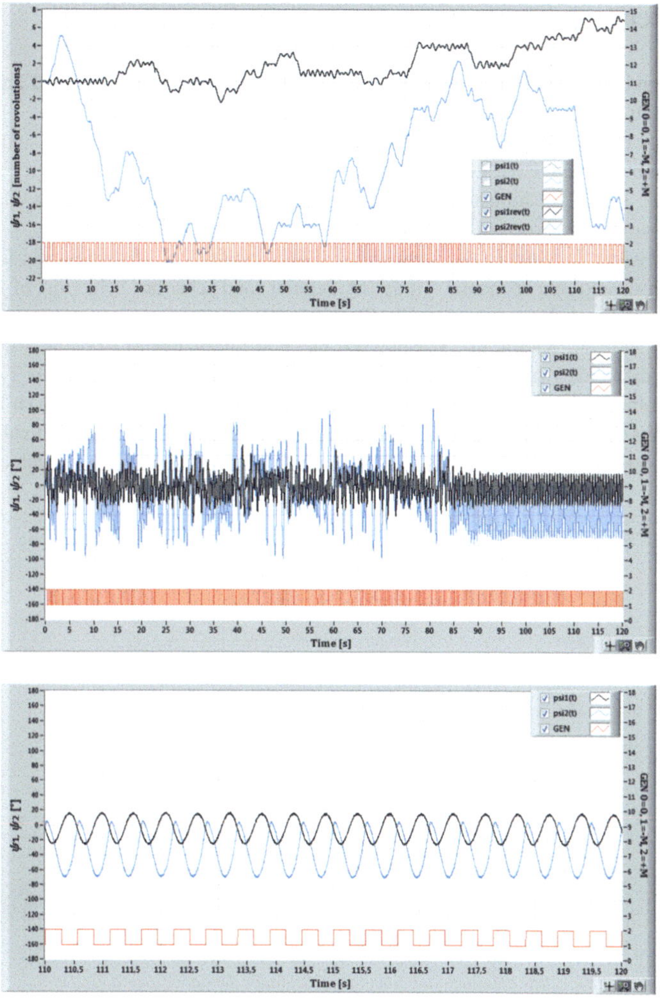

Fig. 7. (*continued*)

4 Conclusions

The constructed magnetic interaction measurement system enables to measure force caused by two repelling permanent magnets. It allows to obtain polynomial approximation curves of moment of force, which will be used for mathematical modeling of the system in further papers. Numerical and experimental analysis of system dynamics will be the purpose of next authors' articles.

The conducted experimental analysis proves that considered system is strongly non-linear. Recorded time series of angular positions show the regular or chaotic motion of the pendulum depending on the forcing frequency.

References

1. Khomeriki, G.: Parametric resonance induced chaos in magnetic damped driven pendulum. Phys. Lett. A **380**, 2382–2385 (2016)
2. Siahmakoun, A., French, V.A., Patterson, J.: Nonlinear dynamics of a sinusoidally driven pendulum in a repulsive magnetic field. Am. J. Phys. **65**, 393 (1997)
3. Tran, V., Brost, E., Johnston, M., Jalkio, J.: Predicting the behavior of a chaotic pendulum with a variable interaction potential. CHAOS **23**, 033103 (2013)
4. Awrejcewicz, J., Supeł, B., Lamarque, C.-H., Kudra, G., Wasilewski, G., Olejnik, P.: Numerical and experimental study of regular and chaotic behavior of triple physical pendulum. Int. J. Bifurc. Chaos **18**(10), 2883–2915 (2008)
5. Awrejcewicz, J., Kudra, G., Wasilewski, G.: Potrójne wahadło fizyczne - wybrane aspekty eksperymentalne i numeryczne. XII Konferencja Naukowa Wibrotechniki i Wibroakustyki, VII Ogólnopolskie Seminarium Wibroakustyka w Systemach Technicznych (2006)
6. Awrejcewicz, J., Wasilewski, G.: Bifurkacje w układzie potrójnego wahadła fizycznego - badania eksperymentalne. In: XXII Sympozjum Mechaniki Eksperymentalnej Ciała Stałego (2006)

Model of Trough-Beam Laser Sensor for Determining the Real Position and Real Response Time

Mirosław Wolski(✉), Tomasz Piątkowski, and Przemysław Osowski

Faculty of Mechanical Engineering, UTP University of Science and Technology, Al. Prof. S. Kaliskiego 7, 85-796 Bydgoszcz, Poland
{Miroslaw.Wolski,Tomasz.Piatkowski, Przemyslaw.Osowski}@utp.edu.pl

Abstract. In almost all industries the use of non-contact sensors to detect objects and their position has become necessary. The laser through-beam sensors is resistant to electromagnetic interference, inexpensive and relatively accurate. The only parameters of trough-beam laser sensor provided by the manufacturer to estimate the object position measurement error are maximal response time and laser beam spot dimension. With such data, the device allows for order of accuracy of 10^{-3} m, which is not sufficient for precision machines. With that in mind, this paper presents a model of laser through beam sensor using a transistor hybrid model developed to improve location detection accuracy. An object in motion was detected in the experiment at the velocity ranging from 0.5 to 10 m/s. The results of the experiment showed good compatibility of the experimental and model output characteristics. The actual object location was determined at the accuracy of $2.5 \cdot 10^{-5}$ m (0.5 m/s) to $2.6 \cdot 10^{-4}$ m (10 m/s).

Keywords: Hybrid transistor equivalent circuit · Object detection · Convolution integral

1 Introduction

In the automated assembly system, including the packing, products sorting, bottles filling, separating, counting, dimensional control operations, the non-contact sensors to detect objects and their position have become necessary. The laser through-beam sensor is resistant to electromagnetic interference, inexpensive and relatively accurate.

The only parameters of through-beam laser sensor provided by the manufacturer to estimate the object-in-motion position measurement error are maximal response time and laser beam spot dimension [1]. With such data, the maximum position error can be estimated as the sum of the laser beam diameter and the product of the object movement velocity as well as response time. When the sensor receiver contains a number of photosensitive components (mostly a CCD), the manufacturer provides also a voltage signal non-linearity error as a function of the depth of interrupted laser beam under quasi-stationary conditions; when the object that intersects the laser beam moves very slowly. The response time is determined by the manufacturers for a condition when a cycle of the interruption of the laser beam in rectangular in shape. The object velocity

J. Świder et al. (Eds.): MECHATRONICS 2017, AISC 934, pp. 465–474, 2019.
https://doi.org/10.1007/978-3-030-15857-6_46

should be infinite to create the rectangular shape of the luminous power amount change for laser beam. The problem of determining the actual sensor response time appears when the object velocity is finite and the light beam is cut gradually under real conditions. In this case, a coupling occurs between the quasi-stationary output signal (voltage or current signal as a function of interrupted laser beam depth) and the step output signal; a step output signal of the sensor is response to the step input signal of laser beam power value).

The paper presents a model of through-beam laser sensor applying a transistor hybrid equivalent circuit developed to enhance the location detection accuracy [2]. The method of experimental model parameters determination and experimental model verification have been demonstrated.

2 Unit Examined

The study involved phototransistor TEPT4400 produced by Vishay [3] (with wavelength at the maximum sensitivity point of 570 nm (Fig. 1a), lens angle of 30°, spectral band range of 440–800 nm) and laser diode (with 3 mW optical power, wavelength of light emitted from 645 nm to 660 nm, light beam diameter 1 m away from the source of 1.32 mm).

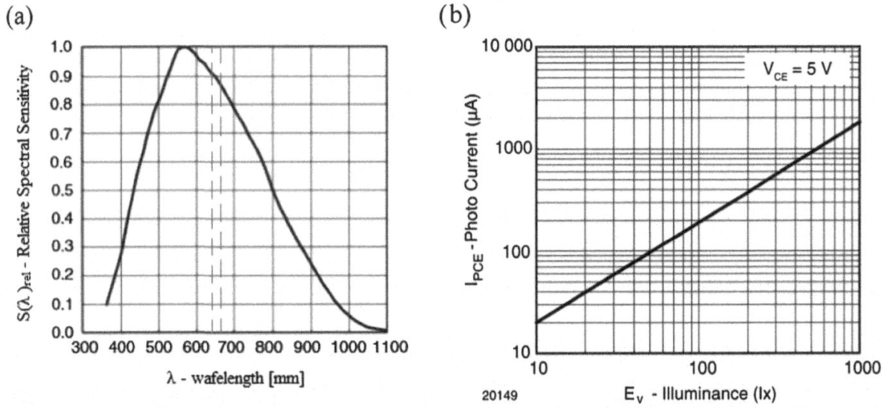

Fig. 1. Basic phototransistor characteristics [3]: (a) spectrum of relative phototransistor sensitivity (black solid line) and the range of the wavelength of light emitted by the laser module (red dashed line), (b) photocurrent as a function of the light intensity for the collector-emitter voltage equal to 5 V.

The phototransistor and the laser module were selected (Fig. 1) in a way to ensure the laser module emission range being close to the maximum relative spectral sensitivity of the phototransistor. Figure 1b shows the current flowing through the phototransistor emitter as a function of the luminous intensity incident on the phototransistor. The sensor receiver was connected in a common emitter system, and the phototransistor – in series with the pull-up resistor to 3.3 V, see Fig. 2.

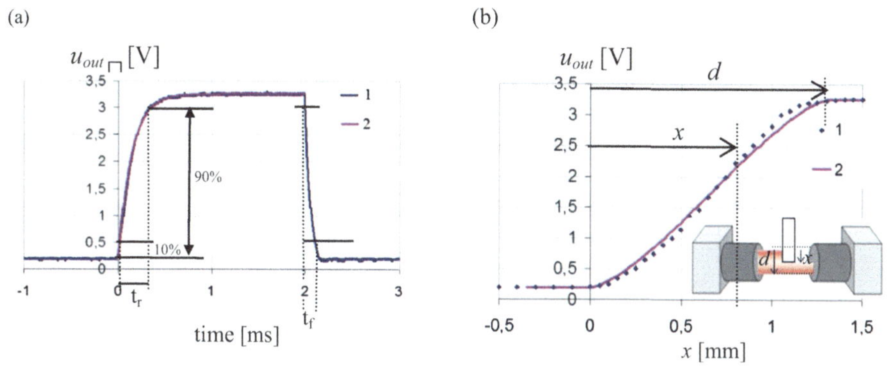

Fig. 2. Basic time characteristics of the optical sensor: (a) step response signal at the sensor output due to the rectangular period of light interruption, 1 – experimental response, 2 – exponential model response of the sensor with *RC* time constant, (b) sensor output voltage versus the interruption depth of laser beam x in quasi-stationary conditions, 1 – experimental characteristics, 2 – theoretical characteristics for assumed proportional dependence between sensor output voltage and surface area $S = f(x)$ of interrupted laser beam

3 Methodology of Research

The documentation provided by the manufacturer contains information about response time t_{resp} and transverse laser beam spot dimension d_b. The sensor response time is defined as rise time t_r (or falling time t_f) of the output signal from 10% to 90% (or 90% to 10%) of the maximum value (Fig. 3a) without a background signal. Response time is determined for the conditions when the light interruption period is rectangular in shape (Fig. 4a). Figure 3b shows a discrete graph of the sensor output voltage versus the interruption depth of laser beam x in quasi-stationary conditions. Quasi-stationary conditions are such conditions when changes in the interruption depth of laser beam over time are so slow that all the transient components of output signal disappear. To measure the sensor output voltage (Fig. 3b), the laser beam was gradually interrupted using a micrometer screw (Fig. 4b). If rise time t_r had equaled zero, the same result as in Fig. 3b could have been obtained for any velocity (without the assumption of quasi-stationarity). The doted curve on the graph (Fig. 3b) could be rectangular in shape in quasi-stationary conditions only when the beam spot size would be an infinitely small light point. Since the point is not infinitely small, its interruption through the object with finite velocity does not satisfy the conditions of the rectangular shape of beam interruption cycle for determining rise time t_r.

Thus, real response time t_{resp_real} should take into account not only the rise time of the output signal for the rectangular input signal of the interrupted laser beam but also the fact that the interruption of the beam does not take place in infinite short time, so the input signal is not rectangular. Based on the manufacturer's data, real response time t_{resp_real} falls within the range of:

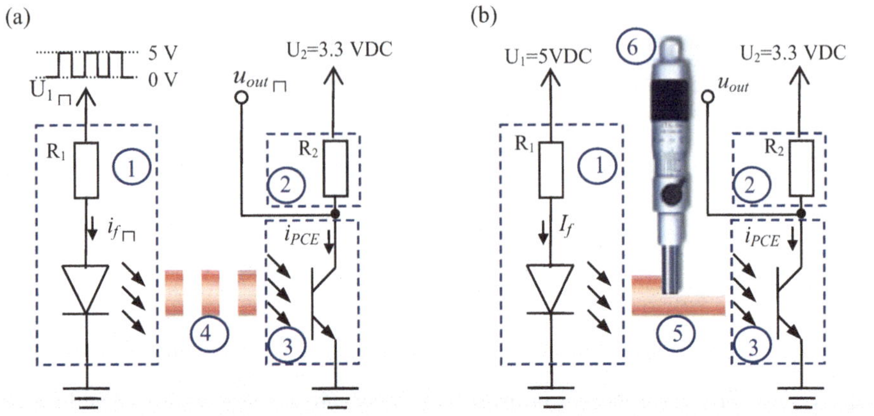

Fig. 3. Diagrams of measurement circuits for characteristics determination presented in Fig. 2. (a) for output step response determination, (b) for determination of output voltage characteristic versus the interruption depth of laser beam x in quasi-stationary conditions, 1 – laser diode module, which consists of laser diode and current limiting resistor, 2 – pull-up resistor of phototransistor, 3 – phototransistor, 4 – laser beam interrupted due to pulse supply to laser diode, 5 – laser beam cutting using micrometer screw 6.

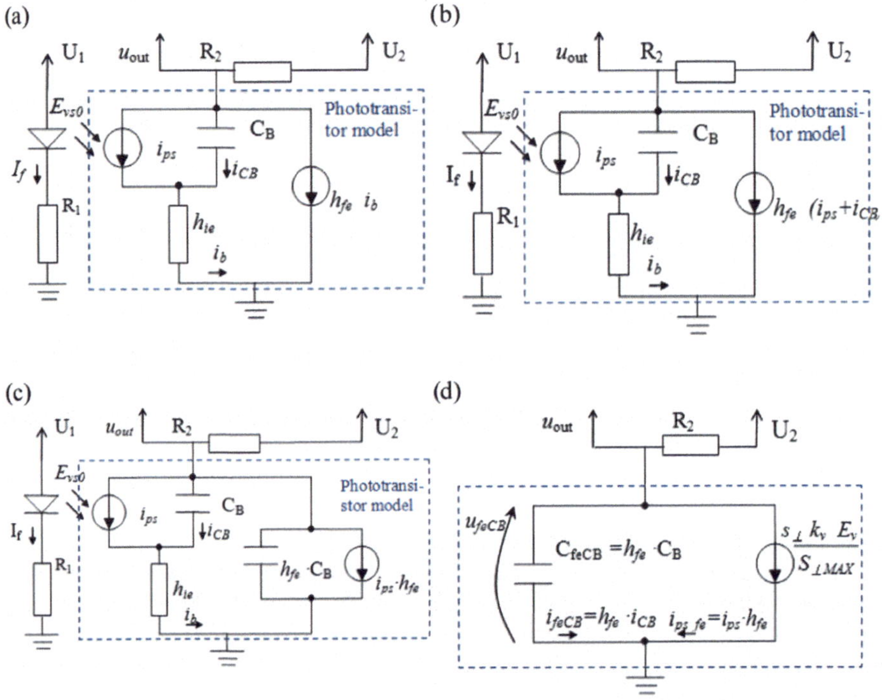

Fig. 4. Hybrid model of phototransistor: (a) – (d) schemes representing the next steps in model simplification.

$$t_{resp} \leq t_{resp\_real} \leq t_{resp} + \frac{d_b}{V} \tag{1}$$

where: t_{resp} – response time for rectangular input signal, t_{resp_real} – real response time for input signal of gradually interrupted laser beam, V – velocity of detected object, d_b – transverse dimension of laser beam.

The real position error can be obtained by multiplying the inequality (1) by object-in-motion velocity V:

$$x_{resp} \leq x_{resp\_real} \leq x_{resp} + d_b \tag{2}$$

where: $x_{resp} = V \cdot t_{resp}$ – object displacement at V velocity before a response of the object detection, assuming that light beam interruption period is rectangular in shape, $x_{resp\_real} = V \cdot t_{resp\_real}$ – real displacement of the object detected.

By calculating the limit of the inequalities, if $V \to \infty$, then $x_{resp\_real} \to x_{resp}$ and if $V \to 0$ then $x_{resp\_real} \to d_b$. However, it is not possible to determine real displacement x_{resp_real} from the manufacturer's data when velocity V is arbitrary.

The step response signal $u_{out\sqcap}$ at the sensor output (Fig. 3a) was obtained according to the conditions as shown in Fig. 4a due to the rectangular voltage supply to the sensor laser diode. As a result, light-sensitive element appears on the phototransistor as rectangular luminous power pulses. The conditions are corresponding to those when the light beam is immediately interrupted all across the surface area of the laser spot, e.g. by a very fast-moving object. The response signal is an exponential function with RC time constant. The RC parameter is described in the next section relating to the sensor model.

When the surface area of the laser beam incident on the phototransistor decreases gradually, the luminous power and the average illuminance are also proportionally reduced. According to the characteristics in Fig. 1b, collector-emitter photocurrent iPCE decreases linearly with the average illuminance of incident laser beam on the phototransistor. With the reduction of the iPCE current, the voltage across resistor R_2 decreases linearly and the output voltage of sensor u_{out} increases linearly.

Figure 2b shows the theoretical characteristics of the output voltage u_{out} versus interrupted laser beam proportional to transversal cut of surface area of disk $S = f(x)$. Characteristics in Fig. 1b are determined for constant voltage on the phototransistor, equal to 5 V. Due to the resistor R_2, in this sensor, the voltage u_{out} across the phototransistor is not constant, so there are some deviations from the linear characteristics of the photocurrent i_{PCE} versus illuminance E_v. The deviations specified in manufacturer's documentation are about 15% in the range of 0.2–3.3 V; hence a deviation between the theoretical and experimental characteristics (Fig. 2b).

4 Through-Beam Laser Sensor Model

The sensor model based on the hybrid model of the transistor allows for finding the sensor output response for any input signal using convolution. Thus, the input signal can also be gradually interrupted laser beam by an object moving with specified velocity $\boldsymbol{V}$. The basic relationship between the phototransistor input and output is described with the equation:

$$I_{PCE} = CTR \cdot I_f \tag{3}$$

where: I_{PCE} – collector-emitter photocurrent, CTR – current transfer ratio of phototransistor, I_f – laser diode current.

The output current of the phototransistor i_{PCE} is also proportional to the light intensity incident on the phototransistor assuming that $I_f = const$ (according to Fig. 1b):

$$I_{PCE} = k_v \cdot E_{v0} \tag{4}$$

where: k_v – photoelectric proportional coefficient [A/lux], E_v – illuminance [lux], E_{v0} – illuminance for $I_f = const$ [lux].

Luminous power Φ_v [lm] can be expressed as [4] (in case of perpendicularity between the light beam and surface of light incidence $S_{\perp}$):

$$\Phi_v = S_{\perp} E_{v0} \tag{5}$$

Dividing Eq. (5) by the maximum cross-sectional area of the laser beam (i.e., when the laser beam does not encounter an obstacle and its full power reaches the phototransistor), we obtain:

$$E_{vs0} = \frac{\Phi_v}{S_{\perp MAX}} = \frac{s_{\perp} E_{v0}}{S_{\perp MAX}} \tag{6}$$

Introduced in the above equation, E_{vs0} represents the equivalent average illuminance reaching the phototransistor, as if the luminous power Φ_v arriving at the phototransistor from the remaining beam on surface area $S_{\perp}$ were distributed over the entire unbroken $S_{\perp MAX}$ beam surface area.

The phototransistor model in Fig. 4 has been presented as a hybrid model of the bipolar transistor after replacing the supply of base current by current coupled with galvanically separated source. The basic relationship between the transistor input and output is described by equation [5]:

$$I_C = h_{fe} \cdot I_B \tag{7}$$

where: I_C – collector current, I_B – base current, h_{fe} – small signal current gain.

Assuming quasi-stationary conditions $I_p = I_B$ and similarity between a transistor and a phototransistor $I_C = I_{PCE}$, comparing the left sides of Eqs. (3) and (7) we obtain:

$$I_p = \frac{CTR \cdot I_f}{h_{fe}} \tag{8}$$

Comparing the left sides of Eqs. (3) and (4) and showing I_f as an independent variable, and then substituting into Eq. (8) we have:

$$I_p = \frac{k_v \cdot E_{v0}}{h_{fe}} \tag{9}$$

For any depth x of interrupted laser beam, substituting E_{v0} in Eq. (9) by E_{vs0} to Eq. (6) we receive:

$$i_{ps} = \frac{k_v}{h_{fe}} \frac{s_{\perp} E_{v0}}{S_{\perp MAX}} \tag{10}$$

Figure 4a shows a phototransistor model and subsequent steps to simplify the model (Figs. 4a–d). The model includes C_B collector-base capacitance and two dependent current sources i_p and $h_{fe}\, I_B$. The transition between Fig. 4a and b consists of replacing i_b with i_{ps} + i_{CB} on the output side. It was further noted that $h_{fe} i_{CB}$ component of current $h_{fe}(i_p + i_{CB})$ can be replaced by a capacitor with $h_{fe}C_B$ capacity on the output side of phototransistor (Fig. 4c). Finally, in Fig. 4d, due to the low value of the current input signal, the input side has been omitted.

The circuit in Fig. 4d can be described by the differential equation:

$$u_{out} = U_2 - R_2 i_{ps\_fe} = R_2 C_{feCB} \frac{du_{feCB}}{dt} + u_{feCB} \tag{11}$$

The equation solved for step input signal marked with by symbol $\sqcap$ and for initial conditions $i_{ps\_fe}(0^-) = I_{ps\_fe\_MAX}$, $i_{ps}(0^+) = 0$, after substituting $t = x/V$ assumed the form:

$$u_{out\sqcap}(x) = U_2 - R_2 I_{ps\_fe\_MAX} e^{-\frac{x}{VR_2 C_{feCB}}} \tag{12}$$

Adjusting the above function to the experimental curve as shown in Fig. 2a allows for determining parameter R_2C_{feCB} (sensor time constant). The impulse response $h(t)$ of the circuit after substitution $t = x/V$ assumed the following form [6]:

$$h(x) = \frac{1}{R_2 I_{ps\_fe\_MAX}} \frac{\partial u_{out\sqcap}(x)}{\partial x} = 1(t)\left[\frac{1}{VR_2 C_{feCB}} e^{-\frac{x}{VR_2 C_{feCB}}}\right] \tag{13}$$

On the other hand, the voltage as a function of depth x of the interrupted laser beam spot was approximated by polynomial $W(x) = ax^3 + bx^2 + d$.

Using the polynomial simplifies the calculation of the convolution integral and better corresponds to the experimental data than the function proportional to the analytical function of the cut of surface area of disk $S = f(x)$. The full quasi-stationary function used for convolution assumes the form:

$$g(x) = \left(U_2 - R_2 I_{pS\_fe\_MAX}\right) 1(-x) + W(x) \cdot [1(x) - 1(x-d)] + U_1 \cdot 1(x-d) \tag{14}$$

After convolution integral according to the following relationship [6]:

$$Q(x) = h(x) * g(x) = \int_{-\infty}^{\infty} h(x)g(x-\chi)d\chi \tag{15}$$

the function is as follows:

$$Q(t) = Q_1(x) + Q_2(x) + Q_3(x) \tag{16}$$

$$Q_1(x) = \left(U_2 - R_2 I_{pS\_fe\_MAX}\right)\left[1(-x) + 1(x)e^{-\frac{x}{VR_2C_{feCB}}}\right] \tag{17}$$

$$Q_2(x) = [1(t)]\left[W_s(x) - W_s(0)e^{-\frac{x}{VR_2C_{feCB}}}\right] - [1(x-d)]\left[W_s(x) - W_s(d)e^{-\frac{x-d}{VR_2C_{feCB}}}\right] \tag{18}$$

$$Q_3(x) = 1(x-x_1)U_1\left(1 - e^{-\frac{(x-x_1)}{VR_2C_{feCB}}}\right) \tag{19}$$

where: $Q_1(x)$ results from the convolution multiplication of the component

$$\begin{aligned}
&\left(U_2 - R_2 I_{ps_fe_MAX}\right) \cdot 1(-x) \text{ function } g(x) \text{ with function } h(x),\\
&Q_1(x) = h(x) * \left(U_2 - R_2 I_{ps_fe_MAX}\right) \cdot 1(-x),\\
&Q_2(x) = h(x) * W(x)[1(x) - 1(x-d)],\\
&Q_3(x) = h(x) * U_2 \cdot 1(x-d).
\end{aligned}$$

Parameters of convolution polynomial $W_s(x) = A_s x^3 + B_s x^2 + D_s x + E_s$ are equal to $A_s = a, B_s = b - 3aCRV, D_s = 6aC^2R^2V^2 - 2bCRV, E_s = -6aC^3R^3V^3 + 2bC^2R^2V^2 + d$. Function is continuous over the entire domain, also at the interval limit of unit step:

$\lim_{x\to d^-} S(x) = \lim_{x\to d^+} S(x) = W(d) + U_2 - [W(0) + U_2 - R_2 I_{pS\_fe\_MAX}]\exp[-d/(CRV)]$ and $\lim_{x\to 0^-} S(x) = \lim_{x\to 0^+} S(x) = U_2 - R_1 I_{pS\_fe\_MAX}$.

It can be shown: if $d/V >> R_2 C_{pS\_fe\_MAX}$ then $S(x) = g(x)$ and if $d/V << R_2 C_{feCB}$ (writing a function $\exp[-d/(CRV)]$ in Taylor series) then $S(x) = u_{out\,\sqcap}(x)$.

5 Comparison of Model and Experiment Results

The experimental verification of the model was performed by recording the output signal when the light beam was interrupted by gear with parameters: t = 5.43 mm – tooth thickness, p = 10.16 mm – transverse pitch, z = 54 – teeth number.

The tooth geometry was included in the calculations. The research has been carried out by the interrupted laser beam frequency: 54.1 Hz (Fig. 5a), 271 Hz (Fig. 5b), 806 Hz

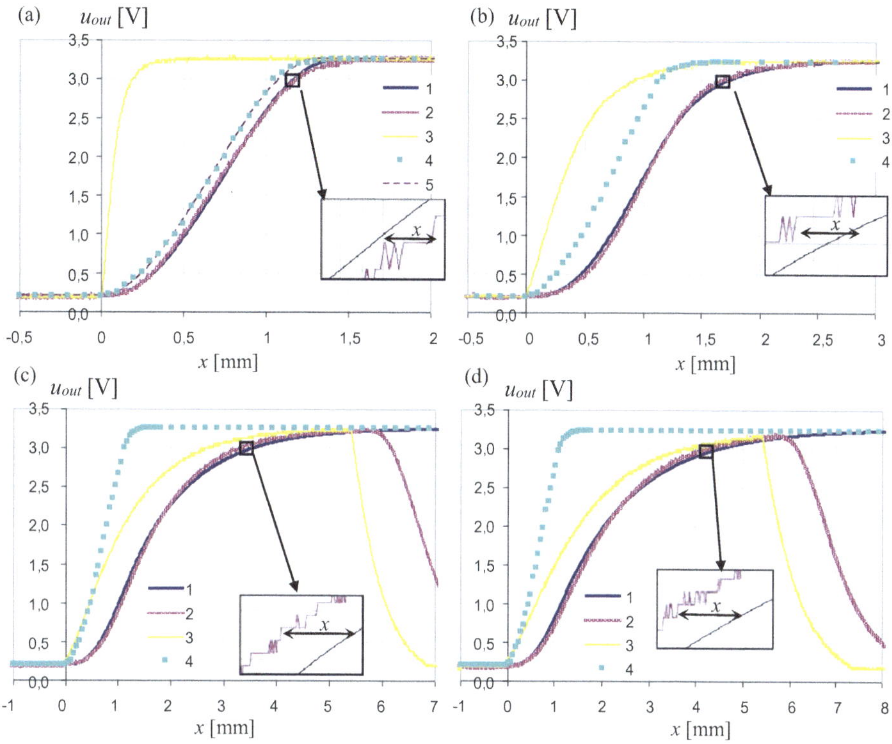

Fig. 5. Sensor output voltage as a function of the object position x for the frequency of the laser beam interruption (a) 54.1 Hz, (b) 271 Hz, (c) 806 Hz, (d) 1070 Hz. 1 – sensor response measured using oscilloscope, 2 –sensor model response, 3 – output signal corresponding to rectangular input signal according to Fig. 3a, 4 – quasi-stationary sensor response determined according to Fig. 3b, 5 – polynomial $W(x)$.

(Fig. 5c), 1070 Hz (Fig. 5d). The frequencies correspond to the respective velocities of the object detected: $V_a = 0.543$ m/s, $V_b = 2.74$ m/s, $V_c = 8.14$ m/s, $V_d = 10.8$ m/s.

In Figs. 5a–d, all the characteristics as a function of time were transformed into characteristics as a function of the position of a moving object according to $x = Vt$. In Fig. 5a, quasi-stationary output signal (characteristic 4) was compiled with polynomial $W(x)$ approximating it.

6 Conclusions

The article provides mathematical relationships (Eqs. 16–19) which facilitate calculating the output voltage characteristics of the sensor as a function of the position where the quasi-stationary sensor response can be approximated by a polynomial not greater than of the third degree. The use of the transistor model helped reducing considerably the measurement errors of the real position and real response time respectively x_{resp_real}

= 0.025 mm and t_{resp_real} = 0.0468 ms (V_a = 0.543 m/s); x_{resp_real} = 0.045 mm and t_{esp_real} = 0.164 ms (V_b = 2.74 m/s); x_{resp_real} = 0.23 mm and t_{esp_real} = 0.0282 ms (V_c = 8.14 m/s); x_{resp_real} = 0.28 mm and t_{resp_real} = 0.259 ms (V_d = 10.8 m/s). Error x_{resp_real} increase with increasing velocity is due to a decreasing value of the first derivative of the position of real and model characteristics. This is particularly noticeable at 90% of the maximal value. Therefore, for comparison, it is much more effective to consider the range where the value of the derivative of the real characteristics is maximal, about 50% of the maximal value. The stationary characteristics of the sensor are much better represented by an approximation polynomial than an analytical function to avoid complications in calculating the convolution integral. Value x in the function of convolution in Eq. (16) is implicit, however, it can be found at a given voltage using known numerical methods. For low-performance controllers is recommended the lookup table for reading the position corrections already calculated from the model for any object velocity V.

References

1. Omron, Notation and Terminology for Phototransistor Output. http://www.omron-ap.co.in/service_support/FAQ/FAQ01138/index.asp. Accessed 30 May 2017
2. Panov, Y.: Small-signal analysis and control design of isolated power supplies with optocoupler feedback. IEEE Trans. Power Electron. **20**(4), 823–832 (2005). https://doi.org/10.1109/tpel.2005.850926
3. Vishay manufacturer. https://www.vishay.com/doc?81341. Accessed 30 May 2017
4. Hänsch, T.W., Träger, F.: Springer Handbook of Lasers and Optics. Springer, Heidelberg (2007). https://doi.org/10.1007/978-3-642-19409-2
5. Bakhsi, U.A., Godse, A.P.: Electronic circuits. Technical Publications Pune, Amit Residency, 412, Shaniwar Peth, Pune 411 030, India (2008)
6. Mayergoyz, I., Lawson, W.: Basic Electric Circuit Theory, 1st edn. Academic Press Hardcover (2012). ISBN: 9780124808652, eBook ISBN: 9780080572284

Application of the Parametric Identification While Modelling the Dynamics of the Electro-Hydraulic Drive

Piotr Wos(✉) and Ryszard Dindorf

Faculty of Mechatronics and Machine Design,
Kielce University of Technology, Kielce, Poland
{wos, dindorf}@tu.kielce.pl

Abstract. The quality of control systems is one of the most important factors leading to proper operation of electro-hydraulic servo-drives. This paper discusses possibilities to apply the dynamic identification with a discrete linear model while assessing the state of the electro-hydraulic drive dynamics. This evaluation is crucial while designing modern power or positional control systems. Experimental data is applied in order to determine the model dynamics of the real system and estimate unknown parameters of an object model. The dependencies were interpreted. The paper includes the selected results of current identification tests of the electro-hydraulic drive system and based on them the discrete parametric object model was derived.

Keywords: System identification · Electro-hydraulic drive · Adaptive forgetting coefficient

1 Introduction

Conventional control systems designed for specific working conditions operate well provided that they are subjected to small deviations from these conditions. In fact, controlled objects do not fulfill the stationary conditions as a result of this interference [7, 9]. Therefore, a problem of incomplete information about the object should not be taken into account. Incomplete information about the object, as far as hydraulic systems are concerned, usually refers to the lack of knowledge of the exact values of all parameters and signals which are applied to determine regulatory quality indicators or other optimization conditions. In the object control structures with incomplete information, algorithms providing the identification of system indeterminateness and change of the controllers' settings might be distinguished. It makes it possible to adjust the control system to variable conditions. Despite considerable advancement in on-line design of the controllers' algorithms, the industry is dominated by periodically-adjusted classic PID controllers. It is generally believed that those controllers are reliable and there are many methods of adjusting them [1]. However, due to the stability of the closed system, the methodology of adjusting the controller in real time is more modern, efficient and reliable for multi-dimensional control systems of high order. The standardization of those algorithms requires additional identification methods for

J. Świder et al. (Eds.): MECHATRONICS 2017, AISC 934, pp. 475–483, 2019.
https://doi.org/10.1007/978-3-030-15857-6_47

continuous systems to be developed which might cooperate with on-line adjustable controllers. Thus, it is essential to build appropriate algorithms of parametric identification as well as appropriate algorithms of state observation. Identification is an approach to build a model of dynamic system which consists in analyzing the dependences between its input and output signals without intruding into the laws of physics governing these dependencies. The identification is applied when the rules governing the model phenomenon are either unknown or too complicated to be used for designing the model construction for control purposes. In such a situation in order to determine the model dynamics of the real system and estimate unknown parameters of an object model, experimental data is applied. The identification procedure in an iterative procedure and encompasses the following stages: planning measurements in order to obtain data, selecting a model structure in terms of its form and row of equations, estimating model parameters and verifying the model correctness by using other measurement data than used for estimating the parameters [3]. Moreover, the identification method should not be sensitive to the nature of input signals. Obtaining data sufficient for identification is only possible through a proper stimulation of system inputs, and such an interaction is not always possible in practice [8].

One method of control systems is adaptive control with model identification [6]. The model identification adaptive control (MIAC) system is illustrated in Fig. 1. Identification of model is determined on the input u and output y. Based on these models optimal controller parameters are calculated and adjusted.

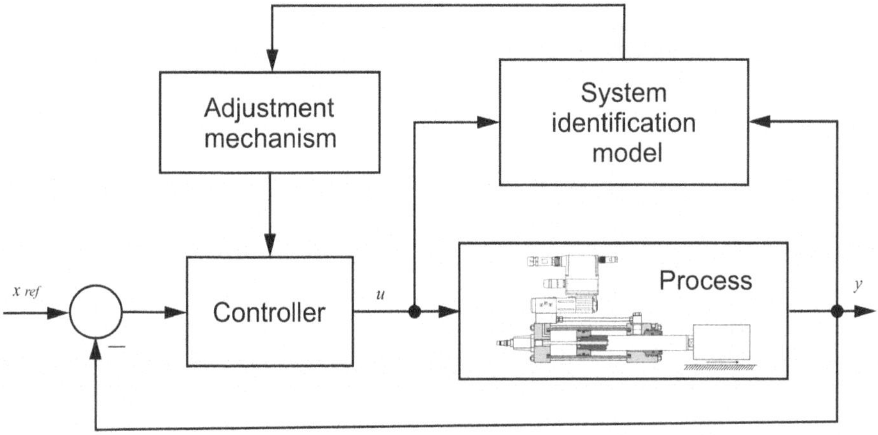

Fig. 1. Model identification adaptive control (MIAC)

2 Mathematical Identification Model

The parametrized model defines precisely the behaviour of the control object under specific conditions which are described by inputs and outputs at present and in the past. The algorithms for estimating the model coefficients were developed with the aim to track the changes of model coefficients. A typical parametric description is to present the properties of the control system in the form of discrete transmittance. The dynamic properties of the control process depend on the parameters' values of the discrete transmittance. In the synthesis of the control algorithms, the choice upon the modelling method depends to a great extent on how the suggested model is interpreted physically and whether it is easy to identify it [4]. Nonparametric models in the form of previously stored samples of the object responses, models formulated in the space of states as well as parametric models in the form of input-output discrete transmittance might be distinguished among many known descriptions [5]. The basis for determining the parametric model is to define the model class. Defining the model class is limited to determining the mathematical expression in the parametric form which describes the studied object. The model is selected from the accepted class by determining the numerical values of its coefficients. As a rule, foreknowledge makes it possible to establish only the model class. However, the coefficient values of the mathematical model are determined as a result of the implementation of the selected identification algorithm which stipulates how the measurement results should be processed in order to determine the values of the model coefficients. Due to the use of measurement data which is of "discrete-in-time" nature, the mathematical identification models are discrete and linear in relation of the desired parameters. They also take into account the fact that the measurement data might include mistakes as a result of introducing random interference into the models. As far as discrete time is concerned, the operator of signal backward shift by k of samples $q^{-kx(n)} = x(n-k)$ is introduced to shorten the record, which gives the equation that is similar to the equations with complex argument z^{-k} in the field of transforms Z.

In order to identify the dynamic model of the electro-hydraulic servo-drive, a deterministic model with the following structure was applied [6]:

$$A(q^{-1})\, y_t = B(q^{-1})\, u_{t-d} + \xi_t \tag{1}$$

where: y_t, u_t – input and control signals, respectively, d – discrete delay, ξ_t – object interference.

Model (1) does not require a priori information about the identification and thus, it is relatively easy and shows no sensitivity to frequently occurring industrial changes of the structure of the control object. The disadvantage of this model is a relatively great number of previously stored samples which, however, can be used to describe objects with complex dynamics. The transmittance model in the discrete form (2) has unknown parameter values of the control object which must be identified:

$$G(z) = \frac{B(z^{-1})}{A(z^{-1})} = \frac{b_1 z^{-1} + b_2 z^{-2} + \ldots + b_n z^{-n}}{1 + a_1 z^{-1} + a_2 z^{-2} + \ldots + a_n z^{-n}} z^{-d} \tag{2}$$

where: $a_1, a_2, \ldots, a_n, b_1, b_2, \ldots, b_n$ – parameters of the control object, n – model order.

3 Algorithm of the Model Identification of the Electro-Hydraulic Drive

The process of identification is a cognitive process, focused on quantitative and qualitative learning about the reality. The basis of this process is experimental research which has limited accuracy influenced by many factors, i.e.: the accuracy of the direct measurement of the object variables, influence of external disturbing factors, choice upon the signal stimulating the object with relation to its ability to identify, properties of the identification algorithm and the class of the adopted model. Both the identification process and the measurement process consist in obtaining the information about the studied object. The basis for determining the coefficients of the identification model are the measurement results of the input and output signals conducted directly on the identified object. An on-line identification algorithm was applied to provide a parametric identification of the electro-hydraulic drive. A solution for the parameters' estimation task with the use of calculation algorithms in accordance with the block diagram shown in Fig. 2 was suggested herein. The identification process consists of a few stages, with the stage of numerical calculations of estimator $\hat{\hat{\theta}}$. Being the most crucial.

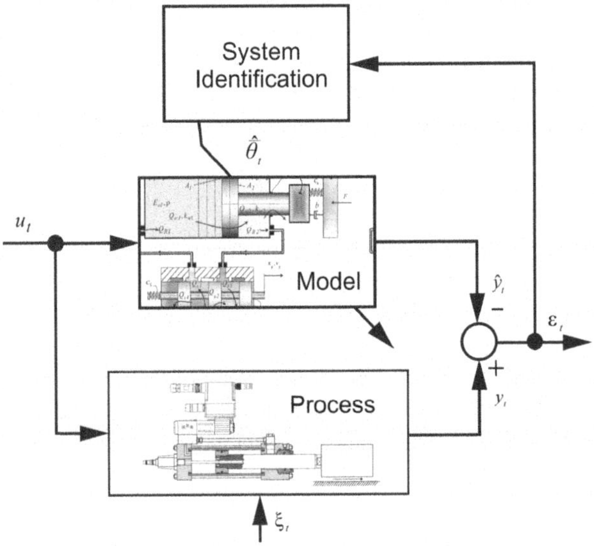

Fig. 2. Block diagram of adjusting the object model: u_t control signal, y_t object output signal, $\widehat{y}_t$ output signal of the object model, $\hat{\hat{\theta}}$ estimate vector of the object model parameters, ε_t model error, ξ_t object interference.

3.1 Systems with Parameters Variable Over Time

Determining the system parameters using the method which takes the whole history of the observation process into account is the average estimation obtained throughout the complete registration period. The applied iterative algorithm does not often follow the changes in the parameters of the objects loaded with random interferences. Moreover, those systems are excessively overloaded by increasing the number of measurements, the size of matrix $\mathbf{P}_t$ and vector $\overline{\varphi}_t$ increase as well at the same time and as a consequence, the demand for the memory of the control system grows. Adjusting the iterative algorithm to tracking the object parameters variable over time leads to the method with exponential forgetfulness. The method based on RLS recursive algorithm with WRLS (*Weighted Recursive Least Squares*) exponential forgetfulness consists in introducing λ_t forgetting coefficient, which decides upon the size of the algorithm memory, into the control equations. If the changes in parameters' values are tracked in an on-line way, estimates $\hat{\theta}$ should be determined on the basis of on-line measurements of input and output variables. Past data has no significant impact on the current object state and should be forgotten. In order to make matrix $\mathbf{P}_t$ independent from the past data, weight parameter λ_t, often called exponential forgetting coefficient, was introduced. Exponential forgetting coefficient λ_t decreases the influence of the past data on the current evaluation determined with covariance matrix $\mathbf{P}_t$:

$$\mathbf{P}_t = \left(\mathbf{P}_{t-1} - \frac{\mathbf{P}_{t-1}\overline{\varphi}_t\overline{\varphi}_t^T\mathbf{P}_{t-1}}{1+\overline{\varphi}_t^T\mathbf{P}_{t-1}\overline{\varphi}_t}\right)\frac{1}{\lambda_t}, \tag{3}$$

Then, the vector of estimated parameters takes the following form:

$$\hat{\hat{\theta}}_t = \hat{\hat{\theta}}_{t-1} + \frac{\mathbf{P}_{t-1}\overline{\varphi}_t}{\lambda_t + \overline{\varphi}_t^T\mathbf{P}_{t-1}\overline{\varphi}_t}\varepsilon_t, \tag{4}$$

$$\varepsilon_t = y_t - \overline{\varphi}_t^T\hat{\hat{\theta}}_{t-1}, \tag{5}$$

Weight coefficient λ_t variable over time is applied in order to accelerate the convergence of the algorithm and is determined with equation: $\lambda_t = \lambda^0\lambda_{t-1} + (1 - \lambda_0)$, where λ_t aims exponentially to 1, while the number of measurement samples increases.

Assuming that $\lambda^0 = \lambda_0 = 1$, the weight coefficient takes the following value: $\lambda = \lambda_t = 1$. This value is applied while tracking low-variable parameters of the control system. For $\lambda = 1$, matrix $\mathbf{P}_t$ depends in the case of every moment t on all previous data. Due to the fact that while adjusting, the parameters were prone to changes, there is no linear dependency between vector elements $\bar{\theta}_t$. Introduction of this parameter results in an exponential limitation of the error signal, which means that older samples of the estimation signal error are taken to total error measurement ε_t with correspondingly lower weight. It should be expected that lower values $\lambda < 1$ will upgrade the model more quickly which is beneficial for a quicker adaptation, and simultaneously lower values λ lead to greater temporary variations of the values of the estimated coefficients which is not desirable. That is why, the selection of forgetting coefficient λ should be well thought over compromise. The value of the forgetting coefficient is selected in accordance with the rate of changes of the identified object parameters as well as the level of interference included in the measurement data (Fig. 3).

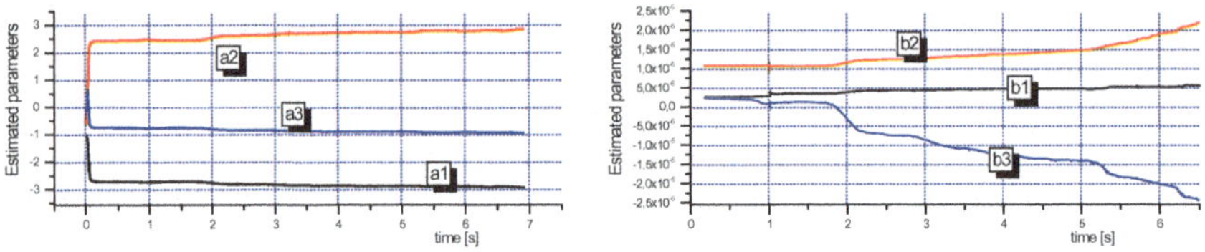

Fig. 3. Time courses of the model parameters of the control object determined using WRLS method

3.2 Robust Estimation of the Parameters of the Electro-Hydraulic Servo-Drive Model

A major challenge for engineers is to create such a model identification algorithm of the control object which will make it possible to accelerate the adaptation of model parameters when changing the reference signal values or the object properties. By estimating the model parameters in real time (*on-line*), a dynamic change of those parameters at the moment of random interference occurrence ξ_t is possible.

A solution for parameters' estimation task with the use of calculation algorithms based on RLS adaptive recursive algorithm with forgetting exponential AWRLS (Adaptive Weighted Recursive Least Squares) [2] was suggested herein. This method consists in constant λ_t value adaptation during the identification process. Adopting $\lambda = const$, as demonstrated by the attempts conducted, is difficult to be implemented in practice. In order to allow tracking the model parameters with quickly variable values, an attempt to improve RLS was made by using the adaptation of exponential forgetting coefficient λ_t so that it would include the change of input and output parameters during the object identification. In this case, exponential coefficient λ_t is updated at any time t and recorded in the following form:

$$\lambda_t = \left\{ 1 + \ln\left(1 + \overline{\varphi}_t^T \mathbf{P}_{t-1} \overline{\varphi}_t\right) + \left[\frac{(\upsilon_t + 1)\frac{\left(y_t - \overline{\varphi}_t^T \hat{\hat{\theta}}_{t-1}\right)^2}{\alpha_t}}{1 + \overline{\varphi}_t^T \mathbf{P}_{t-1} \overline{\varphi}_t + \frac{\left(y_t - \overline{\varphi}_t^T \hat{\hat{\theta}}_{t-1}\right)^2}{\alpha_t}} - 1 \right] \frac{\overline{\varphi}_t^T \mathbf{P}_{t-1} \overline{\varphi}_t}{1 + \overline{\varphi}_t^T \mathbf{P}_{t-1} \overline{\varphi}_t} \right\}^{-1} \tag{6}$$

where:

$$\alpha_t = \lambda_{t-1} \left[\alpha_{t-1} + \frac{\left(y_t - \overline{\varphi}_t^T \hat{\hat{\theta}}_{t-1}\right)^2}{1 + \overline{\varphi}_t^T \mathbf{P}_{t-1} \overline{\varphi}_t} \right], \; \upsilon_t = \lambda_{t-1}(\upsilon_{t-1} + 1)$$

where: y_t – values of output size samples,$\overline{\varphi}_t^T\hat{\bar{\theta}}_t$ – anticipated process output, $\mathbf{P}_t$ – covariance matrix.

Parameter λ_t which is found in Eq. (6) is an adaptive weight parameter, which has a significant influence on the convergence rate of the control servo-drive algorithm. If coefficient λ_t aims to unity, the estimation variance decreases and the process dynamic properties deteriorate by prolonging the memory of previous measurement results. At the same time, in the states determined for a low value of coefficient λ_t, the estimation quality deteriorates as the variance of the generated estimation increases. Constant value tracking of the forgetting coefficient provides great process dynamics (low value of coefficient λ_t) and low random variation of the generated estimation in the event of weak stimulation (the value of coefficient λ_t is close to unity, see Fig. 5a). Figure 4 shows time courses of parameters a_i, b_i for $n = 3$ of the model order of the control object (1), determined using AWRLS while controlling the displacement of the hydraulic actuator for sinusoidal extortion.

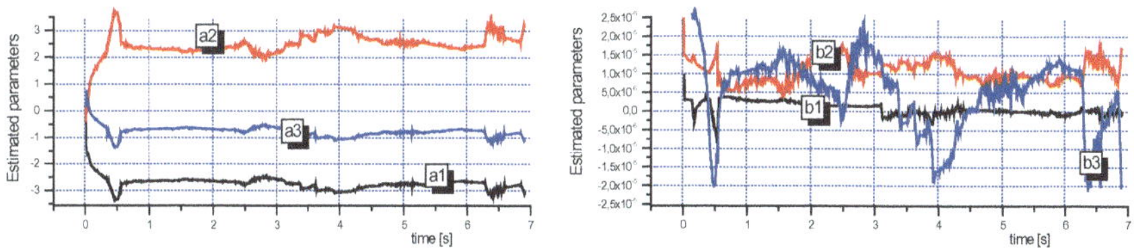

Fig. 4. Courses of identified model parameters (1) a_i, b_i and forgetting coefficient λ_t for the displacement of the actuator piston along sinusoidal trajectory $u_t = 0.5(\sin 2\pi t + \sin 10\pi t + \sin 24\pi t)$

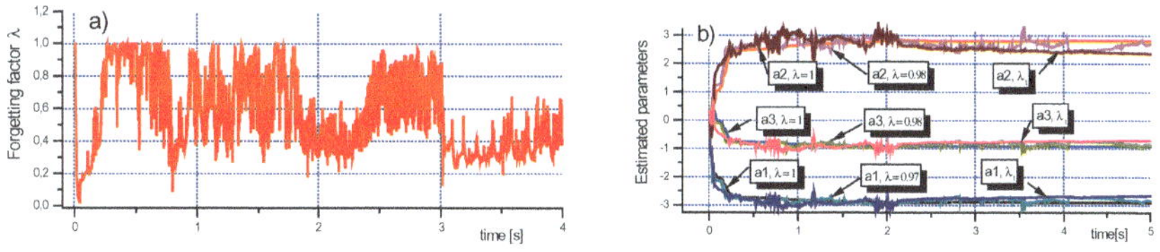

Fig. 5. Change of exponential forgetting factors (a), comparison of changes of parameter of model polynomial $A(q^{-1})$ with adaptation and without adaptation of exponential forgetting factors (b)

4 Comparison of the Efficiency of Estimation Algorithms of the Model Parameters – Model Verification

During the research, an analysis of the time course of remainders ε of adjusting the model to the intended object response was conducted. Due to the random nature of remainders ε, statistical tests were applied which determine the autocorrelation function of remainders $R_x(\tau)$. The autocorrelation function (own correlation) of signal $x(t)$ determines the dependency of the values in two various moments t at the distance of τ from each other. The function definition is stipulated with Eq. (7):

$$R_x(\tau) = \lim_{T \to \infty} \frac{1}{T} \int_0^T x(t)x(t+\tau)d\tau \tag{7}$$

where: τ – displacement over time.

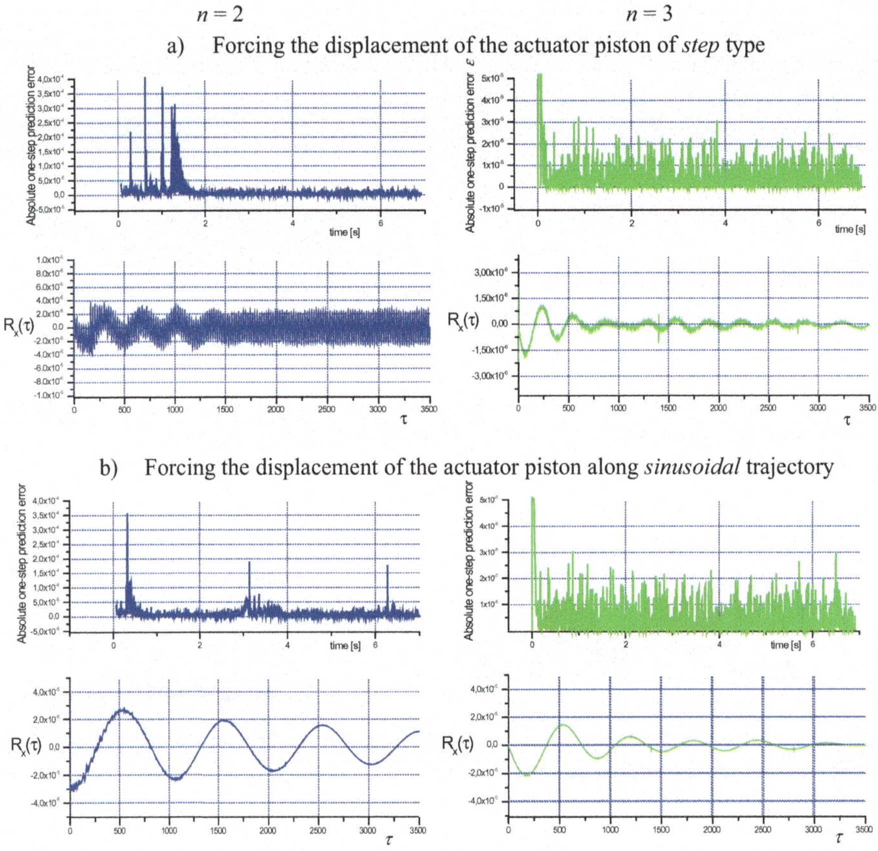

Fig. 6. The courses of absolute error $|\varepsilon|$ and autocorrelation function of remainders $R_x(\tau)$ for reference model $n = 2$ and $n = 3$, AWRLS algorithm: (a) displacement of *step* type, (b) displacement of *sinusoidal* type $u_t = 0.5(\sin 2\pi t + \sin 10\pi t + \sin 24\pi t)$

If the model is matched in a proper way, the properties of remainders ε should be characterized by symmetric distribution and be independent from input signals. They should have the properties of white noise, i.e. a stationary stochastic signal with an average value equaling to zero. This signal does not carry any information about the object reaction to the forced signal. The complete information about the object behaviour should be reproduced form the model and the process output. Figure 6 shows the courses of absolute values $|\varepsilon|$ and autocorrelation function of remainders $R_x(\tau)$ for model (1), type $n = 2$ and $n = 3$.

If the model accuracy is measured with the autocorrelation function of remainders $R_x(\tau)$, then, if properly identified, it will asymptotically equal to zero for each displacement which is different than zero. From the courses presented above (see Fig. 6) it might be concluded that the best match was obtained for model $n = 3$. No proper match with model $n = 2$ was observed. The remainders depend on the input size, and autocorrelation function $R_x(\tau)$ does not tend asymptotically to zero.

5 Summary

An important issue regarding the selection of the adaptive controller parameters is to identify the model of the control object in a proper way. Both the model parameters and the controller parameters are selected in real time (*on-line*). The proper identification of the model parameters determines the quality of the electro-hydraulic servo-drive control. The proper selection of appropriate tools to verify the model match for the intended object response is crucial.

References

1. Ahmadnezhad, M., Soltanpour, M.: Tracking performance evaluation of robust back-stepping control design for a nonlinear electrohydraulic servo system. World Acad. Sci., Eng. Technol. Int. J. Mech. Aerosp. Ind. Mechatron. Manuf. Eng. **9**(7), 1341–1347 (2015)
2. Bobál, V., Böhm, J., Fessl, J., Machácek, J.: Digital Self-tuning Controllers: Algorithms, Implementation and Applications, Technology & Engineering. Springer, Heidelberg (2005). https://doi.org/10.1007/b137409
3. Jelali, M., Kroll, A.: Hydraulic Servo Systems – Modelling, Identification & Control. Springer, Berlin (2003). https://doi.org/10.1007/978-1-4471-0099-7
4. Kheowree, T., Kuntanapreeda, S.: Adaptive dynamic surface control of an electrohydraulic actuator with friction compensation. Asian J. Control. **17**, 855–867 (2015)
5. Rahmat, M.F., Rozali, M.: Modeling and controller design of the electro-hydraulic actuator system. Am. J. Appl. Sci. **7**, 1100–1108 (2010)
6. Takahashi, K., Inoue, M., Ikeo, S.: Application of the model reference adaptive control technique to an electrohydraulic servo system, pp. 68–87. Proc. Int. Conf. Fluid Power Trans. Control, Hangzhou, China (1985)
7. Wos, P., Dindorf, R.: Adaptive control of the electro-hydraulic servo-system with external disturbances. Asian J. Control. **15**, 1065–1080 (2013)
8. Wos, P., Dindorf, R.: Practical parallel position-force controller for electro-hydraulic servo drive using on-line identification. Int. J. Dyn. Control. **4**, 3–9 (2014)
9. Yao, B.: Adaptive robust motion control of the single-rod hydraulic actuators: theory and experiments. IEEE/ASME Trans. Mechatron. **5**(1), 79–91 (2002)

Author Index

J. Świder et al. (Eds.): MECHATRONICS 2017, AISC 934, pp. 485–486, 2019.
https://doi.org/10.1007/978-3-030-15857-6

The manufacturer's authorised representative in the EU is Springer Nature Customer Service Centre GmbH, Europaplatz 3, 69115 Heidelberg, Germany. If you have any concerns regarding our products, please contact ProductSafety@springernature.com

Printed and bound by CPI Group (UK) Ltd, Croydon, CR0 4YY
06/10/2026
02228868-0004